Studies in Economic Theory

Editors

Springer
Berlin
Heidelberg
New York
Barcelona
Hong Kong
London
Milan
Paris
Singapore
Tokyo

Titles in the Series

M. Ali Khan and Nicholas C. Yannelis (Eds.)
Equilibrium Theory in Infinite Dimensional Spaces

Charalambos D. Aliprantis, Kim C. Border and
Wilhelmus A. J. Luxemburg (Eds.)
Positive Operators, Riesz Spaces, and Economics

Donald G. Saari
Geometry of Voting

Charalambos D. Aliprantis and Kim C. Border
Infinite Dimensional Analysis

Jean-Pierre Aubin
Dynamic Economic Theory

Mordecai Kurz (Ed.)
Endogenous Economic Fluctuations

Jean-François Laslier
Tournament Solutions and Majority Voting

Ahmed Alkan, Charalambos D. Aliprantis,
Nicholas C. Yannelis (Eds.)
Theory and Applications

James C. Moore
Mathematical Methods for Economic Theory 1

James C. Moore
Mathematical Methods for Economic Theory 2

James C. Moore

Mathematical Methods for Economic Theory 1

With 19 Figures

Springer

Professor James C. Moore
Purdue University
Krannert School of Management
1310 Krannert Building
West Lafayette, IN 47907-1310
USA

ISBN 3-540-66235-9 Springer-Verlag Berlin Heidelberg New York

Library of Congress Cataloging-in-Publication Data
Die Deutsche Bibliothek - CIP-Einheitsaufnahme
Mathematical methods for economic theory 1 / James C. Moore. - Berlin; Heidelberg; New York; Barcelona; Hong Kong; London; Milan; Paris; Singapore; Tokyo: Springer, 1999
(Studies in economic theory; 9)
ISBN 3-540-66235-9

Printed in Germany

Harcover design: Erich Kirchner, Heidelberg

SPIN 10655637 42/2202-5 4 3 2 1 0 - Printed on acid-free paper

To Donna Rae,

Who is also

My Best Pal . . .

Preface

This is the first of a two-volume work intended to function as a textbook for graduate students in economics as well as a reference work for economic scholars who are either working in theory, or who have a strong interest in microeconomic theory. The mathematics background assumed of the reader is that required of students in their second term in most graduate economics programs; a year of elementary calculus, a semester of linear algebra, and the 'mathematics for economists' course which usually must be taken by students in their first semester of graduate work in economics (for example, a course based upon Novshek [1993]).

Since this is a rather unusual work, a word or two justifying its writing is undoubtedly in order. There are three reasons, essentially, why I have believed it worthwhile to undertake such a writing project. In the first place, I have collected a number of results which are frequently useful in economics, but for which exact statements and proofs are rather difficult to find; for example, a number of results on convex sets and their separation by hyperplanes, some results on correspondences, and some results concerning support functions and their duals. Secondly, the mathematical topics taken up in these two volumes are generally taught somewhere in the mathematics curriculum, but are generally not all done in a two-course sequence as they are arranged here. The order and arrangement of topics here is, I believe, quite logical and efficient from the standpoint of study by graduate students in economics; but the topics have not been 'packaged' in this way in mathematics courses because the needs of students in that area are different. Thirdly, I have tried to present examples of the use of the mathematical tools being developed as they are presented. I have been greatly impressed over the years by how much more quickly students (and I as well, for that matter) learn and understand topics from mathematics if they see how these results can be put to use in their major field.

The material in these two volumes is intended to be studied in the order in which it is presented, and has been used here at Purdue as the main text for a two-semester sequence in mathematical methods for economic theory; volume 1 being the text for the first semester. However, students who have been undergraduate mathematics majors can probably skip most of the content of volume 1; perhaps reading through only Chapters 5 and 6 before moving on to the second volume of this work.

As an aid to studying this material, let me explain a distinction used throughout both volumes. In each section, the reader will find one or more entries labeled 'Examples/Exercises,' as well as exercises at the end of each section. The 'Examples/Exercises' are a part of the text, and if the topics in a given section are new to the student, I would recommend that this material be worked through fairly carefully before going on to the next section. The exercises at the end of the section are there for the usual purpose of providing a self-test of your understanding of the material presented in the section, as well as some practice in making use of the material. I have attempted to arrange these exercises more or less in order of difficulty, and I would recommend that the student work through the first two or three of these exercises at the end of a given section before tackling the next section.

Two further remarks on the organization of these books are in order. First, while the six chapters in this volume are numbered conventionally from 1 to 6, the six chapters in volume 2 are numbered from 7 through 12. Thus, in this volume, when reference is made to 'Chapter 7,' I am speaking of the first chapter in volume 2. Secondly, some sections in each of the two volumes are marked with an asterisk (*). Where this is the case, the material in the section is somewhat peripheral to the main thrust of the text; in fact, the material in the starred sections is never used in the sections which follow it; and thus can quite logically be skipped in a first reading.

This is a textbook much more than a monograph, and apart from Chapter 6, parts of Chapters 3 and 9, and some economic examples elsewhere, my original intent was to be as unoriginal as possible in developing this material. Unfortunately for this resolve, the unusual arrangement and 'packaging' of the topics in these volumes has necessitated somewhat more original work than I had ever intended! However, this comment is essentially a complaint; no mathematician who might happen to read this work will find many mathematical results or proofs which are very strikingly original. On the other hand, in terms of the number of citations to works in mathematics, the book may appear to be rather more original than it is, for the following reason. Where I have found a statement and proof of a mathematical result which I could adapt to use here with little change, my rule of thumb in citing sources was as follows. If I found a statement and proof in at least two mathematics textbooks with no citation of source by either of the two authors, I have assumed that the material was a part of the public domain, and have not cited any source. On the other hand, if I have made use of some material for which I have been unable to find more than one source, I believe that I have always cited the source which I have used.

It is rather embarrassing to have to admit that I spent two Sabbatical leaves working on these volumes. At the same time it is a pleasure to acknowledge the fact that I spent a large portion of a very pleasant and productive visit at the University of Michigan, Ann Arbor, working on this manuscript, as well as nearly all of an equally pleasant and productive semester at the Royal Holloway campus of the University of London. I would particularly like to thank Ted Bergstrom, Charles Brown, and Hal Varian, who arranged the visit to Ann Arbor; and Jeff Frank, who arranged my visit to Royal Holloway. I would also like to thank Deans James Dworkin and Dennis Weidenaar of the Krannert School of Management here at Purdue, who have been quite supportive of my dedication to the writing of these volumes.

I would also like to thank Roko Aliprantis, Ted Bergstrom, Ken Binmore, Larry Blume, Tom Knoblauch, Dan Kovenock, Bill Novshek, Carl Simon, Nicholas Yannelis, and an anonymous referee; all of whom read portions of various versions of the manuscript and made helpful comments and suggestions for its improvement. Special thanks are also due to a number of Research Assistants I have had over the years, who have conscientiously read through versions of this manuscript and helped expunge those dratted 'typos:' Hiro Ishigaki, Eric Munshower, Matt Roelofs, Suddhasatwa Roy, Peter Lee U, and Jason Zimmerman. Thanks are also due to Glen Waddell, who did a splendid job of re-doing my figures and inserting them into this manuscript; to Karen Angstadt, who did her usual efficient and competent job of organizing and preparing the manuscript; and to a number of students in the classes in which I have taught from this manuscript, and who have helped spot errors and 'typos' therein.

A special thanks is also due my teachers at the University of Minnesota, especially John Chipman, Leo Hurwicz, Ket Richter, and Hugo Sonnenschein; all of whom not only taught me a great deal of economics, but also taught me a healthy respect for the 'tools of my trade.' Finally, I would like to thank my wife, Donna, who has had to endure a great deal of inconvenience and frustration because of this project, and without whose support and encouragement this task could never have been completed.

W. Lafayette, IN J. C. M.
May, 1999

Contents

Chapter 1

Set Theory and Properties of $\mathbf{R}^n$

1.1 Set-Theoretic Notation and Concepts

The primary purpose of this section is to provide you with a convenient source of review material, since I am assuming that you already have some familiarity with basic set theory. However, I also want to make sure that we are all using the same elementary definitions and notation. Thus:

$a \in A$ means 'a is an element of the set A.'

$a \notin A$ means 'a is not an element of the set A.'

$A \subseteq B$ means 'A is a subset of B;'

that is, each element of A is also an element of B. The symbol '$\emptyset$' will be used to denote the empty set; that is, $\emptyset$ is the set containing no elements.

The expression

$$A = \{x \in X \mid x \text{ satisfies } \mathbf{P}\}$$

defines A to be the set of all elements of X satisfying the property $\mathbf{P}$. Thus, for example, denoting the set of real numbers by '$\mathbf{R}$,'

$$\mathbf{R}_+ = \{x \in \mathbf{R} \mid x \geq 0\}$$

is the set of nonnegative real numbers; while

$$\mathbf{R}_{++} = \{x \in \mathbf{R} \mid x > 0\}$$

is the set of positive real numbers.

We will use a somewhat nonstandard notation for intervals of real numbers, as is set forth in the following.

1.1. Definitions. Suppose that a and b are real numbers, and that $a \leq b$. Then we define:

1. $[a, b]$ the **closed interval** from a to b, by:

$$[a,b] = \{x \in \mathbf{R} \mid a \leq x \leq b\}$$

2. $]a, b[$ the **open interval** from a to b, by:

$$]a,b[= \{x \in \mathbf{R} \mid a < x < b\}$$

3. the **half-open intervals** from a to b by:

$$]a,b] = \{x \in \mathbf{R} \mid a < x \leq b\}$$

and

$$[a,b[= \{x \in \mathbf{R} \mid a \leq x < b\}.$$

It is more common in the mathematics literature to use, for example, '(a, b)' to denote the open interval from a to b. However, our usage follows a time-honored tradition in economics, having been used by Debreu in *Theory of Value*; and in any case, we will want to reserve the notation '(a, b)' to denote the ordered pair whose first element is a and whose second element is b. If A and B are sets, the **cartesian product** of A and B is defined by

$$A \times B = \{(a,b) \mid a \in A \,\&\, b \in B\}.$$

Thus,

$$\mathbf{R}^2 \stackrel{\text{def}}{=} \mathbf{R} \times \mathbf{R} \equiv \{(x,y) \mid x \in \mathbf{R} \,\&\, y \in \mathbf{R}\},$$

is two-dimensional Euclidean space.[1] In general,

$$\mathbf{R}^n \stackrel{\text{def}}{=} \mathbf{R}^{n-1} \times \mathbf{R} \equiv \{(x_1, x_2, \cdots, x_n) \mid x_i \in \mathbf{R}, \text{ for } i = 1, \cdots, n\}.$$

Extending this idea somewhat, if A_i is a non-empty set, for $i = 1, \ldots, n$, we define the **cartesian product of $\mathbf{A}_1, \ldots, \mathbf{A}_n$**, $\prod_{i=1}^n A_i$, by:

$$\prod_{i=1}^n A_i = \{\mathbf{a} = (a_1, \ldots, a_n) \mid a_i \in A_i \text{ for } i = 1, \ldots, n\}.$$

If A and B are sets, we define:
$A \cup B$, the **union** of A and B, by:

$$A \cup B = \{x \mid x \in A \text{ or } x \in B\},$$

[1] When defining an entity (set, function, whatever) within an equation, I will use the notation '$\stackrel{\text{def}}{=}$' if there might otherwise be a possibility for confusion. If, however, I am repeating a definition, or where it otherwise seems obvious, I will use the simpler notation '$\equiv$' to indicate an identity or definition.

$A \cap B$, the **intersection** of A and B, by:

$$A \cap B = \{x \mid x \in A \text{ and } x \in B\},$$

and $A \setminus B$, the **set-theoretic difference** of A and B, by:

$$A \setminus B = \{x \mid x \in A \text{ and } x \notin B\}.$$

Two non-empty sets, A and B are said to be **disjoint** iff

$$A \cap B = \emptyset.$$

It will generally be the case that all of the sets under discussion will be understood to be subsets of some 'universal set,' S, in which case, for a set A (understood now to be a subset of S), we define the **complement of A**, A^c, by

$$A^c = S \setminus A \stackrel{\text{def}}{=} \{x \in S \mid x \notin A\}.$$

Thus, for example, taking $S = \mathbf{R}$,

$$\mathbf{R}_- \stackrel{\text{def}}{=} \{x \in \mathbf{R} \mid x < 0\} = (\mathbf{R}_+)^c.$$

The following result lists the basic properties of the set-theoretic operations which we have just defined; its proof will be left as an exercise.

1.2. Proposition. *If A, B, and C are subsets of S, then:*

1. $A \subseteq B \Leftrightarrow B^c \subseteq A^c$,
2. $\emptyset^c = S$, $S^c = \emptyset$,
3. $A \cup A^c = S$, $A \cap A^c = \emptyset$,
4. $(A^c)^c = A$,
5. $A \setminus B = A \cap B^c$,
6. $A \cap B = \emptyset \Leftrightarrow A \subseteq B^c \Leftrightarrow B \subseteq A^c$,
7. $A \cap B \subseteq A$ *and* $A \cap B \subseteq B$,
8. $A \cap B = A \Leftrightarrow A \subseteq B$,
9. $A \subseteq A \cup B$ *and* $B \subseteq A \cup B$,
10. $A \cup B = A \Leftrightarrow B \subseteq A$,
11. $A \cup (B \cup C) = (A \cup B) \cup C = A \cup B \cup C$,
12. $A \cap (B \cap C) = (A \cap B) \cap C = A \cap B \cap C$,
13. $A \cap (B \cup C) = (A \cap B) \cup (A \cap C)$,
14. $A \cup (B \cap C) = (A \cup B) \cap (A \cup C)$,
15. $A = (A \cap B) \cup (A \cap B^c) = (A \cap B) \cup (A \setminus B)$.

When we write '$f\colon A \to B$,' we will mean that f is a function with **domain** A and **range space** B; that is, for each a, $f(a)$ is defined, and is an element of B. For such a function, we define the **graph of f**, $\mathcal{G}_f$, by

$$\mathcal{G}_f = \{(x, y) \in A \times B \mid y = f(x)\}.$$

1.3. Definitions. Suppose $f: A \to B$. We shall say that f is:
1. **one-to-one** (abbreviated '1-1') iff, for all $a, a^* \in A$, we have:

$$f(a) = f(a^*) \Rightarrow a = a^*.$$

2. **onto B** iff, for all $b \in B$, there exists $a \in A$ such that $b = f(a)$.[2]

Thus, in Figure 1.1.a, f is neither 1-1 nor onto; in Figure 1.1.b, f is 1-1, but not onto. To complete our inventory, in 1.2.a on the next page, f is onto, but not 1-1; while in Figure 1.2.b, f is both 1-1 and onto.

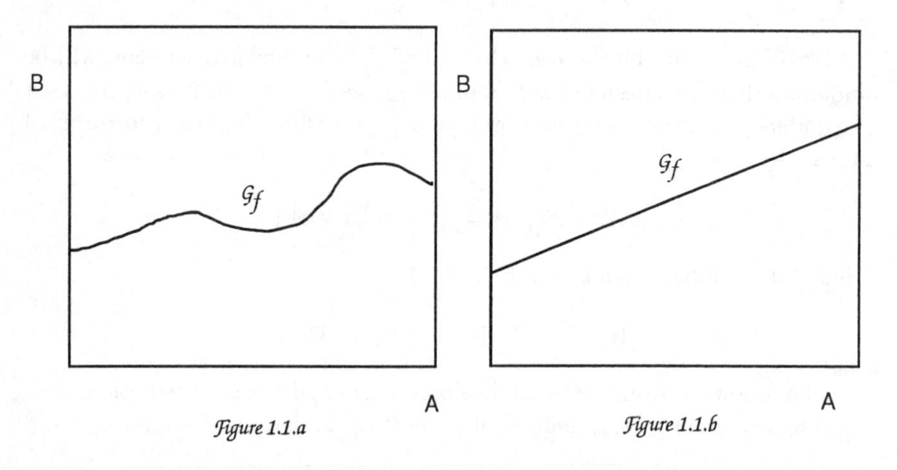

Figure 1.1: Graphs of Functions.

A particular function with which we shall often be working is the **identity function, i**, defined on an arbitrary set, say X, by: $\mathbf{i}(x) = x$. Obviously, **i** is both one-to-one and onto.

We use '$\forall$' and '$\exists$' to denote the universal and existential quantifiers, respectively; and enclose quantifier statements by parentheses, with the end of a chain of quantifier statements being indicated by a colon (:). The colon itself is then generally read as 'such that' or 'it is true that,' depending upon whether the expression immediately preceding it involves $\exists$ or $\forall$, respectively. Thus, the expression:

$$(\forall x \in \mathbf{R})(\exists y \in \mathbf{R}): y > x, \tag{1.1}$$

means 'for every real number, x, there exists a real number, y, which is greater than x;' while the expression:

$$(\exists y \in \mathbf{R})(\forall x \in \mathbf{R}): y > x, \tag{1.2}$$

[2]While we shall not use these terms in this text, the student should be aware that the expressions '**f is injective**' and '**f is surjective**' are often used as synonyms for 'f is 1-1' and 'f is onto,' respectively; while, if f is both 1-1 and onto, one then says that f is **bijective**.

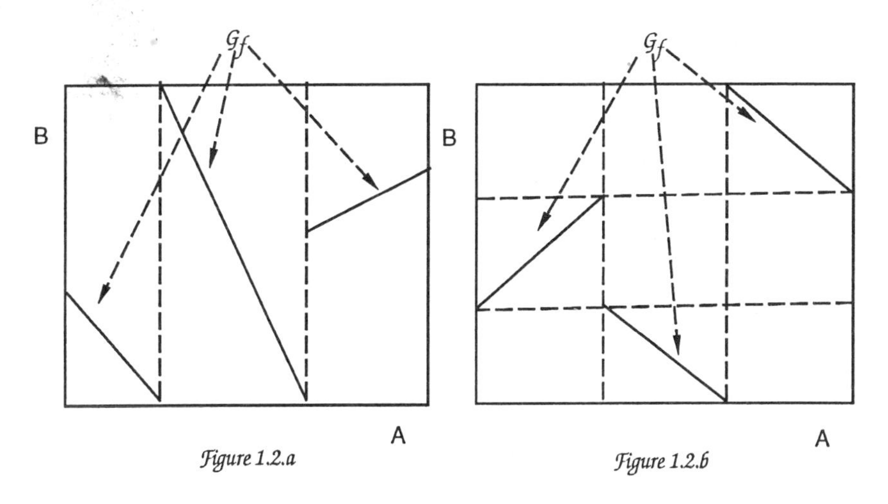

Figure 1.2: Functions Onto B.

means 'there exists a real number y such that for each real number, x, it is true that $y > x$.' Obviously the order in which the quantifiers appear is important; in particular, if the existential quantifier follows a statement containing the universal quantifier, the entity whose existence is being asserted can depend upon the variable appearing in the universal quantifier statement, whereas this is not the case if the existential statement precedes the universal statement. Thus, statement (1.1), above, is true; while statement (1.2) is not. Notice that, with the use of this notation, we can re-write Definitions 1.3.1 and 1.3.2 as:

$$(\forall a, a^* \in A) \colon f(a) = f(a^*) \Rightarrow a = a^*,$$

and

$$(\forall b \in B)(\exists a \in A) \colon b = f(a),$$

respectively. Once you become familiar with this notation, I am confident that you will find it easier to read statements like those just presented when they are written symbolically than when they are written out in English. On the other hand, while you are acquiring this familiarity, I will promise to minimize my use of long expressions involving existential and universal quantifiers.

1.4. Definitions. If $f \colon X \to Y$, and A and B are subsets of X and Y, respectively, we define:

1. **the image of A under f**, denoted '$f(A)$,' by:

$$f(A) = \{y \in Y \mid (\exists x \in A) \colon y = f(x)\}.$$

2. **the inverse image of B under f**, denoted '$f^{-1}(B)$,' by:

$$f^{-1}(B) = \{x \in X \mid f(x) \in B\}.$$

In Figures 1.3.a and 1.3.b, below, we have provided illustrations of these concepts. We will refer to $f(X)$ as the **range of f**; and, of course, f is onto Y if, and only if, Y is the range of f [that is, $Y = f(X)$]. The function $f\colon X \to Y$ is one-to-one iff it satisfies:

$$(\forall y \in f(X))(\exists x \in X)\colon f^{-1}(\{y\}) = \{x\}.$$

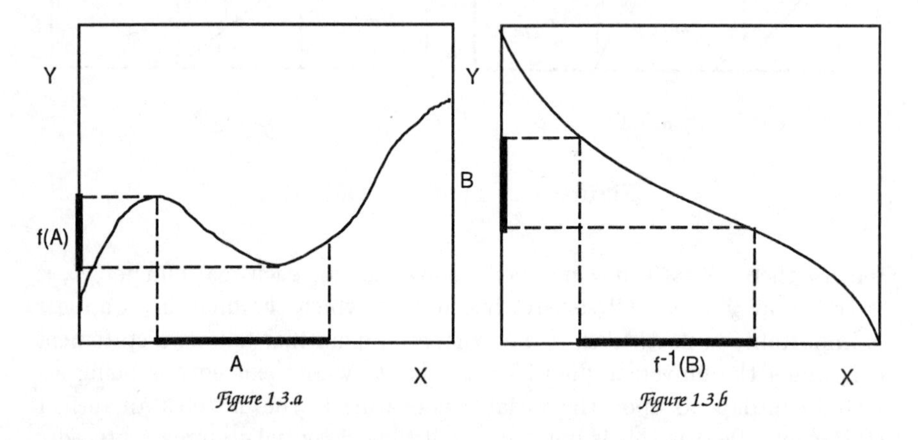

Figure 1.3: Image and Inverse Image.

There are several additional bits of notation which we will find useful in dealing with sets. First of all, given a non-empty set, A, we will denote by '$\mathcal{P}(A)$' the family of all subsets of A (the '**power set of A**'), where we include in the family $\emptyset$ and A itself. Thus, if $A = \{x, y\}$, then

$$\mathcal{P}(A) = \{\emptyset, \{x\}, \{y\}, \{x, y\}\}.$$

Secondly, in dealing with families of sets, it will often be useful to suppose that the family has been labeled with the use of a so-called index set. If A and X are non-empty sets, and $\mathcal{V}$ is a subset of $\mathcal{P}(X)$, we shall say that **A is an index set for $\mathcal{V}$** iff there exists a function $f\colon A \to \mathcal{V}$, which is both one-to-one and onto. In such a case, we will usually denote $f(\alpha)$, for $\alpha \in A$, by something like 'V_α.' We can then consider such entities as:

$$\bigcup_{\alpha \in A} V_\alpha \stackrel{\text{def}}{=} \{x \in X \mid (\exists \alpha \in A)\colon x \in V_\alpha\},$$

and

$$\bigcap_{\alpha \in A} V_\alpha \stackrel{\text{def}}{=} \{x \in X \mid (\forall \alpha \in A)\colon x \in V_\alpha\}.$$

It is important to recognize that the use of an index set and the notation just introduced is purely a matter of convenience, and involves no hidden assumptions or the use of the Axiom of Choice,[3] or whatever. To see that no hidden assumptions are involved, and that there is no loss of generality in using this index set notation, notice that our definition does not stipulate that A be distinct from $\mathcal{V}$. In fact, if X is any set whatsoever, and $\mathcal{V}$ is any subset of $\mathcal{P}(X)$, we can let $A = \mathcal{V}$, and f be the identity function to obtain an index set for $\mathcal{V}$; among other things, this illustrates the fact that we can indicate the union or intersection of the sets in $\mathcal{V}$ in a different (although logically equivalent) way:

$$\bigcup_{V \in \mathcal{V}} V \stackrel{\text{def}}{=} \{x \in X \mid (\exists V \in \mathcal{V})\colon\ x \in V\},$$

and

$$\bigcap_{V \in \mathcal{V}} V \stackrel{\text{def}}{=} \{x \in X \mid (\forall V \in \mathcal{V})\colon\ x \in V\}.$$

I will feel free to use whichever of the two notational devices just introduced is the more convenient in a given context; however, if the equivalence of the two representations of unions and intersections were all there were to the story, we probably wouldn't bother with index sets. As it happens, however, index sets arise naturally and conveniently in many settings. We can probably best illustrate this last point with a couple of examples.

1.5. Examples/Exercises.

1. Let $X = \mathbf{R}$, and A be equal to N, the set of natural numbers:

$$N = \{1, 2, 3, \dots\};$$

and, for each $n \in N$, define V_n by:

$$V_n =]n - 1, n[.$$

In this case we have:

$$\bigcup_{n \in N} V_n = \mathbf{R}_{++} \setminus N$$

while:

$$\bigcap_{n \in N} V_n = \emptyset.$$

This seems a good time to mention two further points about our use of index sets. First, when our index set, A, is equal to N, we will often write:

$$\text{`}\bigcup_{n=1}^{\infty} V_n\text{' and `}\bigcap_{n=1}^{\infty} V_n\text{,'}$$

[3]The reader who is unfamiliar with the Axiom of Choice need not worry about that fact at this juncture; we are, after all, mentioning the Axiom here only to point out that we're not implicitly making use of it.

in place of:

$$\text{`}\bigcup_{n\in N} V_n\text{' and `}\bigcap_{n\in N} V_n\text{,'}$$

respectively. Furthermore, if A is of the form:

$$A = \{1, 2, \ldots, n\},$$

for some $n \in N$, we will usually write:

$$\text{`}\bigcup_{i=1}^{n} V_i\text{' and `}\bigcap_{i=1}^{n} V_i\text{,'}$$

in place of:

$$\text{`}\bigcup_{i\in A} V_i\text{' and `}\bigcap_{i\in A} V_i\text{,'}$$

respectively.

2. Let $X = \mathbf{R}, A = N$, and for each $n \in N$, define X_n by:

$$X_n =]1/n, 3 - 1/n[.$$

Prove that:

$$\bigcup_{n=1}^{\infty} X_n =]0, 3[,$$

and

$$\bigcap_{n=1}^{\infty} X_n =]1, 2[= X_1. \quad \square$$

Using the index set notation just introduced, we can state and prove generalized versions of the associative and distributive laws presented in Proposition 1.2.11–13, as follows.

1.6. Proposition. *If A is a subset of S, and $\mathcal{B} = \{B_\lambda \mid \lambda \in \Lambda\}$ is a family of subsets of S, then:*

1. $A \cap (\cap_{\lambda\in\Lambda} B_\lambda) = \cap_{\lambda\in\Lambda}(A \cap B_\lambda)$ *and* $A \cup (\cup_{\lambda\in\Lambda} B_\lambda) = \cup_{\lambda\in\Lambda}(A \cup B_\lambda)$.
2. $A \cap (\cup_{\lambda\in\Lambda} B_\lambda) = \cup_{\lambda\in\Lambda}(A \cap B_\lambda)$.
3. $A \cup (\cap_{\lambda\in\Lambda} B_\lambda) = \cap_{\lambda\in\Lambda}(A \cup B_\lambda)$.

Proof. I will only prove part 2; leaving parts 1 and 3 as exercises. Accordingly, suppose first that:

$$x \in A \cap \left(\bigcup_{\lambda\in\Lambda} B_\lambda\right).$$

Then from the definitions we have:

$$x \in A \tag{1.3}$$

and

$$x \in \bigcup_{\lambda\in\Lambda} B_\lambda \tag{1.4}$$

From (1.4), there exists $\mu \in \Lambda$ such that

$$x \in B_\mu; \tag{1.5}$$

and, from (1.3) and (1.5), we then have:

$$x \in A \cap B_\mu.$$

Therefore, since $\mu \in \Lambda$,

$$x \in \bigcup_{\lambda \in \Lambda} (A \cap B_\lambda);$$

and we conclude that:

$$A \cap (\bigcup_{\lambda \in \Lambda} B_\lambda) \subseteq \bigcup_{\lambda \in \Lambda} (A \cap B_\lambda).$$

Conversely, suppose now that:

$$x \in \bigcup_{\lambda \in \Lambda} (A \cap B_\lambda).$$

Then for some $\lambda^* \in \Lambda, x \in A \cap B_{\lambda^*}$; so that:

$$x \in A \text{ and } x \in B_{\lambda^*} \subseteq \bigcup_{\lambda \in \Lambda} B_\lambda.$$

Therefore,

$$x \in A \cap (\bigcup_{\lambda \in \Lambda} B_\lambda),$$

and we conclude that:

$$\bigcup_{\lambda \in \Lambda} (A \cap B_\lambda) \subseteq A \cap (\bigcup_{\lambda \in \Lambda} B_\lambda). \quad \square$$

The following is a very fundamental result in set theory, and we will make use of it many times throughout this book.

1.7. Theorem. (*DeMorgan's Laws*). *Let S and A be non-empty sets, and let $\mathcal{X} \equiv \{X_a \mid a \in A\}$ be a family of subsets of S. Then we have:*

1. 'the complement of the union equals the intersection of the complements,' that is:

$$(\bigcup_{a \in A} X_a)^c = \bigcap_{a \in A} (X_a)^c$$

and

2. 'the complement of the intersection equals the union of the complements,' that is:

$$(\bigcap_{a \in A} X_a)^c = \bigcup_{a \in A} (X_a)^c.$$

Proof. I will only prove part 1; leaving the proof of part 2, which is similar, as an exercise. Suppose that

$$x \in (\bigcup_{a\in A} X_a)^c.$$

Then there exists no $a \in A$ such that $x \in X_a$, so that

$$(\forall a \in A) : x \in (X_a)^c;$$

and thus

$$x \in \bigcap_{a\in A} (X_a)^c.$$

Since x was arbitrary, it then follows that:

$$(\bigcup_{a\in A} X_a)^c \subseteq \bigcap_{a\in A} (X_a)^c. \tag{1.6}$$

To prove the converse, let y be an arbitrary element of the intersection of the complements [the right-hand-side of (1.6)]. Then we see that:

$$(\forall a \in A) : y \notin X_a;$$

and thus

$$y \notin \bigcup_{a\in A} X_a.$$

Therefore, since y was arbitrary, it follows that

$$\bigcap_{a\in A} (X_a)^c \subseteq (\bigcup_{a\in A} X_a)^c;$$

and, combining this with (1.6), we see that

$$\bigcap_{a\in A} (X_a)^c = (\bigcup_{a\in A} X_a)^c;$$

as was to be proved. □

Exercises.

1. Prove parts 1 and 3 of Proposition 1.6.
2. Prove Part 2 of Theorem 1.7.
3. Show that if $A \subseteq B$, and $C \subseteq D$, then $A \times C \subseteq B \times D$.
4. Show that, for any sets, A, B, C, and D, we have:

a. $(A \times B) \cup (C \times D) \subseteq (A \cup C) \times (B \cup D)$, and

b. $(A \cap C) \times (B \cap D) = (A \times B) \cap (C \times D)$.

5. Provide an example to show that the inclusion in (a) of exercise 4, above, may be proper; that is, find examples of sets A, B, C, and D, such that:

$$(A \times B) \cup (C \times D) \neq (A \cup C) \times (B \cup D).$$

6. Show that DeMorgan's Laws can be generalized in the following way. Let A and S be non-empty sets, let X be a subset of S, and let $\mathcal{V} = \{V_a \,|\, a \in A\}$ be a non-empty family of subsets of S. Then we have:

a. $X \setminus (\cup_{a\in A} V_a) = \cap_{a\in A}(X \setminus V_a)$, and

b. $X \setminus (\cap_{a\in A} V_a) = \cup_{a\in A}(X \setminus V_a)$.

7. Generalize problem 4b to show the following: if A_i and B_i are subsets of S_i, for $i = 1, \ldots, n$, then

$$\left(\prod_{i=1}^n A_i\right) \cap \left(\prod_{i=1}^n B_i\right) = \prod_{i=1}^n (A_i \cap B_i).$$

1.2 Properties of the Real Numbers

The following is a list of the basic properties of the standard (weak) inequality on $\mathbf{R}$, and of absolute values on $\mathbf{R}$. Since we are treating this material as review, we will simply take these properties as axiomatic.

A. Inequalities. If $a, b, c, d \in \mathbf{R}$ (that is, if a, b, c, and d are any real numbers), then:

1. if $a \le b$ and $c \le d$, then $a + c \le b + d$,
2. if $c \ge 0$, then
   $a \le b \Rightarrow ca \le cb$,
3. if $c \le 0$, then
   $a \le b \Rightarrow cb \le ca$,
4. if $a \ge b > 0$, then
   $0 < 1/a \le 1/b$.

B. Absolute Value.

1. **Definition.** If $x \in \mathbf{R}$, $|x|$, the **absolute value of x**, is defined by:

$$|x| = \left\{ \begin{array}{cl} x & \text{if } x \ge 0 \\ -x & \text{if } x < 0 \end{array} \right\} = \max\{x, -x\},$$

2. for $a, b \in \mathbf{R}$:
   $|a| \le b \iff -b \le a \le b$,
3. for $a \in \mathbf{R}$:
   $|a| = |-a|, |a|^2 = a^2$, and $\sqrt{a^2} = |a|$,
4. for all $a, b \in \mathbf{R}$:
   $|ab| = |a| \cdot |b|$ and $[a^2 \le b^2 \iff |a| \le |b|]$,
5. for $a, b \in \mathbf{R}$:
   $|a + b| \le |a| + |b|$.

Using the above properties of inequalities and absolute value, one can prove the following result, although I will leave the proof as an exercise.

1.8. Proposition. *If x and y are real numbers, then* $\big|\, |x| - |y| \,\big| \le |x - y|$.

1.9. Definition. Let A be a non-empty set of real numbers. A number $\alpha \in \mathbf{R}$ is said to be an **upper bound for A** [respectively, **lower bound for A**] iff we have:

$$(\forall x \in A)\colon x \leq \alpha \text{ [respectively, } (\forall x \in A)\colon x \geq \alpha \text{]}.$$

The set A will be said to be **bounded above** [respectively, **below**] iff there exists $\alpha \in \mathbf{R}$ which is an upper [respectively, lower] bound for A.

1.10. Definition. Let A be a non-empty set of real numbers. A number $\alpha \in \mathbf{R}$ is said to be the **least upper bound**, or **supremum of A**, abbreviated 'l.u.b. of A' or 'sup A,' iff the following two conditions hold:

1. α is an upper bound for A, and
2. if β is an upper bound for A, then $\alpha \leq \beta$.

Notice that a least upper bound for a set A is exactly what it says; that is, it is (1) an upper bound for A, and (2) at least as small as any other upper bound for A. More simply stated, it is the *smallest* upper bound for A (following usual practice, however, we will usually denote it by 'sup A'). In connection with this last point, it should be clear that the supremum of a set is unique (see exercise 1, at the end of this section).

Since the purpose of this section is to present a review of the properties of the real number system, we will not attempt to present a set of axioms defining the real number system; however, because of the importance of the following condition, we will present it formally as an axiom.

The Axiom of Completeness for R. *Any non-empty set of real numbers that is bounded above has a least upper bound (supremum).*

1.11. Definition. If A is a non-empty set of real numbers, a number $\alpha \in \mathbf{R}$ will be said to be the **greatest lower bound for A**, or the **infimum of A**, abbreviated 'g.l.b. of A' or 'inf A,' iff:

1. α is a lower bound for A, and
2. if β is a lower bound for A, then $\alpha \geq \beta$.

Once again, notice how natural the terminology 'greatest lower bound for A' is; on the other hand, we will confuse things slightly again by generally using the expression 'inf A' to denote the greatest lower bound for A. Our first use of the Axiom of Completeness for **R** will be in proving the following result.

1.12. Theorem. *Any non-empty set of real numbers that is bounded below has a greatest lower bound (infimum).*

Proof. Exercise (Hint: consider the set $-A \stackrel{\text{def}}{=} \{x \in \mathbf{R} \mid -x \in A\}$). □

1.13. Definition. A non-empty set $A \subseteq \mathbf{R}$ will be said to be **bounded** iff there exists $\alpha \in \mathbf{R}$ satisfying:

$$(\forall x \in A)\colon |x| \leq \alpha.$$

The following result constitutes an extremely useful characterization of $\sup A$, and we will use it frequently in the remainder of this book.

1.14. Theorem. *Let A be a non-empty set of real numbers. Then $\alpha = \sup A$ if, and only if, for every positive real number, ϵ, we have:*
1. $(\forall x \in A)\colon x < \alpha + \epsilon$, and
2. $(\exists x' \in A)\colon x' > \alpha - \epsilon$.

Proof. Suppose $\alpha = \sup A$, and let $\epsilon > 0$ be given. Obviously we must have

$$(\forall x \in A)\colon x < \alpha + \epsilon,$$

since α is an upper bound for A. Now suppose, by way of obtaining a contradiction, that condition 2 does not hold; that is, that

$$(\forall x \in A)\colon x \leq \alpha - \epsilon. \tag{1.7}$$

Then, defining

$$\beta = \alpha - \epsilon,$$

we see that [by (1.7)], β is an upper bound for A. However, this is impossible, for $\beta < \alpha$, and by assumption α is the *least* upper bound for A. Therefore (1.7) cannot hold, that is, there must exist some $x' \in A$ satisfying:

$$x' > \alpha - \epsilon.$$

I will leave the proof of the converse, that is, that if α satisfies conditions 1 and 2, then $\alpha = \sup A$, as an exercise. □

The following result is of interest as much because of the considerations used in its proof as for the statement of the result itself. Incidentally, having implied that it is particularly important that you read and thoroughly understand the proof of the next result, I suppose I really should remind you that few of the results or proofs in this text are really original with me; in particular, I must reluctantly admit that this proposition is a standard result in analysis texts, and the proof presented here is a standard line of argument as well.

1.15. Proposition. *Suppose that A and B are non-empty sets of real numbers satisfying:*

$$(\forall a \in A)(\forall b \in B)\colon a \geq b.$$

Then $\inf A$ and $\sup B$ both exist, and we have $\inf A \geq \sup B$.

Proof. Let b^* be an arbitrary (fixed) element of B. Then we note that it follows from our hypothesis that b^* satisfies

$$(\forall a \in A) \colon a \geq b^*;$$

and thus b^* is a lower bound for A. It then follows from Theorem 1.12 that $\alpha = \inf A$ exists; and, since α is the *greatest* lower bound for A, that

$$\alpha \geq b^*.$$

Next recall that b^* was an *arbitrary* element of B. Consequently, the argument of the preceding paragraph establishes that α is an upper bound for B; that is, that α satisfies:

$$(\forall b \in B) \colon \alpha \geq b.$$

Therefore it follows from the Axiom of Completeness that $\beta = \sup B$ exists; and, since β is the *least* upper bound for B, we have $\alpha \geq \beta$. □

In economic theory we often have occasion to consider sets which are scalar multiples or sums of other sets. In the case of subsets of the real numbers, we can define these as follows.

1.16. Definition. Let A and B be non-empty sets of real numbers, and let $\lambda \in \mathbf{R}$. Then we define the sets λA and $A + B$ by:

$$\lambda A = \{x \in \mathbf{R} \mid (\exists x' \in A) \colon x = \lambda x'\},$$

and

$$A + B = \{x \in \mathbf{R} \mid (\exists a \in A \,\&\, b \in B) \colon x = a + b\},$$

respectively.

Thus, to take an example, if $A = [0, 1], B = [1, 2]$, and $\lambda = 2$, then

$$\lambda A = [0, 2] \text{ and } A + B = [1, 3]$$

(it is a worthwhile exercise to prove the correctness of these two equalities). If, on the other hand, $A =]0, 1[$ and $B = \{1\}$, then

$$A + B =]1, 2[,$$

and so on. Notice that, for any non-empty set, $A, 0A = \{0\}$.

The supremum of a set which is a scalar multiple of another set, or which is the sum of two other sets, bears a simple relationship to the suprema of the original sets; as is shown in our next two results.

1.17. Proposition. *If A is a non-empty set of real numbers, which is bounded above, and λ is a nonnegative real number ($\lambda \in \mathbf{R}_+$), then* $\sup(\lambda A)$ *exists, and we have*

$$\sup(\lambda A) = \lambda \sup A.$$

Proof. Let us denote the supremum of A by 'α.' We distinguish two cases, according to the value of λ.

1. $\lambda = 0$. In this case, we have

$$\lambda A = \{0\},$$

and obviously

$$0 = \lambda\alpha = \sup(\lambda A).$$

2. $\lambda > 0$. In this case, we notice first that if $x \in \lambda A$, then there exists a point $x' \in A$ satisfying

$$x = \lambda x'.$$

But then, since $\alpha = \sup A$, we have

$$\alpha \geq x';$$

and therefore, since $\lambda > 0$,

$$x = \lambda x' \leq \lambda\alpha.$$

It then follows, since x is an arbitrary element of λA, that $\lambda\alpha$ is an upper bound for λA.

To prove that $\lambda\alpha$ is the *least* upper bound for λA, let $\beta \in \mathbf{R}$ be an upper bound for λA. Then we have, for any $x \in A$,

$$\lambda x \leq \beta;$$

and therefore, since $\lambda > 0$,

$$x \leq \beta/\lambda.$$

Consequently we see that β/λ is an upper bound for A; and, since α is the *least* upper bound for A, it follows that

$$\alpha \leq \beta/\lambda.$$

Therefore, since $\lambda > 0$,

$$\lambda\alpha \leq \beta,$$

and it follows that $\lambda\alpha = \sup(\lambda A)$. □

1.18. Proposition. *If A and B are non-empty sets of real numbers which are bounded above, then $A + B$ is bounded above, and*

$$sup(A + B) = sup\, A + sup\, B.$$

Proof. If we denote the suprema of A and B by 'α' and 'β,' respectively, it is easy to show that the set $A + B$ is bounded above by $\alpha + \beta$; and I will leave this part of the proof as an exercise. To prove that $\alpha + \beta$ is the *least* upper bound for $A + B$, let γ be an upper bound for $A + B$. It then follows from the definition of the set $A + B$ that, if we let b^* be an arbitrary element of B, we will have

$$(\forall a \in A) : a + b^* \leq \gamma,$$

and thus

$$(\forall a \in A) : a \leq \gamma - b^*.$$

But then we see that the number $\gamma - b^*$ is an upper bound for A, so that, since $\alpha = \sup A$,

$$\alpha \leq \gamma - b^*,$$

or

$$b^* \leq \gamma - \alpha.$$

However, b* was an arbitrary element of B, and thus we conclude

$$(\forall b \in B) : b \leq \gamma - \alpha;$$

so that, since $\beta = \sup B$, it then follows that

$$\beta \leq \gamma - \alpha,$$

and thus

$$\alpha + \beta \leq \gamma. \quad \square$$

Another property of the real numbers and/or the positive integers of which we will make frequent use, and also of which, hopefully, you are already aware, is the principle of mathematical induction; which we can state as follows.

Principle of Mathematical Induction. *If a subset, A, of $N \equiv \{1, 2, \ldots\}$ satisfies:*

a. $1 \in A$, *and*

b. for each $k \in A$, we also have $k + 1 \in A$,
then $A = N$.

We will simply take the statement of the principle of mathematical induction as an axiom. We can, however, use it to prove a very useful result, as follows.

Well-Ordering Principle for the Positive Integers. *Every non-empty subset of N has a least element; that is, if $A \subseteq N$ is non-empty, then there exists $k^* \in A$ satisfying:*

$$(\forall n \in A) \colon k^* \leq n. \tag{1.8}$$

Notice that if an element satisfies equation (1.8), then it is unique; that is, if k^* and m^* are elements of A, both of which satisfy (1.8), then $k^* = m^*$. Furthermore, we can equivalently express equation (1.8) as:

$$(\forall n \in A) \colon n \neq k^* \Rightarrow k^* < n. \tag{1.9}$$

We can prove the well-ordering principle as follows. Let A be a non-empty subset of N. If A contains the integer 1, then clearly 1 is a least element for A. Suppose, therefore, that $1 \notin A$; and define the set B by:

$$B = \{n \in N \mid (\forall m \in N) \colon m \leq n \Rightarrow m \notin A\}.$$

Then the set B is non-empty, since $1 \in B$. But then we also see that there must exist an integer, $k \in B$, such that $k + 1 \notin B$; since otherwise it would follow from the principle of mathematical induction that $B = N$, contradicting our assumption that A is non-empty. However, we can now see that k^* defined as

$$k^* = k + 1,$$

is the least element of A. For, first note that we must have $k^* \in A$; for otherwise, for all $n \in N$ such that $n \leq k^*$, we would have $n \notin A$ [recall that $k = k^* - 1$ *is* an element of B], which would imply that $k^* \in B$. Furthermore, if $n \in A$ were such that $n < k^*$, we would have $n \leq k$, and it would then follow that $n \in B$, contradicting the assumption that $n \in A$. Thus, if $n \in A$, we must have $n \geq k^*$. □

Exercises.

1. Show that if $\sup A$ exists, for some subset of $\mathbf{R}$, then it is unique; that is, if $\alpha = \sup A$ and $\beta = \sup A$, then we must have $\alpha = \beta$.
2. Show that, for any non-empty set, A, of real numbers, that A is bounded if, and only if, A is both bounded above and bounded below.
3. Complete the proof of Theorem 1.14.
4. Prove Proposition 1.8.

5. Prove that if A is any non-empty bounded set of real numbers, then there is a *smallest* closed interval containing A; that is, there exists a closed interval of real numbers, I, satisfying:

a. $A \subseteq I$, and

b. if $J \subseteq \mathbf{R}$ is any closed interval such that $A \subseteq J$, then $I \subseteq J$.

6. Formulate and prove a result analogous to Theorem 1.14 which characterizes the infimum of a set A.

7. Formulate and prove results analogous to Proposition 1.17 for:

a. $\inf(\lambda A)$, with $\lambda \geq 0$;

b. $\sup(\lambda A)$, for the case where $\lambda < 0$.

8. Formulate and prove a result analogous to Proposition 1.18 for $\inf(A+B)$.

9. Prove the **Archimedian Property** for the real numbers: if $a \in \mathbf{R}$, and b is a positive real number ($b > 0$), then there exists $n \in N$ such that $a < nb$. [Hint. Consider the set B defined as

$$B = \{nb \,|\, n \in N\}.$$

If it were the case that, for all $n, nb \leq a$, then the set B is bounded above; and, by the Axiom of Completeness, would have a least upper bound, c. You should now have no trouble in establishing that this implies a contradiction.]

10. Prove that if x is a nonnegative real number, then there exists a unique positive integer, n, satisfying

$$n - 1 \leq x < n.$$

1.3 Binary Relations

Whether or not you have encountered a formal definition of what is meant by a 'binary relation,' you certainly have encountered examples of such before this. The weak and strong inequalities on the real number system are both examples of binary relations, as is a consumer preference relation in economic theory (more on this later). Informally, a **binary relation**, R, on a set X, is simply a rule such that for each x and y in X, we can determine whether xRy, yRx, or neither, or both. Thus, for example, for any non-empty set X, we can define the relation E (for equality) by:

$$yEx \iff x = y.$$

To take another example, suppose $f\colon \mathbf{R} \to \mathbf{R}$, and define the relation R by:

$$yRx \iff y \geq f(x).$$

In this section, we will begin by studying the following properties of binary relations. *In the definitions to follow, and throughout the remainder of this section, we will suppose that the set on which the binary relation is defined is non-empty.*

1.19. Definitions. Let G be a binary relation on a set X. We shall say that G is:

1. **total** iff:

$$(\forall x, y \in X) \colon xGy \text{ or } yGx \text{ or } x = y.$$

2. **reflexive** iff:

$$(\forall x \in X) \colon xGx.$$

3. **irreflexive** iff

$$(\forall x \in X) \colon \neg xGx.$$

4. **symmetric** iff:

$$(\forall x, y \in X) \colon xGy \Rightarrow yGx.$$

5. **asymmetric** iff:

$$(\forall x, y \in X) \colon xGy \Rightarrow \neg yGx.$$

6. **antisymmetric** iff:

$$(\forall x, y \in X) \colon [xGy \;\&\; yGx] \Rightarrow x = y.$$

7. **transitive** iff:

$$(\forall x, y, z \in X) \colon [xGy \;\&\; yGz] \Rightarrow xGz.$$

In considering the definitions just stated, it is important to notice that, for example, irreflexivity is *not* the negation of reflexivity; that is, if a relation is not reflexive, it may nonetheless not be irreflexive, and conversely. Similarly, a relation which is not symmetric is not necessarily asymmetric; conversely, a relation may fail to satisfy asymmetry, yet not be symmetric. On the other hand, notice that any asymmetric relation is also antisymmetric and irreflexive. In the examples which follow, we provide illustrations of these points.[4]

1.20. Examples/Exercises.

1. Consider the usual weak inequality relation, $\geq$, on $\mathbf{R}$. Show that $\geq$ is total, reflexive, antisymmetric, and transitive. Incidentally, here is an example of a binary relation which is neither symmetric nor asymmetric.

[4] In this connection, it should be mentioned that the requirement in Definitions 1.19 that the set X be non-empty is there for essentially one reason; namely, that a binary relation on the empty set satisfies all of the conditions, 1–7, of Definition 1.19. Consequently, if we include the empty set in our definitions, the relationships among these conditions become somewhat confused!

2. Show that the usual strict inequality relation, $>$, on the real numbers is total, irreflexive, asymmetric (and thus antisymmetric, since asymmetry implies antisymmetry), and transitive.

3. Let X be the set of all physical objects on the earth at the present time, and define the relation R on X by:

$xRy \iff x$ is at least as heavy as y.

Show that R is total, reflexive, and transitive.

(Actually, this is something of a trick question, since it is really an empirical, and not a mathematical issue. In order to arrive at something which you can prove, assume that mass can be measured to any degree of accuracy that we choose.)

4. Let $f: X \to \mathbf{R}$, where X is any non-empty set, and define E on X by:

$$xEy \iff f(x) = f(y).$$

Show that E is reflexive, symmetric, and transitive. □

Most of the binary relations which we encounter in economic theory are orderings of one type or another, where I will use the term '**ordering**' to refer to any transitive relation. The first particular type of ordering to be considered is defined as follows.

1.21. Definition. Let G be a binary relation on a non-empty set, X. We shall say that G is a **weak order** (or that **G is a weak ordering of X**) iff G is total, reflexive, and transitive.

1.22. Examples/Exercises.

1. It follows immediately from Example 1.20.1–2 that the usual weak inequality, $\geq$, on the real numbers is a weak order, but that the strict inequality for the real numbers, $>$, is not. Since $\geq$ is antisymmetric, it is an example of a more restrictive type of order than a weak order, which we will define shortly.

2. From Example 1.20.3 it follows that the relation 'is at least as heavy as' is a weak order.

3. Let X be any non-empty set, and let $f: X \to \mathbf{R}$. Show that the relation G defined on X by:

$$xGy \iff f(x) \geq f(y),$$

is a weak order on X. □

In working with orderings, it is often useful to divide the relation up by separating it into its so-called asymmetric and symmetric parts, which terminology is justified by the following result (the proof of which is more or less immediate, and will be left as an exercise).

1.23. Proposition. *Let G be a binary relation on a set X, and define P and I on X by:*

$$xPy \iff [xGy \;\&\; \neg yGx], \tag{1.10}$$

and

$$xIy \iff [xGy \;\&\; yGx], \tag{1.11}$$

respectively. Then P is asymmetric and I is symmetric.

1.24. Definitions. If G is a binary relation on X, we define the **asymmetric part of G, P**, and the **symmetric part of G, I**, by equations (1.10) and (1.11), respectively.

1.25. Definition. If X is a non-empty set, and R is a binary relation on X, we shall say that ***R* is an equivalence relation on *X*** iff R is reflexive, symmetric, and transitive.

Using the terminology just introduced, in Exercise 1.20.4 you were asked to show that the relation, E, defined there is an equivalence relation. With the use of the definitions just presented, we can prove the following, although the proof will be left as an exercise.

1.26. Proposition. *If G is a weak order on X, then:*

1. the asymmetric part of G, P, is irreflexive, asymmetric, and transitive, and

2. the symmetric part of G, I, is an equivalence relation.

1.27. Examples/Exercises.

1. In economic theory we usually suppose that a consumer's choice of a 'commodity bundle' is limited to some non-empty subset, X, of $\mathbf{R}^n$, called the **consumption set**; and where for $\mathbf{x} = (x_1, \ldots, x_n) \in X$,

'x_i' denotes the amount of the ith commodity available to the consumer, per unit of time.

In this context, the consumer is supposed to choose according to her/his 'preference relation,' G, defined over the consumption set; where by this we mean the following. Let $\mathcal{B}$ be a family of non-empty subsets of X; for example, if $X \subseteq \mathbf{R}^n$, the family $\mathcal{B}$ would typically be defined as the family of all sets (**budget sets** of the form:

$$b(\mathbf{p}, w) = \{\mathbf{x} \in X \mid \mathbf{p} \cdot \mathbf{x} \leq w\},$$

for $\mathbf{p} \in \mathbf{R}^n_{++}$ and $w \geq \inf\{\mathbf{p} \cdot \mathbf{x} \mid \mathbf{x} \in X\}$. If we then define $h \colon \mathcal{B} \mapsto X$ by:[5]

$$h(B) = \{\mathbf{x} \in B \mid (\forall \mathbf{y} \in B) \colon \mathbf{x}G\mathbf{y}\},$$

[5]The symbol '$\mapsto$,' rather than '$\rightarrow$,' is used here to indicate that the values of h, $h(B)$ for $B \in \mathcal{B}$, will be subsets, instead of single points, of X.

it is assumed that, if the consumer is restricted to a choice from $B \in \mathcal{B}$, then he or she will always choose an element of $h(B)$. Accordingly, we shall refer to the correspondence, h, just defined as the consumer's **choice correspondence**.

Typically one assumes that this preference relation is a weak order (and thus we should perhaps refer to this relation as the 'at least as good as' relation; although it is usually called the **weak preference relation**), and the asymmetric part of G, P, is called the consumer's **strict preference relation**, while the symmetric part of G, I, is called the consumer's **indifference relation**. It of course follows from 1.26 that P is irreflexive, asymmetric, and transitive; while I is an equivalence relation.

2. If G is a weak order on X, and P and I are the asymmetric and symmetric parts of G, respectively, show that P and I satisfy the following conditions. For any w, x, y, and z in X:

a. we have

$$\neg xPy \iff yGx, \tag{1.12}$$

or, equivalently,

$$\neg yGx \iff xPy.$$

b. exactly one of the following conditions holds:

$$xPy, yPx, \text{or } xIy.$$

c. if wGx, xPy, and yGz, then wPz.
d. if xPz, then either xPy or yPz.

Proof of d. Suppose that $\neg xPy$. Then by (1.12), yGx; and, since xPz, it then follows at once from part c that yPz.

3. As we have noted above, it is generally accepted practice to regard the economic theory of consumer behavior as being based upon the idea that, given any two commodity bundles, x and y, a 'consumer' can always tell us (the economic investigator) whether x is at least as good as y, or whether y is at least as good as x, or both. More formally, we take the notion of at least as good as (the binary relation, G) to be a primitive (a basic, undefined, building block) of our theory, and usually assume that it is a weak order. The consumer's strict preference relation, P, and indifference relation, I, are then 'derived concepts' of the theory; that is, they are defined from G.

The foregoing considerations may raise a question in your mind; namely, what happens if we make (strict) preference the basic primitive of the theory? In other words, suppose we begin our theoretical considerations with the assumption that the consumer has a (strict) preference relation, P, which is asymmetric, irreflexive, and transitive; which are the properties which P

has been shown to satisfy in Proposition 1.26. We can then define, G, the at-least-as-good-as relation, from P by (1.12), above. Do we now have an equivalent theory? The answer to this last question is 'no', as is shown by the following (generic) example.

Let X be any non-empty set, let $f\colon X \to \mathbf{R}$ be any real-valued function defined on X, and let $\delta \in \mathbf{R}_{++}$ be a strictly positive real number. If we then define the relation P on X by:

$$xPy \iff f(x) > f(y) + \delta, \tag{1.13}$$

you should have no difficulty in proving that P is irreflexive, asymmetric, and transitive. On the other hand, if we define the relation G from P by (1.12), then the relation G will generally not be transitive (and thus will not be a weak order). For example, let $X = \mathbf{R}_{+}$, let f be given by $\mathbf{i}$ (the identity function), and let $\delta = 1$. Then in this case, the relation G will be given by:

$$xGy \iff y \leq x + 1;$$

and, if we let $x = 0$, $y = 1$, and $z = 2$, we see that:

$$xGy, yGz, \text{but } \neg xGz. \quad \square$$

The additional property which the asymmetric part of a weak order satisfies (additional, that is, to the properties set out in Proposition 1.26.1) is the following.

1.28. Definition. We shall say that a relation, P, on a set X is **negatively transitive** iff for all $x, y, z \in X$, we have:
if xPz, then either xPy or yPz.

While the condition defining negative transitivity undoubtedly appears odd at first reading, notice that, in the case of a (strict) preference ordering, what it says is this: If x is preferred to z, and y is any alternative not preferred to (no better than) z, then x must be preferred to y as well. Using this condition, we can now prove the following result.

1.29. Proposition. *If G is a weak order on X, then P, its asymmetric part, is negatively transitive. Conversely, if P is a relation which is asymmetric and negatively transitive, and we define G on X by (1.12), above, then G is a weak order on X, and P is its asymmetric part.*

Proof. We have already proved that if G is a weak order, then its asymmetric part, P, is negatively transitive; this was 1.27.2.d, above. Suppose now that P is asymmetric and negatively transitive, and let G be defined from P by (1.12). I will leave the proof that G is then total and reflexive as an exercise (see exercise 2 at the end of this section). To prove that G is transitive, let x, y, and z be such that:

$$xGy \text{ and } yGz, \tag{1.14}$$

and suppose, by way of obtaining a contradiction, that $\neg xGz$. Then it follows from (1.12) that zPx. But this is impossible; for it would then follow from negative transitivity that either zPy or yPx, and either of these conditions contradicts (1.14). We conclude, therefore, that G is transitive, and thus is a weak order. The fact that P is the asymmetric part of G then follows from [the negation of] (1.12) [see 1.27.2.a, above]. □

1.30. Corollary. *If P is a binary relation which is asymmetric and negatively transitive, then P is also transitive.*

It should be noted that a binary relation may be irreflexive and negatively transitive without being transitive; as an example, consider the standard inequality relation ($\neq$).

Proposition 1.29 provides the answer to the question addressed earlier: if we begin our development of consumer preference theory with strict preferences as the basic primitive of the theory, and we assume that the strict preference relation, P, is asymmetric and negatively transitive, our theory is equivalent to one which takes the at-least-as-good-as relation, G, as the primitive, and assumes that G is a weak order.

In our discussions of consumer theory for the remainder of this book, we will always take the notion of strict preference, P, as the primitive building block of the theory. Moreover, while we will generally assume that the relation P is asymmetric and transitive (and we will refer to a binary relation satisfying these two properties as an **asymmetric order**),we will try to avoid assuming that is is also negatively transitive. There are several reasons why I would rather not assume, other things being equal, that a consumer's strict preference relation is negatively transitive. Of these I will mention two.

First of all, as was pointed out by W. E. Armstrong [1939] almost sixty years ago, it may be that for a given consumer an alternative, x, is indifferent to y, and y is indifferent to z, yet it may nonetheless be the case that x is preferred to z. The reasoning behind this sort of contention is that the difference between x and y may be too small to notice, and such may also be the case as regards the comparison between y and z; yet the difference between x and z may nonetheless be sufficiently great as to result in the preference of x over z. As a theoretical device to deal with this and similar phenomena in the the area of psychophysics, R. Duncan Luce [1956] developed the notion of a semi-order, an example of which was presented in Example 1.27.3, above. In that example, the relation P was easily seen to be asymmetric and transitive (and thus is an asymmetric order), but it is not negatively transitive.

The second difficulty is that, while we are likely to casually identify the primitive 'consumer' with an individual decision-maker, the 'real world' counterpart with which we would usually identify this theoretical term is the notion of a 'household' as defined by the Bureau of the Census. The difficulty which this fact creates for the theory is that a household typically contains more than one member.

Let's consider for a moment, the simplest sort of non-single-person household; namely one containing just two persons, whom we will, for the sake of convenience (and in case any of my relatives read this book), suppose are husband and wife. Suppose each has a well-defined (weak) preference relation, G_i ($i = 1$ for the wife, and $i = 2$ for the husband), which is a weak order over the household consumption set, X, which we shall suppose is a non-empty subset of $\mathbf{R}^n$. Given a price vector $\mathbf{p} \in \mathbf{R}^n_{++}$, and a household income (or wealth) level, w, the household's budget set, $b(\mathbf{p}, w)$ is given by:

$$b(\mathbf{p}, w) = \{\mathbf{x} \in X \mid \mathbf{p} \cdot \mathbf{x} \leq w\}.$$

The question now is, how does the household choose a 'best point' from this budget set? I do not propose to try to develop a complete theory of household decision-making here, but it certainly does not seem at all likely that the household will end up with a choice $\mathbf{x}^*$, satisfying:

$$\mathbf{x}^* \in b(\mathbf{p}, w) \text{ and } (\forall \mathbf{x} \in b(\mathbf{p}, w)) \colon \mathbf{x}^* G_i \mathbf{x},$$

for either $i = 1$ or $i = 2$; rather, it seems that some sort of compromise solution is likely to be reached. However, regardless of the way in which Ms. 1 and Mr. 2 go about reconciling their diverse preferences, it seems likely that the final choice, $\mathbf{x}^*$, will be such as to satisfy:

$$\mathbf{x}^* \in b(\mathbf{p}, w) \text{ and } (\forall \mathbf{x} \in b(\mathbf{p}, w)) \colon \neg[\mathbf{x} P_1 \mathbf{x}^* \ \&\ \mathbf{x} P_2 \mathbf{x}^*]; \tag{1.15}$$

that is, we would expect that there is no bundle in the budget set which is preferred to $\mathbf{x}^*$ by both Ms. 1 and Mr. 2.

These considerations suggest the following approach. We first define the binary relation, P, on X by:

$$\mathbf{x} P \mathbf{x}' \iff [\mathbf{x} P_1 \mathbf{x}' \ \&\ \mathbf{x} P_2 \mathbf{x}'],$$

where P_i is the strict preference relation derived from G_i, for $i = 1, 2$. We shall refer to P as the 'household's **unanimity ordering**,' and it is an easy exercise to prove that if each P_i is an asymmetric order, then so too is P. Notice also that we can now equivalently express the condition of (1.15) as:

$$\mathbf{x}^* \in b(\mathbf{p}, w) \text{ and } (\forall \mathbf{x} \in X) \colon \mathbf{x} P \mathbf{x}^* \Rightarrow \mathbf{p} \cdot \mathbf{x} > w. \tag{1.16}$$

However, if we define the weak preference relation, G, from P, as in (1.12), above; notice that we can equivalently express the condition (1.16) as:

$$\mathbf{x}^* \in b(\mathbf{p}, w) \text{ and } (\forall \mathbf{x} \in b(\mathbf{p}, w)) \colon \mathbf{x}^* G \mathbf{x}; \tag{1.17}$$

that is, the household's choice must be a bundle from $h(\mathbf{p}, w)$, as defined in Example 1.27.1, above. The thing to keep in mind here, however, is that under the present assumptions, G is not necessarily a weak order. In particular, G may not be transitive.[6]

Of great historical importance in the economic theory of consumer behavior is the notion of an ordinal utility function. In this connection, the following is the key concept.

1.31. Definition. If G is a binary relation on X, we shall say that a function $f: X \to \mathbf{R}$ **represents G on X** iff for all x and y in X, we have:

$$xGy \iff f(x) \geq f(y). \tag{1.18}$$

(In the special case in which G is a consumer preference relation, we shall say that a function f which represents G on X is an **[ordinal] utility function for G**.) In the case in which a function f exists which represents G on X, we shall say that **there exists a representation for G**, or that G **admits of a real-valued representation.**

You should have no difficulty in proving that a function, f, represents a binary relation, G, if and only if, for all $x, y \in X$,

$$xPy \iff f(x) > f(y), \tag{1.19}$$

where 'P' denotes the asymmetric part of G. It is often more convenient to use (1.19) instead of (1.18) in dealing with representations of weak orders; and when this is the case, I will feel free to do so.

In the special case in which X is a finite set, it is very easy to set out necessary and sufficient conditions for the existence of a function f which represents a binary relation, G. In fact, we have the following.

1.32. Proposition. *Suppose that X is a finite set, and that G is a binary relation on X. Then there exists $f: X \to \mathbf{R}$ such that f represents G if, and only if, G is a weak order.*

Proof. It follows at once from 1.22.3 that if there exists a function representing G, then G must be a weak order. To prove the converse, suppose G is a weak order, and define the real-valued function f on X by:

$$f(x) = \#\{y \in X \mid xGy\} = \#xG,$$

where we define xG by:

$$xG = \{z \in X \mid xGz\}$$

[6] I hope it goes without saying that the approach taken here should not be regarded as definitive. For an alternative approach to the reconciliation of individual and household preferences, see Samuelson [1956], Chipman and Moore [1979], and Dow and Sonnenschein [1986].

and where we use the notation '$\#A$' to denote the number of elements in the set A (for a finite set, A); thus, $f(x)$ is the number of elements, y, of X such that xGy. To prove that f represents G, suppose first that xGy. Then if $z \in X$ is such that yGz, it follows from the transitivity of G that xGz as well. Therefore,

$$xG \equiv \{z \in X \mid xGz\} \supseteq \{v \in X \mid yGv\} \equiv yG,$$

and it follows that:

$$f(x) \equiv \#xG \geq \#yG \equiv f(y).$$

Conversely, suppose that x' and y' are such that $x'Py'$. Then

$$x'Gy', \tag{1.20}$$

and

$$\neg x'Gy'. \tag{1.21}$$

From (1.20) and the transitivity of G, it is easy to see that

$$y'G \subseteq x'G; \tag{1.22}$$

while from the reflexivity of G and (1.21), we see that:

$$x' \in x'G \text{ and } x' \notin y'G. \tag{1.23}$$

From (1.22) and (1.23) it follows that:

$$f(x') \equiv \#x'G > \#y'G \equiv f(y').$$

Thus we have shown that:

$$x'Py' \Rightarrow f(x') > f(y'),$$

which is equivalent, given that G is total, to:

$$f(y') \geq f(x') \Rightarrow y'Gx'. \quad \square$$

The above result makes it clear that we cannot make use of the convenience that utility functions provide if we are basing our consumer preference theory upon the assumption that the consumer's strict preference relation is an asymmetric order. However, we can nonetheless prove that the consumer's choice correspondence will be well-defined in this situation. We will establish this fact in a very general context in Chapter 8, and we will here prove a special case which will be used as a lemma for our proof in Chapter 8. We begin by introducing the following definition, which is often used in the revealed preference literature.

1.33. Definition. Let P be a binary relation on a non-empty set, X. We say that P is **cyclic** iff, for some positive integer, n, there exist points, $x_1, x_2, \ldots, x_n \in X$ such that:

$$x_1 P x_2 \ \& \ x_2 P x_3 \ \& \ldots x_{n-1} P x_n,$$

but $x_n P x_1$. If no such cycle exists, we will say that P is **acyclic**.

You should have no difficulty in finding examples of binary relations which are acyclic, but not transitive; on the other hand, a transitive and irreflexive binary relation is easily shown to be acyclic, and an acyclic binary relation is also irreflexive and asymmetric. It is easily seen that an asymmetric order satisfies the hypotheses of the following result.

1.34. Proposition. *Suppose P is an acyclic binary relation on a nonempty set, X, let $\mathcal{B}$ be the family of finite non-empty subsets of X, and let G be defined from P by:*

$$xGx' \iff \neg[x'Px].$$

Then, given any $B \in \mathcal{B}$, we have:

$$h(B) = \{x \in B \mid (\forall y \in B)\colon xGy\} = \{x \in B \mid (\forall y \in X)\colon yPx \Rightarrow y \notin B\},.$$

is non-empty.

Proof. We will prove the contrapositive of the conclusion. Accordingly, suppose there exists $B^* \in \mathcal{B}$ such that $h(B^*) = \emptyset$. We will show that in this case, P must be cyclic.

Since B^* is finite, we may suppose that $\#B^* = k$, where k is a positive integer. Since $h(B^*) = \emptyset$, it follows at once that if $k = 1$, then P must be reflexive, and thus is cyclic. Suppose now that $k > 1$, and that P is irreflexive (if it is not irreflexive, then it is necessarily cyclic). Picking any element of B^* to label 'x_1,' we see that there must exist a second element of B^*, distinct from x_1 (since P is irreflexive) such that $x_2 P x_1$. Moreover, if $k > 2$, there must exist a third element of B^*, x_3, such that $x_3 P x_2$; and in fact, generalizing this argument, there must exist a labeling of the elements of B^*, $x_1, \ldots, x_k$, such that:

$$x_k P x_{k-1} \ \& \ldots \ \& \ x_3 P x_2 \ \& \ x_2 P x_1.$$

However, since $h(B^*) = \emptyset$, there must exist $j \in \{1, \ldots, k-1\}$ such that $x_j P x_k$. But we then have:

$$x_k P x_{k-1} \ \& \ldots \ \& \ x_{j+1} P x_j \ \text{ and } \ x_j P x_k,$$

and thus we see that P is cyclic. □

Returning to the problem of determining conditions sufficient to ensure that a weak order admits of a real-valued representation, it should be immediately noted that the issue becomes much more complicated when X is not a finite set; although it is easy to see that it is still necessary that G be a weak order in order that a representation exist. We will not pursue the problem of obtaining sufficient conditions for the existence of a real-valued representation further;[7] however, we will shortly present an example of an order which does not admit of a real-valued representation. Our example will actually satisfy conditions stronger than those defining a weak order; so before presenting the example, suppose we first turn our attention to defining some other types of orderings.

1.35. Definitions. Suppose G is a binary relation on a set X. We shall say that G is a:

1. **preorder** (or a **preordering**) iff G is reflexive and transitive.
2. **partial order** iff G is reflexive, anti-symmetric, and transitive.
3. **linear** (or **simple**) **order** iff G is total, reflexive, anti-symmetric, and transitive.

Thus a preorder which is also antisymmetric is a partial order, and a partial order which is also total (and thus not reasonably called 'partial') is a linear order. The chart on the next page, which is an adaptation of that appearing on p. 8 of Debreu (1959) shows the relationships among the types of orderings which we have introduced to this point.

1.36. Examples/Exercises.

1. The most familiar example of a linear order is the usual weak inequality, $\geq$, on $\mathbf{R}$. Moreover its asymmetric part, $>$, is almost a linear order itself; being antisymmetric (since it is asymmetric), transitive, and total.

2. Let $f\colon X \to \mathbf{R}$, and define G on X by:

$$xGy \iff f(x) \geq f(y) \quad \text{for } x, y \in X.$$

Then, as pointed out earlier (Exercise 1.22.3), G is a weak order. Supposing that X contains at least two distinct elements, under what condition(s), if any, will G be a linear order?

3. Let $X = \mathbf{R}^2_+$, and define $>_L$, the **lexicographic order**, on X by:

$$(x_1, x_2) >_L (y_1, y_2) \iff \left\{ \begin{array}{lll} x_1 > y_1, & \text{or} & \\ x_1 = y_1 & \text{and} & x_2 > y_2. \end{array} \right\} \tag{1.24}$$

It is easy to show that $>_L$ is total and asymmetric (and thus antisymmetric). To prove that $>_L$ is negatively transitive, suppose $\mathbf{x}, \mathbf{y}$, and $\mathbf{z}$ are elements of X, that

$$\mathbf{x} >_L \mathbf{z},$$

[7] For an excellent general discussion of this issue, see Fishburn [1970], pp. 1–39.

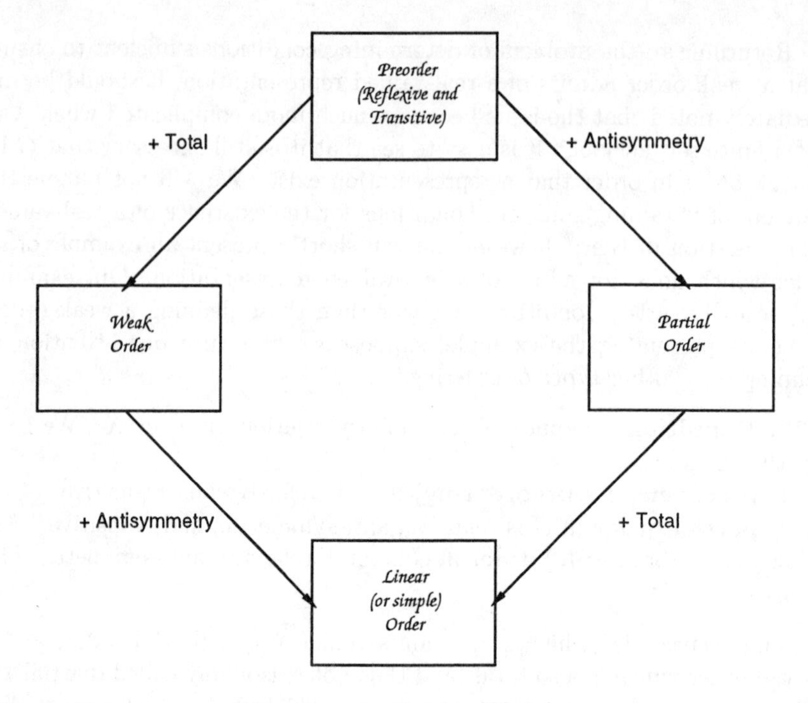

Figure 1.4: Types of Orders.

and that $\neg \mathbf{x} >_L \mathbf{y}$. Then it follows that we must have:

$$y_1 > x_1 \text{ or } [y_1 = x_1 \ \& \ y_2 \geq x_2]. \tag{1.25}$$

However, by assumption, $\mathbf{x} >_L \mathbf{z}$, and thus:

$$x_1 > z_1 \text{ or } [x_1 = z_1 \ \& \ x_2 > z_2.] \tag{1.26}$$

Upon comparing (1.25) and (1.26), it is easily seen that we must in any case have $\mathbf{y} >_L \mathbf{z}$.

Having proved that $>_L$ is negatively transitive, it follows at once from 1.29 (and our earlier observations regarding the asymmetry of $>_L$) that its negation, the relation $\geq_L$ defined by:

$$\mathbf{x} \geq_L \mathbf{y} \iff \neg \mathbf{y} >_L \mathbf{x},$$

is a weak order; and, in fact, since $\geq_L$ is obviously antisymmetric, we see that it is actually a linear order. The following argument (which is due to Debreu [1959, pp. 72-3]) depends upon some cardinal number concepts which you may not have previously encountered. However, I think you will find that the basic argument is clear enough, nonetheless.

Suppose, by way of obtaining a contradiction, that $\geq_L$ admits of a real-valued representation; so that there exists a function $f \colon X \to \mathbf{R}$ satisfying:

$$(\forall \mathbf{x}, \mathbf{y} \in \mathbf{R}^2_+) : \mathbf{x} >_L \mathbf{y} \iff f(\mathbf{x}) > f(\mathbf{y});$$

and, for the sake of convenience in the remainder of our argument, let us use the generic notation '(x, y)' to denote elements of $\mathbf{R}_+^2$. Then, for each $x \in \mathbf{R}_+$, we can define real numbers a_x and b_x by:

$$a_x = f(x, 0) \text{ and } b_x = \sup_{y \in \mathbf{R}_+} f(x, y). \tag{1.27}$$

Moreover, it is easy to see that, since for each $x \in \mathbf{R}_+$, $(x, 1) >_L (x, 0)$, we must have:

$$a_x \equiv f(x, 0) < f(x, 1) \leq b_x;$$

while, for $x, x^* \in \mathbf{R}_+$, such that $x^* > x$, similar considerations establish that:

$$b_x \leq a_{x^*}.$$

Thus we see that the family $\mathcal{J}$ given by:

$$\mathcal{J} = \{[a_x, b_x[\,|\, x \in \mathbf{R}_+\},$$

is a family of disjoint, non-degenerate intervals of real numbers; a distinct such interval for each nonnegative real number, x. But this is impossible, because there are only a countable number of such intervals; whereas there are an uncountable number of nonnegative real numbers. More crudely put, there are simply too many nonnegative real numbers to obtain a non-degenerate interval for each such that no two such intervals have any points in common if the corresponding x's are not equal![8] □

1.37. Definitions. For $\mathbf{x}, \mathbf{y} \in \mathbf{R}^n$, we define:

1. $\mathbf{x} \geq \mathbf{y}$ [read '$\mathbf{x}$ is greater than or equal to $\mathbf{y}$'] iff

$$x_i \geq y_i \quad \text{for } i = 1, \ldots, n.$$

2. $\mathbf{x} > \mathbf{y}$ [read '$\mathbf{x}$ is semi-greater than $\mathbf{y}$'] iff $\mathbf{x} \geq \mathbf{y}$, but $\mathbf{y} \not\geq \mathbf{x}$.

3. $\mathbf{x} \gg \mathbf{y}$ [read '$\mathbf{x}$ is strictly greater than $\mathbf{y}$'] iff

$$x_i > y_i \quad \text{for } i = 1, \ldots, n.$$

As you are probably already well aware, if $n = 1$, the distinction between $>$ and $\gg$ disappears. On the other hand, for $n \geq 2$, there is a real difference between the two. The weak inequality relation for $\mathbf{R}^n$ is a partial order, but is not a linear order if $n > 1$, as is stated in the following result. The proof of Theorem 1.38 is fairly easy, and will be left as an exercise. In 1.39 we introduce further examples of preorders which are not partial orders, and of partial orders which are not linear orders.

[8] You may find that it clears things up a bit if you work through Exercise 6, at the end of this section.

1.38. Theorem. *The weak inequality ($\geq$) for $\mathbf{R}^n$ is a partial order (that is, it is reflexive, antisymmetric, and transitive). However, $\geq$ is not total, for $n \geq 2$ (and thus it is not a linear order, for $n \geq 2$).*

1.39. Examples/Exercises.

1. Let $f\colon X \to \mathbf{R}^n$, and define G on X by

$$xGy \iff f(x) \geq f(y).$$

Show that G is a preorder on X. Can you state a condition under which G will be a partial order? Can you state a condition sufficient to insure that G is a linear order?

2. Show that both the semi-greater-than ($>$) and strict inequality ($\gg$) relations defined in 1.36.2 and 1.36.3, respectively, are irreflexive, asymmetric, and transitive. However, neither is total, if $n \geq 2$. □

We conclude this section with some definitions related to the weak inequality just defined for $\mathbf{R}^n$, and with some results stating further properties of $\geq$. The student should provide proofs of all the results stated here.

1.40. Theorem. *(Properties of $\geq$ on $\mathbf{R}^n$) For any $\mathbf{x}, \mathbf{y}, \mathbf{z} \in \mathbf{R}^n$, and any $\lambda \in \mathbf{R}$, we have:*

1. $\mathbf{x} \geq \mathbf{y} \Rightarrow (\mathbf{x} + \mathbf{z} \geq \mathbf{y} + \mathbf{z})$.
2. $\mathbf{x} \geq \mathbf{y} \Rightarrow |\lambda|\mathbf{x} \geq |\lambda|\mathbf{y}$.

1.41. Definitions. We define $\mathbf{R}^n_+$, the **nonnegative orthant** in $\mathbf{R}^n$, as the set of all nonnegative vectors in $\mathbf{R}^n$; that is

$$\mathbf{R}^n_+ = \{\mathbf{x} \in \mathbf{R}^n \mid \mathbf{x} \geq \mathbf{0}\};$$

and $\mathbf{R}^n_{++}$, the **strictly positive orthant** in $\mathbf{R}^n$, by:

$$\mathbf{R}^n_{++} = \{\mathbf{x} \in \mathbf{R}^n \mid \mathbf{x} \gg \mathbf{0}\},$$

where '$\mathbf{0}$' denotes the origin in $\mathbf{R}^n$ in each of the above statements.

In general, you should be careful to distinguish between the strictly positive orthant and the 'semi-positive orthant:'

$$\mathbf{R}^n_+ \setminus \{\mathbf{0}\} = \{\mathbf{x} \in \mathbf{R}^n \mid \mathbf{x} > \mathbf{0}\};$$

although in $\mathbf{R}$ (that is, for the case in which $n = 1$), the two sets coincide. We will refer to the elements of $\mathbf{R}^n_+$, $\mathbf{R}^n_+ \setminus \{\mathbf{0}\}$, and $\mathbf{R}^n_{++}$ as **nonnegative**, **semi-positive**, and **(strictly) positive vectors**, respectively.

Exercises.

1. Show that if H is *any* asymmetric relation on X, then the relation R is total and reflexive; where we define R on X by:

$$xRy \iff \neg yHx.$$

2. Show that the inequality $>$ on $\mathbf{R}^2$ is *not* negatively transitive.

3. Show that if $f\colon X \to \mathbf{R}$ represents the weak order G, and $F\colon f(X) \to \mathbf{R}$ is any strictly increasing function, then the composition of F and f, the function g defined by:

$$g(x) = F[f(x)] \quad \text{for } x \in X,$$

also represents G. It is because of this consideration that such representations are called **ordinal utility functions** in consumer demand theory.

4. Let $f\colon X \to X$, and define the relation R_f on X by:

$$xR_f y \iff x \geq f(y).$$

Which of the properties defined in 1.19 will this relation satisfy? Can you think of a condition on the function, f, under which R_f will be transitive? reflexive? total? [You may wish to begin by considering the problem of developing a sufficient condition for R_f to be reflexive.]

5. Let the sets N_0 and X be defined by:

$$N_0 = \{0, 1, 2, \dots\} = N \cup \{0\} \text{ and } X = N_0 \times \mathbf{R}_+,$$

respectively. Show that the lexicographic order, $\geq_L$, is representable on X in this case; in fact, it can be represented by the function f defined by:

$$f(x, y) = x + y/(1 + y) \quad \text{for } (x, y) \in X.$$

Show also that the numbers a_x and b_x defined in equation (1.27), above, are here given by:

$$a_x = x, \text{ and } b_x = x + 1,$$

respectively.

Actually, this example can be generalized as follows: let m be any positive integer, and define

$$Y = \{x \in \mathbf{R} \mid (\exists n \in N_0) : x = n/m\},$$

and

$$X = Y \times \mathbf{R}_+.$$

Show that $\geq_L$ is representable on this set X.

1.4 Inner Product, Norm, and Euclidean Metric

We begin this section by reviewing the properties of the inner product on $\mathbf{R}^n$, which is defined as follows.

1.42. Definition. For $\mathbf{x}, \mathbf{y} \in \mathbf{R}^n$, we define $\mathbf{x} \cdot \mathbf{y}$, the **inner product of x and y**, by:

$$\mathbf{x} \cdot \mathbf{y} = \sum_{i=1}^{n} x_i \cdot y_i.$$

The inner product is of particular interest in economics, because it is the natural representation of such things as expenditure and profit. For example, if within the context of the theoretical framework set out in 1.27.1, a consumer is considering the purchase of a commodity bundle $\mathbf{x} \in X$, given the vector of prices $\mathbf{p} \in \mathbf{R}^n$, then the necessary expenditure is given by $\mathbf{p} \cdot \mathbf{x}$. The following result lists the basic properties of the inner product on $\mathbf{R}^n$; its proof, which follows directly from the definition, will be left as an exercise.

1.43. Proposition. *(Basic properties of the inner product). If* $\mathbf{x}, \mathbf{y}$, *and* $\mathbf{z}$ *are elements of* $\mathbf{R}^n$, *and* $a \in \mathbf{R}$, *then:*

1. $\mathbf{x} \cdot \mathbf{x} \geq 0$ *and* $[\mathbf{x} \cdot \mathbf{x} = 0 \Leftrightarrow \mathbf{x} = \mathbf{0}]$.
2. $\mathbf{x} \cdot \mathbf{y} = \mathbf{y} \cdot \mathbf{x}$.
3. $\mathbf{x} \cdot (a\mathbf{y}) = a(\mathbf{x} \cdot \mathbf{y}) = (a\mathbf{x}) \cdot \mathbf{y}$.
4. $\mathbf{x} \cdot (\mathbf{y} + \mathbf{z}) = \mathbf{x} \cdot \mathbf{y} + \mathbf{x} \cdot \mathbf{z}$.

Our next result, while it may appear somewhat out of place in this section, is very useful in proving some of the properties of the euclidean norm.

1.44. Lemma. *If a, b, and c are real numbers satisfying:*

$$(\forall \lambda \in \mathbf{R}) \colon a\lambda^2 + 2b\lambda + c \geq 0, \tag{1.28}$$

then we must have:

$$a \geq 0, c \geq 0, \textit{and}\, ac \geq b^2.$$

Proof. "Clearly" we must have

$$a \geq 0 \;\&\; c \geq 0.$$

In order to prove that $ac \geq b^2$, we distinguish two cases.

1. $a = 0$. In this case we must also have $b = 0$; for if not, then taking

$$\lambda = -(1 + c)/2b,$$

we have:

$$a\lambda^2 + 2b\lambda + c = -c - 1 + c = -1 < 0.$$

Since this contradicts the hypothesis stated as equation (1.28), it follows that if $a = 0$, then $b = 0$; and thus in this case, that

$$ac = 0 = b^2.$$

2. $a > 0$. In this case, letting

$$\lambda = -b/a,$$

we have:

$$0 \le a\lambda^2 + 2b\lambda + c = a\left(\frac{-b}{a}\right)^2 + 2b\left(\frac{-b}{a}\right) + c = \frac{b^2}{a} - \frac{2b^2}{a} + c = c - \frac{b^2}{a}.$$

Therefore,

$$c - \frac{b^2}{a} \ge 0,$$

so that, recalling that $a > 0$, we obtain:

$$ac \ge b^2. \quad \square$$

1.45. Definition. For $\mathbf{x} \in \mathbf{R}^n$, we define $\|\mathbf{x}\|$, the **euclidean norm of x**, by:

$$\|\mathbf{x}\| = \left(\sum_{i=1}^{n} x_i^2\right)^{1/2} = (\mathbf{x} \cdot \mathbf{x})^{1/2}.$$

The following result establishes an inequality which is of use in a truly amazing variety of situations, and will be invaluable to us in our further investigation of euclidean space. Our first application of the inequality will be to establish the basic properties of the euclidean norm.

1.46. Theorem. (Cauchy-Schwarz Inequality). *For all* $\mathbf{x}, \mathbf{y} \in \mathbf{R}^n$, *we have* $|\mathbf{x} \cdot \mathbf{y}| \le \|\mathbf{x}\| \cdot \|\mathbf{y}\|$.

Proof. Let $\mathbf{x}, \mathbf{y} \in \mathbf{R}^n$, and let $\lambda \in \mathbf{R}$ be arbitrary. Then:

$$0 \le (\lambda\mathbf{x} + \mathbf{y}) \cdot (\lambda\mathbf{x} + \mathbf{y}) = \lambda^2\mathbf{x} \cdot \mathbf{x} + (2\mathbf{x} \cdot \mathbf{y})\lambda + \mathbf{y} \cdot \mathbf{y}.$$

Since this inequality holds for all $\lambda \in \mathbf{R}$, it follows from Lemma 1.44 that:

$$(\mathbf{x} \cdot \mathbf{x})(\mathbf{y} \cdot \mathbf{y}) \ge (\mathbf{x} \cdot \mathbf{y})^2.$$

Taking square roots of both sides, we then have:

$$|\mathbf{x} \cdot \mathbf{y}| \le \|\mathbf{x}\| \cdot \|\mathbf{y}\|. \quad \square$$

1.47. Theorem. *The euclidean norm,* $\|\cdot\|$, *satisfies, for each* $\mathbf{x}, \mathbf{y}, \mathbf{z} \in \mathbf{R}^n$, *and each* $\lambda \in \mathbf{R}$:

1. $\|\mathbf{x}\| \ge 0 \ \& \ \left[\|\mathbf{x}\| = 0 \Leftrightarrow \mathbf{x} = \mathbf{0}\right]$,
2. *(homogeneity):* $\|\lambda\mathbf{x}\| = |\lambda| \cdot \|\mathbf{x}\|$,
3. *(triangle inequality):* $\|\mathbf{x} + \mathbf{y}\| \le \|\mathbf{x}\| + \|\mathbf{y}\|$.

Proof. I will leave the proof of parts 1 and 2 of the conclusion as exercises; providing only a skeletal proof of part 3 here (the student should provide the justification for the steps in the following argument).

$$\|\mathbf{x}+\mathbf{y}\|^2 = (\mathbf{x}+\mathbf{y})\cdot(\mathbf{x}+\mathbf{y}) = \mathbf{x}\cdot\mathbf{x} + 2\mathbf{x}\cdot\mathbf{y} + \mathbf{y}\cdot\mathbf{y}$$

$$\leq \mathbf{x}\cdot\mathbf{x} + 2|\mathbf{x}\cdot\mathbf{y}| + \mathbf{y}\cdot\mathbf{y} \leq \mathbf{x}\cdot\mathbf{x} + 2\|\mathbf{x}\|\cdot\|\mathbf{y}\| + \mathbf{y}\cdot\mathbf{y} = (\|\mathbf{x}\| + \|\mathbf{y}\|)^2. \quad \square$$

Notice that it follows from property 2 of the list in 1.47 that, for any $\mathbf{x}\in\mathbf{R}^n$, $\|-\mathbf{x}\| = \|\mathbf{x}\|$.

The euclidean norm can be regarded as providing a measure of the length, or magnitude of a vector. We can use this norm to define the euclidean metric, which can be regarded as a measure of the distance between two points in $\mathbf{R}^n$, as follows.

1.48. Definition. We define the **euclidean metric**, usually denoted by 'd(·),' on $\mathbf{R}^n \times \mathbf{R}^n$ by:

$$d(\mathbf{x},\mathbf{y}) = \left[\sum\nolimits_{i=1}^{n}(x_i - y_i)^2\right]^{1/2} = [(\mathbf{x}-\mathbf{y})\cdot(\mathbf{x}-\mathbf{y})]^{1/2} = \|\mathbf{x}-\mathbf{y}\|. \quad (1.29)$$

While the sum of squares term in the above equation (the term following the first equality in the expression) is the basic definition, and the most useful of the above formulas when it comes to actually calculating the euclidean distance between two points, the last expression appearing in equation (1.29) is often more useful when developing a theoretical argument, as you will see in the next proof.

1.49. Theorem. *The euclidean metric satisfies the following properties: for each* $\mathbf{x},\mathbf{y},\mathbf{z}\in\mathbf{R}^n$, *and each* $\lambda\in\mathbf{R}$:

1. $d(\mathbf{x},\mathbf{y}) \geq 0$ *and* $[d(\mathbf{x},\mathbf{y}) = 0 \Leftrightarrow \mathbf{x}=\mathbf{y}]$,
2. *(symmetry):* $d(\mathbf{x},\mathbf{y}) = d(\mathbf{y},\mathbf{x})$,
3. *(triangle inequality):* $d(\mathbf{x},\mathbf{z}) \leq d(\mathbf{x},\mathbf{y}) + d(\mathbf{y},\mathbf{z})$,
4. *(homogeneity):* $d(\lambda\mathbf{x},\lambda\mathbf{y}) = |\lambda| d(\mathbf{x},\mathbf{y})$,
5. *(translation invariance)* $d(\mathbf{x}+\mathbf{z},\mathbf{y}+\mathbf{z}) = d(\mathbf{x},\mathbf{y})$.

Proof. Exercise. (You should have little trouble with this exercise if you remember that $d(\mathbf{x},\mathbf{y}) = \|\mathbf{x}-\mathbf{y}\|$, and use Theorem 1.47.) $\square$

We say that a subset of $\mathbf{R}^n$ is bounded if there exists a real number such that the distance between any two of its points is necessarily less than this number. More formally, we define a bounded set as follows.

1.50. Definition. A subset, A, of $\mathbf{R}^n$ is said to be **bounded** iff there exists a positive real number, α, satisfying:

$$(\forall \mathbf{x},\mathbf{y}\in A)\colon d(\mathbf{x},\mathbf{y}) < \alpha.$$

Hopefully, you are already quite familiar with the notion of the neighborhood of a point in euclidean space. In any case, we will make a great deal of use of the following definitions.

1.51. Definitions. For $\mathbf{x} \in \mathbf{R}^n$ and a positive real number δ, we define the **(euclidean) neighborhood of x with radius** δ, $N(\mathbf{x}, \delta)$, by:

$$N(\mathbf{x}, \delta) = \{\mathbf{y} \in \mathbf{R}^n \mid d(\mathbf{x}, \mathbf{y}) < \delta\};$$

and the **deleted neighborhood of x with radius** δ, $N'(\mathbf{x}, \delta)$, by:

$$N'(\mathbf{x}, \delta) = \{\mathbf{y} \in \mathbf{R}^n \mid 0 < d(\mathbf{x}, \mathbf{y}) < \delta\};$$

Notice that

$$N(\mathbf{x}, \delta) = N'(\mathbf{x}, \delta) \cup \{\mathbf{x}\};$$

or, equivalently,

$$N'(\mathbf{x}, \delta) = N(\mathbf{x}, \delta) \setminus \{\mathbf{x}\}.$$

We will often refer to these neighborhoods as 'the δ-neighborhood of $\mathbf{x}$' and 'the deleted δ-neighborhood of $\mathbf{x}$,' respectively.

An alternative way to define boundedness is to say that a set is bounded if it is contained in some euclidean neighborhood of the origin. As it turns out, this definition is equivalent to the one given in 1.50, as is established in the following result.

1.52. Proposition. *A subset, A, of $\mathbf{R}^n$ is bounded if, and only if, there exists $\epsilon \in \mathbf{R}_{++}$ such that:*

$$A \subseteq N(\mathbf{0}, \epsilon). \tag{1.30}$$

Proof. Suppose A is bounded, let $\alpha \in \mathbf{R}_{++}$ be such that:

$$(\forall \mathbf{x}, \mathbf{y} \in A) : d(\mathbf{x}, \mathbf{y}) < \alpha,$$

let $\mathbf{x}^*$ be an arbitrary element of A (to be held fixed in our argument), and define:

$$\epsilon = \|\mathbf{x}^*\| + \alpha.$$

Then if $\mathbf{x} \in A$, we have:

$$d(\mathbf{x}, \mathbf{0}) = \|\mathbf{x}\| = \|\mathbf{x} - \mathbf{x}^* + \mathbf{x}^*\| \leq \|\mathbf{x} - \mathbf{x}^*\| + \|\mathbf{x}^*\| < \alpha + \|\mathbf{x}^*\| \equiv \epsilon;$$

and thus we see that

$$A \subseteq N(\mathbf{0}, \epsilon).$$

Conversely, suppose equation (1.30) holds, for some $\epsilon \in \mathbf{R}_{++}$, and define

$$\alpha = 2\epsilon.$$

Then for $\mathbf{x}, \mathbf{y} \in A$, we have:

$$d(\mathbf{x}, \mathbf{y}) = \|\mathbf{x} - \mathbf{y}\| \leq \|\mathbf{x}\| + \|-\mathbf{y}\| = \|\mathbf{x}\| + \|\mathbf{y}\| < \epsilon + \epsilon = \alpha;$$

and we see that A is bounded. □

We will often have occasion to deal with the cartesian product of two or more euclidean spaces, and in doing so, it will often be necessary to consider relationships between the normal euclidean metric on the cartesian product and the metrics for the original spaces. The following result sets forth the relationships among the norms. In reading this result, be sure you understand the fact (and the reasons) that if

$$\sum_{i=1}^{n} m_i = k, \qquad \text{where } m_i \geq 1, \text{ for } i = 1, \ldots, n,$$

then we can consider $\mathbf{R}^k$ to be the same thing as $\mathbf{R}^{m_1} \times \mathbf{R}^{m_2} \times \cdots \times \mathbf{R}^{m_n}$ (and we will, whenever it is convenient to do so). When considering $\mathbf{R}^k$ to be the cartesian product, we will denote points in the space by:

$$\mathbf{x} = (\mathbf{x}_1, \ldots, \mathbf{x}_n) \text{ where } \mathbf{x}_i \in \mathbf{R}^{m_i} \text{ for } i = 1, \ldots, n;$$

and, where necessary for the sake of clarity, we shall denote the origin in $\mathbf{R}^{m_i}$ by '$\mathbf{0}_i$' (for $i = 1, \ldots, n$). Furthermore, we will, as in the following result, denote the euclidean norm for $\mathbf{R}^{m_i}$ by '$\|\cdot\|_i$,' for $i = 1, \ldots, n$.

1.53. Proposition. *Given the constructions of the previous paragraph, and denoting the euclidean norm on $\mathbf{R}^k$ by '$\|\cdot\|$,' for all $\mathbf{x} = (\mathbf{x}_1, \ldots, \mathbf{x}_n) \in \mathbf{R}^k$, we have:*

$$\|\mathbf{x}_i\|_i \leq \|\mathbf{x}\| \;\; \text{for } i = 1, \ldots, n; \tag{1.31}$$

and

$$\|\mathbf{x}\| \leq \sum_{i=1}^{n} \|\mathbf{x}_i\|_i. \tag{1.32}$$

Proof. The inequalities (1.31) are immediate, or nearly so, and I will leave the details of the proof as an exercise. To prove (1.32), we note first that we can write:

$$\mathbf{x} = (\mathbf{x}_1, \ldots, \mathbf{x}_n) = \sum_{i=1}^{n} \mathbf{y}_i, \tag{1.33}$$

where we define $\mathbf{y}_i$ by:

$$\mathbf{y}_i = (\mathbf{0}_1, \ldots, \mathbf{0}_{i-1}, \mathbf{x}_i, \mathbf{0}_{i+1}, \ldots, \mathbf{0}_n) \text{ for } i = 1, \ldots, n; \tag{1.34}$$

in other words, $\mathbf{y}_i$ is the vector having all its coordinates equal to zero, except those corresponding to the ith sub-vector, which is equal to $\mathbf{x}_i$. Notice that we have:

$$\|\mathbf{y}_i\| = \|\mathbf{x}_i\|_i \text{ for } i = 1, \ldots, n; \tag{1.35}$$

so that, combining this fact with (1.33) and the triangle inequality, it follows that:

$$\|\mathbf{x}\| \leq \sum_{i=1}^{n} \|\mathbf{y}_i\| = \sum_{i=1}^{n} \|\mathbf{x}_i\|_i. \quad \square$$

It is apparent from the above result that, for all $\mathbf{x} = (x_1, \ldots, x_n) \in \mathbf{R}^n$, we have:

$$|x_i| \leq \|\mathbf{x}\| \text{ for } i = 1, \ldots, n,$$

and

$$\|\mathbf{x}\| \leq \sum_{i=1}^{n} |x_i|.$$

The following corollary is also immediate, once we recall the relationship between the euclidean norm and metric.

1.54. Corollary. *Under the hypotheses of 1.53, and denoting the euclidean metric for $\mathbf{R}^{m_i}$ by '$d_i(\cdot)$,' for $i = 1, \ldots, n$, and that for $\mathbf{R}^k$ by '$d(\cdot)$,' we have, for all $\mathbf{x}$ and $\mathbf{y}$ in $\mathbf{R}^k$:*

$$d_i(\mathbf{x}_i, \mathbf{y}_i) \leq d(\mathbf{x}, \mathbf{y}) \text{ for } i = 1, \ldots, n,$$

while:

$$d(\mathbf{x}, \mathbf{y}) \leq \sum_{i=1}^{n} d_i(\mathbf{x}_i, \mathbf{y}_i).$$

Exercises.

1. Show that if $\mathbf{x}, \mathbf{y} \in \mathbf{R}^n_+$, then:

$$\mathbf{x} \cdot \mathbf{y} \geq 0.$$

2. Show that if $\mathbf{x}, \mathbf{y}, \mathbf{z} \in \mathbf{R}^n$, and $\mathbf{x} \geq \mathbf{0}$, then:

$$\mathbf{y} \geq \mathbf{z} \Rightarrow \mathbf{x} \cdot \mathbf{y} \geq \mathbf{x} \cdot \mathbf{z}.$$

3. Show that if $\mathbf{x}, \mathbf{y} \in \mathbf{R}^n$ satisfy:

$$\mathbf{x} \gg \mathbf{0} \text{ and } \mathbf{y} \geq \mathbf{0},$$

then

$$\mathbf{x} \cdot \mathbf{y} \geq 0 \text{ and } [\mathbf{x} \cdot \mathbf{y} = 0 \Leftrightarrow \mathbf{y} = \mathbf{0}].$$

4. Show that if $\mathbf{x}, \mathbf{y}, \mathbf{z} \in \mathbf{R}^n$ satisfy:

$$\mathbf{x} \gg \mathbf{0} \text{ and } \mathbf{y} \geq \mathbf{z},$$

then

$$\mathbf{x} \cdot \mathbf{y} \geq \mathbf{x} \cdot \mathbf{z} \text{ and } [\mathbf{x} \cdot \mathbf{y} = \mathbf{x} \cdot \mathbf{z} \Leftrightarrow \mathbf{y} = \mathbf{z}].$$

5. Show that if A is a subset of $\mathbf{R}^n$ such that there exist $\mathbf{a}, \mathbf{b} \in \mathbf{R}^n$ satisfying:

$$(\forall \mathbf{x} \in A) : \mathbf{a} \leq \mathbf{x} \leq \mathbf{b},$$

then A is bounded. [Hint: you may find it helpful to begin by using exercise 2 to establish that, for any $\mathbf{x} \in A$, $d(\mathbf{x}, \mathbf{a}) \leq d(\mathbf{b}, \mathbf{c})$.]

1.5 Open and Closed Sets in $\mathbf{R}^n$

Let's begin our review of the definitions and properties of open and closed sets in $\mathbf{R}^n$ by defining an interior point of a set. The collection of all interior points of a set is called the interior of the set, and this notion is then used to define an open set.

1.55. Definition. Let $\mathbf{x} \in \mathbf{R}^n$, and let X be a subset of $\mathbf{R}^n$. We shall say that **x is an interior point of *X*** iff there exists a positive real number ϵ such that:

$$N(\mathbf{x}, \epsilon) \subseteq X.$$

We shall denote the set of all interior points of X by '$intX$,' which set will be called the **interior of *X***.

Notice that for any subset, X, of $\mathbf{R}^n$, we must have:

$$intX \subseteq X.$$

[This is true for the empty set, $\emptyset$, also for the simple reason that $int\emptyset = \emptyset$; that is, $\emptyset$ has no interior points, and therefore $int\emptyset$ is empty.] Our next proposition presents a class of sets, X, for which $X \subseteq intX$ as well; but before turning to this result, it will be useful to consider a lemma, the proof of which I will leave as an exercise.

1.56. Lemma. *If $\mathbf{x}^*$ is an element of $\mathbf{R}^n$, $\delta \in \mathbf{R}_{++}$, and $\mathbf{y} \in N(\mathbf{x}^*, \delta)$, and if we define*

$$\epsilon = \delta - d(\mathbf{x}^*, \mathbf{y}),$$

then:

$$N(\mathbf{y}, \epsilon) \subseteq N(\mathbf{x}^*, \delta).$$

1.57. Proposition. *Let $\mathbf{x}^*$ be an element of $\mathbf{R}^n$, and let δ be a positive real number. Then $int[N(\mathbf{x}^*, \delta)] = N(\mathbf{x}^*, \delta)$.*

Proof. We recall from our discussion above the fact that we necessarily have

$$int[N(\mathbf{x}^*, \delta)] \subseteq N(\mathbf{x}^*, \delta). \tag{1.36}$$

In order to prove the reverse inclusion, let $\mathbf{y}$ be an arbitrary element of $N(\mathbf{x}^*, \delta)$, and define

$$\epsilon = \delta - d(\mathbf{x}^*, \mathbf{y}).$$

By Lemma 1.56 we have

$$N(\mathbf{y}, \epsilon) \subseteq N(\mathbf{x}^*, \delta);$$

and therefore $\mathbf{y} \in int[N(\mathbf{x}^*, \delta)]$. It follows that

$$N(\mathbf{x}^*, \delta) \subseteq int[N(\mathbf{x}^*, \delta)];$$

and, combining this with (1.36), we see that $int[N(\mathbf{x}^*, \delta)] = N(\mathbf{x}^*, \delta)$. □

1.58. Definition. A subset, X, of $\mathbf{R}^n$ is said to be **open** iff $int X = X$.

Notice that $\emptyset$, the empty set, is open; for obviously the interior of $\emptyset$ is empty, that is

$$int(\emptyset) = \emptyset!$$

The following sets forth the basic facts regarding unions and intersections of open sets in $\mathbf{R}^n$.

1.59. Theorem. *In $\mathbf{R}^n$:*

1. $\emptyset$ and $\mathbf{R}^n$ are open sets.

2. if $\mathfrak{U} = \{U_a \mid a \in A\}$ is any family of subsets of $\mathbf{R}^n$ satisfying:

$$(\forall a \in A)\colon U_a \textit{ is an open subset of } \mathbf{R}^n,$$

then $U \equiv \cup_{a \in A} U_a$ is an open subset of $\mathbf{R}^n$.

3. if V_i is an open subset of $\mathbf{R}^n$, for $i = 1, \ldots, m$, then

$$V \stackrel{def}{=} \bigcap_{i=1}^{m} V_i$$

is an open subset of $\mathbf{R}^n$.

Proof. It should be obvious that both $\emptyset$ and $\mathbf{R}^n$ are open sets. To prove part 2, suppose that $\mathcal{U} = \{U_a \mid a \in A\}$ is a family of open sets, and let

$$\mathbf{x} \in U \stackrel{\text{def}}{=} \bigcup_{a \in A} U_a.$$

Then there exists some particular element of A, call it a*, such that $\mathbf{x} \in U_{a^*}$; and, since U_{a^*} is an open set, there exists a positive number, δ, such that:

$$N(\mathbf{x}, \delta) \subseteq U_{a^*}.$$

But then we see that

$$N(\mathbf{x}, \delta) \subseteq U \equiv \bigcup_{a \in A} U_a,$$

so that $\mathbf{x} \in int(U)$; and, since $\mathbf{x}$ was an arbitrary element of U, it follows that U is an open set.

Now suppose that V_i is an open set, for $i = 1, \ldots, m$, and let

$$\mathbf{y} \in V \equiv \bigcap_{i=1}^{m} V_i.$$

Then, since $\mathbf{y}$ is an element of each V_i, and each V_i is an open set, we see that for each i, there exists δ_i such that:

$$N(\mathbf{y}, \delta_i) \subseteq V_i \quad \text{for } i = 1, \ldots, m. \tag{1.37}$$

If we then define the positive real number δ by:

$$\delta = \min\{\delta_1, \ldots, \delta_m\},$$

we see that

$$N(\mathbf{y}, \delta) \subseteq N(\mathbf{y}, \delta_i) \quad \text{for } i = 1, \ldots, m;$$

and it then follows from (1.37) that

$$N(\mathbf{y}, \delta) \subseteq V \equiv \bigcap_{i=1}^{m} V_i.$$

Thus we see that V is an open set. □

1.60. Examples/Exercises.

1. Let A be a closed interval of real numbers of the form

$$A = [a, b],$$

where

$$0 < a \leq b;$$

and, for each $\alpha \in A$, define the set V_α, a subset of $\mathbf{R}^n$, by

$$V_\alpha = N(\mathbf{0}, \alpha).$$

If we then define

$$\mathcal{V} = \{V_\alpha \mid \alpha \in A\},$$

it is easy to see that

$$\bigcup_{V \in \mathcal{V}} V = \bigcup_{\alpha \in A} V_\alpha = N(\mathbf{0}, b) = \{\mathbf{x} \in \mathbf{R}^n \mid 0 \leq \|\mathbf{x}\| < b\},$$

while

$$\bigcap_{V \in \mathcal{V}} V = \bigcap_{\alpha \in A} V_\alpha = N(\mathbf{0}, a) = \{\mathbf{x} \in \mathbf{R}^n \mid 0 \leq \|\mathbf{x}\| < a\},$$

In particular, notice that the union of all the sets in the family is an open set. In this case, the intersection of all the sets in the family is also open; however, this need not be true, as is shown by our next example.

2. For each $n \in N$, define the open interval of real numbers, X_n, by:

$$X_n =]1 - 1/n, 2 + 1/n[.$$

Prove that:

$$\bigcup_{n=1}^{\infty} X_n =]0, 3[\text{ and } \bigcap_{n=1}^{\infty} X_n = [1, 2].$$

1.61. Definition. A subset of $\mathbf{R}^n$ is said to be **closed** iff its complement is open.

The following result sets forth the basic properties of unions and intersections of closed sets in $\mathbf{R}^n$. The student should compare this result with Theorem 1.59.

1.62. Theorem. *In* $\mathbf{R}^n$*:*

1. *$\emptyset$ and $\mathbf{R}^n$ are closed sets.*
2. *if C_i is a closed subset of $\mathbf{R}^n$, for $i = 1, \ldots, m$, then:*

$$C \stackrel{def}{=} \bigcup_{i=1}^{m} C_i,$$

is a closed set.

3. *if $\{D_a \mid a \in A\}$ is a family of closed subsets of $\mathbf{R}^n$, then:*

$$D \stackrel{def}{=} \bigcap_{a \in A} D_a$$

is a closed set.

Proof. To prove part 1, notice that it follows from Theorem 1.59.1 and the definition of a closed set that both $\emptyset$ and $\mathbf{R}^n$ are closed sets.

Now suppose C_i is closed, for $i = 1, \ldots, m$, and define

$$U_i = C_i^c = \mathbf{R}^n \setminus C_i \text{ for } i = 1, \ldots, m.$$

Then U_i is open, for each i; and thus by Theorem 1.59.3,

$$U \equiv \bigcap_{i=1}^{m} U_i,$$

is an open set. But then

$$U^c = \mathbf{R}^n \setminus U$$

is closed, and by Theorem 1.7.2,

$$U^c = \left[\bigcap_{i=1}^{m} U_i\right]^c = \bigcup_{i=1}^{m} U_i^c = \bigcup_{i=1}^{m} C_i,$$

where the last equality is by definition of the U_i.

Finally, suppose that, for each $a \in A$, C_a is a closed set, and define:

$$U_a = C_a^c \text{ for each } a \in A.$$

Then U_a is open, for each $a \in A$, so that by Theorem 1.59.2,

$$U \equiv \bigcup_{a \in A} U_a,$$

is open. But then we see that U^c is closed, and by Theorem 1.7.1,

$$U^c \equiv \left[\bigcup_{a \in A} U_a\right]^c = \bigcap_{a \in A} U_a^c = \bigcap_{a \in A} C_a. \quad \square$$

The following example shows that the union of an infinite number of closed sets may not itself be a closed set.

1.63. Examples/Exercises. Define the family $\mathcal{X}$, of subsets of $\mathbf{R}$ by letting:

$$X_n = [1/n, 3 - 1/n],$$

for each $n \in N$. Prove that

$$\bigcup_{n=1}^{\infty} X_n =]0, 3[\text{ and } \bigcap_{n=1}^{\infty} X_n = [1, 2].$$

1.64. Definition. Let A be a subset of $\mathbf{R}^n$. A point $\mathbf{x} \in \mathbf{R}^n$ is said to be a **point of closure of** $\boldsymbol{A}$ iff we have, for each positive real number ϵ,

$$N(\mathbf{x}, \epsilon) \cap A \neq \emptyset.$$

We denote the set consisting of all points of closure of A by '$\overline{A}$,' and call this set the **closure of** $\boldsymbol{A}$.

1.65. Proposition. *If A is a subset of $\mathbf{R}^n$, then:*

1. $A \subseteq \overline{A}$,
2. $\overline{(\overline{A})} = \overline{A}$, *and*
3. $\overline{A}$ *is closed.*

Proof. It should be obvious that each element of A is a point of closure of A, which is what is stated in the first conclusion.

To prove part 2, notice first that, from part 1:

$$\overline{A} \subseteq \overline{(\overline{A})}.$$

To prove the converse, let

$$\mathbf{x}^* \in \overline{(\overline{A})}.$$

Then, given an arbitrary $\epsilon > 0$, we see that there exists a point $\mathbf{y}$ such that:

$$\mathbf{y} \in \overline{A} \cap N(\mathbf{x}^*, \epsilon).$$

If we then define

$$\delta = \epsilon - d(\mathbf{x}^*, \mathbf{y}),$$

it follows from the fact that $\mathbf{y}$ is a point of closure of A that there exists a point $\mathbf{z}$ such that:

$$\mathbf{z} \in A \cap N(\mathbf{y}, \delta).$$

However, since by Lemma 1.56:

$$N(\mathbf{y}, \delta) \subseteq N(\mathbf{x}^*, \epsilon),$$

it then follows that:

$$N(\mathbf{x}^*, \epsilon) \cap A \neq \emptyset;$$

and thus, since ϵ was arbitrary, that:

$$\mathbf{x}^* \in \overline{A}.$$

But then, since $\mathbf{x}^*$ is an arbitrary element of $\overline{(\overline{A})}$, we conclude that

$$\overline{(\overline{A})} \subseteq \overline{A}.$$

Finally, in order to prove that the closure of A is a closed set, let

$$\mathbf{x}^* \in (\overline{A})^c.$$

It then follows from part 2 that $\mathbf{x}^*$ is not a point of closure of $\overline{A}$, so that there exists a positive real number ϵ^* such that:

$$N(\mathbf{x}^*, \epsilon^*) \cap \overline{A} = \emptyset.$$

We then have from Proposition 1.2.1 that:

$$N(\mathbf{x}^*, \epsilon^*) \subseteq (\overline{A})^c,$$

and, since $\mathbf{x}^*$ was arbitrary, we can conclude that the complement of $\overline{A}$ is open; and thus that $\overline{A}$ is closed. □

1.66. Theorem. *A subset, A, of $\mathbf{R}^n$ is closed if, and only if $A = \overline{A}$; that is, a set is closed if, and only if, it is equal to its own closure.*

Proof. The fact that A is closed if it is equal to its own closure is simply a re-statement of 1.65.3. To prove the converse, suppose that A is closed. Then A^c is open, so that, given any $\mathbf{x}^* \in A^c$, there exists $\epsilon > 0$ such that $N(\mathbf{x}^*, \epsilon) \subseteq A^c$; and thus:

$$N(\mathbf{x}^*, \epsilon) \cap A = \emptyset.$$

However, it then follows that $\mathbf{x}^*$ is not a point of closure of A; and thus, since $\mathbf{x}^*$ was an arbitrary element of A^c, that

$$A^c \subseteq (\overline{A})^c,$$

which is equivalent to $\overline{A} \subseteq A$. Since we always have $A \subseteq \overline{A}$, it follows that:

$$A = \overline{A}. \quad \square$$

1.67. Definition. Let A be a subset of $\mathbf{R}^n$. A point $\mathbf{x}^* \in \mathbf{R}^n$ is said to be a **boundary point of A** iff, for each $\epsilon \in \mathbf{R}_{++}$, we have:

$$N(\mathbf{x}^*, \epsilon) \cap A \neq \emptyset \text{ and } N(\mathbf{x}^*, \epsilon) \cap A^c \neq \emptyset.$$

The set of all boundary points of A is called the **boundary of A**, and is denoted by '$B(A)$.'

Thus we see that a boundary point of A is a point of closure of both A and its complement; and conversely, if a point is in the closure of both sets, then it is a boundary point of A.

1.68. Proposition. *If A is a subset of $\mathbf{R}^n$, then:*
1. *$int(A) \cap B(A) = \emptyset$, and*
2. *$\overline{A} = int(A) \cup B(A)$.*

Proof. Part 1 of our conclusion should be obvious. To prove part 2, we first note that it is apparent that

$$B(A) \subseteq \overline{A},$$

and it then follows at once that:

$$int(A) \cup B(A) \subseteq \overline{A}.$$

Conversely, suppose $\mathbf{x}^*$ is a point of closure of A, so that for each $\epsilon > 0$,

$$N(\mathbf{x}^*, \epsilon) \cap A \neq \emptyset.$$

Then we either have

$$(\forall \epsilon \in \mathbf{R}_{++}) : N(\mathbf{x}^*, \epsilon) \cap A^c \neq \emptyset$$

as well, in which case $\mathbf{x}^* \in B(A)$; or

$$(\exists \epsilon^* \in \mathbf{R}_{++}) : N(\mathbf{x}^*, \epsilon^*) \cap A^c = \emptyset,$$

so that $N(\mathbf{x}^*, \epsilon^*) \subseteq A$, and thus $\mathbf{x}^* \in int(A)$. Therefore,

$$\overline{A} \subseteq int(A) \cup B(A),$$

and our result follows. □

The preceding result provides us with the basic distinction between a point of closure and a boundary point: a boundary point of A is a point of closure of both A and A^c; whereas an interior point of A is a point of closure of A, but not of A^c, and therefore is not a boundary point of A. On the other hand, every boundary point of A is, of course, a point of closure of A. The student should be careful to distinguish both points of closure and boundary points from limit points, or points of accumulation, which we define as follows.

1.69. Definition. If A is a non-empty subset of $\mathbf{R}^n$, a point $\mathbf{x}^*$ of $\mathbf{R}^n$ will be said to be a **limit point** (or **point of accumulation** or **cluster point**) **of A** iff, for each $\epsilon > 0$, we have:[9]

$$N'(\mathbf{x}^*, \epsilon) \cap A \neq \emptyset.$$

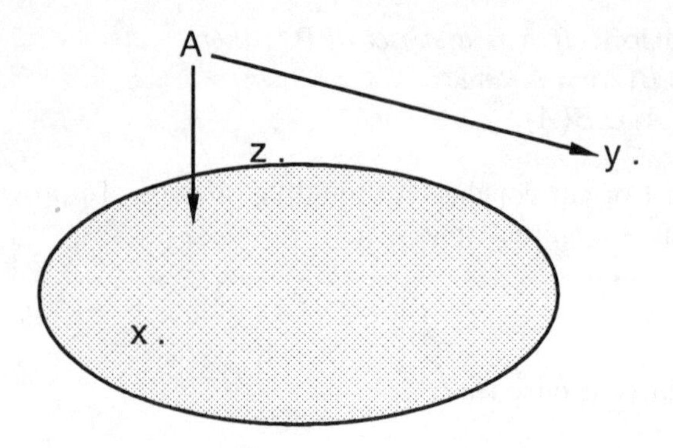

Figure 1.5: Limit Points and Points of Closure.

In the figure above, $\mathbf{x}$ is a point of closure of A which is not a boundary point of A, while $\mathbf{y}$ is a boundary point of A which is not a limit point of A, and $\mathbf{z}$ is a limit point of A. (I should also note that, while it is probably not at all apparent from the diagram, I am taking $\mathbf{y}$ to be an element of A.)

Limit points, points of closure, and boundary points all arise naturally in some standard parts of economic theory, as we will illustrate in the following example.

1.70. Example. Consumer Preference Theory in Economics. We can illustrate many of the concepts we have been studying in this section within the context of consumer preference theory, the basic elements of which were set out in Examples 1.27. As in that example, we shall denote the consumer's strict preference relation by 'P,' and its negation, the consumer's weak preference relation by 'G.'

Notice that, if G is reflexive, each $\mathbf{x}$ is necessarily a point of closure of the at-least-as-good-as set, $G\mathbf{x}$. An assumption of some economic significance, and which is obviously not satisfied by all asymmetric orders, P, is that $\mathbf{x}$ is also a point of closure of the preferred-to-$\mathbf{x}$ set, $P\mathbf{x}$, defined by:

$$P\mathbf{x} = \{\mathbf{y} \in X \mid \mathbf{y}P\mathbf{x}\}.$$

We shall say that P is **locally non-saturating at $\mathbf{x}^*$** iff

$$\mathbf{x}^* \in \overline{P\mathbf{x}^*},$$

and we shall say that ***P* is locally non-saturating on *X***, or simply that ***P* is locally non-saturating**, if P is locally non-saturating at each $\mathbf{x} \in X$. Similarly, if we begin with the consumer's weak preference relation (always assumed to be reflexive, at least), we shall say that G is locally non-saturating iff its asymmetric part, P, is locally non-saturating.

[9] As you will probably recall, '$N'(\mathbf{x}^*, \epsilon)$' denotes the deleted neighborhood of $\mathbf{x}^*$ with radius ϵ (see Definition 1.51).

Notice that P will be locally non-saturating at each $\mathbf{x}^*$ if, and only if, for each $\epsilon \in \mathbf{R}_{++}$, there exists $\mathbf{y} \in N(\mathbf{x}^*, \epsilon)$ such that $\mathbf{y}P\mathbf{x}^*$. Consequently, since P is asymmetric, it is locally non-saturating at a point, $\mathbf{x}^*$, iff $\mathbf{x}^*$ is a *limit point* of $P\mathbf{x}^*$. Thus in the standard textbook indifference map, as illustrated in Figure 1.6.a, below, P is locally non-saturating at $\mathbf{x}^*$; whereas in Figure 1.6.b, P is not locally non-saturating at $\mathbf{x}^*$. In fact, the situation in the latter diagram illustrates the case in which P is defined by:

$$\mathbf{x}P\mathbf{y} \iff f(\mathbf{x}) > f(\mathbf{y}) + \delta, \tag{1.38}$$

where $f\colon X \to \mathbf{R}$, and δ is a strictly positive real number; in other words, if P is a semi-order.

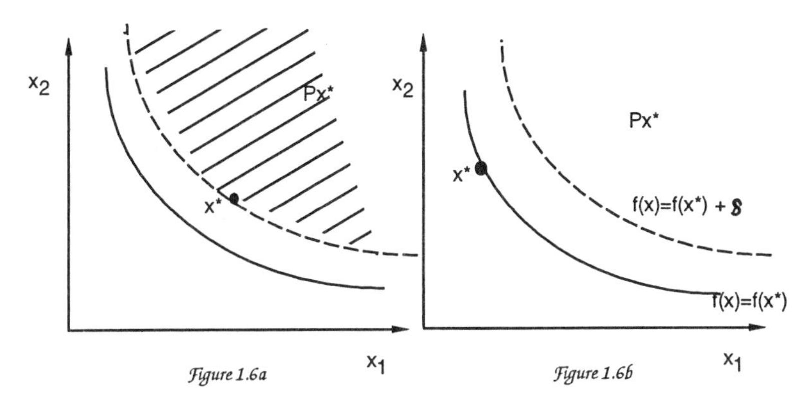

Figure 1.6: Local Non-Saturation.

Local non-saturation is a fairly abstract notion; however, it is implied by a couple of more intuitively satisfying and fairly plausible conditions. In order to introduce them, let's begin by defining the following: we shall say that the consumption set X is **neoclassical** iff X is a closed subset of $\mathbf{R}^n$, and satisfies:

$$(\forall \mathbf{x} \in X)\colon \mathbf{x} + \mathbf{R}^n_+ \subseteq X. \tag{1.39}$$

In the form in which it is presented, condition (1.39) may appear to be fairly obscure itself! Notice, however, that it is equivalent to the following statement:

$$\text{if } \mathbf{x} \in X, \text{and } \mathbf{y} \in \mathbf{R}^n \text{ is such that } \mathbf{y} \geq \mathbf{x}, \text{ then } \mathbf{y} \in X; \tag{1.40}$$

that is, if $\mathbf{x}$ is contained in X, and $\mathbf{y}$ is any other element of $\mathbf{R}^n$ which is at least as large as $\mathbf{x}$ in each coordinate, then $\mathbf{y}$ is also contained in X. In the following three definitions, suppose that X is a neoclassical consumption set, and that P is a binary (strict preference) relation on X. Then we shall say that P is:

1. **non-decreasing** iff P satisfies:

$$(\forall \mathbf{x}, \mathbf{y} \in X) \colon \mathbf{x} P \mathbf{y} \Rightarrow \mathbf{y} \not\geq \mathbf{x}.$$

2. **increasing** iff P is non-decreasing, and also satisfies:

$$(\forall \mathbf{x}, \mathbf{y} \in X) \colon \mathbf{y} \gg \mathbf{x} \Rightarrow \mathbf{y} P \mathbf{x}.$$

3. **strictly increasing** iff P satisfies:

$$(\forall \mathbf{x}, \mathbf{y} \in X) \colon \mathbf{y} > \mathbf{x} \Rightarrow \mathbf{y} P \mathbf{x}.$$

Once again we shall say that a reflexive (weak) preference relation, G, satisfies one of the above conditions if its asymmetric part satisfies the condition.

Obviously, if P is strictly increasing, then it is increasing; and if P is increasing, then it is non-decreasing. In other words, the above definitions are presented in order of decreasing generality. The student should have no difficulty in proving that if P is increasing, then it is locally non-saturating on X (obviously this will also be true if P is strictly increasing). □

Exercises.

1. Show that a subset of $\mathbf{R}^n$ is closed if, and only if, it contains its boundary.

2. Can the boundary of a subset of $\mathbf{R}^n$ be empty? If your answer is 'no,' can you provide a proof? If your answer is 'yes,' provide an example.

3. Prove Lemma 1.56.

4. Show that conditions (1.39) and (1.40), which were introduced in connection with the definition of a neoclassical consumption set, are equivalent.

5. Show that the lexicographic ordering, $>_L$, is locally non-saturating.

1.6 Relatively Open and Closed Sets

In economic theory we often deal with functions whose domains are proper subsets of $\mathbf{R}^n$, and with relations which are defined only on proper subsets of $\mathbf{R}^n$ as well (for example, a preference relation on a consumption set). In such situations, it is very useful to introduce the notion of openness relative to the underlying set with which we are working. We define this as follows.

1.71. Definition. Let X be a non-empty subset of $\mathbf{R}^n$. We shall say that a subset, A, of X, is **open relative to X** (or simply **relatively open**, if X is clear from the context) iff there exists an open subset of $\mathbf{R}^n$, U, such that:

$$A = X \cap U.$$

Obviously, if a subset of X is open in $\mathbf{R}^n$, then it is also open relative to X. Furthermore, it follows at once from the properties of open sets in $\mathbf{R}^n$ that if X is open, then a subset, A, of X, is open relative to X if, and only if, A is open (in $\mathbf{R}^n$). On the other hand, in economics we often take the consumption set, X, to be the nonegative orthant in $\mathbf{R}^n$; that is, $X = \mathbf{R}^n_+$. Since $\mathbf{R}^n_+$ is closed in $\mathbf{R}^n$, a set may be open relative to $\mathbf{R}^n_+$, yet not be open in $\mathbf{R}^n$. Thus, for example, in the diagram below, the area lying below the curve through $\mathbf{x}^*$ is easily seen to be open relative to $\mathbf{R}^2_+$, even though it is not open in $\mathbf{R}^2$. [The point $\mathbf{y}$, for example, is an element of $\mathbf{x}^*P$, but is not an interior point of $\mathbf{x}^*P$.]

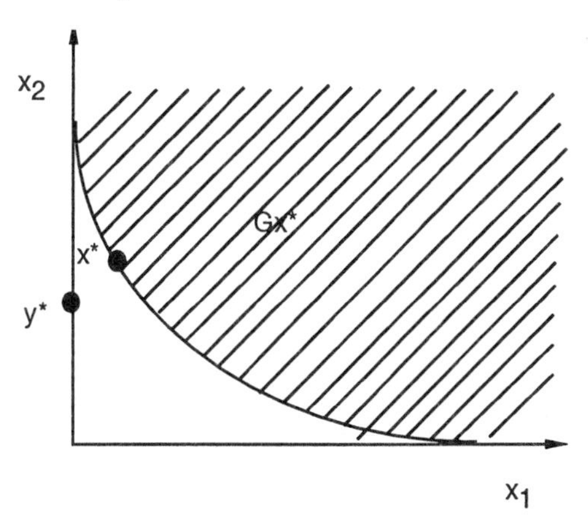

Figure 1.7: A Relatively Open Set.

In principle, determining which sets are open relative to X is no more difficult than determining which sets are open in $\mathbf{R}^n$; as is shown in the following.

1.72. Theorem. *Let X be a subset of $\mathbf{R}^n$, and let A be a subset of X. Then A is open relative to X if, and only if, for each $\mathbf{x} \in A$, there exists $\epsilon \in \mathbf{R}_{++}$ such that:*

$$N(\mathbf{x}, \epsilon) \cap X \subseteq A.$$

Proof. Suppose first that A is open relative to X, and let $\mathbf{x}^* \in A$. Since A is open relative to X, there exists an open subset, U, of $\mathbf{R}^n$, such that:

$$A = X \cap U.$$

Consequently, $\mathbf{x}^* \in U$, and therefore, since U is open, there exists a positive number ϵ such that:

$$N(\mathbf{x}^*, \epsilon) \subseteq U.$$

Thus we see that:

$$N(\mathbf{x}^*, \epsilon) \cap X \subseteq U \cap X = A.$$

Now suppose A satisfies the stated condition. Then, for each $\mathbf{x} \in A$, there exists $\epsilon_{\mathbf{x}} > 0$ such that:

$$N(\mathbf{x}^*, \epsilon_{\mathbf{x}}) \cap X \subseteq A. \tag{1.41}$$

But then, if we define

$$U = \bigcup_{\mathbf{x} \in A} N(\mathbf{x}, \epsilon_{\mathbf{x}}),$$

it follows that U (since it is the union of open sets) is an open set (in $\mathbf{R}^n$); and by (1.41) we have:

$$U \cap X = \Big[\bigcup_{\mathbf{x} \in A} N(\mathbf{x}, \epsilon_{\mathbf{x}})\Big] \cap X = \bigcup_{\mathbf{x} \in A} [X \cap N(\mathbf{x}, \epsilon_{\mathbf{x}})] \subseteq A. \tag{1.42}$$

On the other hand, we obviously have

$$A \subseteq \bigcup_{\mathbf{x} \in A} [X \cap N(\mathbf{x}, \epsilon_{\mathbf{x}})] \equiv X \cap U;$$

and, combining this with (1.42), it follows that:

$$A = X \cap U. \quad \square$$

Not only are relatively open sets as easy to identify, in principle, as are open sets, they satisfy the same properties with respect to union and intersection; as we will now demonstrate.

1.73. Theorem. *(Properties of Relatively Open Sets.*[10]*) If X is a subset of $\mathbf{R}^n$, then:*

1. $\emptyset$ and X are open relative to X.

2. if $A_i \subseteq X$ is open relative to X, for $i = 1, \ldots, m$, then

$$A \stackrel{def}{=} \bigcap_{i=1}^{m} A_i$$

is open relative to X.

3. if $U_\beta \subseteq X$ is open relative to X, for each $\beta \in B$, then:

$$U \stackrel{def}{=} \bigcup_{\beta \in B} U_\beta,$$

is open relative to X.

[10] Compare Theorem 1.59.

Proof. We note first that, since

$$\emptyset = \emptyset \cap X \text{ and } X = X \cap \mathbf{R}^n,$$

it is immediate that both sets are open relative to X.

To prove part 2, let $A_i \subseteq X$ be relatively open, for $i = 1, \dots, m$. Then there exist open sets $U_i \subseteq \mathbf{R}^n$ such that:

$$A_i = X \cap U_i \quad \text{for } i = 1, \dots, m.$$

But then:

$$A \equiv \bigcap_{i=1}^{m} A_i = \bigcap_{i=1}^{m} (X \cap U_i) = X \cap \left[\bigcap_{i=1}^{m} U_i\right];$$

and it then follows at once from Theorem 1.59 and Definition 1.71 that A is open relative to X.

I will leave the proof of part 3 as an exercise. □

1.74. Definition. Let X be a subset of $\mathbf{R}^n$. We shall say that a subset, C, of X, is **closed relative to X** iff $A \equiv X \setminus C$, is open relative to X.

1.75. Theorem. *(Properties of Relatively Closed Sets.*[11]*) If X is a subset of $\mathbf{R}^n$, then:*

1. *$\emptyset$ and X are closed relative to X.*
2. *if $C_i \subseteq X$ is closed relative to X, for $i = 1, \dots, m$, then*

$$C \stackrel{def}{=} \bigcup_{i=1}^{m} C_i$$

is closed relative to X.

3. *if $C_\alpha \subseteq X$ is closed relative to X, for each $\alpha \in A$, then:*

$$C \stackrel{def}{=} \bigcap_{\alpha \in A} C_\alpha$$

is closed relative to X.

Proof. This result follows from Theorem 1.73 and DeMorgan's Laws (Theorem 1.7) by an argument essentially identical to that used to prove Theorem 1.62. The details will be left as an exercise. □

1.76. Proposition. *Let $A \subseteq X \subseteq \mathbf{R}^n$. Then A is closed relative to X if, and only if, there exists a closed set, C (closed in $\mathbf{R}^n$) such that $A = X \cap C$.*

Proof. Suppose A is closed relative to X. Then $X \setminus A$ is open relative to X, and thus there exists a set U which is open in $\mathbf{R}^n$ and such that:

$$X \setminus A = X \cap U. \tag{1.43}$$

[11] Compare Theorem 1.62.

Now, if we define

$$C = U^c,$$

then C is closed. Furthermore, since $C \cup U = \mathbf{R}^n$, we necessarily have

$$X = (X \cap C) \cup (X \cap U);$$

and, since C and U are disjoint, it then follows from (1.43) that

$$A = X \cap C.$$

Now suppose $A = X \cap C$, where C is closed in $\mathbf{R}^n$; and define:

$$U = C^c \equiv \mathbf{R}^n \setminus C.$$

Then, as in part 1 of our proof, it follows that

$$X \setminus A = X \cap U;$$

so that $X \setminus A$ is open relative to X, and thus A is closed relative to X. □

Notice that it follows immediately from Proposition 1.76 that if a subset of X, A, is closed in $\mathbf{R}^n$, then A is closed relative to X. Furthermore, if X itself is closed in $\mathbf{R}^n$, then A is closed relative to X if, and only if, A is closed (in $\mathbf{R}^n$).

1.77. Theorem. *If $A \subseteq X \subseteq \mathbf{R}^n$, then A is closed relative to X if, and only if, $A = \overline{A} \cap X$.*

Proof. If $A = \overline{A} \cap X$, then, since $\overline{A}$ is a closed set, it follows at once from Proposition 1.76 that A is closed relative to X.

Now suppose A is closed relative to X. Then by Proposition 1.76, there exists a closed subset of $\mathbf{R}^n$, C, such that:

$$A = X \cap C.$$

But then we have

$$A \subseteq C,$$

so that, since C is closed,

$$\overline{A} \subseteq \overline{C} = C.$$

Thus

$$C \cap X = A = A \cap X \subseteq \overline{A} \cap X \subseteq C \cap X,$$

and we conclude that

$$A = \overline{A} \cap X. \quad \square$$

We can now use the concepts just introduced to define some very useful continuity conditions for irreflexive orders, as follows.

1.78. Definitions. If P is a binary relation on a non-empty subset of $\mathbf{R}^n$, X, we shall say that P is:

1. **upper semi-continuous (on X)** iff

$$(\forall \mathbf{x} \in X)\colon \mathbf{x}P \stackrel{\text{def}}{=} \{\mathbf{y} \in X \mid \mathbf{x}P\mathbf{y}\},$$

is open relative to X.

2. **lower semi-continuous (on X)** iff

$$(\forall \mathbf{x} \in X)\colon P\mathbf{x} \stackrel{\text{def}}{=} \{\mathbf{y} \in X \mid \mathbf{y}P\mathbf{x}\}$$

is open relative to X.

3. **continuous (on X)** iff P is both upper and lower semi-continuous on X.[12]

In the chapters to follow we will find a great many applications for these conditions; they will be particularly useful in our study of demand correspondences. These conditions are necessary for the existence of a continuous utility function representing the preference relation, and have been used in the proof of the existence of such a utility function as well (Debreu [1959, pp. 56–59]). In connection with the question of sufficient conditions for the existence of a real-valued representation of G, it is suggestive to note that the lexicographic order, $>_L$, introduced in Example 1.35.4 is *neither* upper nor lower semi-continuous.

The following two examples provide very simple illustrations of the fact that an asymmetric order may be upper, but not lower semi-continuous; or lower, but not upper semi-continuous. In defining these two examples, we will use for the first time a technique which is often quite useful in defining binary relations; specifically, we define the relation by defining $\mathbf{x}P$, the set of points inferior to $\mathbf{x}$ (or lower countour set for $\mathbf{x}$), for each $\mathbf{x} \in X$.

1.79. Examples.

1. Let $X = \mathbf{R}^2_+$, and define P on X by:

$$\mathbf{x}P = \begin{cases} \mathbf{R}^2_+ \setminus \{\mathbf{x}\} & \text{for } \mathbf{x} = (1,1), \text{ and} \\ \emptyset & \text{for } \mathbf{x} \neq (1,1), \end{cases}$$

that is, the point $\mathbf{x}^* = (1,1)$ is preferred to everything else in X, but no other point is preferred to any other point. It is easy to see that P is asymmetric and negatively transitive (and therefore the negation of P is a weak order), although I will leave the proof of this as an exercise. Since the lower contour sets are open in X, for each $\mathbf{x} \in X$, it is clear that P

[12] Once again, we shall say that a reflexive relation, G, satisfies one of these three conditions iff its asymmetric part satisfies the condition.

is upper semi-continuous. On the other hand, for any $\mathbf{x} \in X$ such that $\mathbf{x} \neq \mathbf{x}^* \equiv (1,1)$, we have:

$$P\mathbf{x} = \{\mathbf{x}^*\},$$

which is not open relative to X. Therefore, P is *not* lower semi-continuous.

2. Let $X = \mathbf{R}^n_+$, and define P on X by:

$$\mathbf{x}P = \begin{cases} \{\mathbf{0}\} & \text{for } \mathbf{x} \neq \mathbf{0}, \text{and} \\ \emptyset & \text{for } \mathbf{x} = \mathbf{0}, \end{cases}$$

where '$\mathbf{0}$' denotes the origin in $\mathbf{R}^n$. Here everything is preferred to the origin ('anything is preferred to having nothing at all'), but there are no other bundles such that one is preferred to the other. Once again it is easy to show that P is asymmetric and negatively transitive, and it is also easy to show that P is lower, but not upper semi-continuous. □

Exercises.

1. Complete the proof of Theorem 1.73.
2. Prove Theorem 1.75.
3. Show that the relation P defined in Example 1.79.1 is negatively transitive; and is *nowhere* locally non-saturating.
4. Show that the relation, P, defined in Example 1.79.2, above, is:
   a. locally non-saturating only at $\mathbf{0}$, and
   b. lower, but not upper semi-continuous.
5. Show that if G is a total and reflexive binary relation on a subset, X, of $\mathbf{R}^n$, then G is:

a. upper semi-continuous if, and only if, the **no-worse-than-x-set**, defined by:

$$G\mathbf{x} = \{\mathbf{y} \in X \mid \mathbf{y}G\mathbf{x}\},$$

is closed relative to X, for each $\mathbf{x} \in X$.

b. lower semi-continuous if, and only if, the **no-better-than-x-set**, defined by:

$$\mathbf{x}G = \{\mathbf{y} \in X \mid \mathbf{x}G\mathbf{y}\},$$

is closed relative to X, for each $\mathbf{x} \in X$.

1.7 Some Linear Space Properties of $\mathbf{R}^n$

We will conclude this chapter with a brief review of some of the linear space properties of $\mathbf{R}^n$; primarily those connected with linear functions on $\mathbf{R}^n$. In order to make it easier to distinguish matrices from sets of real numbers, I will use bold-face type, **A**, **B**, etc., as generic notation for matrices [as well as continuing to distinguish vectors from scalars (real numbers) by using bold-face type for the former; **x**, **y**, etc.].

1.80. Definition. A function $\mathbf{f}\colon \mathbf{R}^n \to \mathbf{R}^m$ is **linear** iff, for all $\mathbf{x}$, $\mathbf{y} \in \mathbf{R}^n$, and all $a \in \mathbf{R}$, we have:

1. $\mathbf{f}(\mathbf{x}+\mathbf{y}) = \mathbf{f}(\mathbf{x}) + \mathbf{f}(\mathbf{y})$, and
2. $\mathbf{f}(a\mathbf{x}) = a\mathbf{f}(\mathbf{x})$.

It is easy to show that the above two-part definition of a linear function is equivalent to the single condition: for all $\mathbf{x}, \mathbf{y} \in \mathbf{R}^n$, and for all $a, b \in \mathbf{R}$, we have:

$$\mathbf{f}(a\mathbf{x} + b\mathbf{y}) = a\mathbf{f}(\mathbf{x}) + b\mathbf{f}(\mathbf{y}). \tag{1.44}$$

It is also easy to show (by a simple induction argument) that if $\mathbf{f}$ is linear, then the condition in equation (1.44) holds not only for pairs of vectors; as is stated in the folowing proposition.

1.81. Proposition. *Suppose* $\mathbf{f}\colon \mathbf{R}^n \to \mathbf{R}^m$ *is linear, that* $\mathbf{x}_1, \ldots, \mathbf{x}_q \in \mathbf{R}^n$, *and that* $a_1, \ldots, a_q$ *are real numbers. Then:*

$$\mathbf{f}\left(\sum\nolimits_{i=1}^{q} a_i\mathbf{x}_i\right) = \sum\nolimits_{i=1}^{q} a_i\mathbf{f}(\mathbf{x}_i).$$

Our next result sets out the basic relationship between linear functions and matrices.

1.82. Proposition. *A function* $\mathbf{f}\colon \mathbf{R}^n \to \mathbf{R}^m$ *is linear if, and only if, there exists an* $m \times n$ *matrix,* $\mathbf{A}$, *such that for every* $\mathbf{x} \in \mathbf{R}^n$*:*

$$\mathbf{f}(\mathbf{x}) = \mathbf{A}\mathbf{x}. \tag{1.45}$$

Proof. Suppose $\mathbf{f}$ is linear, and define the vectors $\mathbf{a}_{\cdot j} \in \mathbf{R}^m$ by:

$$\mathbf{a}_{\cdot j} = f(\mathbf{e}_j) \ \text{ for } j = 1, \ldots, n,$$

where $\mathbf{e}_j$ is the jth unit coordinate vector in $\mathbf{R}^n$; and define the matrix $\mathbf{A}$ by:

$$\mathbf{A} = [\mathbf{a}_{\cdot 1}, \ldots, \mathbf{a}_{\cdot n}]$$

(that is, $\mathbf{A}$ is the $m\times n$ matrix whose j^{th} column is $\mathbf{a}_{\cdot j}$). If $\mathbf{x} = (x_1, \ldots, x_n) \in \mathbf{R}^n$, we have

$$\mathbf{x} = \sum\nolimits_{j=1}^{n} x_j\mathbf{e}_j,$$

so that, using Proposition 1.81 and the definition of $\mathbf{A}$, we have:

$$\mathbf{f}(\mathbf{x}) = \mathbf{f}\left(\sum\nolimits_{j=1}^{n} x_j\mathbf{e}_j\right) = \sum\nolimits_{j=1}^{n} x_j\mathbf{f}(\mathbf{e}_j) = \sum\nolimits_{j=1}^{n} \mathbf{a}_{\cdot j}x_j = \mathbf{A}\mathbf{x}.$$

(The student having trouble understanding the above equality may find it helpful to look at 1.83.2, below.)

It is easy to show that if $\mathbf{f}$ is of the form (1.45), above, then $\mathbf{f}$ is linear; and I will leave this part of the proof as an exercise. □

Notice that, if we regard $\mathbf{x}$ and $\mathbf{y}$ as column (as opposed to row) vectors (and we will whenever it makes a difference), then their inner product is also a matrix-vector product:

$$\mathbf{x} \cdot \mathbf{y} = \mathbf{x}'\mathbf{y},$$

where $\mathbf{x}'$ is the transpose of $\mathbf{x}$; that is, the row vector whose i^{th} coordinate is x_i. Having noted this fact, we can now see that an immediate implication of Proposition 1.82 is that, if $f \colon \mathbf{R}^n \to \mathbf{R}$ is linear, then there exists a (fixed) vector $\mathbf{a} \in \mathbf{R}^n$ such that, for all $\mathbf{x} \in \mathbf{R}^n$, we have:

$$f(\mathbf{x}) = \mathbf{a} \cdot \mathbf{x}.$$

The following lists a number of other useful (and, hopefully, familiar) facts regarding linear functions (or linear transformations, a term we will use as a synonym for 'linear functions') on $\mathbf{R}^n$. I will feel free to use these facts and properties throughout the remainder of the book.

1.83. Facts regarding linear transformations on $\mathbf{R}^n$.
1. If $\mathbf{f} \colon \mathbf{R}^n \to \mathbf{R}^m$ is linear, so that $\mathbf{f}$ can be written in the form

$$\mathbf{f}(\mathbf{x}) = \mathbf{A}\mathbf{x} \quad \text{for } \mathbf{x} \in \mathbf{R}^n,$$

then the matrix $\mathbf{A}$ is called the **matrix of the transformation.**
2. If $\mathbf{A}$ is an $m \times n$ matrix, and $\mathbf{x} \in \mathbf{R}^n$, we can write:

$$\mathbf{A}\mathbf{x} = \sum_{j=1}^{n} \mathbf{a}_{\cdot j} x_j = \begin{pmatrix} \mathbf{a}_{1\cdot} \cdot \mathbf{x} \\ \mathbf{a}_{2\cdot} \cdot \mathbf{x} \\ \vdots \\ \mathbf{a}_{m\cdot} \cdot \mathbf{x} \end{pmatrix}$$

where '$\mathbf{a}_{i\cdot}$' denotes the row vector, $\mathbf{a}_{i\cdot} = (a_{i1}, \ldots, a_{in})$, and '$\mathbf{a}_{\cdot j}$' denotes the column vector:

$$\mathbf{a}_{\cdot j} = \begin{pmatrix} a_{1j} \\ a_{2j} \\ \vdots \\ a_{mj} \end{pmatrix}.$$

3. Given two $m \times n$ matrices, $\mathbf{A}$ and $\mathbf{B}$, and the associated transformations,

$$\mathbf{f} \colon \mathbf{R}^n \to \mathbf{R}^m \ \& \ \mathbf{g} \colon \mathbf{R}^n \to \mathbf{R}^m,$$

the matrix $\mathbf{C} = \mathbf{A} + \mathbf{B}$ is the matrix of the transformation $\mathbf{f} + \mathbf{g}$.

4. Given an $m \times n$ matrix, $\mathbf{A}$, and a scalar $\alpha \in \mathbf{R}$, the matrix $\alpha\mathbf{A}$ is the matrix of the transformation $\alpha\mathbf{f}$, where $\mathbf{f}$ is defined by:

$$\mathbf{f}(\mathbf{x}) = \mathbf{A}\mathbf{x} \quad \text{for } \mathbf{x} \in \mathbf{R}^n.$$

5. Let $\mathbf{f}\colon \mathbf{R}^n \to \mathbf{R}^m$, and $\mathbf{g}\colon \mathbf{R}^m \to \mathbf{R}^p$ be defined by:

$$\mathbf{f}(\mathbf{x}) = \mathbf{A}\mathbf{x} \quad \text{for } \mathbf{x} \in \mathbf{R}^n,$$

and

$$\mathbf{g}(\mathbf{y}) = \mathbf{B}\mathbf{y} \quad \text{for } \mathbf{y} \in \mathbf{R}^m,$$

where $\mathbf{A}$ and $\mathbf{B}$ are $m \times n$ and $p \times m$ matrices, respectively. Then the matrix

$$\mathbf{C} = \mathbf{BA},$$

is the matrix of the composite transformation $\mathbf{h} = \mathbf{g} \circ \mathbf{f}$:

$$\mathbf{h}(\mathbf{x}) = \mathbf{g}[\mathbf{f}(\mathbf{x})] = \mathbf{B}(\mathbf{A}\mathbf{x}) = (\mathbf{BA})\mathbf{x} \quad \text{for } \mathbf{x} \in \mathbf{R}^n.$$

6. The **row rank** of an $m \times n$ matrix $\mathbf{A}$ is the number of linearly independent row vectors of $\mathbf{A}$, while the **column rank** of $\mathbf{A}$ is the number of linearly independent column vectors of $\mathbf{A}$; however, for any $\mathbf{A}$, the row rank is necessarily equal to the column rank of $\mathbf{A}$, and we call this common value the **rank of A**, which we shall denote by '$r(\mathbf{A})$.' Notice that if $\mathbf{A}$ is an $m \times n$ matrix, we necessarily have:

$$r(\mathbf{A}) \leq \min\{m, n\}.$$

An $m \times n$ matrix $\mathbf{A}$ is said to have **full rank** iff $r(\mathbf{A}) = \min\{m, n\}$.

7. A square matrix of full rank is said to be **non-singular** (otherwise it is said to be **singular**). An $n \times n$ matrix is non-singular if, and only if, $|\mathbf{A}| \neq 0$, where '$|\mathbf{A}|$' denotes the determinant of $\mathbf{A}$.

8. If $\mathbf{A}$ is an $m \times n$ matrix of rank r, then the **kernel** of the associated transformation, the set S defined as:

$$S = \{\mathbf{x} \in \mathbf{R}^n \mid \mathbf{A}\mathbf{x} = \mathbf{0}\},$$

is a linear subspace of dimension $n - r$.

9. Suppose $\mathbf{f}\colon \mathbf{R}^n \to \mathbf{R}^m$ is linear, and let $\mathbf{A}$ be the matrix of the transformation. Then $\mathbf{f}$ is one-to-one if, and only if, $r(\mathbf{A}) = n$.

10. Suppose $\mathbf{f}\colon \mathbf{R}^n \to \mathbf{R}^m$ is linear, and let $\mathbf{A}$ be the matrix of the transformation. Then $\mathbf{f}$ is onto $\mathbf{R}^m$ if, and only if, $r(\mathbf{A}) = m$.

11. Suppose $\mathbf{f}\colon \mathbf{R}^n \to \mathbf{R}^m$ is linear, and let $\mathbf{A}$ be the matrix of the transformation. Then $\mathbf{f}$ is both one-to-one and onto $\mathbf{R}^m$ if, and only if, $m = n$ and $r(\mathbf{A}) = n$ [and thus $\mathbf{A}$ is non-singular].

12. For an arbitrary positive integer, n, the **n × n identity matrix**, denoted by '$\mathbf{I}_n$,' is defined by:

$$\mathbf{I}_n = \begin{bmatrix} 1 & 0 & 0 & \dots & 0 \\ 0 & 1 & 0 & \dots & 0 \\ 0 & 0 & 1 & \dots & 0 \\ \dots & \dots & \dots & \dots & \dots \\ 0 & 0 & 0 & \dots & 1 \end{bmatrix}.$$

[However, if the dimension of $\mathbf{I}_n$ is clear from the context, we shall generally write simply '$\mathbf{I}$' in place of '$\mathbf{I}_n$.']

13. If $\mathbf{A}$ is an $m \times n$ matrix, then there exists an $n \times m$ matrix $\mathbf{B}$ satisfying:

$$\mathbf{BA} = \mathbf{I}_n,$$

if, and only if, $r(\mathbf{A}) = n$.

14. If $\mathbf{A}$ is an $m \times n$ matrix, then there exists an $n \times m$ matrix $\mathbf{C}$ satisfying:

$$\mathbf{AC} = \mathbf{I}_m,$$

if, and only if, $r(\mathbf{A}) = m$.

15. If $\mathbf{A}$ is an $m \times n$ matrix, then there exist $n \times m$ matrices $\mathbf{B}$ and $\mathbf{C}$ satisfying

$$\mathbf{BA} = \mathbf{I}_n \text{ and } \mathbf{AC} = \mathbf{I}_m, \tag{1.46}$$

if, and only if, $m = r(\mathbf{A}) = n$. Furthermore, if there exist matrices $\mathbf{B}$ and $\mathbf{C}$ satisfying equation (1.46), then $\mathbf{B} = \mathbf{C}$, and we call $\mathbf{B}$ the **inverse of A**, and write:

$$\mathbf{B} = \mathbf{A}^{-1}.$$

16. If $\mathbf{A}$ is non-singular, then the associated linear transformation, f, given by:

$$\mathbf{f}(\mathbf{x}) = \mathbf{Ax} \quad \text{for } \mathbf{x} \in \mathbf{R}^n,$$

is one-to-one and onto (onto $\mathbf{R}^n$ itself, in this case), and $\mathbf{A}^{-1}$ is the matrix of the inverse transformation, $\mathbf{f}^{-1}$.

Chapter 2

Sequences and Infinite Series

2.1 Sequences of Real Numbers

In this chapter, we will review the basic theory of sequences and limits. The material we will be studying is extremely important for a thorough understanding of functions and continuity of functions, among other things. We will also be examining the theory of infinite series; material which is critical to the development of many dynamic models in economics.

We begin with a bit of notation, which we will be using throughout the remainder of this book. **We will denote the set of natural numbers by 'N;'** that is,

$$N = \{1, 2, 3, \ldots\}.$$

2.1. Definition. An (infinite) **sequence** of real numbers is a function $f\colon N \to \mathbf{R}$.

If $f\colon N \to \mathbf{R}$ is a sequence, we denote $f(n)$, for $n \in N$, by

$$f(n) = x_n \quad \text{for } n = 1, 2, \ldots ;$$

and we will generally use the notation '$\langle x_n \rangle$' to denote the sequence itself. Moreover, if A is a subset of $\mathbf{R}$, we will write

$$\langle x_n \rangle \subseteq A,$$

as shorthand for the statement:

$$(\forall n \in N) : \; x_n \in A.$$

Note. Until further notice, when we say that $\langle x_n \rangle$ is a sequence, we will always mean that $\langle x_n \rangle$ is a sequence of real numbers.

Some examples of sequences are $\langle x_n \rangle$, $\langle y_n \rangle$, and $\langle z_n \rangle$, defined by:

$$x_n = n, y_n = (-1)^n \text{ and } z_n = 1/n \quad \text{for } n = 1, 2, \ldots ,$$

respectively. Since this material is review for you, you probably are already aware that the sequence $\langle z_n \rangle$ differs from the other two in that it is convergent; a concept formally defined as follows.

2.2. Definition. Let $\langle x_n \rangle$ be a sequence, and let $x^* \in \mathbf{R}$. The sequence $\langle x_n \rangle$ is said to **converge to x*** iff

$$(\forall \epsilon > 0)(\exists m \in N)(\forall n \geq m) \colon |x_n - x^*| < \epsilon.$$

In this case, x^* is said to be the **limit of the sequence** $\langle x_n \rangle$ (and $\langle x_n \rangle$ is said to be **convergent**), and we write:

$$\lim_{n \to \infty} x_n = x^*, \text{ or } x_n \to x^*.$$

Notice that we can paraphrase the definition of convergence by the statement, "$x_n \to x^*$ iff every ϵ-neighborhood of x^* contains all but a finite number of terms of the sequence." Consequently, the following facts should be obvious.

1. The limit of a sequence, if it exists, is unique.
2. The alteration of a finite number of terms of a sequence has no effect on convergence, divergence, or limit.
3. If all but a finite number of terms of a sequence are equal to some constant, then the sequence converges to that constant.

2.3. Definition. If a sequence converges to some real number $x \in \mathbf{R}$ it is said to be **convergent**; otherwise it is said to be **divergent**.

Actually, in the terminology which we will be following here, a sequence may approach a limit without being convergent. The following definition should resolve this apparent paradox.

2.4. Definition. A sequence $\langle x_n \rangle$, is said to have the **limit** $+\infty$ [respectively, $-\infty$], or to **diverge to** $+\infty$, iff for every $\alpha \in \mathbf{R}$, there exists $m \in N$ such that:

$$(\forall n \geq m) : \ x_n > \alpha$$

[respectively, $(\forall n \geq m) \colon x_n < \alpha$]; in which case we write:

$$\lim_{n \to \infty} x_n = +\infty, \text{ or } x_n \to +\infty$$

[respectively, $\lim_{n \to \infty} x_n = -\infty$, or $x_n \to -\infty$].

2.5. Examples/Exercises. [1]

1. Consider the sequence $\langle x_n \rangle = \langle 1/2, 1/4, 1/6, 1/8, \ldots \rangle$. Here it is easy to see that the general term of the sequence is given by

$$x_n \equiv f(n) = 1/2n \quad \text{for } n = 1, 2, \ldots .$$

[1]Examples 1 and 2 are undoubtedly presented in much greater detail than most of you will need; the detail is provided to help refresh your memory regarding the basics of working with sequences.

Intuitively it is apparent that $x_n \to 0$. We can formally prove this as follows.

Let $\epsilon > 0$ be given. Then there exists m large enough (see Exercise 9, at the end of Sec. 1.2) so that

$$m > 1/2\epsilon. \tag{2.1}$$

But then, if $n > m$, we have by equation (2.1) that

$$2n > 2m > 1/\epsilon,$$

and thus

$$|(1/2n) - 0| = |1/2n| = 1/2n < \epsilon.$$

Therefore, $x_n \to 0$.

2. Consider the sequence $\langle x_n \rangle = \langle 1/2, 2/5, 3/8, 4/11, \dots \rangle$. Here we can write the general term as

$$x_n = n/d_n, \tag{2.2}$$

where "d_n" denotes the (as-yet-to-be-determined) denominator of the nth term of the sequence. Thus we wish to find the general term of the sequence

$$\langle d_n \rangle = \langle 2, 5, 8, 11, \dots \rangle.$$

Notice that, for the terms shown,

$$d_{n+1} - d_n = 3,$$

which suggests that we try the formula

$$d_n = 3n + k, \tag{2.3}$$

where k is a constant term (independent of n). If the formula for the general term of $\langle d_n \rangle$ takes the form shown in equation (2.3), then, by substituting $n = 1$ into (2.3), we see that we must have:

$$2 = 3 \cdot 1 + k,$$

or

$$k = -1.$$

Checking the formula (2.3) for $n = 2, 3, 4$, and with $k = -1$, we find that indeed

$$d_n = 3n - 1 \quad \text{for } n = 1, 2, \dots ;$$

and thus from (2.2),

$$x_n = n/(3n - 1) \quad \text{for } n = 1, 2, \dots .$$

Now, the next question is, does this sequence converge? To answer this, let's begin by dividing both numerator and denominator of the terms x_n by n to obtain:

$$x_n = 1/[3 - (1/n)] \quad \text{for } n = 1, 2, \ldots .$$

Since $1/n \to 0$, it then appears that

$$x_n \to 1/3. \tag{2.4}$$

We can prove this as follows.

Notice that for $n \in N$,

$$|x_n - 1/3| = \left|\frac{n}{3n-1} - \frac{1}{3}\right| = \left|\frac{3n-(3n-1)}{3(3n-1)}\right| = \frac{1}{3(3n-1)}.$$

Thus we see that, for given $\epsilon > 0$, we will have

$$|x_n - 1/3| < \epsilon,$$

if, and only if

$$\frac{1}{3(3n-1)} < \epsilon,$$

or

$$3n > (1/3\epsilon) + 1 = (3\epsilon + 1)/3\epsilon;$$

which is equivalent to

$$n > (3\epsilon + 1)/9\epsilon. \tag{2.5}$$

Now, given $\epsilon > 0$, there exists a unique $n \in N$ satisfying:[2]

$$m - 1 \leq (3\epsilon + 1)/9\epsilon < m. \tag{2.6}$$

Therefore, for $n \geq m$, it follows by reversing the steps whereby we obained equation (2.5), that

$$|x_n - 1/3| = \frac{1}{9n-3} \leq \frac{1}{9m-3} < \frac{1}{9[(3\epsilon+1)/9\epsilon] - 3} = \frac{1}{1/\epsilon} = \epsilon.$$

It follows, therefore, that equation (2.4) is correct.

3. Show that the sequence $\langle x_n \rangle$ defined by

$$x_n = (-1)^n \quad \text{for } n = 1, 2, \ldots ,$$

is divergent. [Hint. Suppose that $x_n \to x$, and consider the positive real number ϵ^* defined by $\epsilon^* = \max\{|x-1|, |x+1|\}/2$]. □

---

[2] See Exercise 10, at the end of Section 1.2.

2.6. Definition. A sequence, $\langle x_n \rangle$, is said to be **bounded below** [respectively, **above**] iff there exists $\alpha \in \mathbf{R}$ satisfying

$$x_n \geq \alpha \text{ [respectively, } x_n \leq \alpha] \quad \text{for } n = 1, 2, \ldots .$$

The sequence $\langle x_n \rangle$ is said to be **bounded** iff it is bounded both above and below.

2.7. Proposition. *A sequence $\langle x_n \rangle$ is bounded if, and only if, there exists $\alpha \in \mathbf{R}$ satisfying*

$$|x_n| < \alpha \quad \textit{for } n = 1, 2, \ldots .$$

Proof. Exercise. □

Example 2.5.3 shows that a sequence may be bounded, yet nonetheless not be convergent. On the other hand, a moment's reflection will probably serve to convince you that a convergent sequence must necessarily be bounded, and our next result verifies this conjecture.

2.8. Proposition. *If the sequence $\langle x_n \rangle$ converges, then it is bounded.*

Proof. Suppose $x_n \to x$. Then there exists $m \in N$ such that:

$$(\forall n \geq m) \colon |x_n - x| < 1. \tag{2.7}$$

Define

$$\alpha = \max\{|x_1| + 1, |x_2| + 1, \ldots, |x_{m-1}| + 1, |x| + 1\}.$$

Then, for $1 \leq n \leq m - 1$, we obviously have

$$|x_n| < \alpha;$$

whereas, for $n \geq m$, we have, using equation (2.7) and the definition of α:

$$|x_n| = |x_n - x + x| \leq |x_n - x| + |x| < 1 + |x| \leq \alpha. \quad \square$$

One of the beauties of the theory of sequences is that, if we operate on two or more convergent sequences arithmetically, adding corresponding terms, or multiplying each term by a constant, or whatever, the new sequence so obtained is again convergent. Furthermore, the limit of the new sequence is obtained by the corresponding arithmetic operations on the limits of the original sequences! This is all set out more precisely in the next several results.

2.9. Theorem. *Let $\langle x_n \rangle$ and $\langle y_n \rangle$ be convergent sequences, and define*

$$\lim_{n\to\infty} x_n = x \quad \textit{and} \quad \lim_{n\to\infty} y_n = y.$$

Then we have:

1. $\lim_{n\to\infty}(x_n + y_n) = x + y$.
2. $(\forall \alpha \in \mathbf{R})\colon \alpha x_n \to \alpha x$.
3. $\lim_{n\to\infty} x_n \cdot y_n = x \cdot y$.

Proof.
1. Define

$$z_n = x_n + y_n \quad \text{for } n = 1, 2, \ldots, \quad z = x + y,$$

and let $\epsilon > 0$ be given. Then there exist $n_1, n_2 \in N$ such that:

$$(\forall n \geq n_1)\colon |x_n - x| < \epsilon/2,$$

and

$$(\forall n \geq n_2)\colon |y_n - y| < \epsilon/2.$$

Consequently, defining

$$m = \max\{n_1, n_2\},$$

we have, for $n \geq m$,

$$|z_n - z| = |x_n + y_n - x - y| \leq |x_n - x| + |y_n - y| < \epsilon/2 + \epsilon/2 = \epsilon.$$

2. Exercise.
3. We begin by noting that, since $y_n \to y$, it follows from Proposition 2.8 that there exists $\alpha > 0$ such that

$$|y_n| < \alpha \quad \text{for } n = 1, 2, \ldots. \tag{2.8}$$

Now, let $\epsilon > 0$ be given. Then, using the convergence of $\langle x_n \rangle$ and $\langle y_n \rangle$, there exist n_1 and n_2 such that:

$$(\forall n \geq n_1)\colon |x_n - x| < \epsilon/2\alpha$$

[where α is from (2.8)], and

$$(\forall n \geq n_2)\colon |y_n - y| < \epsilon/2(|x| + 1).$$

Defining $m = \max\{n_1, n_2\}$, we then have, for $n \geq m$:

$$|x_n \cdot y_n - x \cdot y| = |x_n \cdot y_n - x \cdot y_n + x \cdot y_n - x \cdot y|$$

$$\leq |y_n(x_n - x)| + |x(y_n - y)| = |y_n| \cdot |x_n - x| + |x| \cdot |y_n - y|$$

$$< \alpha \cdot \left(\frac{\epsilon}{2\alpha}\right) + |x| \cdot \left(\frac{\epsilon}{2(|x|+1)}\right) < \frac{\epsilon}{2} + \frac{\epsilon}{2} = \epsilon. \quad \square$$

For an example of the meaning and use of the above result, consider the sequence $\langle x_n \rangle$ given by:

$$x_n = 1/n \quad \text{for } n = 1, 2, \ldots .$$

We have, for each n,

$$x_n = 2[1/2n];$$

and thus, since we have shown in Example 2.5.1 that the sequence $\langle y_n \rangle$ defined by

$$y_n = 1/2n \quad \text{for } n = 1, 2, \ldots ,$$

converges to zero, it follows at once from Theorem 2.9.2 that $\langle x_n \rangle$ converges to zero as well.

As a second example, consider the sequence $\langle z_n \rangle$ given by:

$$z_n = \frac{2n-1}{n} \quad \text{for } n = 1, 2, \ldots .$$

Since we can write

$$z_n = 2 - 1/n \quad \text{for } n = 1, 2, \ldots ,$$

it follows immediately from Theorem 2.9.1 and the example just given that $\langle z_n \rangle$ converges to 2.

Our next two results are both corollaries of 2.9, and follow easily from that theorem. Consequently, I will leave the proof of the first corollary as an exercise (it is actually a corollary of the proof of 2.9), and provide only an outline of the proof of the second.

2.10. Corollary. *If $\langle x_n \rangle$ and $\langle y_n \rangle$ are sequences such that $x_n \to 0$, and $\langle y_n \rangle$ is bounded, then $x_n \cdot y_n \to 0$.*

2.11. Proposition. *Suppose we have m convergent sequences $\langle x_{in} \rangle$ ($i = 1, \ldots, m$), and that*

$$x_{in} \to \bar{x}_i \quad \text{for } i = 1, \ldots, m.$$

Then, given any m real numbers α_i ($i = 1, \ldots, m$), if we define the sequence $\langle z_n \rangle$ by

$$z_n = \sum_{i=1}^{m} \alpha_i x_{in} \quad \text{for } n = 1, 2, \ldots ,$$

we have:

$$z_n \to \sum_{i=1}^{m} \alpha_i \bar{x}_i.$$

Proof. (Outline). We proceed by induction on m, as follows.

1. Suppose $m = 1$. Then it follows at once from Theorem 2.9.2 that

$$\alpha_1 x_{in} \to \alpha_1 \bar{x}_i.$$

2. Suppose the result holds for $m = k$. Then for $m = k + 1$, we have:

$$z_n \stackrel{\text{def}}{=} \sum_{i=1}^{k+1} \alpha_i x_{in} = \left(\sum_{i=1}^{k} \alpha_i x_{in}\right) + \alpha_{k+1} x_{k+1,n}$$

$$\to \lim_{n\to\infty} \left(\sum_{i=1}^{k} \alpha_i x_{in}\right) + \lim_{n\to\infty} \alpha_{k+1} x_{k+1,n}$$

$$= \sum_{i=1}^{k} \alpha_i \bar{x}_i + \alpha_{k+1} \bar{x}_{k+1}. \quad \square$$

As yet we have not considered the arithmetic operation of division in connection with sequences. The next two results remedy this situation.

2.12. Lemma. *If $\langle x_n \rangle$ is a sequence having no terms equal to zero, and $x_n \to x^* \neq 0$, then $1/x_n \to 1/x^*$.*

Proof. Since $x_n \to x^*$, and $x^* \neq 0$, there exists $m \in N$ such that

$$(\forall n \geq m)\colon \ ||x_n| - |x^*|| \leq |x_n - x^*| < |x^*|/2;$$

and thus, for all $n \geq m$,

$$-|x^*|/2 < |x_n| - |x^*| < |x^*|/2,$$

or

$$|x^*|/2 < |x_n| < 3|x^*|/2.$$

If we then define $\alpha > 0$ by

$$\alpha = \min\{|x_1|, |x_2|, \dots, |x_{m-1}|, |x^*|/2\},$$

we see that

$$|x_n| \geq \alpha \quad \text{for } n = 1, 2, \dots. \tag{2.9}$$

Now, let $\epsilon > 0$ be given. Once again using the convergence of $\langle x_n \rangle$ and the fact that $x^* \neq 0$, we see that there exists $p \in N$ such that

$$(\forall n \geq p)\colon |x_n - x^*| < (\alpha \cdot |x^*|) \cdot \epsilon. \tag{2.10}$$

But then, using (2.9) and (2.10) in turn, it follows that, if $n \geq p$:

$$\left|\frac{1}{x_n} - \frac{1}{x^*}\right| = \left|\frac{x^* - x_n}{x_n \cdot x^*}\right| = \frac{|x_n - x^*|}{|x_n| \cdot |x^*|} \leq \frac{|x_n - x^*|}{\alpha \cdot |x^*|} < \epsilon. \quad \square$$

The following result is now an immediate consequence of Theorem 2.9 and Lemma 2.12. Details of the proof will be left to the reader.

2.13. Theorem. (Corollary). *If $\langle x_n \rangle$ and $\langle y_n \rangle$ are sequences satisfying:*

1. $(\forall n \in N) : x_n \neq 0$,
2. $x_n \to x^* \neq 0$, *and*
3. $y_n \to y^*$,

then $y_n/x_n \to y^*/x^*$.

We conclude this section with several propositions, the proofs of which will be left as exercises. Each of these propositions is of some importance in and of itself. However, it is particularly important to notice that, collectively, they provide a rigorous justification for the following sort of argument. "Consider the sequence $\langle x_n \rangle$ defined by

$$x_n = \frac{n^2 + 12n + 23}{n^2 + 6n + 5} \quad \text{for } n = 1, 2, \ldots .$$

We have

$$\left| \frac{n^2 + 12n + 23}{n^2 + 6n + 5} - 1 \right| = \left| \frac{n^2 + 12n + 23 - n^2 - 6n - 5}{n^2 + 6n + 5} \right|$$

$$= 6 \cdot \left| \frac{n + 3}{(n + 5)(n + 1)} \right| \leq \frac{6}{n} \to 0.$$

Therefore $x_n \to 1$."

As a hint to help you in doing these proofs, I should mention that it is particularly convenient to prove Propositions 2.14–2.18 in the order in which they are presented.

2.14. Proposition. *If $\langle x_n \rangle$ is a sequence, and x^* is a real number, we have $x_n \to x^*$ if, and only if, $|x_n - x^*| \to 0$.*

2.15. Proposition. *Suppose $\langle x_n \rangle$ is a sequence satisfying:*

$$x_n \geq 0 \quad \text{for } n = 1, 2, \ldots .$$

Then if $\langle x_n \rangle$ converges, we must have:

$$\lim_{n \to \infty} x_n \geq 0.$$

2.16. Proposition. *If $\langle x_n \rangle$ and $\langle y_n \rangle$ satisfy:*

$$x_n \to x^*, y_n \to y^*,$$

and

$$x_n \geq y_n \quad \text{for } n = 1, 2, \ldots ,$$

then

$$x^* \geq y^*.$$

2.17. Proposition. *If $\langle x_n \rangle$ and $\langle y_n \rangle$ satisfy:*

$$0 \leq x_n \leq |y_n| \quad \text{for } n = 1, 2, \dots ,$$

and $y_n \to 0$, then $x_n \to 0$ as well.

2.18. Proposition. *If $\langle x_n \rangle$, $\langle y_n \rangle$, and $\langle z_n \rangle$ satisfy:*

$$x_n \leq y_n \leq z_n \quad \text{for } n = 1, 2, \dots ,$$

$x_n \to x^*$, and $z_n \to x^*$, then $y_n \to x^*$ as well.

Exercises.

1. Find the formula for the general term of the sequence,

$$3/2, 5/4, 7/6, 9/8, \dots ;$$

then (a) find the limit of the sequence, and (b) prove that the sequence converges to the limit which you have found.

2. Show that the general term of the sequence

$$\langle x_n \rangle = \langle -1/2, 1/8, -1/18, 1/32, -1/50, 1/72, \dots \rangle$$

is given by

$$x_n = (-1)^n / 2n^2 \quad \text{for } n = 1, 2, \dots ;$$

and show that

$$x_n \to 0.$$

3. Prove Propositions 2.14–2.18.

4. Use Propositions 2.14–2.18 to justify the steps used to establish the convergence of $\langle x_n \rangle$ in the example immediately preceding the statement of Proposition 2.14.

5. Does the sum (difference, product, quotient) of two *divergent* sequences also diverge? Prove or provide a counterexample.

6. If $x_n \to x^*$, does $|x_n| \to |x^*|$? How about the converse? Prove or provide counterexamples.

7. Show that if $x > 1$, then the sequence $\langle x_n \rangle$ defined by

$$x_n = (1/x)^n \quad \text{for } n = 1, 2, \dots ,$$

converges to zero. [Hint: write $x = 1 + \alpha$, for $\alpha > 0$, and then see if you can find a sequence $\langle y_n \rangle$ satisfying $x^n = (1 + \alpha)^n > y_n$ and such that $1/y_n \to 0$.]

2.2 Subsequences and Cauchy Sequences

While the section heading lists two apparently unrelated topics to be covered in this section, they fit together fairly well (or at least they can both be covered in one lecture). Of course, just to confuse things, we begin with a topic not even mentioned in the heading: monotonic sequences.

2.19. Definition. A sequence $\langle x_n \rangle$ is said to be **monotone** or **monotonic** iff either

1. $x_1 \leq \cdots \leq x_n \leq x_{n+1} \leq \cdots$ [that is, $x_n \leq x_{n+1}$ for $n = 1, 2, \ldots$], or
2. $x_1 \geq \cdots \geq x_n \geq x_{n+1} \geq \cdots$ [that is, $x_n \geq x_{n+1}$ for $n = 1, 2, \ldots$].

In the first case, the sequence is said to be **monotonically increasing** or simply **increasing** [**strictly increasing** if a strict inequality holds in each case, that is, if $x_{n+1} > x_n$, for $n = 1, 2, \ldots$]; and the sequence is said to be **monotonically decreasing**, or simply **decreasing** in the second case [**strictly decreasing** if $x_{n+1} < x_n$, for $n = 1, 2, \ldots$].

Thus, for example, the sequence $\langle x_n \rangle$ defined by

$$x_n = 1/n \quad \text{for } n = 1, 2, \ldots ,$$

is strictly decreasing; while the sequence $\langle y_m \rangle$ defined by

$$y_m = 2m \quad \text{for } m = 1, 2, \ldots ,$$

is strictly increasing. One of the particularly useful and interesting things about monotonic sequences is that, if they are bounded, then they always converge.[3]

2.20. Theorem. *Every bounded monotonic sequence of real numbers converges.*

Proof. Suppose $\langle x_n \rangle$ satisfies

$$x_n \leq x_{n+1} \quad \text{for } n = 1, 2, \ldots , \tag{2.11}$$

and consider the set A consisting of all of the values of $\langle x_n \rangle$; that is, define A by

$$A = \{x_1, x_2, x_3, \ldots\}.$$

Since $\langle x_n \rangle$ is bounded, $\sup A$ exists, and we define

$$x^* = \sup A.$$

Now let $\epsilon > 0$ be given. It follows easily from Theorem 1.14 that:

$$(\exists m \in N) : x_m > x^* - \epsilon.$$

[3]Actually, if a monotone sequence is unbounded, it has either $+\infty$ or $-\infty$ as a limit; so that a monotone sequence always has a limit.

But then, using (2.11) and the definition of the supremum, it follows that:

$$x^* - \epsilon < x_n < x^* + \epsilon \quad \text{for } n = m, m+1, \ldots ;$$

so that

$$(\forall n \geq m) \colon |x_n - x^*| < \epsilon.$$

Therefore $x_n \to x^* \equiv \sup A$; and a similar proof establishes convergence (with $\inf A$ replacing $\sup A$) for the case in which $\langle x_n \rangle$ is decreasing. □

2.21. Definition. If $\langle x_n \rangle$ and $\langle y_n \rangle$ are sequences, $\langle y_n \rangle$ is said to be a **subsequence of** $\langle x_n \rangle$ iff there exists a sequence $\langle n_k \rangle$ of positive integers satisfying:

1. $(\forall i, j \in N) \colon i > j \Rightarrow n_i > n_j$, and
2. $(\forall i \in N) \colon y_i = x_{n_i}$.

(In such a case, the subsequence is usually denoted by '$\langle x_{n_i} \rangle$' rather than '$\langle y_i \rangle$.')

Notice that condition 1 of the above definition requires that $\langle n_i \rangle$ be a *strictly increasing* sequence of positive integers; not just a monotonically increasing sequence. The significance of this distinction is illustrated in the first of the following examples.

2.22. Examples/Exercises.

1. Consider the sequences $\langle x_n \rangle = \langle 1, 1/2, 1/8, 1/16, \ldots \rangle$, and $\langle y_m \rangle = \langle 1, 1, 1, 1, \ldots \rangle$. Is $\langle y_m \rangle$ a subsequence of $\langle x_n \rangle$?

Solution. Notice that if we define the sequence $\langle n_i \rangle$ by:

$$n_i = 1 \quad \text{for } i = 1, 2, \ldots , \tag{2.12}$$

then we have

$$y_i = x_{n_i} \quad \text{for } i = 1, 2, \ldots . \tag{2.13}$$

However, $\langle n_i \rangle$ is *not* a strictly increasing sequence of positive integers. Thus with $\langle n_i \rangle$ defined as in (2.12), part 2 of Definition 2.21 is satisfied, but not part 1. Moreover, it is easy to see that no other sequence $\langle n_i \rangle$ will satisfy (2.13) in this case, and therefore $\langle y_i \rangle$ is *not* a subsequence of $\langle x_n \rangle$.

2. Consider the sequences $\langle x_m \rangle$, $\langle y_n \rangle$, and $\langle z_q \rangle$, where

$$\langle x_m \rangle = \langle 0, 3/4, 0, 7/8, 0, 15/16, 0, 31/32, 0, 63/64, \ldots \rangle,$$

$$y_n = 1 - (1/2)^{n+1} \quad \text{for } n = 1, 2, \ldots ,$$

and

$$z_q = 0 \quad \text{for } q = 1, 2, \ldots .$$

If we define $\langle m_i \rangle$ and $\langle n_j \rangle$ by

$$m_i = 2i \quad \text{for } i = 1, 2, \ldots ,$$

and

$$n_j = 2j - 1 \quad \text{for } j = 1, 2, \dots ,$$

then

$$y_i = 1 - (1/2)^{i+1} = x_{m_i} \quad \text{for } i = 1, 2, \dots ,$$

and

$$z_j = x_{n_j} \quad \text{for } j = 1, 2, \dots .$$

Therefore, both $\langle y_n \rangle$ and $\langle z_q \rangle$ are subsequences of $\langle x_m \rangle$.

3. Consider the sequences $\langle x_m \rangle$ and $\langle y_n \rangle$, where

$$\langle x_m \rangle = \langle 1, -1, 2, 8, 3, 1/2, 1/4, 1/8, 1/16, \dots \rangle,$$

and

$$y_n = (1/2)^n \quad \text{for } n = 1, 2, \dots .$$

Is $\langle y_n \rangle$ a subsequence of $\langle x_m \rangle$? Explain.

4. Consider the sequence $\langle x_n \rangle$ defined by:

$$x_n = n(-1)^{n+1}/(3n - 1) \quad \text{for } n = 1, 2, \dots .$$

Show that the sequences $\langle y_m \rangle$ and $\langle z_q \rangle$ defined by

$$y_m = (2m - 1)/(6m - 4) \quad \text{for } m = 1, 2, \dots ,$$

and

$$z_q = -2q/(6q - 1) \quad \text{for } q = 1, 2, \dots ,$$

are both subsequences of $\langle x_n \rangle$, *and* that both converge. □

If $\langle x_n \rangle$ is a convergent sequence, then *any* subseqence of $\langle x_n \rangle$ converges to $\lim x_n$, as is shown in the next result. We will leave the proof as an exercise.

2.23. Proposition. *If $\langle x_n \rangle$ is a sequence such that $x_n \to x^*$, and $\langle x_{n_i} \rangle$ is any subsequence of $\langle x_n \rangle$, then $x_{n_i} \to x^*$.*

In Example 2.22.4, above, the two subsequences considered converge to $1/3$ and $-1/3$, respectively. Thus it follows from Proposition 2.23 that $\langle x_n \rangle$ is divergent.

Subsequences of real numbers are intimately related to cluster points of the sequence, which are defined as follows.

2.24. Definition. We shall say that $x^* \in \mathbf{R}$ is a **cluster point** (or **accumulation point**) of the sequence $\langle x_n \rangle$ iff, given any $\epsilon > 0$, and any positive integer, m, there exists $n > m$ such that

$$|x_n - x^*| < \epsilon.$$

Intuitively, x^* is a cluster point of $\langle x_n \rangle$ if, and only if, each neighborhood of x^* contains infinitely many of the terms of the sequence. Notice the distinction between this statement and the earlier verbal statement "x^* is the limit of $\langle x_n \rangle$ iff every neighborhood of x^* contains all but a finite number of the terms of the sequence." Thus, if $\langle x_n \rangle$ converges to x^*, then x^* is also a cluster point of the sequence; but on the other hand, if x^* is a cluster point of the sequence, this does not necessarily mean that $\langle x_n \rangle$ converges to x^*. If you take another look at Example 2.22.2, it should be immediately apparent that zero is a cluster point of the sequence $\langle x_m \rangle$ considered there, but clearly zero is not the limit of the sequence. Notice, however, that a subsequence of $\langle x_m \rangle$ does converge to zero in that example [namely, the subsequence $\langle z_q \rangle$ defined there]. This is typical of the general case; in fact, x^* is a cluster point of $\langle x_n \rangle$ if, and only if, a subsequence of $\langle x_n \rangle$ converges to x^*, as we shall now prove.

2.25. Proposition. *The real number x^* is a cluster point of the sequence $\langle x_n \rangle$ if, and only if, there exists a subsequence of $\langle x_n \rangle$, $\langle x_{n_i} \rangle$, such that $x_{n_i} \to x^*$.*

Proof. Suppose first that x^* is a cluster point of $\langle x_n \rangle$. Then the set N_1 defined by

$$N_1 = \{n \in N \mid |x_n - x^*| < 1\},$$

is non-empty, and we may define

$$n_1 = \min N_1$$

[recall that any non-empty set of positive integers has a smallest element]. Similarly, and again using the fact that x^* is a cluster point of $\langle x_n \rangle$, we see that the set N_2 defined by

$$N_2 = \{n \in N \mid n > n_1 \;\&\; |x_n - x^*| < 1/2\},$$

is non-empty; and we can define

$$n_2 = \min N_2.$$

Proceeding in this way, we note that, if we have defined n_{i-1}, for $i \geq 2$, then the set

$$N_i \equiv \{n \in N \mid n > n_{i-1} \;\&\; |x_n - x^*| < 1/i\},$$

is necessarily non-empty, and we can then define

$$n_i = \min N_i.$$

Notice that we have in this way defined (inductively) a strictly increasing sequence of positive integers, $\langle n_i \rangle$, and that the subsequence, $\langle x_{n_i} \rangle$ of $\langle x_n \rangle$, which it defines, satisfies

$$|x_{n_i} - x^*| < 1/i \quad \text{for } i = 1, 2, \dots ,$$

and therefore it is obvious that $x_{n_i} \to x^*$.

Conversely, if x^* is such that there exists a subsequence of $\langle x_n \rangle$, $\langle x_{n_i} \rangle$, such that

$$x_{n_i} \to x^*,$$

then it is easy to show that x^* is a cluster point of $\langle x_n \rangle$. Details will be left as an exercise. □

As an immediate corollary of 2.25 and 2.23 we then have the following.

2.26. Corollary. *If the sequence $\langle x_n \rangle$ converges to $x^* \in \mathbf{R}$, then x^* is a cluster point; in fact the* only *cluster point, of $\langle x_n \rangle$.*

Our next result is of fundamental importance in analysis, and we will have many occasions to make use of it in the subsequent chapters of this book.

2.27. Theorem. (Bolzano-Weierstrass Theorem for R). *If $\langle x_n \rangle$ is a bounded sequence, then there exists a subsequence, $\langle x_{n_i} \rangle$, of $\langle x_n \rangle$, and a number $z \in \mathbf{R}$, such that*

$$\lim_{i \to \infty} x_{n_i} = z.$$

Proof. It follows from 2.25 that we need only establish that $\langle x_n \rangle$ has a cluster point, which is what we will do. We begin by noting that, since $\langle x_n \rangle$ is bounded, there exists a positive number, a, satisfying

$$-a < x_n < a \quad \text{for } n = 1, 2, \ldots .$$

Thus the set of real numbers, A, defined by

$$A = \{x \in \mathbf{R} \mid x_n \geq x \text{ for an infinite number of terms of } \langle x_n \rangle\}$$

$$= \{x \in \mathbf{R} \mid (\forall m \in N)(\exists n \geq m) : x_n \geq x\},$$

is non-empty (since it certainly contains $-a$) and bounded above by a. Therefore A has a (finite) supremum; and we define

$$z = \sup A. \tag{2.14}$$

Now let $\epsilon > 0$ and $m \in N$ be given. Then we note first that the number $z + \epsilon/2$ is not an element of A, and thus only a finite number of terms of $\langle x_n \rangle$ can exceed $z + \epsilon/2$. Consequently, there exists an integer, q, such that for all $n \geq q$,

$$x_n < z + \epsilon. \tag{2.15}$$

Next, we note that if there were to exist some integer, p, such that for all $n \geq p$,

$$x_n < z - \epsilon/2,$$

then it would follow that every element of A is less than or equal to $z - \epsilon/2$; which, since $\epsilon > 0$, contradicts (2.14). Thus it must be the case that there exists $n > \max\{m, q\}$ such that

$$x_n \geq z - \epsilon/2 > z - \epsilon. \tag{2.16}$$

Combining (2.15) and (2.16), we see that z is a cluster point of $\langle x_n \rangle$. □

In our proof of the Bolzano-Weierstrass Theorem, we actually found not just a cluster point of the sequence, but the *largest* cluster point of the sequence, as we will demonstrate presently. In the meantime, we consider a condition developed by the great French mathematician Augustin-Louis Cauchy, and for which we will find a great deal of use.

2.28. Definition. A sequence $\langle x_n \rangle$ is called a **Cauchy sequence** (or is said to satisfy the **Cauchy condition**) iff, for every $\epsilon > 0$, there exists $p \in N$ such that, for all $m, n \geq p$,

$$|x_n - x_m| < \epsilon.$$

We will demonstrate shortly that a sequence converges if, and only if, it is a Cauchy sequence. In order to do this, we begin by establishing the following, the proof of which I will leave as an exercise.

2.29. Proposition. *If $\langle x_n \rangle$ is a Cauchy sequence, then $\langle x_n \rangle$ is bounded.*

2.30. Theorem. *A sequence $\langle x_n \rangle$ is convergent if, and only if, it is a Cauchy sequence.*

Proof. Suppose first that $x_n \to x^*$, and let $\epsilon > 0$ be given. Then there exists $p \in N$ such that

$$(\forall n \geq p)\colon |x_n - x^*| < \epsilon/2. \tag{2.17}$$

But then we see that it follows from (2.17) and the triangle inequality that, if $m, n \geq p$, then

$$|x_n - x_m| \leq |x_n - x^*| + |x_m - x^*| < \epsilon/2 + \epsilon/2 = \epsilon.$$

Therefore, $\langle x_n \rangle$ is a Cauchy sequence.

Now suppose that $\langle x_n \rangle$ is a Cauchy sequence. By Proposition 2.29, $\langle x_n \rangle$ is bounded, and thus by the Bolzano-Weierstrass Theorem, there exists a subsequence, $\langle x_{n_i} \rangle$, and a real number, x^*, such that

$$x_{n_i} \to x^*. \tag{2.18}$$

Now let $\epsilon > 0$ be given. Then by (2.18), there exists $j \in N$ such that for all $i \geq j$,

$$|x_{n_i} - x^*| < \epsilon/2. \tag{2.19}$$

Furthermore, since $\langle x_n \rangle$ is a Cauchy sequence, there exists $q \in N$ such that, for all $m, n \geq q$,

$$|x_m - x_n| < \epsilon/2; \tag{2.20}$$

and, since $\langle n_i \rangle$ is strictly increasing, there exists $k \geq j$ [where j is from (2.19)] such that, for all $i \geq k$,

$$n_i \geq q.$$

Thus it follows from (2.19) and (2.20) that if $n \geq q$, then

$$|x_n - x^*| \leq |x_n - x_{n_k}| + |x_{n_k} - x^*| < \epsilon/2 + \epsilon/2 = \epsilon.$$

Therefore $x_n \to x^*$. □

One particularly useful aspect of the Cauchy condition is that we can use it to establish convergence of a sequence in situations where we do not know the limit of the sequence. In particular, therefore, we will find this condition especially useful in theoretical arguments; in which the sequence may not be numerically defined.

Our next concern will be to show, essentially, that any bounded sequence has both a largest and a smallest cluster point, and to develop one way of finding these points. We begin with the following.

2.31. Proposition. *Let $\langle x_n \rangle$ be a bounded sequence, and define the sequences $\langle y_m \rangle$ and $\langle z_q \rangle$ by:*

$$y_m = \inf\{x_m, x_{m+1}, \dots\} = \inf_{n \geq m} x_n \quad \textit{for}\, m = 1, 2, \dots,$$

and

$$z_q = \sup\{x_q, x_{q+1}, \dots\} = \sup_{n \geq q} x_n \quad \textit{for}\, q = 1, 2, \dots,$$

respectively. Then the sequences $\langle y_m \rangle$ and $\langle z_q \rangle$ both converge, and we have:

$$\lim_{q \to \infty} z_q \geq \lim_{m \to \infty} y_m.$$

Proof. Since

$$\{x_{m+1}, x_{m+2}, \dots\} \subseteq \{x_m, x_{m+1}, x_{m+2}, \dots\},$$

it follows that, for each $m \in N$:

$$y_m \equiv \inf\{x_m, x_{m+1}, \dots\} \leq y_{m+1} \equiv \inf\{x_{m+1}, x_{m+2}, \dots\}.$$

Thus we see that $\langle y_m \rangle$ is increasing, and similar considerations establish that $\langle z_q \rangle$ is decreasing. Moreover, for each m we have

$$y_m \equiv \inf\{x_m, x_{m+1}, \dots\} \leq \sup\{x_m, x_{m+1}, \dots\} \equiv z_m;$$

and therefore,

$$y_1 \le y_m \le z_m \le z_1 \quad \text{for } m = 1, 2 \ldots. \tag{2.21}$$

It follows from (2.21) that both $\langle y_m \rangle$ and $\langle z_q \rangle$ are bounded; and thus, from Theorem 2.20, that both are convergent. Furthermore, from (2.21) and Proposition 2.16, we see that

$$\lim_{q\to\infty} \sup_{n\ge q} x_n = \lim_{q\to\infty} z_q \ge \lim_{m\to\infty} y_m = \lim_{m\to\infty} \inf_{n\ge m} x_n. \quad \square$$

2.32. Definition. The limits of $\langle y_m \rangle$ and $\langle z_q \rangle$, as defined in the above result are denoted by "lim inf x_n" and "lim sup x_n," respectively.

To illustrate these definitions, consider the sequence $\langle x_n \rangle$ defined by

$$x_n = (1/2)^n \quad \text{for } n = 1, 2, \ldots.$$

In this case, it should be clear that, for each m:

$$y_m \stackrel{\text{def}}{=} \inf\{(1/2)^m, (1/2)^{m+1}, \ldots\} = 0,$$

and

$$z_m \stackrel{\text{def}}{=} \sup\{(1/2)^m, (1/2)^{m+1}, \ldots\} = (1/2)^m;$$

and thus in this case,

$$\liminf x_n = \limsup x_n = 0.$$

2.33. Proposition. *Suppose $\langle x_n \rangle$ is a bounded sequence, and define*

$$C = \{x \in \mathbf{R} \mid x \text{ is a cluster point of } \langle x_n \rangle\}.$$

Then we have:

1. C is bounded and non-empty.

2. $x^* \stackrel{def}{=} \sup C$ and $x_* \stackrel{def}{=} \inf C$ are both finite, and both are cluster points of $\langle x_n \rangle$ (so that both are elements of C, $x^* = \max C$, and $x_* = \min C$).

3. $x^* = \limsup x_n$ and $x_* = \liminf x_n$.

Proof. The fact that C is non-empty is simply a re-statement of the Bolzano-Weierstrass Theorem; and the boundedness of C will follow from the next part of our proof.

In order to prove part 2, we begin by noting that a re-reading of the proof of Theorem 2.27 makes it easy to see that, letting 'z' denote the supremum of the set A considered there:

$$z \in C, \tag{2.22}$$

and, given any $\epsilon > 0$, there exists $m \in N$ such that

$$x_n < z + \epsilon \quad \text{for } n = m+1, m+2, \ldots. \tag{2.23}$$

It is an easy consequence of (2.23) that, if x is a cluster point of $\langle x_n \rangle$, then $x \leq z$; so that we see that z is an upper bound for C. Since z is also an element of C [by (2.22)], it follows at once that

$$z = \sup C \equiv x^*; \tag{2.24}$$

which establishes the portion of part 2 of our conclusion which deals with $\sup C$. We also note that we can now re-state (2.23) as: given any $\epsilon > 0$, there exists $m \in N$ such that, for all $n \geq m$,

$$x_n < x^* + \epsilon. \tag{2.25}$$

Now, as in Proposition 2.31, define

$$z_q = \sup\{x_q, x_{q+1}, \ldots\} = \sup\{x_n \mid n \geq q\} \quad \text{for } q = 1, 2, \ldots;$$

and suppose, by way of obtaining a contradiction, that for some m,

$$z_m < x^*.$$

Then it follows immediately from the definition of $\langle z_q \rangle$ that x^* cannot be a cluster point of $\langle x_n \rangle$; contradicting what we have established in the previous paragraph. Therefore we have

$$z_q \geq x^* \quad \text{for } q = 1, 2, \ldots,$$

and it then follows by Proposition 2.16 that

$$\limsup x_n = \lim_{z \to \infty} z_q \geq x^*. \tag{2.26}$$

Furthermore, it is an immediate consequence of (2.25) that, given any $\epsilon > 0$, there exists m such that, for all $q \geq m$,

$$z_q \leq x^* + \epsilon;$$

from which it follows easily that

$$\limsup x_n = \lim_{q \to \infty} z_q \leq x^*. \tag{2.27}$$

Combining (2.26) and (2.27) establishes that

$$\limsup x_n = \lim_{q \to \infty} z_q = x^*.$$

Similar considerations establish the fact that x_* is a cluster point of $\langle x_n \rangle$, and that

$$\liminf x_n = x_*.$$

Details will be left as an exercise. □

We observed earlier that, if $\langle x_n \rangle$ is convergent, then $\lim_{n\to\infty} x_n$ is the unique cluster point of the sequence. Notice that it follows at once from this consideration and 2.33 that a sequence $\langle x_n \rangle$ converges if, and only if,

$$\liminf x_n = \limsup x_n;$$

in which case, the common value is the limit of the sequence. It is also an easy consequence of Propositions 2.25 and 2.33 that if z is the limit of a subsequence of $\langle x_n \rangle$, then we must have

$$\liminf x_n \leq z \leq \limsup x_n.$$

These facts will be of use to us sufficiently often that we will state them formally, as follows.

2.34. Proposition. *If $\langle x_n \rangle$ is a bounded sequence, then:*

1. $\langle x_n \rangle$ converges if, and only if,

$$\liminf x_n = \limsup x_n,$$

in which case,

$$\lim_{n\to\infty} x_n = \limsup x_n;$$

and

2. if z is the limit of a subsequence of $\langle x_n \rangle$, then

$$\liminf x_n \leq z \leq \limsup x_n.$$

Exercises.

1. Find the general term of the following sequences, and in each case, determine whether or not the sequence converges (you may find it necessary and/or more convenient to define separate subsequences, rather than to provide one general formula).

a. $1/2, -1/7, 1/8, -1/14, 1/14, \ldots$.

b. $3/5, -7/5, 11/5, -3, 19/5, \ldots$.

c. $2/3, 2/5, 5/9, 5/11, 8/15, 8/17, \ldots$.

2. Does a bounded sequence necessarily converge? Prove or provide a counterexample.

3. Show that if a monotone sequence has a convergent subsequence, then the original sequence is convergent.

4. Prove Propositions 2.23 and 2.29.

2.3 Infinite Series

In this section we will look at the basic elements of the theory of infinite series. For us, an infinite series is perhaps best considered as an indirectly-defined sequence; most often, a sequence which is itself defined from a sequence. Thus, for example, suppose that a producer has a choice of two production/pricing policies:

1. policy A, which will yield a profit of $\$x_0$ to be paid now, $\$\ x_1$, which he will receive one year from now, and so on, with $\$x_n$ being the profit he (or his heirs) will receive n years from now; or

2. policy B, under which he will receive $\$y_0$ now, $\$y_1$ a year from now, and so on.

There are, in fact, many ways in which one might devise indices, or whatever, to choose between policies A and B, but the one most often considered in economics is the criterion of maximizing **present value**. The idea here is that, if the relevant rate of interest is 'i' (expressed as a decimal), and interest is compounded annually, then $\$p$ available now could be invested so as to be worth

$$q = (1 + i)p$$

one year from now. Thus the amount

$$p_1 \equiv [1/(1 + i)]x_1$$

is the maximum amount which the producer should be just willing to pay now in order to receive the $\$x_1$ profit obtainable under policy A one year from now. Similarly, if our producer were to invest $\$p$ now and leave it for two years, the investment would, at the end of that time, be worth

$$q = (1 + i)^2 p;$$

so that the $\$x_2$ profits which policy A pays at the end of two years has a present value of

$$p_2 = [1/(1 + i)^2]x_2 = [1/(1 + i)]^2 x_2.$$

The general principle should now be clear: if we define the **discount rate**, δ, by

$$\delta = 1/(1 + i),$$

then the profit obtainable under policy A at the end of n years has a present value of

$$p_n \stackrel{\text{def}}{=} \delta^n \cdot x_n.$$

Thus the total profits paid out under policy A through the end of n years from now have a present value of

$$S_n = \sum_{m=0}^{n} \delta^m x_m; \tag{2.28}$$

while the present value of the total profits paid out under policy B through the first n years is given by

$$T_n = \sum_{m=0}^{n} \delta^m y_m. \tag{2.29}$$

Thus, assuming that both the sequences, $\langle S_n \rangle$ and $\langle T_n \rangle$ converge, policy A yields a present value higher than policy B if, and only if,

$$\lim_{n \to \infty} S_n > \lim_{n \to \infty} T_n.$$

Hopefully, with this background (which you may well have already had), the reason for our interest in the following definitions will be apparent.

2.35. Definition. If $\langle a_n \rangle$ is a sequence, the expression

$$\sum_{n=1}^{\infty} a_n. \tag{2.30}$$

is called an **infinite series**; the numbers $a_1, a_2, \ldots, a_n, \ldots$, being the **terms of the series.** The sequence $\langle s_n \rangle$ defined by:

$$s_n = a_1 + a_2 + \cdots + a_n = \sum_{i=1}^{n} a_i \quad \text{for } n = 1, 2, \ldots, \tag{2.31}$$

is called the **sequence of partial sums of the series**; and the series is said to **converge** or **diverge** according to whether $\langle s_n \rangle$ converges or diverges. In the case where $s_n \to s$, we write

$$s = \sum_{n=1}^{\infty} a_n,$$

and say that the **series converges to s.**

Another expression which will be of interest is '$S_{n,p}$' given by:

$$S_{n,p} \equiv \sum_{m=n+1}^{n+p} a_m = s_{n+p} - s_n.$$

From the Cauchy criterion for the convergence of a sequence (Theorem 2.30), we then obtain the following result, the proof of which is immediate.

2.36. Proposition. *The infinite series, $\sum_{n=1}^{\infty} a_n$, converges if, and only if, for each $\epsilon > 0$, there exists $n \in N$ such that, for all $p \in N$,*

$$|S_{n,p}| < \epsilon.$$

We will sometimes denote the first term in the sequence $\langle a_n \rangle$, which defines the series, by 'a_0,' instead of 'a_1;' and, correspondingly, define s_n by:

$$s_n = \sum_{m=0}^{n} a_m \quad \text{for } n = 0, 1, 2, \ldots; \tag{2.32}$$

rather than as in (2.31). While this may sound confusing, you probably will not find it so in practice; just be careful to notice whether the first term in the defining sequence is subscripted by a '0' or by '1!'

Before proceeding further with our general discussion of the theory of infinite series, suppose we consider some examples.

2.37. Examples/Exercises.

1. From Proposition 2.36, it follows at once that if the series,

$$\sum_{n=1}^{\infty} a_n \tag{2.33}$$

converges, then $a_n \to 0$. Thus, consider a series of the form of (2.33) where $\langle a_n \rangle$ is given by

$$a_n = (n-1)/n \quad \text{for } n = 1, 2, \ldots .$$

In this case, we see at once that the series does not converge, since

$$a_n \to 1 \neq 0 \text{ as } n \to \infty.$$

Moreover, suppose a series converges to, say $s^* \in \mathbf{R}$, and, for each n, define r_n by

$$r_n = \lim_{p \to \infty} S_{n,p}.$$

It follows at once from the definition of $S_{n,p}$ that, for each n, we have

$$|S_{n,p} - (s^* - s_n)| = |S_{n,p} + s_n - s^*| = |s_{n+p} - s^*|.$$

Since s_{n+p} converges to s^*, we then see that r_n is well-defined, for each n, and that

$$r_n = \lim_{p \to \infty} S_{n,p} = \sum_{m=1}^{\infty} a_m - s_n = s^* - s_n, \quad \text{for } n = 1, 2, \ldots .$$

Thus, if the series converges, then we see that the sequence $\langle r_n \rangle$ is well defined [the term r_n is often called the **remainder after n terms**]; and, in fact, it follows from Proposition 2.36 that the sequence converges, and that

$$r_n \to 0.$$

2. Consider the series $\sum_{n=0}^{\infty} a_n$, where

$$a_n = (1/2)^n \quad \text{for } n = 0, 1, 2, \ldots,$$

so that

$$\sum_{n=0}^{\infty} a_n = \sum_{n=0}^{\infty} (1/2)^n = 1 + 1/2 + 1/4 + 1/8 + \ldots .$$

Here it is easy to verify that, for any n,

$$s_n - (1/2)s_n = 1 - (1/2)^{n+1};$$

so that, solving for s_n, we obtain:

$$s_n = 2 - (1/2)^n \quad \text{for } n = 0, 1, 2, \ldots .$$

Thus we see that $s_n \to 2$, and it follows that

$$\sum_{n=0}^{\infty} a_n = \sum_{n=0}^{\infty} (1/2)^n = 2.$$

3. Generalizing the previous example, consider the series in which a_n is given by:

$$a_n = \alpha^n \quad \text{for } n = 0, 1, 2, \ldots,$$

where α is a (fixed) real number; which is the **geometric series**. If $|\alpha| \geq 1$, then

$$\lim_{n \to \infty} \alpha^n \neq 0,$$

and we see that the series does not converge. On the other hand, if $|\alpha| < 1$, then proceeding as in the previous example, we see that in this case,

$$s_n - \alpha s_n = 1 - \alpha^{n+1};$$

and thus

$$s_n = 1/(1 - \alpha) - \alpha^{n+1}/(1 - \alpha).$$

It then follows easily that the series converges in this case; and, still supposing $|\alpha| < 1$, that

$$\sum_{n=0}^{\infty} \alpha^n = 1/(1 - \alpha).$$

Notice also that, since $\alpha^0 = 1$, it is apparent from the work of the previous paragraph and Example 2.37.1, above, that:

$$s_n^* \equiv \sum_{m=1}^{n} \alpha^m = \alpha/(1 - \alpha) - \alpha^{n+1}/(1 - \alpha);$$

and that

$$\lim_{n \to \infty} s_n^* = \sum_{m=1}^{\infty} \alpha^m = [1/(1 - \alpha)] - 1 = \alpha/(1 - \alpha).$$

4. Suppose we return to the present value calculations considered at the beginning of this section, and consider the present value of a discounted income stream given by:

$$I_n = \sum_{m=0}^{n} \delta^m x_m;$$

where

$$\delta = 1/(1 + i),$$

'i' denotes the current rate of interest, and 'x_m' denotes the income recieved in period m. If we suppose that the income to be received is the same in each period, that is,

$$x_m = x_1 \equiv x \quad \text{for } m = 0, 1, 2, \ldots,$$

then it follows from the previous example (see also Theorem 2.38, which follows) that

$$I_n = x[1/(1-\delta) - \delta^{n+1}/(1-\delta)] = [x(1+i)/i] \cdot [1 - \delta^{n+1}],$$

and that

$$\lim_{n \to \infty} I_n = x(1+i)/i.$$

Moreover, in the more usual case in which the first income payment is not received until the end of the first period, the present value formula becomes:

$$I_n^* = \sum_{m=1}^{n} \delta^m x,$$

which, from the analysis of the previous example, simplifies to:

$$I_n^* = x\left[1/(1-\delta) - \delta^{n+1}/(1-\delta) - 1\right] = x\left[\delta/(1-\delta) - \delta^{n+1}/(1-\delta)\right]$$
$$= (x/i) \cdot (1 - \delta^n);$$

and we have:

$$\lim_{n \to \infty} I_n^* = x/i.$$

5. Consider the **harmonic series**

$$\sum_{n=1}^{\infty} 1/n = 1 + 1/2 + 1/3 + \dots. \tag{2.34}$$

In this case, we see that $a_n \to 0$; as required for the convergence of the series. However, given any n, we also see that:

$$\begin{aligned} S_{n,n} &= \sum_{m=n+1}^{2n} (1/m) \\ &= 1/(n+1) + 1/(n+2) + \cdots + 1/2n \geq n(1/2n) = 1/2. \end{aligned} \tag{2.35}$$

Consequently, it follows readily from Proposition 2.36 that the harmonic series is divergent. On the other hand, the sequence of partial sums, $\langle s_n \rangle$, does have a *limit* in this case; that is, $s_n \to \infty$. □

It is an immediate consequence of the definition of the convergence of an infinite series and Theorem 2.9 that, if we form a new series from two convergent infinite series by adding corresponding terms, then the new series is also convergent. Similarly, if we multiply each term of a convergent series by a constant, the new series is also convergent. We formalize these observations in the following. Details of the proof will be left as an exercise.

2.38. Theorem. *Suppose $\sum_{n=1}^{\infty} a_n$ and $\sum_{n=1}^{\infty} b_n$ are both convergent. If we form a new series by letting:*

1. $c_n = a_n + b_n$ for $n = 1, 2, \dots$,

then $\sum_{n=1}^{\infty} c_n$ also converges, and we have

$$\sum_{n=1}^{\infty} c_n = \left(\sum_{n=1}^{\infty} a_n\right) + \left(\sum_{n=1}^{\infty} b_n\right).$$

2. $d_n = \alpha a_n$ *for* $n = 1, 2, \ldots,$
where α *is any real number, then* $\sum_{n=1}^{\infty} d_n$ *also converges, and we have:*

$$\sum_{n=1}^{\infty} d_n = \alpha \left(\sum_{n=1}^{\infty} a_n \right).$$

For some purposes, the distinction between a series which is absolutely convergent, and one which is conditionally convergent, is critical. We will consider some of the reasons for the importance of this distinction later; in the meantime, the formal definitions are as follows.

2.39. Definition. If the series $\sum_{n=1}^{\infty} |a_n|$ converges, then we say $\sum_{n=1}^{\infty} a_n$ is **absolutely convergent**. If $\sum_{n=1}^{\infty} a_n$ is convergent, but $\sum_{n=1}^{\infty} |a_n|$ is not, we say that the series is **conditionally convergent**.

The terminology of the definition probably seems a bit strange, but perhaps our next result will clear it up a bit. The proposition is an easy consequence of Proposition 2.36 and the definition of convergence, and the proof will be left as an exercise.

2.40. Proposition. *If* $\sum_{n=1}^{\infty} a_n$ *is absolutely convergent, then it is convergent.*

Obviously, if each term in the sequence $\langle a_n \rangle$ is nonnegative, then the series formed from this sequence is convergent if, and only if, it is absolutely convergent. Moreover, using 2.38.2, it then is easy to see that a series in which all the terms are non-positive is convergent if, and only if, it is absolutely convergent. Consequently, it is apparent that, for real-valued series (which is our concern here), the distinction between convergence and absolute convergence can only be significant if the terms in $\langle a_n \rangle$ alternate in sign. We will return to the distinction between conditional and absolute convergence later in this section; in the meantime, we present a very simple criterion for the convergence of nonnegative series, which we now see also provides a criterion for the absolute convergence of a series. The proof will be left as an exercise; the result is a nearly immediate consequence of Theorem 2.20. Before turning to the result, we will present a couple of definitions, which are sufficiently natural as to almost need no formal statement.

2.41. Definitions. We shall say that the series $\sum_{n=1}^{\infty} a_n$ is:

1. **nonnegative** iff $a_n \geq 0$, for $n = 1, 2, \ldots$.
2. **bounded** iff the sequence of partial sums, $\langle s_n \rangle$, is bounded (as per Definition 2.6).

2.42. Proposition. *A nonnegative series,* $\sum_{n=1}^{\infty} a_n$*, is convergent if, and only if, it is bounded.*

Of the results concerning infinite series which we will study in this section, our next theorem is probably the one you will use most frequently in attempting to establish the convergence of a series.

2.43. Theorem. (The Comparison Test). *Suppose $\sum_{n=1}^{\infty} b_n$ is a non-negative series, and let $\sum_{n=1}^{\infty} a_n$ be a second series such that there exists a constant, $\alpha \geq 0$, and an integer m satisfying:*

$$|a_n| \leq \alpha b_n \quad \text{for } n = m+1, m+2, \ldots. \tag{2.36}$$

Then we have the following:

1. *if $\sum_{n=1}^{\infty} b_n$ is convergent, then $\sum_{n=1}^{\infty} a_n$ is absolutely convergent.*
2. *if $\sum_{n=1}^{\infty} |a_n|$ is divergent, then $\sum_{n=1}^{\infty} b_n$ is also divergent.*

Proof. Notice that part 2 of our conclusion is simply the contrapositive of part 1. Consequently, we need only prove part 1; and only for the case in which $\alpha > 0$, since the case in which $\alpha = 0$ is obvious.

Accordingly, let $\epsilon > 0$ be given. Since $\sum_{n=1}^{\infty} b_n$ is convergent, there exists $m' \geq m$ [where m is from (2.36)] such that, for all $n \geq m'$ and all $p \in N$:

$$\left|\sum_{q=n+1}^{n+p} b_q\right| = \sum_{q=n+1}^{n+p} b_q < \epsilon/\alpha. \tag{2.37}$$

Using (2.36) and (2.37) it then follows that, for $n \geq m'$, and any $p \in N$:

$$\left|\sum_{q=n+1}^{n+p} |a_q|\right| = \sum_{q=n+1}^{n+p} |a_q| \leq \sum_{q=n+1}^{n+p} \alpha b_q < \alpha(\epsilon/\alpha) = \epsilon;$$

and thus it follows from 2.36 that $\sum_{n=1}^{\infty} a_n$ is absolutely convergent. □

2.44. Examples/Exercises.

1. Suppose α is a real number less than or equal to one. Then we have, for any $n \geq 1, n^{\alpha} \leq n$. Consequently, letting

$$a_n = 1/n \text{ and } b_n = n^{-\alpha} \quad \text{for } n = 1, 2, \ldots,$$

we have

$$|a_n| = a_n \equiv 1/n \leq n^{-\alpha} \equiv b_n \quad \text{for } n = 1, 2, \ldots;$$

and, since we have already shown that the harmonic series, $\sum_{n=1}^{\infty} 1/n = \sum_{n=1}^{\infty} |1/n|$, diverges (Example 2.37.5), it follows that $\sum_{n=1}^{\infty} n^{-\alpha}$ diverges, for $\alpha \leq 1$.

2. Now consider the series $\sum_{n=1}^{\infty} n^{-\alpha} = \sum_{n=1}^{\infty} (1/n)^{\alpha}$, for $\alpha > 1$. If we define the new series, $\sum_{q=0}^{\infty} b_q$, by:

$$b_q = \sum_{n=2^q+1}^{2^{q+1}} (1/n)^{\alpha} \quad \text{for } q = 0, 1, 2, \ldots,$$

it should be clear that the original series converges if, and only if, the new series converges. However, for each q, we have (since $2^{q+1} - 2^q = 2^q$):

$$b_q \leq 2^q \cdot \left(\frac{1}{2^q}\right)^{\alpha} = 2^{q-\alpha q} = \left[\left(\frac{1}{2}\right)^{\alpha-1}\right]^q \equiv \beta^q,$$

where β is defined by:

$$\beta = (1/2)^{\alpha-1}.$$

Since $\alpha > 1$, we see that $0 < \beta < 1$, and therefore the geometric series, $\sum_{q=0}^{\infty} \beta^q$, converges (2.37.3). Consequently, it follows from 2.43 that $\sum_{q=0}^{\infty} b_q$ converges, and thus that $\sum_{n=1}^{\infty} a_n$ converges as well.

3. In the economic theory of general equilibrium, it is common to represent a firm's technological production possibilities by a production set, Y, taken to be a subset of, say $\mathbf{R}^m$, with the interpretation:

$$\mathbf{y} \in Y \text{ iff the vector } \mathbf{y} \text{ is technologically feasible for the firm,}$$

where $\mathbf{y}$ represents an input-output vector (or '**production vector**'), with the interpretation, for each $j \in \{1, \ldots, m\}$:

$y_j \geq 0$ indicates the firm is producing y_j units of the jth commodity, per unit of time,

$y_j < 0$ indicates the firm is using the jth commodity as an input, in the amount $|y_j| = -y_j$ per unit of time.

With these interpretations, notice that if the prices faced by the firm are given by a vector $\mathbf{p} \in \mathbf{R}_+^m$, where

$$p_j \text{ is the price of the } j^{th} \text{ commodity, for } j = 1, \ldots, m,$$

then the firm's profit (gross profit, if the firm has fixed costs) if it chooses $\mathbf{y}^* \in Y$, is given by:

$$\pi(\mathbf{p}, \mathbf{y}^*) = \mathbf{p} \cdot \mathbf{y}^*.$$

In the theory of the competitive firm, we assume that the firm takes the price vector, $\mathbf{p}$, as given; and, given $\mathbf{p}$, tries to choose $\mathbf{y}^* \in Y$ so as to maximize profit. That is, we assume the firm chooses a production vector, $\mathbf{y}^* \in Y$ satisfying:

$$(\forall \mathbf{y} \in Y)\colon\ \mathbf{p} \cdot \mathbf{y} \leq \mathbf{p} \cdot \mathbf{y}^*.$$

It is easy to show that the behavior of the firm[4] is invariant with respect to proportional changes in all prices; that is, if $\mathbf{y}^*$ is the firm's profit-maximizing choice when the price vector is $\mathbf{p}^*$, and $\mathbf{p}'$ is a price vector such that

$$\mathbf{p}' = \theta \mathbf{p}^* \quad \text{for some } \theta > 0,$$

then $\mathbf{y}^*$ is also the profit-maximizing choice when the price vector is $\mathbf{p}'$. Consequently, it is usual in general equilibrium theory to 'normalize' prices; assuming that the price vectors under consideration are all elements of the unit simplex in $\mathbf{R}^m$, Δ_m, defined by

$$\Delta_m = \left\{ \mathbf{p} \in \mathbf{R}_+^m \mid \sum_{j=1}^{m} p_j = 1 \right\}. \tag{2.38}$$

[4]That is, the profit-maximizing choice of the firm.

What I would like to do in this example is to consider some aspects of a generalization of this theory to the case in which the firm is choosing a production plan over time.

Accordingly, suppose now that a firm is considering a choice over production vectors of the form

$$\mathbf{y} = \langle \mathbf{y}_t \rangle, \tag{2.39}$$

where

$$\mathbf{y}_t \in \mathbf{R}^m \quad \text{for } t = 0, 1, 2, \ldots; \tag{2.40}$$

and where $\mathbf{y}_t$ is to be the production vector for the tth period from the present. Suppose further that the goal of the firm is to maximize the present value of its stream of profits, given the (expected) sequence of prices,

$$\mathbf{p} = \langle \mathbf{p}_t \rangle,$$

where $\mathbf{p}_t \in \Delta_m$ is the vector of prices for the tth period, and given the rate of discount (recall our discussion at the beginning of this section), δ, where

$$0 < \delta < 1.$$

We will suppose further that the firm's (infinite horizon) production set is given by $\mathbf{Y}$, where $\mathbf{Y}$ is the collection of all $\mathbf{y}$ satisfying (2.39) and (2.40), and which are technologically feasible choices for the firm. We will also suppose that $\mathbf{Y}$ satisfies one further condition; namely, that if $\mathbf{y} \in \mathbf{Y}$, then $\mathbf{y}$ is a bounded sequence; that is, there exists $\alpha \in \mathbf{R}_{++}$ (independent of t) such that

$$\|\mathbf{y}_t\| < \alpha \quad \text{for } t = 0, 1, 2, \ldots. \tag{2.41}$$

Thus our firm seeks to choose $\mathbf{y}^* \in \mathbf{Y}$ such that for all $\mathbf{y} \in \mathbf{Y}$, we have:

$$\pi(\mathbf{p}, \mathbf{y}^*) \equiv \sum_{t=0}^{\infty} \delta^t \mathbf{p}_t \cdot \mathbf{y}_t^* \geq \sum_{t=0}^{\infty} \delta^t \mathbf{p}_t \cdot \mathbf{y}_t \equiv \pi(\mathbf{p}, \mathbf{y}). \tag{2.42}$$

In a later chapter we will consider the question of deriving sufficient conditions for the existence of a production sequence, $\mathbf{y}^*$, satisfying (2.42); at present, our concern is the simpler one of determining whether the problem is well-formulated. The issue which occupies our attention here is simply this: if it is the case that for all $\mathbf{y} \in \mathbf{Y}$, we have

$$\pi(\mathbf{p}, \mathbf{y}) = +\infty,$$

then the issue of maximizing present value is fairly meaningless; or to put it another way, the assumption that the firm attempts to maximize present value has no predictive power whatsoever under these conditions. As it turns out, however, we can show that the present value expressions defined in (2.42) are always finite; that is, the series always converge.

We can show this as follows. Consider the tth term in the series defining $\pi(\mathbf{p}, \mathbf{y})$, for some $\mathbf{y} \in \mathbf{Y}$. We have:

$$|\delta^t \mathbf{p}_t \cdot \mathbf{y}_t| = |\delta^t| \cdot |\mathbf{p}_t \cdot \mathbf{y}_t| \leq \delta^t \cdot \|\mathbf{p}_t\| \cdot \|\mathbf{y}_t\| \leq (\alpha \cdot \|\mathbf{p}_t\|) \cdot \delta^t, \tag{2.43}$$

where we have used the Cauchy-Schwarz Inequality [equation (1.46)] and (2.41), in turn, to derive the inequalities in (2.43). Moreover, by Proposition 1.53[5] we have (recall that $\mathbf{p}_t$ is an element of Δ_m):

$$\|\mathbf{p}_t\| \leq \sum_{j=1}^{m} |p_{tj}| = \sum_{j=1}^{m} p_{tj} = 1. \tag{2.44}$$

Substituting (2.44) into (2.43), we obtain:

$$|\delta^t \mathbf{p}_t \cdot \mathbf{y}_t| \leq \alpha \cdot \delta^t \quad \text{for } t = 0, 1, 2, \ldots, \tag{2.45}$$

and, since it follows from the fact that $0 < \delta < 1$ that the geometric series,

$$\sum_{t=1}^{\infty} \delta^t,$$

converges, it follows from 2.43 that the infinite series defining $\pi(\mathbf{p}, \mathbf{y})$ converges as well.

4. Use the comparison test to show that if $\langle a_n \rangle$ is bounded, and $\sum_{n=1}^{\infty} b_n$ is absolutely convergent, then the series $\sum_{n=1}^{\infty} a_n b_n$ is absolutely convergent.

5. Use the comparison test to prove the following. If $\langle a_n \rangle$ is a sequence such that there exists $p > 1$ satisfying:

$$\lim_{n \to \infty} n^p a_n = 0,$$

then $\sum_{n=1}^{\infty} a_n$ is absolutely convergent.

6. Suppose $\sum_{n=1}^{\infty} |a_n|^p$ converges, for some $p \geq 1$. Show that for $q > p$, $\sum_{n=1}^{\infty} |a_n|^q$ is also convergent. □

Infinite series involving terms which are products arise very naturally in economic problems involving choices over time. However, one has to be a bit careful in dealing with such series, for they do not necessarily converge! We will presently consider an example of an infinite series of a form very much like that in the example just considered, but which does not converge; but in the meantime, we will first consider another general test for convergence, and then turn to some further considerations regarding series whose terms can be conveniently expressed as products.

---

[5] Notice also the discussion immediately following the proof of that result.

2.45. Proposition. (D'Alembert's Ratio Test) *Suppose* $a_n > 0$, *for* $n = 1, 2, \dots$, *and that*

$$\limsup(a_{n+1}/a_n) = r < 1. \tag{2.46}$$

Then $\sum_{n=1}^{\infty} a_n$ *converges. If, on the other hand,*

$$\liminf(a_{n+1}/a_n) = s > 1, \tag{2.47}$$

then the series diverges.

Proof. Suppose $r \equiv \limsup(a_{n+1}/a_n)$ is less than one, and define

$$\alpha = (1 + r)/2.$$

Then $r < \alpha < 1$, and it follows from Proposition 2.33 that there exists an integer, m, such that for all $n \geq m$, we will have:

$$\frac{a_{n+1}}{a_n} < \alpha. \tag{2.48}$$

Defining

$$\beta = a_m,$$

it then follows from (2.48) via an easy inductive argument, that

$$a_{m+p} < \beta\alpha^p \quad \text{for } p = 1, 2, \dots .$$

Consequently, it follows from the comparison test (Theorem 2.43) that $\sum_{n=1}^{\infty} a_n$ converges.

A similar argument, based upon the contrapositive part of the comparison principle, shows that if (2.47) holds, then the series diverges. Details will be left as an exercise. □

2.46. Example. Consider the series, $f(x)$, defined by:

$$f(x) = \sum_{n=0}^{\infty} \frac{x^n}{n!} \quad \text{for } x \in \mathbf{R}$$

(recall that, by definition, $0! = 1$). Obviously the series converges if $x = 0$. Otherwise, we have

$$\frac{|a_{n+1}|}{|a_n|} = \left(\frac{|x^{n+1}|}{(n+1)!}\right) \div \left(\frac{|x^n|}{n!}\right) = \frac{|x|}{n+1}.$$

Since, for any $x \in \mathbf{R}$,

$$\lim_{n\to\infty} \left(\frac{|x|}{n+1}\right) = 0,$$

it follows from the ratio test that $f(x)$ is absolutely convergent, and thus is convergent, for any $x \in \mathbf{R}$. As a matter of fact, you are already familiar with this series, in some sense, since as it happens,

$$f(x) = \exp(x). \quad \square$$

D'Alembert's test gives an almost complete criterion for convergence. Unfortunately, if neither (2.46) nor (2.47) holds, then the series may either converge or diverge. In fact, in the special case of this in which

$$\lim_{n\to\infty}\left(\frac{a_{n+1}}{a_n}\right) = 1,$$

[recall Proposition 2.34], the series may either converge or diverge. You can prove this to yourself by considering the series of the form

$$\sum_{n=1}^{\infty} n^{-\alpha},$$

for $\alpha \in \mathbf{R}_+$.

In order to study the results in which we are interested regarding series whose terms are products, we begin by considering the following proposition regarding finite sums.

2.47. Proposition. (Abel's Inequality) *If $\langle a_n\rangle$ and $\langle b_n\rangle$ are sequences such that, for some $m \in N$, and $\alpha \in \mathbf{R}_+$, we have*

$$b_1 \geq \cdots \geq b_{m-1} \geq b_m \geq 0;$$

and

$$|s_n| \equiv \left|\sum_{i=1}^{n} a_i\right| \leq \alpha \quad \textit{for}\, n = 1, 2, \ldots, m, \tag{2.49}$$

then we have

$$\left|\sum_{n=1}^{m} a_n b_n\right| \leq \alpha b_1.$$

Proof. We have from the definition of s_n $(n = 1, 2, \ldots, m)$:

$$\left|\sum_{n=1}^{m} a_n b_n\right| = \left|s_1 b_1 + \sum_{n=2}^{m}(s_n - s_{n-1})b_n\right|.$$

Moreover,

$$\left|s_1 b_1 + \sum_{n=2}^{m}(s_n - s_{n-1})b_n\right| = \left|s_1(b_1 - b_2) + \cdots + s_{m-1}(b_{m-1} - b_m) + s_m b_m\right|$$

$$\leq |s_1|\cdot|b_1 - b_2| + \cdots + |s_{m-1}|\cdot|b_{m-1} - b_m| + |s_m|\cdot|b_m|$$

$$= |s_1|\cdot(b_1 - b_2) + \cdots + |s_{m-1}|\cdot(b_{m-1} - b_m) + |s_m|\cdot b_m$$

$$\leq \alpha\cdot(b_1 - b_2) + \cdots + \alpha\cdot(b_{m-1} - b_m) + \alpha\cdot b_m = \alpha\cdot b_1;$$

where the last inequality is by (2.49) □

The following corollary of Abel's inequality is quite often useful. The proof will be left as an exercise.

2.48. Corollary. *If $\langle b_n \rangle$ is a sequence such that, for some $m \in N$ we have:*

$$b_1 \geq b_2 \geq \cdots \geq b_m \geq 0,$$

then:

$$\left| \sum_{n=1}^{m} (-1)^n \cdot b_n \right| \leq b_1.$$

Abel's inequality and its corollary can be used to derive several results regarding the convergence of series whose terms are products. We present one such here, and another as an exercise at the end of this section.

2.49. Proposition. (Dirichlet's Test) *Suppose $\langle a_n \rangle$ is a sequence such that the corresponding sequence of partial sums, $\langle s_n \rangle$, is bounded; and let $\langle b_n \rangle$ be a nonnegative monotone decreasing sequence such that $b_n \to 0$. Then the series, $\sum_{n=1}^{\infty} a_n b_n$, converges.*

Proof. Let $\alpha > 0$ be such that

$$|s_n| < \alpha \quad \text{for } n = 1, 2, \dots,$$

and let $\epsilon > 0$ be given. Then, since $b_n \to 0$, there exists $m \in N$ such that, for all $n \geq m$,

$$|b_n| = b_n < \epsilon / \alpha.$$

But then, we see that, for any $p \in N$,

$$\left| \sum_{n=m}^{m+p} a_n b_n \right| \leq \alpha \cdot b_m < \alpha \cdot \left(\frac{\epsilon}{\alpha} \right) = \epsilon;$$

where the first inequality is by Abel's inequality. It now follows from 2.36 that the series is convergent. □

2.50. Examples/Exercises.

1. Let $\alpha \in]0, 1]$, and consider the series

$$\sum_{n=1}^{\infty} (-1)^n \cdot n^{-\alpha}.$$

This series is of the form $\sum_{n=1}^{\infty} a_n b_n$, where $\langle a_n \rangle$ and $\langle b_n \rangle$ are given by:

$$a_n = (-1)^n \quad \text{and} \quad b_n = n^{-\alpha} \quad \text{for } n = 1, 2, \dots,$$

respectively. Since

$$\left| \sum_{n=1}^{m} (-1)^n \right| \leq 1 \quad \text{for } m = 1, 2, \dots;$$

while $\langle b_n \rangle$ is a nonnegative decreasing sequence such that

$$b_n \equiv n^{-\alpha} \to 0,$$

it follows from Dirichlet's test that the series is convergent.

2. The following example provides an object lesson regarding the dangers of getting careless in the use of Dirichlet's test. Consider the series $\sum_{n=1}^{\infty} a_n b_n$, where

$$a_n = (-1)^n \cdot n^{-1/2} \quad \text{and} \quad b_n = (-1)^n \cdot (1/n)^{1/2} \quad \text{for } n = 1, 2, \ldots,$$

respectively. It follows from the previous example that $\sum_{n=1}^{\infty} a_n$ is convergent (and thus that the corresponding sequence of partial sums is bounded), and clearly $b_n \to 0$ (in fact, $\sum_{n=1}^{\infty} b_n$ converges). Nonetheless,

$$\sum_{n=1}^{\infty} a_n b_n = \sum_{n=1}^{\infty} 1/n,$$

which we know to be divergent by 2.37.5! Why doesn't this contradict Proposition 2.49? □

In 2.50 we showed that the series $\sum_{n=1}^{\infty}(-1)^n(1/n)$ is conditionally convergent, although it is not absolutely convergent, since we showed earlier that

$$\sum_{n=1}^{\infty} |(-1)^n(1/n)| = \sum_{n=1}^{\infty} 1/n,$$

is divergent. We noted earlier that there are important differences between absolutely convergent and conditionally convergent series, other than the definitional distinction. One of these differences is that, if we re-arrange the terms in an absolutely convergent series, then the new series is still convergent, and converges to the same limit.[6] On the other hand, if we re-arrange the terms in a conditionally convergent series, then the new series may not be convergent; or if it is still convergent, it may approach a limit different from that of the original series. For a more complete discussion of this issue, I particularly recommend that the student consult Binmore [1982, pp. 61 – 3]; in fact, those of you who might like a bit more leisurely and complete introduction to the topic of infinite series than we have provided will find Chapter 6, pp 54–64, of Binmore's book to be very illuminating. Those of you who might be interested in a treatment of infinite series which is both more thorough and a bit more sophisticated than that presented here will find Apostol [1974], Eggleston [1962], or Shilov [1973], to be very useful.

Exercises.

1. Prove that $\sum_{n=1}^{\infty}(1/2)^n = 1$.

2. Prove that, if α is any real number such that $0 < |\alpha| < 1$, and β is any real number, then

$$\sum_{n=0}^{\infty} \beta \cdot \alpha^n = \frac{\beta}{1-\alpha}.$$

[6] Another important difference can be illustrated by the comparison of Examples 2.44.4 and 2.50.2.

3. Suppose $\langle a_n \rangle$ is a bounded sequence, and that α is a real number satisfying $0 < |\alpha| < 1$. Show that the series $\sum_{n=1}^{\infty} a_n \cdot \alpha^n$ is convergent. In particular, this shows that if we take $\alpha = 1/10$ and

$$a_n \in \{0, 1, \dots, 9\} \quad \text{for } n = 0, 1, 2, \dots;$$

then such a series always converges. What is such a series called in this latter case?

4. Supposing that $\langle a_n \rangle$ satisfies

$$a_n \geq a_{n+1} \geq 0 \quad \text{for } n = 1, 2, \dots m-1,$$

show that

$$\sum_{n=1}^{m} (-1)^{n+1} \cdot a_n \leq a_1.$$

5. Let $\langle a_m \rangle$ be a sequence, and $\langle m_n \rangle (n = 0, 1, \dots)$ be a strictly increasing sequence of positive integers such that $m_0 = 1$. Show that, if we define the sequence $\langle b_n \rangle$ by:

$$b_n = \sum_{m=m_{n-1}}^{m_n - 1} a_m \quad \text{for } n = 1, 2, \dots ,$$

then

a. if $\sum_{m=1}^{\infty} a_m$ converges, then so does $\sum_{n=1}^{\infty} b_n$.

b. if $\langle a_m \rangle$ is a nonnegative sequence, and $\sum_{n=1}^{\infty} b_n$ converges, then so does $\sum_{m=1}^{\infty} a_m$; and

$$\sum_{m=1}^{\infty} a_m = \sum_{n=1}^{\infty} b_n.$$

6. Prove Theorem 2.38.

7. Suppose $\langle a_n \rangle$ is a non-increasing sequence such that $a_n \to 0$. Show that

$$\sum_{n=1}^{\infty} (-1)^{n-1} \cdot a_n \tag{2.50}$$

converges. Suppose now that $\langle a_n \rangle$ is a non-decreasing sequence such that $a_n \to 0$. Does (2.50) still converge?

8. Show that if $\sum_{n=1}^{\infty} a_n$ and $\sum_{n=1}^{\infty} b_n$ are both absolutely convergent, then the series $\sum_{n=1}^{\infty} c_n$ formed by setting:

$$c_n = a_n + b_n \quad \text{for } n = 1, 2, \dots ,$$

is also absolutely convergent (see Theorem 2.38 and proposition 2.42). Is it the case that we necessarily have:

$$\sum_{n=1}^{\infty} |c_n| = \sum_{n=1}^{\infty} |a_n| + \sum_{n=1}^{\infty} |b_n| ?$$

9. Show that if $\sum_{n=1}^{\infty} a_n$, $\sum_{n=1}^{\infty} b_n$ and $\sum_{n=1}^{\infty} c_n$ are all absolutely convergent, then so is $\sum_{n=1}^{\infty} (a_n + b_n) c_n$ (see Example 2.44.4 and Exercise 8, above).

2.4 Efficient Allocation in a Simple Growth Model

In this section we will be considering a growth model in which there is one commodity which can, in each period, either be consumed or used as an input to produce a stock of the commodity to be available at the beginning of the next period. We will be considering some of the problems involved in obtaining a decentralized (over time) solution to the problem of efficient intertemporal allocation of the commodity between production and consumption. In our treatment, we will be borrowing heavily from Majumdar [1974, 1988] and Hurwicz and Majumdar [1988].

The basic notation and structure of the model to be considered is as follows. The stock of the commodity available at the beginning of the tth period is denoted by 'y_t', with the initial stock given by

$$y_0 = \bar{y}$$

(the 0th period being the present). In each period, the commodity stock is divided between consumption, c_t, and production input (or capital), x_t; so that we require

$$y_t = c_t + x_t \quad \text{for } t = 0, 1, 2, \ldots. \tag{2.51}$$

For each period, we will suppose there is given a production function, $f_t \colon \mathbf{R}_+ \to \mathbf{R}_+$, with the technological requirement that:

$$0 \leq y_{t+1} \leq f_{t+1}(x_t) \quad \text{for } t = 0, 1, 2, \ldots. \tag{2.52}$$

We will always suppose that each f_t satisfies the following two conditions:

A.1. $f_t(0) = 0$,

A.2. f_t is (strictly) increasing and continuous on $\mathbf{R}_+$;

although we will often impose further assumptions as well.

We will employ the generic notation, '$\mathbf{x}$','$\mathbf{y}$', and '$\mathbf{c}$', to denote sequences,

$$\mathbf{x} = \langle x_t \rangle, \mathbf{y} = \langle y_t \rangle, \text{ and } \mathbf{c} = \langle c_t \rangle,$$

respectively; and, similarly, we will use the notation '$\mathbf{f}$' to denote the sequence of production functions, $\mathbf{f} = \langle f_t \rangle$. We will denote the set of all real-valued infinite sequences by '$\mathbf{R}^\infty$,' and, consistent with the idea of thinking of a sequence as an infinite tuple of real numbers, we will use the kind of notation for inequalities involving sequences that you are already used to employing in $\mathbf{R}^n$; that is, for $\mathbf{x}, \mathbf{x}' \in \mathbf{R}^\infty$, we write:

$$\mathbf{x} \geq \mathbf{x}' \iff [x_t \geq x_t' \quad \text{for } t = 0, 1, 2, \ldots],$$

$$\mathbf{x} > \mathbf{x}' \iff [\mathbf{x} \geq \mathbf{x}' \;\&\; \mathbf{x}' \not\geq \mathbf{x}],$$

and

$$\mathbf{x} \gg \mathbf{x}' \iff [x_t > x_t' \quad \text{for } t = 0, 1, 2, \ldots].$$

Denoting the origin in $\mathbf{R}^\infty$ by '$\mathbf{0}$,' we then define

$$\mathbf{R}^\infty_+ = \{\mathbf{x} \in \mathbf{R}^\infty \mid \mathbf{x} \geq \mathbf{0}\},$$

that is, '$\mathbf{R}^\infty_+$' will denote the set of all infinite sequences of nonnegative real numbers, and

$$\mathbf{R}^\infty_{++} = \{\mathbf{x} \in \mathbf{R}^\infty \mid \mathbf{x} \gg \mathbf{0}\},$$

so that $\mathbf{R}^\infty_{++}$ is the set of all infinite sequences of positive real numbers. In terms of this notation, then, we will be considering input, output, and consumption sequences, $\mathbf{x}$,$\mathbf{y}$, and $\mathbf{c}$, respectively, such that

$$\mathbf{x}, \mathbf{y}, \mathbf{c} \in \mathbf{R}^\infty_+.$$

2.51. Definition. Given a sequence of production functions, $\mathbf{f}$, and an initial stock, $\bar{y} > 0$, we shall say that a triple ($\mathbf{x}$, $\mathbf{y}$, $\mathbf{c}$) is **feasible for** $(\mathbf{f}, \bar{y})$ iff $\{\mathbf{x}, \mathbf{y}, \mathbf{c}\} \subseteq \mathbf{R}^\infty_+$,

$$y_0 = \bar{y}, \tag{2.53}$$

and

$$\left.\begin{array}{l} y_t = c_t + x_t \\ 0 \leq y_{t+1} \leq f_{t+1}(x_t) \end{array}\right\} \quad \text{for } t = 0, 1, 2, \ldots; \tag{2.54}$$

and, in this case we shall also say that $(\mathbf{x}, \mathbf{y})$ is a **feasible production program**, and that $\mathbf{c} = \langle c_t \rangle$ is a **feasible consumption program** for $(\mathbf{f}, \bar{y})$.

2.52. Definition. If $(\mathbf{x}^*, \mathbf{y}^*, \mathbf{c}^*)$ is a feasible program for $(\mathbf{f}, \bar{y})$, we will say that $(\mathbf{x}^*, \mathbf{y}^*, \mathbf{c}^*)$ is **intertemporally efficient for** $(\mathbf{f}, \bar{y})$ iff there is no other feasible program for $(\mathbf{f}, \bar{y})$, $(\mathbf{x}, \mathbf{y}, \mathbf{c})$, such that $\mathbf{c} > \mathbf{c}^*$.

The second of the above definitions tends to be just a little bit confusing, since we are generally more used to dealing with a notion of pure production efficiency, at least in the static case. Thus, consider the following example.

2.53. Example. (*Program of pure accumulation*) (Hurwicz and Majumdar [1988], Example 3.1, p. 237). Given $(\mathbf{f}, \bar{y})$, we define the program $(\hat{\mathbf{x}}, \hat{\mathbf{y}}, \hat{\mathbf{c}})$ by:

$$\hat{y}_0 = \bar{y}, \hat{x}_t = \hat{y}_t, \hat{c}_t = 0, \text{ and } \hat{y}_{t+1} = f_{t+1}(\hat{x}_t) \quad \text{for } t = 0, 1, 2, \ldots. \tag{2.55}$$

If one examines Definition 2.52 carefully, it is easy to see that $(\hat{\mathbf{x}}, \hat{\mathbf{y}}, \hat{\mathbf{c}})$ is *not* efficient for $(\mathbf{f}, \bar{y})$, assuming, that is, that $\bar{y} > 0$. □

It is common in this literature to consider intertemporal profit maximization of some sort or another. The criterion which we will consider is defined in the following.

2.54. Definition. We will say that a feasible program $(\mathbf{x}^*, \mathbf{y}^*, \mathbf{c}^*)$, is **competitive for** $(\mathbf{f}, \bar{y})$ iff there exists $\mathbf{p} \in \mathbf{R}^{\infty}_{++}$ such that for all t $(t = 0, 1, \ldots)$, and for all $(x, y) \in \mathbf{R}^2_+$ such that $y \leq f_{t+1}(x)$, we have:

$$p_{t+1} f_{t+1}(x_t^*) - p_t x_t^* \geq p_{t+1} y - p_t x. \tag{2.56}$$

From the static (more correctly, the finite horizon) case, we are accustomed to a duality between competitive profit maximization and efficiency. In the infinite horizon case, things do not work out quite so simply, however. In the first place, a feasible program may be competitive, yet not be efficient. For example, suppose we return to the program of pure accumulation defined in 2.53,and suppose now that $\mathbf{f}$ satisfies the **Inada conditions**; that is, in addition to A.1 and A.2, suppose the following condition is also satisfied:

A.3. For each t, f_t is twice differentiable at each $x > 0$, with $f_t'(x) > 0$ and $f_t''(x) < 0$, and

$$\lim_{x \to 0+} f_t'(x) > 1, \quad \text{and} \quad 0 < \lim_{x \to \infty} f'(t) < 1.$$

Given these assumptions, if we define the price sequence $\mathbf{p}$ by

$$p_0 = 1, \quad \text{and} \quad p_{t+1} = p_t / f_{t+1}'(\hat{x}_t) \quad \text{for } t = 0, 1, 2, \ldots, \tag{2.57}$$

it is easy to show (exercise) that $\mathbf{p}$ satisfies (2.56) for the pure accumulation program. Consequently, it follows that a program may be competitive without being efficient, in the sense in which we are using the term here. As a matter of fact, there are a number of problems connected with defining competitive prices to support an efficient program; some of which we will be considering in a later chapter. In the meantime, let's take note of one aspect of a competitive program; if it were true that a competitive program is necessarily efficient, then the use of competitive prices would allow decentralization over time. In other words, if a competitive program is efficient, then (given competitive prices) an efficient program can be attained by individual period-by-period profit maximization in which the agents in time period t need only know p_t, p_{t+1}, and f_{t+1}. We will consider some aspects of this decentralization of decisions for this general model later on in this book, but in the meantime we consider a special case of the model.

In the remainder of this section we will suppose that the production function in each period is *linear*, so that, for each t, there exists ρ_t such that

$$f_t(x) = \rho_t x \quad \text{for } t = 1, 2, \ldots. \tag{2.58}$$

In this case, the sequence of production functions is fully specified by the positive real sequence $\langle \rho_t \rangle$, and thus we will substitute $\boldsymbol{\rho} = \langle \rho_t \rangle$ for $\mathbf{f}$ in our definitions of feasibility and optimality. For the sake of reference, let's also note that our technology constraint for feasibility can now be written as:

$$y_{t+1} \leq \rho_{t+1} x_t \quad \text{for } t = 0, 1, 2, \ldots, \tag{2.59}$$

in place of (2.53); and that assumption A.2 implies that

$$\rho_t > 0 \quad \text{for } t = 1, 2, \ldots. \tag{2.60}$$

Because of (2.60), we can make use of an alternative specification of production technology in this case; defining the **input requirement function**, g_t, for each t, by:

$$g_t(y) = (1/\rho_t)y \quad \text{for } t = 1, 2, \ldots. \tag{2.61}$$

With this definition, we can equivalently express the technology requirement (2.59) by:

$$g_{t+1}(y_{t+1}) = (1/\rho_{t+1})y_{t+1} \leq x_t \text{ (and } y_{t+1} \geq 0) \quad \text{for } t = 1, 2, \ldots. \tag{2.62}$$

In our work in the remainder of this section, we will also find it convenient to make use of the sequence $\langle \sigma_t \rangle$ defined by:

$$\sigma_0 = 1 \quad \text{and} \quad \sigma_t = \prod_{j=1}^{t} \rho_j \quad \text{for } t = 1, 2, \ldots. \tag{2.63}$$

Majumdar has established the following interesting and useful criteria for this case.

2.55. Proposition. *(Majumdar [1974]). Suppose $\rho \in \mathbf{R}^{\infty}_{++}$ and $\bar{y} > 0$. If $(\mathbf{x}^*, \mathbf{y}^*, \mathbf{c}^*)$ is feasible for $(\rho, \bar{y})$, then*

$$\sum_{t=0}^{\infty} (c_t^*/\sigma_t) \leq \bar{y}; \tag{2.64}$$

and, in addition, $(\mathbf{x}^*, \mathbf{y}^*, \mathbf{c}^*)$ is efficient for $(\rho, \bar{y})$ if, and only if,

$$\sum_{t=0}^{\infty} (c_t^*/\sigma_t) = \bar{y}. \tag{2.65}$$

Proof. Suppose $(\mathbf{x}, \mathbf{y}, \mathbf{c})$ is feasible for $(\rho, \bar{y})$.
Part 1. By the definition of feasibility,

$$c_1 = y_1 - x_1,$$

and thus

$$(1/\rho_1)c_1 = (1/\rho_1)y_1 - (1/\rho_1)x_1 \leq x_0 - (1/\rho_1)x_1 = \bar{y} - c_0 - (1/\rho_1)x_1,$$

where we have used the technological constraint (2.62) to obtain the inequality. Therefore,

$$c_0 + (1/\rho_1)c_1 \leq \bar{y} - (1/\rho_1)x_1;$$

or, using the definitions in (2.63),

$$(1/\sigma_0)c_0 + (1/\sigma_1)c_1 \leq \bar{y} - (1/\rho_1)x_1 \leq \bar{y}. \tag{2.66}$$

Now suppose that for $0 \leq T \leq s$, we have

$$\sum_{t=0}^{T} (1/\sigma_t)c_t \leq \bar{y} - (1/\sigma_T)x_T \leq \bar{y}. \tag{2.67}$$

Then, for $T = s + 1$, we have, using our inductive hypothesis, feasibility, and (2.63), and technological feasibility (2.62), in turn:

$$\sum_{t=0}^{s+1} (1/\sigma_t)c_t = \sum_{t=0}^{s} (1/\sigma_t)c_t + (1/\sigma_{s+1})c_{s+1}$$

$$\leq \bar{y} - (1/\sigma_s)x_s + (1/\sigma_{s+1})c_{s+1} = \bar{y} - (1/\sigma_s)x_s + (1/\sigma_s)[(1/\rho_{s+1})(y_{s+1} - x_{s+1})]$$

$$\leq \bar{y} - (1/\sigma_s)x_s + (1/\sigma_s)[x_s - (1/\rho_{s+1})x_{s+1}] = \bar{y} - (1/\sigma_{s+1})x_{s+1} \leq \bar{y}.$$

Thus we have shown that the sequence of partial sums, $\langle \sum_{t=0}^{T} (1/\sigma_t)c_t \rangle$ is a non-decreasing sequence which is bounded above by $\bar{y}$. Therefore, the series converges, and

$$\sum_{t=0}^{\infty} (1/\sigma_t)c_t \leq \bar{y}.$$

Part 2. Now suppose $(\mathbf{x}, \mathbf{y}, \mathbf{c})$ satisfies

$$\sum_{t=0}^{\infty} (1/\sigma_t)c_t = \bar{y}; \tag{2.68}$$

and suppose, by way of obtaining a contradiction, that there exists an alternative feasible program for $(\boldsymbol{\rho}, \bar{y}), (\mathbf{x}', \mathbf{y}', \mathbf{c}')$, for which

$$\mathbf{c}' > \mathbf{c}. \tag{2.69}$$

Then, since $(\mathbf{x}', \mathbf{y}', \mathbf{c}')$ is feasible, $\sum_{t=0}^{\infty} (1/\sigma_t)c_t'$ converges, and, by part 1, we have:

$$\sum_{t=0}^{\infty} (1/\sigma_t)c_t' \leq \bar{y}. \tag{2.70}$$

However, since $1/\sigma_t > 0$ for each t, it follows from (2.69) that

$$\sum_{t=0}^{\infty} (1/\sigma_t)c_t' > \sum_{t=0}^{\infty} (1/\sigma_t)c_t;$$

which, given (2.68) and (2.70), yields a contradiction.

Part 3. Now suppose $(\mathbf{x}, \mathbf{y}, \mathbf{c})$ satisfies:

$$\sum_{t=0}^{\infty} (1/\sigma_t)c_t = \bar{y} - \eta, \tag{2.71}$$

where $\eta > 0$. We will demonstrate that in this case there exists a feasible program for $(\boldsymbol{\rho}, \bar{y}), (\mathbf{x}', \mathbf{y}', \mathbf{c}')$, which dominates $(\mathbf{x}, \mathbf{y}, \mathbf{c})$. To do this, we begin by noting that the series $\sum_{t=1}^{\infty} (1/\sigma_t)c_t$ converges; and, from (2.71):

$$\sum_{t=1}^{\infty} (1/\sigma_t)c_t = \bar{y} - c_0 - \eta. \tag{2.72}$$

Moreover, for an arbitrary $t \geq 1$ and $T > t$, the sequence of partial sums, $\sum_{s=t}^{T}(1/\sigma_s)c_s$, is increasing; and, since

$$\sum\nolimits_{s=t}^{T}(1/\sigma_s)c_s \leq \sum\nolimits_{s=0}^{T}(1/\sigma_s)c_s \leq \bar{y};$$

it is clear that, for each t, the series $\sum_{s=t}^{\infty}(1/\sigma_s)c_s$ converges. But then it follows from Theorem 2.38.2 that, for each t, the series

$$\sum\nolimits_{j=t+1}^{\infty}(\sigma_t/\sigma_j)c_j = (1/\rho_{t+1})c_{t+1} + [1/(\rho_{t+1} \cdot \rho_{t+2})]c_{t+2} + \ldots,$$

also converges. We make use of these facts in defining an alternative feasible program, $(\mathbf{x}', \mathbf{y}', \mathbf{c}')$, as follows:

$$c_0' = c_0 + \eta, x_0' = \bar{y} - c_0 - \eta,$$

$$c_t' = c_t,\ x_t' = \sum\nolimits_{j=t+1}^{\infty}(\sigma_t/\sigma_j)c_j,\ y_t' = c_t' + x_t' \quad \text{for } t = 1, 2, \ldots.$$

since we have, by definition of c_0' and x_0', that

$$x_0' + c_0' = \bar{y} - c_0 - \eta + c_0 + \eta = \bar{y};$$

it follows from the definition of y_t, for $t \geq 1$, that to show that $(\mathbf{x}', \mathbf{y}', \mathbf{c}')$ is feasible, we need only show that the program satisfies the production constraint:

$$(1/\rho_{t+1})y_{t+1}' \leq x_t' \quad \text{for } t = 0, 1, 2, \ldots$$

(notice that it follows from the definition that $y_t \geq 0$ for each $t \geq 1$). However, for $t = 0$,

$$(1/\rho_1)y_1' = (1/\rho_1)c_1 + (1/\rho_1)x_1' = (1/\rho_1)c_1 + (1/\rho_1)\left[\sum_{j=2}^{\infty}(\sigma_1/\sigma_j)c_j\right]$$

$$= (1/\rho_1)c_1 + \sum_{j=2}^{\infty}(1/\sigma_j)c_j = \sum_{t=1}^{\infty}(1/\sigma_t)c_t = \bar{y} - c_0 - \eta = x_0',$$

where the last two equalities are by (2.72) and the definition of x_0', respectively. Similarly, for $t \geq 1$, we have:

$$\left(\frac{1}{\rho_{t+1}}\right)y_{t+1}' = \left(\frac{1}{\rho_{t+1}}\right)c_{t+1}' + \left(\frac{1}{\rho_{t+1}}\right)x_{t+1}'$$

$$= \left(\frac{1}{\rho_{t+1}}\right)c_{t+1} + \left(\frac{1}{\rho_{t+1}}\right)\left[\sum\nolimits_{j=t+2}^{\infty}\left(\frac{\sigma_{t+1}}{\sigma_j}\right)c_j\right]$$

$$= \left(\frac{1}{\rho_{t+1}}\right)c_{t+1} + \sum_{j=t+2}^{\infty}\left(\frac{\sigma_t}{\sigma_j}\right)c_j = \left(\frac{\sigma_t}{\sigma_{t+1}}\right)c_{t+1} + \sum_{j=t+2}^{\infty}\left(\frac{\sigma_t}{\sigma_j}\right)c_j$$

$$= \sum\nolimits_{j=t+1}^{\infty}\left(\frac{\sigma_t}{\sigma_j}\right)c_j = x_t'.$$

We have shown, therefore, that $(\mathbf{x}', \mathbf{y}', \mathbf{c}')$ is feasible; and, since

$$c_0' = c_0 + \eta > c_0, \quad \text{while} \quad c_t' = c_t \quad \text{for } t = 1, 2, \ldots,$$

it follows that $(\mathbf{x}, \mathbf{y}, \mathbf{c})$ is *not* efficient. Consequently, we see that if a feasible program, $(\mathbf{x}, \mathbf{y}, \mathbf{c})$, *is* efficient, we must have

$$\sum_{t=0}^{\infty} (1/\sigma_t) c_t = \bar{y}. \quad \square$$

We will conclude this section with an extended example from Hurwicz and Majumdar [1988].

2.56. Example. Here we will consider a particular efficient program for the model which is the subject of Proposition 2.55, above. Let $\theta \in \,]0, 1[$, and consider the program $(\mathbf{x}^*, \mathbf{y}^*, \mathbf{c}^*)$ defined by:

$$\left.\begin{array}{ll} y_0^* = \bar{y} & \\ c_t^* = \theta y_t^*, x_t^* = (1-\theta) y_t^*, \, y_{t+1}^* = \rho_{t+1} x_t^* & \text{for } t = 0, 1, \ldots. \end{array}\right\} \tag{2.73}$$

It is apparent in this case that $(\mathbf{x}^*, \mathbf{y}^*, \mathbf{c}^*)$ is a feasible program.

In order to prove that $(\mathbf{x}^*, \mathbf{y}^*, \mathbf{c}^*)$ is efficient, we begin by noting that

$$y_t^* = \sigma_t (1-\theta)^t \bar{y} \quad \text{for } t = 1, 2, \ldots, \tag{2.74}$$

where σ_t is from (2.63). To prove (2.74), we note first that the formula is obviously true for $t = 1$. Suppose it holds for $t = 1, \ldots, s$. Then we have

$$y_{s+1}^* = \rho_{s+1} x_s^* = \rho_{s+1}(1-\theta) y_s^* = \rho_{s+1}(1-\theta)\sigma_s(1-\theta)^s \bar{y} = \sigma_{s+1}(1-\theta)^{s+1}\bar{y}.$$

Next we note that

$$c_0/\sigma_0 = \theta \bar{y};$$

while for $t \geq 1$:

$$c_t^*/\sigma_t = \theta y_t^*/\sigma_t = \theta(1-\theta)^t \bar{y} = (1-\theta)^t(\theta \bar{y}).$$

Therefore, we see that for an arbitrary $T \geq 1$,

$$\sum_{t=0}^{T} (c_t^*/\sigma_t) = (\theta \bar{y}) \cdot \left[\sum_{t=0}^{T} (1-\theta)^t\right];$$

and it follows from 2.37.3 and 2.38.2 that

$$\sum_{t=0}^{\infty} (c_t^*/\sigma_t) = (\theta \bar{y}) \cdot \left[\sum_{t=0}^{\infty} (1-\theta)^t\right] = (\theta \bar{y}) \cdot \left(1/[1-(1-\theta)]\right) = \bar{y},$$

and thus from Proposition 2.55 that $(\mathbf{x}^*, \mathbf{y}^*, \mathbf{c}^*)$ is efficient.

Notice that the program $(\mathbf{x}^*, \mathbf{y}^*, \mathbf{c}^*)$, while very simple, is also completely temporally decentralized. At each time period, t, only x_{t-1}^*, ρ_t, and θ needs to be known at the outset of the period; from these values, y_t^*, c_t^*, and x_t^* can be calculated. $\square$

2.5 Sequences and Series in $\mathbf{R}^n$

In this section we will be dealing with sequences $\langle \mathbf{x}_m \rangle$, each term of which is an element, $\mathbf{x}_m = (x_{1m}, \ldots, x_{nm})$, of $\mathbf{R}^n$. Thus, a sequence $\langle \mathbf{x}_m \rangle$ in $\mathbf{R}^n$ defines n sequences of real numbers,

$$\langle x_{im} \rangle \subseteq \mathbf{R} \quad \text{for } i = 1, 2, \ldots, n;$$

to which we will refer as the **n coordinate sequences defined by** $\langle \mathbf{x}_m \rangle$. Conversely, of course, n coordinate sequences [of real numbers] define a sequence in $\mathbf{R}^n$; and because of these basic relationships, it turns out that we will encounter nothing very new in our study of sequences in $\mathbf{R}^n$, as you will see. In fact, in this section we will study only enough of the theory of sequences in $\mathbf{R}^n$ to enable you to see that all of the basic results follow readily from the theory of sequences of real numbers. Incidentally, in this section, when we say that $\langle \mathbf{x}_m \rangle$ is a sequence, we will always mean that

$$\mathbf{x}_m \in \mathbf{R}^n \quad \text{for } m = 1, 2, \ldots.$$

2.57. Definition. We shall say that the sequence $\langle \mathbf{x}_m \rangle$ **converges to $\mathbf{x}^* \in \mathbf{R}^n$** iff, given any $\epsilon > 0$, there exists $p \in N$ such that

$$(\forall m \geq p) \colon d(\mathbf{x}_m, \mathbf{x}^*) \equiv \|\mathbf{x}_m - \mathbf{x}^*\| < \epsilon.$$

In this case, $\mathbf{x}^*$ is said to be the **limit of the sequence** $\langle \mathbf{x}_m \rangle$, and we write:

$$\lim_{m \to \infty} \mathbf{x}_m = \mathbf{x}^*, \text{ or } \mathbf{x}_m \to \mathbf{x}^*,$$

or sometimes simply '$\lim \mathbf{x}_m = \mathbf{x}^*$.'

The following lemma will enable us to prove quite easily that a sequence $\langle \mathbf{x}_m \rangle$ converges if, and only if, each of its n coordinate sequences converges. The lemma itself is an immediate consequence of Corollary 1.54.

2.58. Lemma. *(Cf. Corollary 1.54) Suppose* $\mathbf{x}, \mathbf{y} \in \mathbf{R}^n$. *Then we have:*

1. $|x_i - y_i| \leq d(\mathbf{x}, \mathbf{y}) \equiv \|\mathbf{x} - \mathbf{y}\|$ *for* $i = 1, 2, \ldots, n$,

and

2. $d(\mathbf{x}, \mathbf{y}) \leq \sum_{i=1}^{n} |x_i - y_i|$.

If we utilize Lemma 2.58, it is then easy to prove that a sequence in $\mathbf{R}^n$ converges if, and only if, each of its coordinate sequences converges. I will leave the proof of the formal statement of this fact as an exercise.

2.59. Theorem. *A sequence* $\langle \mathbf{x}_m \rangle$ *converges to* $\mathbf{x}^* \in \mathbf{R}^n$ *if, and only if,*

$$\lim_{m \to \infty} x_{im} = x_i^* \quad \text{for } i = 1, \ldots, n.$$

Having established 2.59, it now follows readily from 2.9 and 2.11 that if we perform standard arithmetic operations on convergent sequences, we will obtain a new convergent sequence, and we can obtain the limit of the new sequence by performing the same arithmetic operations on the limits of the original sequences. This is set out more intelligibly in the following result, and once again I will leave the proof as an (easy) exercise.

2.60. Theorem. *If $\langle \mathbf{x}_m \rangle$ and $\langle \mathbf{y}_m \rangle$ are sequences, and $\mathbf{x}^*, \mathbf{y}^* \in \mathbf{R}^n$ are such that $\mathbf{x}_m \to \mathbf{x}^*$ and $\mathbf{y}_m \to \mathbf{y}^*$, then:*

1. $\lim(\mathbf{x}_m + \mathbf{y}_m) = \mathbf{x}^* + \mathbf{y}^*$,
2. *for any* $\alpha \in \mathbf{R}$, $\lim(\alpha \mathbf{x}_m) = \alpha \mathbf{x}^*$, *and*
3. $\lim(\mathbf{x}_m \cdot \mathbf{y}_m) = \mathbf{x}^* \cdot \mathbf{y}^*$.

The following generalizes Conclusion 2 of Theorem 2.60 somewhat, and is often useful in applications.

2.61. Proposition. *Suppose $\langle a_m \rangle$ is a sequence of real numbers which converges to $a \in \mathbf{R}$, and let $\langle \mathbf{y}_m \rangle \subseteq \mathbf{R}^n$ be a sequence which converges to $\mathbf{y}^* \in \mathbf{R}^n$. Then the sequence $\langle \mathbf{z}_m \rangle$ defined by:*

$$\mathbf{z}_m = a_m \mathbf{y}_m \quad \text{for } m = 1, 2, \ldots,$$

converges to $a\mathbf{y}^*$.

Proof. Let $\epsilon > 0$ be given. Since $\langle a_m \rangle$ converges, it is bounded; thus there exists $\alpha \in \mathbf{R}$ such that:

$$|a_m| < \alpha \quad \text{for } m = 1, 2, \ldots.$$

Moreover, since both $\langle a_m \rangle$ and $\langle \mathbf{y}_m \rangle$ are convergent, there exist integers p and q such that for all $m \geq p$:

$$|a_m - a| < \epsilon / \big[2(1 + \|\mathbf{y}^*\|)\big],$$

and for all $m \geq q$:

$$\|\mathbf{y}_m - \mathbf{y}^*\| < \epsilon / 2\alpha.$$

But then if $m \geq \max\{p, q\}$, we have:

$$\begin{aligned} \|a_m \mathbf{y}_m - a\mathbf{y}^*\| &\leq \|a_m \mathbf{y}_m - a_m \mathbf{y}^*\| + \|a_m \mathbf{y}^* - a\mathbf{y}^*\| \\ &= |a_m| \cdot \|\mathbf{y}_m - \mathbf{y}^*\| + |a_m - a| \cdot \|\mathbf{y}^*\| < \epsilon. \quad \square \end{aligned}$$

As has been the case with all of the definitions we have looked at so far, bounded sequences and Cauchy sequences in $\mathbf{R}^n$ are defined in ways which are very natural generalizations of the corresponding definitions for sequences of real numbers, as I hope you will agree. The proof of Proposition 2.64, which follows the definitions, will be left as an exercise.

2.62. Definition. A sequence $\langle \mathbf{x}_m \rangle$ is said to be **bounded** iff there exists $\alpha \in \mathbf{R}$ satisfying:

$$(\forall m \in N)\colon \|\mathbf{x}_m\| < \alpha.$$

2.63. Definition. A sequence $\langle \mathbf{x}_m \rangle$ is called a **Cauchy sequence** iff, for every $\epsilon > 0$, there exists $q \in N$ such that:

$$(\forall m, p \geq q)\colon d(\mathbf{x}_m, \mathbf{x}_p) \equiv \|\mathbf{x}_m - \mathbf{x}_p\| < \epsilon.$$

2.64. Proposition. *The sequence $\langle \mathbf{x}_m \rangle$ is a Cauchy sequence if, and only if, each coordinate sequence $\langle x_{im} \rangle$ $(i = 1, \ldots, n)$, is a Cauchy sequence.*

2.65. Proposition. *If $\langle \mathbf{x}_m \rangle$ is a Cauchy sequence, then $\langle \mathbf{x}_m \rangle$ is bounded.*

Proof. This follows immediately from 2.29, 2.58, and 2.64. □

2.66. Theorem. *A sequence $\langle \mathbf{x}_m \rangle$ is convergent if, and only if, it is a Cauchy sequence.*

Proof. This follows immediately from 2.30, 2.59, and 2.64. □

2.67. Theorem. (Bolzano-Weierstrass Theorem for $\mathbf{R}^n$) *If $\langle \mathbf{x}_m \rangle$ is a bounded sequence, then there exists a subsequence of $\langle \mathbf{x}_m \rangle, \langle \mathbf{x}_{m_i} \rangle$, and a point $\mathbf{x}^* \in \mathbf{R}^n$ such that $\mathbf{x}_{m_i} \to \mathbf{x}^*$.*

Proof. We will prove this result for the case in which $n = 2$. The same basic argument is valid for an arbitrary positive integer, n, but the notation gets very (or perhaps I should say even more) messy!

1. We first note that, since $\langle \mathbf{x}_m \rangle$ is bounded, it follows readily from Lemma 2.58 that the jth coordinate sequence, $\langle x_{jm} \rangle$ is bounded as well, for $j = 1, 2$.

2. Since $\langle x_{1m} \rangle$ is bounded, it follows from the Bolzano-Weierstrass Theorem for $\mathbf{R}$ (Theorem 2.27) that there exists a subsequence of $\langle x_{1m} \rangle, \langle x_{1m_i} \rangle$, and a point $x_1^* \in \mathbf{R}$ such that:

$$x_{1m_i} \to x_1^*. \tag{2.75}$$

3. Consider the sequence $\langle x_{2m_i} \rangle$, where $\langle m_i \rangle$ is the strictly increasing sequence of positive integers obtained in the previous paragraph. Obviously, $\langle x_{2m_i} \rangle$ is a subsequence of $\langle x_{2m} \rangle$; and, since $\langle x_{2m} \rangle$ is bounded, $\langle x_{2m_i} \rangle$ is bounded as well. Thus it follows from the Bolzano-Weierstrass Theorem for $\mathbf{R}$ that there exists a subsequence of $\langle x_{2m_i} \rangle, \langle x_{2m_{i_j}} \rangle$ [with $\langle m_{i_j} \rangle$ a subsequence of $\langle m_i \rangle$], and a number x_2^* such that:

$$x_{2m_{i_j}} \to x_2^*. \tag{2.76}$$

4. By (2.75) and 2.23, $\langle x_{1m_{i_j}} \rangle$, which is a subsequence of $\langle x_{1m_i} \rangle$, converges to x_1^*. Thus, using (2.75), it follows from Proposition 2.23 that

$$\mathbf{x}_{m_{i_j}} \to \mathbf{x}^*,$$

where we define $\mathbf{x}^* = (x_1^*, x_2^*)$. □

We can use the results regarding sequences in $\mathbf{R}^n$ to define infinite series in $\mathbf{R}^n$; in fact, we can proceed in pretty much the same way as in dealing with infinite series of real numbers, beginning with the following.

2.68. Definition. If $\langle \mathbf{x}_m \rangle$ is a sequence from $\mathbf{R}^n$, we say that the series, $\sum_{m=1}^{\infty} \mathbf{x}_m$ **converges** iff the sequence of partial sums, $\langle \mathbf{s}_t \rangle$, defined by:

$$\mathbf{s}_t = \sum_{m=1}^{t} \mathbf{x}_m \quad \text{for } t = 1, 2, \ldots,$$

converges. In other words, we shall say that the series converges if, and only if, there exists $\mathbf{x}^* \in \mathbf{R}^n$ such that, for all $\epsilon > 0$, there exists T such that, for all $t \geq T$, $\|\mathbf{s}_t - \mathbf{x}^*\| < \epsilon$.

Having formally stated the definition of convergence, however, it should be apparent from our discussion of sequences in $\mathbf{R}^n$ that a series will converge if, and only if, each of the n coordinate series of which it is composed also converge. A number of other results then become more or less immediate implications of the work which we have done so far. Thus, as an example, if we define, for each pair of positive integers, p and q, the sum

$$\mathbf{S}_{p,q} = \sum_{m=p+1}^{p+q} \mathbf{x}_m = \mathbf{s}_{p+q} - \mathbf{s}_p,$$

the following is a very easy consequence of Theorem 2.66.

2.69. Proposition. *The infinite series, $\sum_{m=1}^{\infty} \mathbf{x}_m$, converges if, and only if, for each $\epsilon > 0$, there exists $p \in N$ such that, for all $q \in N$, $\|\mathbf{S}_{p,q}\| < \epsilon$.*

In fact, most of the results on series which we studied in Section 3 have useful re-statements for series in $\mathbf{R}^n$. We list two of the more useful of these results as our next propositions;[7] the proofs of which will be left as exercises.

2.70. Theorem. *Suppose $\sum_{m=1}^{\infty} \mathbf{x}_m$ and $\sum_{m=1}^{\infty} \mathbf{y}_m$ are both convergent. If we form a new series by letting:*

1. $\mathbf{z}_m = \mathbf{x}_m + \mathbf{y}_m$ for $m = 1, 2, \ldots,$

then $\sum_{m=1}^{\infty} \mathbf{z}_m$ also converges, and we have:

$$\sum_{m=1}^{\infty} \mathbf{z}_m = \sum_{m=1}^{\infty} \mathbf{x}_m + \sum_{m=1}^{\infty} \mathbf{y}_m.$$

2. $\mathbf{w}_m = \alpha \mathbf{x}_m$ for $m = 1, 2, \ldots,$

where α is any real number, then $\sum_{m=1}^{\infty} \mathbf{w}_m$ also converges, and we have:

$$\sum_{m=1}^{\infty} \mathbf{w}_m = \alpha \cdot \left(\sum_{m=1}^{\infty} \mathbf{x}_m \right).$$

[7] Of course, all of the convergence 'tests' of Section 3 can be applied to the individual coordinate series. The results which we list here are probably the most useful of those which can be stated without mention of coordinates.

2.71. Proposition. *A nonnegative series from $\mathbf{R}^n$ is convergent if, and only if, it is bounded.*

The fact that a sequence of points in $\mathbf{R}^n$, $\langle \mathbf{x}_m \rangle$, converges if, and only if, its n coordinate sequences converge provides us with one way of applying the results of Section 3 to test for the convergence of series in $\mathbf{R}^n$. The following result provides us with a second such avenue.

2.72. Proposition. *If the series $\sum_{m=1}^{\infty} \|\mathbf{x}_m\|$ converges, then the series $\sum_{m=1}^{\infty} \mathbf{x}_m$ converges as well.*

Proof. Let $\epsilon > 0$ be given. Then, since $\sum_{m=1}^{\infty} \|\mathbf{x}_m\|$ is a convergent (real) series, it follows from Proposition 2.36 that there exists $n \in N$ such that, for all $p \in N$:

$$\left| \sum_{q=n+1}^{n+p} \|\mathbf{x}_q\| \right| = \sum_{q=n+1}^{n+p} \|\mathbf{x}_q\| < \epsilon.$$

But then we have, for all $p \in N$:

$$\|S_{n,p}\| \stackrel{\text{def}}{=} \left\| \sum_{q=n+1}^{n+p} \mathbf{x}_q \right\| \leq \sum_{q=n+1}^{n+p} \|\mathbf{x}_q\| < \epsilon,$$

and it follows from Proposition 2.69 that $\sum_{m=1}^{\infty} \mathbf{x}_m$ converges. □

One illustration of the usefulness of Proposition 2.72 is provided by the following result; the proof of which will be left as an exercise.

2.73. Proposition. *Suppose $\langle a_m \rangle$ is a sequence of real numbers such that $\sum_{m=1}^{\infty} a_m$ is absolutely convergent, and let $\langle \mathbf{y}_m \rangle \subseteq \mathbf{R}^n$ be a bounded sequence. Then the series $\sum_{m=1}^{\infty} a_m \mathbf{y}_m$ converges.*

Our main order of business for the remainder of this section will be to examine a generalization to n commodities of the linear growth model of Section 2.4. The result we will consider here is from Majumdar [1974], and in developing it, we will make considerable use of the following result.

2.74. Proposition. *Suppose $\mathbf{A}_1, \ldots, \mathbf{A}_t$ are $n \times n$ matrices such that each column of $\mathbf{A}_s$ is semi-positive,[8] for $s = 1, \ldots, t$, and suppose that $\mathbf{y} \in \mathbf{R}^n$ is also semi-positive. Then the matrix-vector product:*

$$\mathbf{A}_1 \mathbf{A}_2 \ldots \mathbf{A}_t \mathbf{y},$$

is semi-positive as well.

Proof. The matrix-vector product $\mathbf{A}_t \mathbf{y}$ can be expressed as (see 1.83.2):

$$\mathbf{x} = \sum_{j=1}^{n} \mathbf{a}^t_{\cdot j} y_j;$$

[8] Recall the terminology and vector inequalities from Section 1.3

which is clearly semi-positive, since all of the products, $\mathbf{a}^t_{.j} y_j$, are nonnegative, and at least one must be semi-positive (since $\mathbf{y}$ has at least one positive coordinate). By the same reasoning, we see that

$$\mathbf{A}_{t-1}(\mathbf{A}_t \mathbf{y})$$

is semi-positive, and so on. □

In our growth model we will suppose that n commodities are produced in each period, and that the stocks of these commodities can then either be used as inputs in producing output for the next period, or they can be consumed. Once again we will denote output, input, and consumption in period t by '$\mathbf{y}_t$','$\mathbf{x}_t$', and '$\mathbf{c}_t$', respectively; but this time we will assume that

$$\mathbf{y}_t, \mathbf{x}_t, \mathbf{c}_t \in \mathbf{R}^n_+ \quad \text{for } t = 0, 1, 2, \ldots; \tag{2.77}$$

and, as before, we will require that:

$$\mathbf{x}_t + \mathbf{c}_t = \mathbf{y}_t \quad \text{for } t = 0, 1, 2, \ldots. \tag{2.78}$$

We will once again denote the initial stock by '$\overline{\mathbf{y}}$,' but now we will be assuming that:

$$\overline{\mathbf{y}} \in \mathbf{R}^n_+ \setminus \{\mathbf{0}\}. \tag{2.79}$$

As suggested earlier, we will suppose that the production technology is linear; in fact, we will make use of an extension of the input-requirement function introduced earlier, supposing that the input of the ith commodity required to produce a unit of the jth commodity is given by the nonnegative number a_{ij} $(i, j = 1, 2, \ldots, n)$. Thus, defining the **input-requirement matrix**, $\mathbf{A}$, by:

$$\mathbf{A} = [a_{ij}],$$

we will suppose that an input-output pair $(\mathbf{x}, \mathbf{y}) \in \mathbf{R}^{2n}_+$ is technologically feasible iff:

$$\mathbf{A}\mathbf{y} \leq \mathbf{x}; \tag{2.80}$$

and we will also suppose throughout our discussion that:

$$\mathbf{a}_{.j} > \mathbf{0} \quad \text{for } j = 1, 2, \ldots, n, \tag{2.81}$$

where $\mathbf{a}_{.j}$ is the jth column of the input-requirement matrix, $\mathbf{A}$. Notice that the vector

$$\mathbf{a}_{.j} = \begin{pmatrix} a_{1j} \\ a_{2j} \\ \vdots \\ a_{nj} \end{pmatrix}$$

is the input vector required to produce a unit output of the jth commodity. Thus the requirement expressed in equation (2.81) simply means that a positive quantity of at least one commodity is needed to produce a unit of the jth commodity. While one would normally make stronger assumptions in a linear growth model, the assumptions expressed in (2.80) and (2.81) will be adequate for our present purposes, and would appear to be totally innocuous.

In the remainder of this section, we will denote the collection of all infinite sequences, $\langle \mathbf{x}_m \rangle$, from $\mathbf{R}^n$, by '$\mathbf{R}^{n\infty}$.' As before, we will denote elements of $\mathbf{R}^{n\infty}$ by $\mathbf{x}, \mathbf{y}, \mathbf{c}$, etc., and we define $\mathbf{R}^{n\infty}_{+}$ and $\mathbf{R}^{n\infty}_{++}$ in the obvious way. We are then interested in the following definitions.

2.75. Definition. Given a sequence of $n \times n$ input-requirement matrices, $\langle \mathbf{A}_t \rangle$, and an initial stock of commodities, $\overline{\mathbf{y}} \in \mathbf{R}^n_+ \setminus \{\mathbf{0}\}$, we shall say that a triple $(\mathbf{x}, \mathbf{y}, \mathbf{c})$ is **feasible for** $(\langle \mathbf{A}_t \rangle, \overline{\mathbf{y}})$ iff $\{\mathbf{x}, \mathbf{y}, \mathbf{c}\} \subseteq \mathbf{R}^{n\infty}_+$,

$$\mathbf{y}_0 = \overline{\mathbf{y}}, \tag{2.82}$$

and

$$\left.\begin{array}{rcl} \mathbf{y}_t & = & \mathbf{c}_t + \mathbf{x}_t \\ \mathbf{A}_{t+1}\mathbf{y}_{t+1} & \leq & \mathbf{x}_t \end{array}\right\} \quad \text{for } t = 0, 1, 2, \ldots; \tag{2.83}$$

and, in this case we shall also say that $(\mathbf{x}, \mathbf{y})$ is a **feasible production program**, and that $\mathbf{c} = \langle \mathbf{c}_t \rangle$ is a **feasible consumption program for** $(\langle \mathbf{A}_t \rangle, \overline{\mathbf{y}})$.

2.76. Definition. Given a sequence of $n \times n$ input-requirement matrices, $\langle \mathbf{A}_t \rangle$, and an initial stock of commodities, $\overline{\mathbf{y}} \in \mathbf{R}^n_+ \setminus \{\mathbf{0}\}$, we shall say that a feasible program $(\mathbf{x}, \mathbf{y}, \mathbf{c})$ is **intertemporally efficient for** $(\langle \mathbf{A}_t \rangle, \overline{\mathbf{y}})$ iff there is no alternative feasible program for $(\langle \mathbf{A}_t \rangle, \overline{\mathbf{y}})$, $(\mathbf{x}', \mathbf{y}', \mathbf{c}')$, such that $\mathbf{c}' > \mathbf{c}$.

In the remainder of this section, we will study a result due to Majumdar[9] which characterizes feasibility and efficiency for this model. In our developement, we will need the following definitions. We define the sequence of $n \times n$ matrices, $\langle \mathbf{B}_t \rangle$, by:

$$\mathbf{B}_0 = \mathbf{I} \quad \text{and} \quad \mathbf{B}_t = \mathbf{B}_0 \mathbf{A}_1 \ldots \mathbf{A}_t \quad \text{for } t = 1, 2, \ldots;$$

and, for each pair of positive integers, (t, T), satisfying $t \geq 1$ and $T > t$, we define the matrix $\mathbf{B}_{t,T}$ by

$$\mathbf{B}_{t,T} = \mathbf{A}_t \mathbf{A}_{t+1} \ldots \mathbf{A}_T.$$

With this notation, we can state the result to be proved, as follows.

[9]Majumdar's result is stated and proved for the case in which:

$$\mathbf{A}_t = \mathbf{A}_1 \equiv \mathbf{A} \quad \text{for } t = 1, 2, \ldots;$$

but, as he notes in his article, there is no real difficulty in generalizing the result slightly by dropping this restriction.

2.77. Proposition. (Majumdar [1974]) *Suppose $\langle \mathbf{A}_t \rangle$ is a sequence of $n \times n$ input-requirement matrices, and $\overline{\mathbf{y}} \in \mathbf{R}^n_+ \backslash \{\mathbf{0}\}$ is an initial stock of commodities. If $(\mathbf{x}, \mathbf{y}, \mathbf{c})$ is feasible for $(\langle \mathbf{A}_t \rangle, \overline{\mathbf{y}})$, then*

$$\sum_{t=0}^{\infty} \mathbf{B}_t \mathbf{c}_t \leq \overline{\mathbf{y}}; \tag{2.84}$$

and, in addition, $(\mathbf{x}, \mathbf{y}, \mathbf{c})$ is intertemporally efficient for $(\langle \mathbf{A}_t \rangle, \overline{\mathbf{y}})$ if, and only if:

$$\sum_{t=0}^{\infty} \mathbf{B}_t \mathbf{c}_t = \overline{\mathbf{y}} \tag{2.85}$$

Proof.[10] Suppose $(\mathbf{x}, \mathbf{y}, \mathbf{c})$ is feasible for $(\langle \mathbf{A}_t \rangle, \overline{\mathbf{y}})$.

Part 1. By the definition of feasibility,

$$\mathbf{c}_1 = \mathbf{y}_1 - \mathbf{x}_1,$$

and thus

$$\mathbf{A}_1 \mathbf{c}_1 = \mathbf{A}_1 \mathbf{y}_1 - \mathbf{A}_1 \mathbf{x}_1 \leq \mathbf{x}_0 - \mathbf{A}_1 \mathbf{x}_1 = \overline{\mathbf{y}} - \mathbf{c}_0 - \mathbf{A}_1 \mathbf{x}_1.$$

Therefore,

$$\mathbf{c}_0 + \mathbf{A}_1 \mathbf{c}_1 = \mathbf{I} \mathbf{c}_0 + \mathbf{A}_1 \mathbf{c}_1 \equiv \mathbf{B}_0 \mathbf{c}_0 + \mathbf{B}_1 \mathbf{c}_1 \leq \overline{\mathbf{y}} - \mathbf{B}_1 \mathbf{x}_1.$$

Suppose now that we have established that, for $1 \leq T \leq s$, we have:

$$\sum_{t=0}^{T} \mathbf{B}_t \mathbf{c}_t \leq \overline{\mathbf{y}} - \mathbf{B}_T \mathbf{x}_T.$$

Then for $T = s + 1$, we have, using our inductive hypothesis, feasibility and the definition of $\mathbf{B}_{s+1}$, and technological feasibility in turn, that:

$$\sum_{t=0}^{s+1} \mathbf{B}_t \mathbf{c}_t = \sum_{t=0}^{s} \mathbf{B}_t \mathbf{c}_t + \mathbf{B}_{s+1} \mathbf{c}_{s+1} \leq \overline{\mathbf{y}} - \mathbf{B}_s \mathbf{x}_s + \mathbf{B}_{s+1} \mathbf{c}_{s+1}$$

$$= \overline{\mathbf{y}} - \mathbf{B}_s \mathbf{x}_s + \mathbf{B}_s [\mathbf{A}_{s+1} (\mathbf{y}_{s+1} - \mathbf{x}_{s+1})] \leq \overline{\mathbf{y}} - \mathbf{B}_s \mathbf{x}_s + \mathbf{B}_s (\mathbf{x}_s - \mathbf{A}_{s+1} \mathbf{x}_{s+1})$$

$$= \overline{\mathbf{y}} - \mathbf{B}_s \mathbf{x}_s + \mathbf{B}_s \mathbf{x}_s - \mathbf{B}_{s+1} \mathbf{x}_{s+1} = \overline{\mathbf{y}} - \mathbf{B}_{s+1} \mathbf{x}_{s+1} \leq \overline{\mathbf{y}}.$$

It follows, therefore, that the sequence of partial sums, $\langle \sum_{t=0}^{T} \mathbf{B}_t \mathbf{c}_t \rangle$, is a nonnegative, nondecreasing sequence which is bounded above by $\overline{\mathbf{y}}$. Therefore, by 2.71, the series converges, and

$$\sum_{t=0}^{\infty} \mathbf{B}_t \mathbf{c}_t \leq \overline{\mathbf{y}}.$$

Part 2. Now suppose $(\mathbf{x}, \mathbf{y}, \mathbf{c})$ satisfies:

$$\sum_{t=0}^{\infty} \mathbf{B}_t \mathbf{c}_t = \overline{\mathbf{y}}; \tag{2.86}$$

[10] A comparison of the proof provided here with the proof of Proposition 2.55 will reveal that the reasoning is virtually identical. I have chosen to accentuate this similarity by keeping the wording of the two proofs as similar as is practicable.

and suppose, by way of obtaining a contradiction, that there exists an alternative feasible program for $(\langle \mathbf{A}_t \rangle, \overline{\mathbf{y}}), (\mathbf{x}', \mathbf{y}', \mathbf{c}')$, such that $\mathbf{c}' > \mathbf{c}$. Then, since $(\mathbf{x}', \mathbf{y}', \mathbf{c}')$ is feasible, we have from part 1 that $\sum_{t=0}^{\infty} \mathbf{B}_t \mathbf{c}'_t$ converges, and

$$\sum_{t=0}^{\infty} \mathbf{B}_t \mathbf{c}'_t \leq \overline{\mathbf{y}}. \tag{2.87}$$

However, it follows from the fact that $\mathbf{c}' > \mathbf{c}$ that, for some integer, $q, \mathbf{c}'_q > \mathbf{c}_q$, and from Proposition 2.74 and the definition of $\mathbf{B}_q$, it then follows that:

$$\mathbf{B}_q \mathbf{c}'_q > \mathbf{B}_q \mathbf{c}_q;$$

and thus

$$\sum_{t=0}^{\infty} \mathbf{B}_t \mathbf{c}'_t > \sum_{t=0}^{\infty} \mathbf{B}_t \mathbf{c}_t;$$

which, given (2.86) and (2.87), yields a contradiction.

Part 3. Now suppose that

$$\sum_{t=0}^{\infty} \mathbf{B}_t \mathbf{c}_t < \overline{\mathbf{y}}; \tag{2.88}$$

so that, defining

$$\mathbf{u} = \overline{\mathbf{y}} - \sum_{t=0}^{\infty} \mathbf{B}_t \mathbf{c}_t,$$

we have

$$\sum_{t=0}^{\infty} \mathbf{B}_t \mathbf{c}_t = \overline{\mathbf{y}} - \mathbf{u}. \tag{2.89}$$

We will demonstrate that in this case there exists a feasible program for $(\langle \mathbf{A}_t \rangle, \overline{\mathbf{y}}), (\mathbf{x}', \mathbf{y}', \mathbf{c}')$, which dominates $(\mathbf{x}, \mathbf{y}, \mathbf{c})$. To do this, we begin by noting that the series $\sum_{t=1}^{\infty} \mathbf{B}_t \mathbf{c}_t$ converges; in fact, from (2.89), we have that:

$$\sum_{t=1}^{\infty} \mathbf{B}_t \mathbf{c}_t = \overline{\mathbf{y}} - \mathbf{c}_0 - \mathbf{u}. \tag{2.90}$$

Next, we note that it follows from the definition of feasibility that, given any $t \geq 1$:

$$\mathbf{A}_{t+1} \mathbf{c}_{t+1} = \mathbf{A}_{t+1}(\mathbf{y}_{t+1} - \mathbf{x}_{t+1}) \leq \mathbf{x}_t - \mathbf{A}_{t+1} \mathbf{x}_{t+1}.$$

Now suppose that, for $t + 1 \leq T \leq s$, we have:

$$\sum_{j=t+1}^{T} \mathbf{B}_{t+1,\, j} \mathbf{c}_j \leq \mathbf{x}_t - \mathbf{B}_{t+1,\, T} \mathbf{x}_T.$$

Then, for $T = s + 1$, we have:

$$\begin{aligned}\sum_{j=t+1}^{s+1} \mathbf{B}_{t+1,\, j} \mathbf{c}_j &= \sum_{j=t+1}^{s} \mathbf{B}_{t+1,\, j} \mathbf{c}_j + \mathbf{B}_{t+1,\, s+1} \mathbf{c}_{s+1} \\ &\leq \mathbf{x}_t - \mathbf{B}_{t+1,\, s} \mathbf{x}_s + \mathbf{B}_{t+1,\, s}[\mathbf{A}_{s+1}(\mathbf{y}_{s+1} - \mathbf{x}_{s+1})]\end{aligned}$$

$$\leq \mathbf{x}_t - \mathbf{B}_{t+1,\,s}\mathbf{x}_s + \mathbf{B}_{t+1,\,s}(\mathbf{x}_s - \mathbf{A}_{s+1}\mathbf{x}_{s+1}) = \mathbf{x}_t - \mathbf{B}_{t+1,\,s+1}\mathbf{x}_{s+1} \leq \mathbf{x}_t.$$

Thus it follows, using Proposition 2.71, that for each $t \geq 1$, the series $\sum_{j=t+1}^{\infty} \mathbf{B}_{t+1,\,j}\mathbf{c}_j$ converges, and

$$\sum\nolimits_{j=t+1}^{\infty} \mathbf{B}_{t+1,\,j}\mathbf{c}_j \leq \mathbf{x}_t \quad \text{for } t = 1, 2, \ldots.$$

We are now in a position to define our alternative feasible program, $(\mathbf{x}', \mathbf{y}', \mathbf{c}')$, as follows. Let:

$$\mathbf{c}_0' = \mathbf{c}_0 + \mathbf{u}, \mathbf{c}_t' = \mathbf{c}_t \quad \text{for } t = 1, 2, \ldots,$$

$$\mathbf{x}_0' = \overline{\mathbf{y}} - \mathbf{c}_0 - \mathbf{u}, \mathbf{x}_t' = \sum\nolimits_{j=t+1}^{\infty} \mathbf{B}_{t+1,\,j}\mathbf{c}_j \quad \text{for } t = 1, 2, \ldots,$$

and

$$\mathbf{y}_t' = \mathbf{c}_t' + \mathbf{x}_t' \quad \text{for } t = 1, 2, \ldots.$$

From the definition of $(\mathbf{x}', \mathbf{y}', \mathbf{c}')$ it is easy to see that, in order to prove that it is feasible, we need only establish technological feasibility, for each t. However, for $t = 0$, we have, using the definitions of $\mathbf{B}_{2,\,j}$ and $\mathbf{B}_j$, Theorem 2.70, and equation (2.90),[11]

$$\begin{aligned}\mathbf{A}_1\mathbf{y}_1' &= \mathbf{A}_1\mathbf{c}_1' + \mathbf{A}_1\mathbf{x}_1' = \mathbf{A}_1\mathbf{c}_1 + \mathbf{A}_1\left(\sum\nolimits_{j=2}^{\infty} \mathbf{B}_{2,\,j}\mathbf{c}_j\right) \\ &= \mathbf{A}_1\mathbf{c}_1 + \sum\nolimits_{j=2}^{\infty} \mathbf{A}_1\mathbf{B}_{2,\,j}\mathbf{c}_j = \mathbf{B}_1\mathbf{c}_1 + \sum\nolimits_{t=2}^{\infty} \mathbf{B}_t\mathbf{c}_t = \overline{\mathbf{y}} - \mathbf{c}_0 - \mathbf{u} \equiv \mathbf{x}_0'.\end{aligned}$$

Similarly, for $t \geq 1$,

$$\begin{aligned}\mathbf{A}_{t+1}\mathbf{y}_{t+1}' &= \mathbf{A}_{t+1}\mathbf{c}_{t+1}' + \mathbf{A}_{t+1}\mathbf{x}_{t+1}' \\ &= \mathbf{A}_{t+1}\mathbf{c}_{t+1} + \mathbf{A}_{t+1}\left(\sum\nolimits_{j=t+2}^{\infty} \mathbf{B}_{t+2,\,j}\mathbf{c}_j\right) \\ &= \mathbf{A}_{t+1}\mathbf{c}_{t+1} + \sum\nolimits_{j=t+2}^{\infty} \mathbf{A}_{t+1}\mathbf{B}_{t+2,\,j}\mathbf{c}_j \\ &= \mathbf{A}_{t+1}\mathbf{c}_{t+1} + \sum\nolimits_{j=t+2}^{\infty} \mathbf{B}_{t+1,\,j}\mathbf{c}_j = \sum\nolimits_{j=t+1}^{\infty} \mathbf{B}_{t+1,\,j}\mathbf{c}_j = \mathbf{x}_t',\end{aligned}$$

where we have used Theorem 2.70 in obtaining the third of the above equalities.

We have shown, therefore, that $(\mathbf{x}', \mathbf{y}', \mathbf{c}')$ is feasible; and, since

$$\mathbf{c}_0' = \mathbf{c}_0 + \mathbf{u} > \mathbf{c}_0$$

while

$$\mathbf{c}_t' = \mathbf{c}_t \quad \text{for } t = 1, 2, \ldots;$$

[11]For a justification of the second equality in the present expression in the text, and of the third equality in the next expression, see also Exercise 5, at the end of this section.

it follows that $(\mathbf{x}, \mathbf{y}, \mathbf{c})$ is *not* intertemporally efficient. Consequently, we see that if $(\mathbf{x}, \mathbf{y}, \mathbf{c})$ *is* intertemporally efficient, then (2.88) cannot hold; whence it follows from part 1 that:

$$\sum_{t=0}^{\infty} \mathbf{B}_t \mathbf{c}_t = \overline{\mathbf{y}}. \quad \square$$

Not only is this last theorem essentially the same, both in statement and in the structure of its proof, as Proposition 2.55, we can also construct an intertemporally efficient program in the same way as it was done for the one-commodity case in Example 2.56. This is shown in the following.

2.78. Example. Suppose $\langle \mathbf{A}_t \rangle$ is such that $\mathbf{A}_t$ is **nonnegatively invertible**, for $t = 1, 2, \ldots$; that is, suppose that for each t, A_t^{-1} exists and is a nonnegative matrix.[12] Given this assumption, we will construct a balanced growth program for this case in essentially the same way in which we proceeded in Example 2.56.

Let $\theta \in]0, 1[$, and define the program $(\mathbf{x}^*, \mathbf{y}^*, \mathbf{c}^*)$ by:

$$\mathbf{y}_0^* = \overline{\mathbf{y}},$$
$$\mathbf{c}_t^* = \theta \mathbf{y}_t^*, \mathbf{x}_t^* = (1-\theta)\mathbf{y}_t^*, \mathbf{y}_{t+1}^* = \mathbf{A}_{t+1}^{-1}\mathbf{x}_t^* \quad \text{for } t = 1, 2, \ldots.$$

It is apparent that $(\mathbf{x}^*, \mathbf{y}^*, \mathbf{c}^*)$ is a feasible program. We can use an argument almost identical to that employed in Example 2.56 to show that $(\mathbf{x}^*, \mathbf{y}^*, \mathbf{c}^*)$ is intertemporally efficient. It is easy to prove, using a simple induction argument, that:

$$\mathbf{y}_t^* = (1-\theta)^t \mathbf{A}_t^{-1} \cdots \mathbf{A}_1^{-1} \overline{\mathbf{y}} \quad \text{for } t = 1, 2, \ldots.$$

Consequently:

$$\begin{aligned} \mathbf{B}_t \mathbf{c}_t^* = \theta \mathbf{B}_t \mathbf{y}_t^* &= \theta(1-\theta)^t \mathbf{A}_1 \cdots \mathbf{A}_t \cdot \mathbf{A}_t^{-1} \cdots \mathbf{A}_1^{-1} \overline{\mathbf{y}} \\ &= \theta(1-\theta)^t \overline{\mathbf{y}} \quad \text{for } t = 1, 2, \ldots. \end{aligned}$$

Since $\mathbf{B}_0 \mathbf{c}_0^* = \theta \overline{\mathbf{y}} = \theta(1-\theta)^0 \overline{\mathbf{y}}$ as well, we see that, for each $T \geq 1$:

$$\sum_{t=0}^{T} \mathbf{B}_t \mathbf{c}_t^* = \left(\theta \left[\sum_{t=0}^{T} (1-\theta)^t\right]\right) \overline{\mathbf{y}}.$$

However, we know that:

$$\theta \left[\sum_{t=0}^{T} (1-\theta)^t\right] \to 1;$$

and thus, from Proposition 2.61:

$$\lim_{T \to \infty} \left(\sum_{t=0}^{T} \mathbf{B}_t \mathbf{c}_t^*\right) \equiv \sum_{t=0}^{\infty} \mathbf{B}_t \mathbf{c}_t^* = \overline{\mathbf{y}}. \quad \square$$

[12] Of course, it must actually be semi-positive, since it is also invertible.

Exercises.

1. Prove (a) Theorem 2.59, (b) Theorem 2.60, and (c) Proposition 2.64, concerning sequences in $\mathbf{R}^n$.

2. Prove (a) Proposition 2.69, (b) Theorem 2.70, (c) Proposition 2.71, and (d) Proposition 2.73, concerning series in $\mathbf{R}^n$.

3. Show that if $\langle \mathbf{x}_m^i \rangle \subseteq \mathbf{R}^n$, for $i = 1, \ldots, k$, $\langle \mathbf{a}_m \rangle \subseteq \mathbf{R}^k$, $\mathbf{x}^i \in \mathbf{R}^n$ ($i = 1, \ldots, k$) and $\mathbf{a} \in \mathbf{R}^k$ are such that:

$$\lim_{m\to\infty} \mathbf{x}_m^i = \mathbf{x}^i \quad \text{for } i = 1, \ldots, k \text{ and } \lim_{m\to\infty} \mathbf{a}_m = \mathbf{a},$$

then:

$$\lim_{m\to\infty} \left[\sum_{i=1}^k a_{im} \mathbf{x}_m^i \right] = \sum_{i=1}^k a_i \mathbf{x}^i.$$

4. Complete the induction argument for the proof of Proposition 2.74.

5. Show that if $\sum_{t=1}^\infty \mathbf{x}_t$ converges to $\mathbf{y} \in \mathbf{R}^n$, and $\mathbf{A}$ is any $m \times n$ matrix, then the series $\sum_{t=1}^\infty \mathbf{z}_t$ defined by

$$\mathbf{z}_t = \mathbf{A}\mathbf{x}_t \quad \text{for } t = 1, 2, \ldots,$$

converges to $\mathbf{A}\mathbf{y}$.

Chapter 3

Continuity

3.1 Continuous Vector-Valued Functions

While I am sure that you have had an introduction to the theory of continuous real-valued functions, you may not have been introduced to the theory of vector-valued functions; and, as it turns out, this sort of function is extremely important in economic theory. Not to worry, however; dealing with a continuous function mapping into m-dimensional Euclidean space involves almost nothing, conceptually, which is not already familiar to you from the theory of real-valued functions.

3.1. Definition. Let $f\colon X \to \mathbf{R}^m$, where X is a non-empty subset of $\mathbf{R}^n$. We say that **f is continuous at $\mathbf{x}^* \in X$** iff, for every $\epsilon \in \mathbf{R}_{++}$, there exists $\delta \in \mathbf{R}_{++}$ such that:

$$(\forall \mathbf{x} \in N(\mathbf{x}^*, \delta) \cap X)\colon \|f(\mathbf{x}) - f(\mathbf{x}^*)\| < \epsilon.$$

The function f will be said to be **continuous on X** (or simply **continuous**, if X is understood to be the entire domain of f) iff f is continuous at each $\mathbf{x} \in X$.

Notice that if $f\colon X \to \mathbf{R}^m$, we can write:[1]

$$f(\mathbf{x}) = (f_1(\mathbf{x}), f_2(\mathbf{x}), \ldots, f_m(\mathbf{x})) \quad \text{for } \mathbf{x} \in X,$$

where

$$f_i\colon X \to \mathbf{R} \quad \text{for } i = 1, \ldots, m.$$

It is because of this that we encounter nothing very new in generalizing the notion of continuity from real-valued functions to vector-valued funtions, as is demonstrated by the following result.

[1] It would be more consistent, given our practice of distinguishing between real numbers and vectors by using bold-face type, to denote the function f by '$\mathbf{f}$;' however, in practice, I don't believe that you will find the present notation to be confusing.

3.2. Theorem. *If $f\colon X \to \mathbf{R}^m$, where X is a non-empty subset of $\mathbf{R}^n$, then f is continuous at a point $\mathbf{x}^* \in X$ if, and only if, each f_i is continuous at $\mathbf{x}^*$.*

Proof. Since this result is an almost immediate consequence of Lemma 2.58, I will present only a sketch of a proof. You should have no trouble filling in the details.

The fact that the continuity of f at $\mathbf{x}^* \in X$ implies that each f_i is continuous at $\mathbf{x}^*$ follows easily from the observation that, for each i, and each $\mathbf{x} \in X$, we have:

$$|f_i(\mathbf{x}) - f_i(\mathbf{x}^*)| \leq \|f(\mathbf{x}) - f(\mathbf{x}^*)\|.$$

Similarly, the proof of the converse statement is a nearly immediate consequence of the fact that:

$$\|f(\mathbf{x}) - f(\mathbf{x}^*)\| \leq \sum\nolimits_{i=1}^{m} |f_i(\mathbf{x}) - f_i(\mathbf{x}^*)|. \quad \square$$

As is the case with real-valued functions, we can characterize the continuity of vector-valued functions in terms of what happens to the value of the function as its arguments approach some limit. First, however, we must define what we mean by the limit of a function.

3.3. Definition. Let $f\colon X \to \mathbf{R}^m$, where X is a non-empty subset of $\mathbf{R}^n$, and let $\mathbf{x}^* \in \mathbf{R}^n$ be a limit point of X. We say that $\mathbf{y} \in \mathbf{R}^m$ is the **limit of f as $\mathbf{x}$ approaches $\mathbf{x}^*$ ($\mathbf{x} \to \mathbf{x}^*$)** iff, for each $\epsilon > 0$, there exists $\delta \in \mathbf{R}_{++}$ such that:

$$(\forall \mathbf{x} \in N'(\mathbf{x}^*, \delta) \cap X)\colon \|f(\mathbf{x}) - \mathbf{y}\| < \epsilon,$$

and we then write '$\lim_{\mathbf{x}\to\mathbf{x}^*} f(\mathbf{x}) = \mathbf{y}$,' or '$f(\mathbf{x}) \to \mathbf{y}$ as $\mathbf{x} \to \mathbf{x}^*$.'

From the same considerations as were used in the proof of Theorem 3.2, it should be clear that if $f(\mathbf{x}) \to \mathbf{y}^* \in \mathbf{R}^m$ as $\mathbf{x} \to \mathbf{x}^*$, then:

$$f_i(\mathbf{x}) \to y_i^* \quad \text{for } i = 1, \ldots, m;$$

and conversely. We shall also be interested in infinite limits for real-valued functions, defined as follows.

3.4. Definition. If $f\colon X \to \mathbf{R}$ where X is a non-empty subset of $\mathbf{R}^n$, and $\mathbf{x}^*$ is a limit point of X, we say that the **limit of f as $\mathbf{x}$ approaches $\mathbf{x}^*$** is $+\infty$ (respectively, $-\infty$), and write

$$\lim_{\mathbf{x}\to\mathbf{x}^*} f(\mathbf{x}) = +\infty \text{ [respectively, } \lim_{\mathbf{x}\to\mathbf{x}^*} f(\mathbf{x}) = -\infty],$$

iff, for every $a \in \mathbf{R}_+$, there exists $\delta \in \mathbf{R}_{++}$ such that:

$$(\forall \mathbf{x} \in N'(\mathbf{x}^*, \delta) \cap X)\colon f(\mathbf{x}) \geq a \text{ [respectively, } f(\mathbf{x}) \leq -a].$$

3.5. Examples/Exercises.

1. Let $f\colon X \to \mathbf{R}$, where $X \subseteq \mathbf{R}$ is given by:

$$X = [0, 1],$$

and f is defined by:

$$f(x) = \begin{cases} 1 & \text{for } 0 \leq x < 1, \\ 4 & \text{for } x = 1. \end{cases}$$

Then $\lim_{x \to 1} f(x) = 1$.

2. Let $X = [0, 1] \subseteq \mathbf{R}$, and let $f\colon X \to \mathbf{R}$ be defined by:

$$f(x) = \begin{cases} 1/x & \text{for } x > 0, \\ 0 & \text{for } x = 0. \end{cases}$$

Then $\lim_{x \to 0} f(x) = +\infty$.

3. Let $X = \mathbf{R}_+$, and let f be defined on X by:

$$f(x) = \begin{cases} \sin(1/x) & \text{for } x > 0, \\ 1 & \text{for } x = 0. \end{cases}$$

Here $\lim_{x \to 0} f(x)$ does not exist.

4. If we modify the last example by letting f be defined by:

$$f(x) = \begin{cases} x \cdot \sin(1/x) & \text{for } x > 0, \\ 1 & \text{for } x = 0, \end{cases}$$

then we have: $\lim_{x \to 0} f(x) = 0$. □

We can now characterize continuous functions in terms of limits, as follows. The proof will be left as an exercise.

3.6. Proposition. *Let $f\colon X \to \mathbf{R}^m$, where X is a non-empty subset of $\mathbf{R}^n$, and let $\mathbf{x}^* \in X$ be a limit point of X. Then f is continuous at $\mathbf{x}^*$ iff* $\lim_{\mathbf{x} \to \mathbf{x}^*} f(\mathbf{x}) = f(\mathbf{x}^*)$.

Notice that, in terms of Definition 3.1, if $f\colon X \to \mathbf{R}^m$, and $\mathbf{x}^* \in X$ is *not* a limit point of X, then f is necessarily continuous at $\mathbf{x}^*$ (that is, *any* such function, f, is continuous at $\mathbf{x}^*$). Thus, for all practical purposes, Proposition 3.6 provides a complete characterization[2] of the continuity of a function from $\mathbf{R}^n$ into $\mathbf{R}^m$.

3.7. Definition. Let $f\colon X \to \mathbf{R}^m$, where X is a non-empty subset of $\mathbf{R}^n$, and let A be a subset of X. We shall say that **f is bounded on A** iff there exists $a \in \mathbf{R}_+$ such that:

$$(\forall \mathbf{x} \in A)\colon \|f(\mathbf{x})\| \leq a.$$

If $A = X$, then we shall usually simply say that **f is bounded**.

[2]That is, both necessary *and* sufficient conditions.

A function which approaches a finite limit at a limit point of its domain is necessarily bounded in a neighborhood of that point, as we shall now demonstrate.

3.8. Proposition. *If $f: X \to \mathbf{R}^m$, where X is a subset of $\mathbf{R}^n$, $\mathbf{x}^*$ is a limit point of X, and $\mathbf{y} \in \mathbf{R}^m$ is such that $\lim_{\mathbf{x}\to\mathbf{x}^*} f(\mathbf{x}) = \mathbf{y}$, then there exists a positive real number, δ, such that f is bounded on $N(\mathbf{x}^*, \delta) \cap X$.*

Proof. Using the fact that $\lim_{\mathbf{x}\to\mathbf{x}^*} f(\mathbf{x}) = \mathbf{y}$, let $\delta \in \mathbf{R}_{++}$ be such that:

$$(\forall \mathbf{x} \in N'(\mathbf{x}^*, \delta) \cap X)\colon \|f(\mathbf{x}) - \mathbf{y}\| < 1.$$

We then have, for all $\mathbf{x} \in N(\mathbf{x}^*, \delta) \cap X$

$$\|f(\mathbf{x})\| = \|f(\mathbf{x}) - \mathbf{y} + \mathbf{y}\| \leq \|f(\mathbf{x}) - \mathbf{y}\| + \|\mathbf{y}\| < 1 + \|\mathbf{y}\|.$$

Consequently, if we define $\alpha \in \mathbf{R}_{++}$ by:

$$\alpha = \begin{cases} \max\{\|\mathbf{y}\| + 1, \|f(\mathbf{x}^*)\|\} & \text{if } \mathbf{x}^* \in X, \\ \|\mathbf{y}\| + 1 & \text{if } \mathbf{x}^* \notin X; \end{cases}$$

we have:

$$(\forall \mathbf{x} \in N'(\mathbf{x}^*, \delta) \cap X)\colon \|f(\mathbf{x})\| \leq \alpha. \quad \square$$

One of the reasons for the importance of continuous functions is that arithmetic operations on continuous functions preserve continuity. This fact leads to some very simple methods for establishing the continuity, or lack thereof, of functions which we may encounter. Moreover, it leads to some very important abstract properties of continuous functions which we shall study later. The fundamental facts yielding all of these consequences are derived from the behavior of the limits of functions, which, not very surprisingly, is much like the behavior of limits of convergent sequences.

3.9. Theorem. *Suppose $f: X \to \mathbf{R}^m$ and $g: X \to \mathbf{R}^m$, where X is a subset of $\mathbf{R}^n$, let $\mathbf{x}^* \in \mathbf{R}^n$ be a limit point of X, and suppose $\mathbf{y}, \mathbf{z} \in \mathbf{R}^m$ are such that:*

$$\lim_{\mathbf{x}\to\mathbf{x}^*} f(\mathbf{x}) = \mathbf{y} \quad \textit{and} \quad \lim_{\mathbf{x}\to\mathbf{x}^*} g(\mathbf{x}) = \mathbf{z}.$$

Then we have:

1. $\lim_{\mathbf{x}\to\mathbf{x}^*} [f(\mathbf{x}) + g(\mathbf{x})] = \mathbf{y} + \mathbf{z}$,
2. *for any* $\alpha \in \mathbf{R}$, $\lim_{\mathbf{x}\to\mathbf{x}^*} \alpha f(\mathbf{x}) = \alpha \mathbf{y}$,

and

3. $\lim_{\mathbf{x}\to\mathbf{x}^*} f(\mathbf{x}) \cdot g(\mathbf{x}) = \mathbf{y} \cdot \mathbf{z}$.

Proof. Since this result can be proved with an argument very much like that used to prove the similar result for sequences [2.9 and 2.60], I will leave the proofs of parts 1 and 2 as exercises. To prove part 3, define the function $h: X \to \mathbf{R}$, by:

$$h(\mathbf{x}) = f(\mathbf{x}) \cdot g(\mathbf{x}) \quad \text{for } \mathbf{x} \in X;$$

and suppose

$$\lim_{\mathbf{x}\to\mathbf{x}^*} f(\mathbf{x}) = \mathbf{y} \quad \text{and} \quad \lim_{\mathbf{x}\to\mathbf{x}^*} g(\mathbf{x}) = \mathbf{z}.$$

We wish to prove that:

$$\lim_{\mathbf{x}\to\mathbf{x}^*} h(\mathbf{x}) = \beta,$$

where the real number β is defined by:

$$\beta = \mathbf{y}\cdot\mathbf{z}.$$

We begin by noting that, since $g(\mathbf{x}) \to \mathbf{z}$ as $\mathbf{x} \to \mathbf{x}^*$, it follows from Proposition 3.8 that there exist positive real numbers, α and δ_1 such that:

$$(\forall \mathbf{x} \in N(\mathbf{x}^*, \delta_1) \cap X)\colon \|g(\mathbf{x})\| \leq \alpha. \tag{3.1}$$

Let $\epsilon > 0$ be given. Then by (3.1), there exist $\delta_2, \delta_3 \in \mathbf{R}_{++}$ such that:

$$(\forall \mathbf{x} \in N'(\mathbf{x}^*, \delta_2) \cap X)\colon \|f(\mathbf{x}) - \mathbf{y}\| < \epsilon/2\alpha, \tag{3.2}$$

and

$$(\forall \mathbf{x} \in N'(\mathbf{x}^*, \delta_3) \cap X) : \|g(\mathbf{x}) - \mathbf{z}\| < \epsilon/2(\|\mathbf{y}\| + 1). \tag{3.3}$$

But then if we define:

$$\delta = \min\{\delta_1, \delta_2, \delta_3\},$$

we see, using the Cauchy-Schwarz inequality, and (3.1) - (3.3) in turn, that, for $\mathbf{x} \in N'(\mathbf{x}^*, \delta) \cap X$, we have:

$$|h(\mathbf{x}) - \beta| = |f(\mathbf{x})\cdot g(\mathbf{x}) - \mathbf{y}\cdot\mathbf{z}\,| = |f(\mathbf{x})\cdot g(\mathbf{x}) - \mathbf{y}\cdot g(\mathbf{x}) + \mathbf{y}\cdot g(\mathbf{x}) - \mathbf{y}\cdot\mathbf{z}\,|$$

$$\leq \Big|g(\mathbf{x})[f(\mathbf{x}) - \mathbf{y}]\Big| + \Big|\mathbf{y}\cdot[g(\mathbf{x}) - \mathbf{z}]\Big| \leq \|g(\mathbf{x})\|\cdot\|f(\mathbf{x}) - \mathbf{y}\,\| + \|\mathbf{y}\,\|\cdot\|g(\mathbf{x}) - \mathbf{z}\,\|$$

$$< \alpha\left(\frac{\epsilon}{2\alpha}\right) + \|\mathbf{y}\|\cdot\left[\frac{\epsilon}{2(\|\mathbf{y}\| + 1)}\right] < \frac{\epsilon}{2} + \frac{\epsilon}{2} = \epsilon. \quad \square$$

Before leaving the topic of limits of functions, I should mention that, in the special case in which $f\colon X \to \mathbf{R}^n$, with $X \subseteq \mathbf{R}$, we will sometimes wish to consider 'one-sided' limits, as follows. First, if $x \in X$ is such that, for some $\eta > 0$, x is a limit point of $X \cap [x, x+\eta\,[$, we define the **limit of f as y approaches x from above**, denoted by '$\lim_{y\to x^+} f(y)$' or '$\lim_{y\searrow x} f(y)$,' by, effectively, treating the domain of f as if it were the set Y defined by

$$Y = X \cap\,]x, x+\eta\,[;$$

that is,

$$\mathbf{z} = \lim_{y\to x^+} f(y) = \lim_{y\searrow x} f(y)$$

if, and only if, for each $\epsilon > 0$, there exists $\delta > 0$ such that:

$$(\forall y \in N(x, \delta) \cap Y)\colon \|f(y) - \mathbf{z}\| < \epsilon.$$

We define the **limit of f as y approaches x from below**, denoted by '$\lim_{y \to x^-} f(y)$' or '$\lim_{y \nearrow x} f(y)$,' in a symmetric fashion. Details will be left to the student.

As I have already mentioned, the fact that arithmetic operations are preserved by limits of functions implies that continuity is preserved by such operations as well. The following theorem (corollary) is a viturally immediate consequence of Theorem 3.9.

3.10. Theorem. (Corollary) *Suppose $f\colon X \to \mathbf{R}^m$ and $g\colon X \to \mathbf{R}^m$, where X is a non-empty subset of $\mathbf{R}^n$. If f and g are both continuous at $\mathbf{x}^* \in X$, then we have:*

1. *$f + g$ is continuous,*
2. *$(\forall \alpha \in \mathbf{R})\colon \alpha f$ is continuous,*
3. *$f \cdot g$ is continuous,*

at $\mathbf{x}^*$.

Our next result, while dealing only with real-valued functions, will often be useful to us. The proof will be left as an exercise, since it can be done in very much the same way as we proved Theorem 2.13.

3.11. Theorem. *Suppose $f\colon X \to \mathbf{R}$ and $g\colon X \to \mathbf{R}$, where X is a non-empty subset of $\mathbf{R}^n$, and that both f and g are continuous at $\mathbf{x}^* \in X$. If $g(\mathbf{x}^*) \neq 0$, then the function*

$$h(\mathbf{x}) = f(\mathbf{x})/g(\mathbf{x}),$$

is well-defined in a neighborhood of $\mathbf{x}^*$, and is continuous at $\mathbf{x}^*$.

Another result which is extremely useful in establishing the continuity of complicated functions is the following.

3.12. Theorem. *Let $f\colon X \to \mathbf{R}^n$ and $g\colon Y \to \mathbf{R}^p$, where*

$$X \subseteq \mathbf{R}^m \quad \textit{and} \quad f(X) \subseteq Y \subseteq \mathbf{R}^n;$$

and suppose f is continuous at $\mathbf{x}^* \in X$, and that g is continuous at $\mathbf{y}^* \equiv f(\mathbf{x}^*) \in Y$. Then the composite function,[3] $h = g \circ f$, is continuous at $\mathbf{x}^*$.

---

[3]Recall that the composite function, h, is defined on X by

$$h(\mathbf{x}) = g[f(\mathbf{x})] \quad \text{for } \mathbf{x} \in X.$$

Proof. Let $\epsilon > 0$ be given. Then there exists $\eta > 0$ such that

$$(\forall \mathbf{y} \in N(\mathbf{y}^*, \eta) \cap Y) \colon \|g(\mathbf{y}) - g(\mathbf{y}^*)\| < \epsilon. \tag{3.4}$$

However, since f is continuous at $\mathbf{x}^*$, there exists $\delta > 0$ such that:

$$(\forall \mathbf{x} \in N(\mathbf{x}^*, \delta) \cap X) \colon \|f(\mathbf{x}) - f(\mathbf{x}^*)\| < \eta. \tag{3.5}$$

Combining (3.4) and (3.5), and recalling that $f(X) \subseteq Y$, we see that:

$$(\forall \mathbf{x} \in N(\mathbf{x}^*, \delta) \cap X) : \|h(\mathbf{x}) - h(\mathbf{x}^*)\| \equiv \Big\| g[f(\mathbf{x})] - g[f(\mathbf{x}^*)] \Big\| < \epsilon. \quad \square$$

The following set of examples demonstrates the way in which Theorems 3.10 – 3.12 can be utilized to build up the stock of functions which we know to be continuous. You can usefully employ the results of these examples in Exercises 4 and 5, at the end of this section.

3.13. Examples/Exercises.

1. If we define $f \colon \mathbf{R}^n \to \mathbf{R}^n$ by:

$$f(\mathbf{x}) = \mathbf{x} \quad \text{for } \mathbf{x} \in \mathbf{R}^n,$$

then f is continuous on $\mathbf{R}^n$.

2. If we define $f \colon \mathbf{R}^n \to \mathbf{R}$ by:

$$f(\mathbf{x}) = \mathbf{b} \cdot \mathbf{x},$$

where $\mathbf{b} \in \mathbf{R}^n$ is a fixed vector, then f is continuous on $\mathbf{R}^n$.

3. If $n \in N$, and we define $f \colon \mathbf{R} \to \mathbf{R}$ by:

$$f(x) = x^n \quad \text{for } x \in \mathbf{R},$$

then f is continuous on $\mathbf{R}$. [This is easy to prove by an induction argument, and making use of Theorem 3.10.]

4. Suppose k is a positive integer, and that $f_i \colon X \to \mathbf{R}^m$ is continuous, for $i = 1, \ldots, k$, where X is a non-empty subset of $\mathbf{R}^n$. Then, if $a_i \in \mathbf{R}$, for $i = 1, \ldots, k$, the function f defined on X by:

$$f(\mathbf{x}) = \sum\nolimits_{i=1}^{k} a_i f_i(\mathbf{x}),$$

is continuous on X.

5. The function $f \colon \mathbf{R} \to \mathbf{R}$ defined by:

$$f(x) = \sum\nolimits_{q=0}^{k} a_q x^q,$$

where $k \in N$ and $a_q \in \mathbf{R}$ for $q = 0, 1, \ldots, k$ is continuous on $\mathbf{R}$.

6. Suppose $f\colon X_1 \to \mathbf{R}^m$ is continuous, where X_1 is a non-empty subset of $\mathbf{R}^n$, let X_2 be a subset of $\mathbf{R}^p$, and define the subset X of $\mathbf{R}^{n+p}$ and $F\colon X \to \mathbf{R}^m$ by:

$$X = X_1 \times X_2 \quad \text{and} \quad F(x_1, x_2) = f(x_1),$$

respectively. Then F is continuous on X. [Notice that we can consider F to be the composition of the function $g(x_1, x_2) = x_1$ with f.]

7. Suppose $f\colon X \to \mathbf{R}^n$ and $g\colon X \to \mathbf{R}^p$ are both continous on X, where X is a subset of $\mathbf{R}^m$. Show that the function $F\colon X \to \mathbf{R}^{n+p}$ defined by:

$$F(\mathbf{x}) = (f(\mathbf{x}), g(\mathbf{x})) \quad \text{for } \mathbf{x} \in X,$$

is continuous on X.

8. Suppose $f_i\colon X_i \to \mathbf{R}^m$, where $X_i \subseteq \mathbf{R}^{n_i}$, is continuous, for $i = 1, 2$; and define f on $X = X_1 \times X_2$ by

$$f(\mathbf{x}) = f(\mathbf{x}_1, \mathbf{x}_2) = f_1(\mathbf{x}_1) \cdot f_2(\mathbf{x}_2).$$

Then f is continuous on X. □

We can also characterize the continuity of functions in terms of the behavior of sequences, as is shown by the following result.

3.14. Proposition. *Suppose $f\colon X \to \mathbf{R}^m$, where X is a non-empty subset of $\mathbf{R}^n$, and let $\mathbf{x}^* \in X$. Then f is continuous at $\mathbf{x}^*$ if, and only if, for every sequence, $\langle \mathbf{x}_q \rangle \subseteq X$, we have:*

$$\mathbf{x}_q \to \mathbf{x}^* \Rightarrow f(\mathbf{x}_q) \to f(\mathbf{x}^*). \tag{3.6}$$

Proof. It is an immediate consequence of Proposition 3.6 that if f is continuous at $\mathbf{x}^*$, then equation (3.6) holds. To prove the converse, suppose f is *not* continuous at $\mathbf{x}^*$. Then, using the definition of continuity, there exists $\epsilon^* > 0$ such that:

$$(\forall \delta \in \mathbf{R}_{++})(\exists \mathbf{x}^\delta \in N(\mathbf{x}^*, \delta) \cap X)\colon \|f(\mathbf{x}^\delta) - f(\mathbf{x}^*)\| \geq \epsilon^*.$$

But then it follows that there exists a sequence, $\langle \mathbf{x}_q \rangle$ in X such that:

$$\mathbf{x}_q \in N(\mathbf{x}^*, 1/q) \cap X \quad \text{and} \quad \|f(\mathbf{x}_q) - f(\mathbf{x}^*)\| \geq \epsilon^* \quad \text{for } q = 1, 2, \ldots;$$

and thus:

$$\mathbf{x}_q \to \mathbf{x}^*, \text{ but } f(\mathbf{x}_q) \nrightarrow f(\mathbf{x}^*).$$

Consequently, we see that if f is not continuous at $\mathbf{x}^*$, then (3.6) does not hold; and it follows that if (3.6) *does* hold, then f is continuous at $\mathbf{x}^*$. □

We often give a verbal statement of continuity something like the following: "... if a function is continuous on its domain, and if the independent variable is very close to some point, $\mathbf{x}^*$, in the domain, then we know that the value of the function is very close to its value at $\mathbf{x}^*$ [that is, to $f(\mathbf{x}^*)$]." That this statement is at best misleading (and probably should be said to be just plain incorrect) is show by Exercise 1 at the end of this section. Consequently, the verbal statement is more correctly given as something like "... given any degree of closeness to $f(\mathbf{x}^*)$ which is desired, we can determine how close $\mathbf{x}$ needs to be to $\mathbf{x}^*$ in order to insure that $f(\mathbf{x})$ is as close as desired to $f(\mathbf{x}^*)$." Notice, however, that how close we need to be to $\mathbf{x}^*$ may be a function of $\mathbf{x}^*$ itself. Thus, consider the following example.

3.15. Example. Consider the function, $f\colon \mathbf{R}_{++} \to \mathbf{R}_{++}$ defined by:

$$f(x) = 1/x \quad \text{for } x \in \mathbf{R}_{++}.$$

Notice, first of all, that it follows at once from Theorem 3.11 and Example 3.13.1 that f is continuous on $\mathbf{R}_{++}$. Let $x^* = 1$, let $\epsilon = 1$, and let's see if we can find a value of δ which will allow us to verify the continuity of f at x^* via our ϵ-δ definition of continuity. If we take $\delta = 1/2$, then for any $x \in \mathbf{R}$ satisfying

$$|x - x^*| = |x - 1| < \delta = 1/2,$$

we have $1/2 < x < 3/2$, so that

$$f(3/2) = 2/3 < f(x) < f(1/2) = 2;$$

and therefore

$$|f(x) - f(x^*)| = |f(x) - 1| < 1 \equiv \epsilon.$$

On the other hand, if we instead take $x^* = 1/2$, then you should have no difficulty in finding a value of x' such that, for the same values of ϵ and δ,

$$|x' - x^*| < \delta \text{ but } |f(x') - f(x^*)| > \epsilon^*.$$

In fact, we can expand upon the idea of the choice of x' in the above paragraph in the following way. Let a positive value of δ^* be chosen as small as you please, but in particular, small enough so that $0 < \delta^* < 1$; and let η be as large a number as you please, but in particular, large enough so that:

$$\eta \cdot \delta^* > 4. \tag{3.7}$$

Given (3.7), we see that if we define x^* by

$$x^* = \delta^*/(\eta \cdot \delta^* - 2),$$

then we will have $x^* > 0$. Furthermore, it also follows easily from (3.7) that

$$3(\eta \cdot \delta^* - 2) > 4,$$

so that

$$\frac{\delta^*}{2} > \frac{2\delta^*}{3(\eta \cdot \delta^* - 2)} = \frac{2x^*}{3}. \tag{3.8}$$

Consequently, if we define x' by:

$$x' = \frac{x^*}{3} = \frac{\delta^*}{3(\eta \cdot \delta^* - 2)}, \tag{3.9}$$

it is clear, using (3.8) and the definition of x', that:

$$0 < x' < x^* \quad \text{and} \quad \delta^*/2 > x^* - x'.$$

However, we also have:

$$|f(x') - f(x^*)| = \frac{1}{x'} - \frac{1}{x^*} = \frac{2}{x^*} = \frac{2\eta \cdot \delta^* - 4}{\delta^*} > \frac{\eta \cdot \delta^*}{\delta^*} = \eta,$$

where the last inequality is by (3.7) □

In the example, we have just seen that the size of the allowable deviation in the value of the argument which is commensurate with a given deviation in the value of a continuous function may be highly dependent upon the starting point in the domain. However, we shall see that if the function is *uniformly continuous*, things are quite different.

3.16. Definition. Let $f\colon X \to \mathbf{R}^m$, where X is a non-empty subset of $\mathbf{R}^n$. We shall say that f is **uniformly continuous on X** iff, given any $\epsilon > 0$, there exists $\delta > 0$ such that for all $\mathbf{x}^*, \mathbf{x}' \in X$, we have:

$$d(\mathbf{x}', \mathbf{x}^*) < \delta \Rightarrow \|f(\mathbf{x}') - f(\mathbf{x}^*)\| < \epsilon.$$

In our discussion in Example 3.15, we showed that the function f defined there was *not* uniformly continuous (even though it was continuous). On the other hand, it should be obvious that any constant function on $\mathbf{R}^n$ and mapping into $\mathbf{R}^m$, is uniformly continuous. We can develop a bit more interesting class of functions which is uniformly continuous, as follows.

Let $\mathbf{a} \in \mathbf{R}^n$ be a (fixed) vector, and define $f\colon \mathbf{R}^n \to \mathbf{R}$ by:

$$f(\mathbf{x}) = \mathbf{a} \cdot \mathbf{x} \quad \text{for } \mathbf{x} \in \mathbf{R}^n.$$

If $\mathbf{a} = \mathbf{0}$, then the function is identically zero; and thus is obviously uniformly continuous. Suppose, therefore, that $\mathbf{a}$ is non-zero, let $\epsilon > 0$ be given, and define δ by:

$$\delta = \epsilon/\|\mathbf{a}\|.$$

Then, given that $\mathbf{x}'$ and $\mathbf{x}^*$ satisfy

$$d(\mathbf{x}', \mathbf{x}^*) = \|\mathbf{x}' - \mathbf{x}^*\| < \delta,$$

we have:

$$|f(\mathbf{x}') - f(\mathbf{x}^*)| = |\mathbf{a} \cdot \mathbf{x}' - \mathbf{a} \cdot \mathbf{x}^*| \leq \|\mathbf{a}\| \cdot \|\mathbf{x}' - \mathbf{x}^*\| < \|\mathbf{a}\| \cdot (\epsilon/\|\mathbf{a}\|) = \epsilon$$

[where the first inequality is by the Cauchy-Schwarz inequality (Theorem 1.46)]. Thus, recalling Proposition 1.82, we see that we have just established that any real-valued linear function on $\mathbf{R}^n$ is uniformly continuous.

Exercises.

1. Let x^* be a positive real number, say $x^* = 1$ for definiteness, and let δ and ϵ be any positive real numbers [which can be as 'small' and as 'large,' respectively, as we wish]. Show that there exists a continuous function, $f\colon \mathbf{R}_+ \to \mathbf{R}$, and a point $x' \in \mathbf{R}_+$ such that:

$$|x^* - x'| < \delta \quad \text{and} \quad f(x^*) - f(x') > \epsilon.$$

[Hint. Try to find a linear function satisfying the desired inequalities.]

2. Suppose $f\colon X \to \mathbf{R}$ is continuous at $x^* \in X$, where X is a subset of $\mathbf{R}^n$, and that $f(x^*) \neq 0$. Show that there exists $\delta > 0$ such that:

$$(\forall x \in N(x^*, \delta) \cap X)\colon f(x) \cdot f(x^*) > 0.$$

Can you generalize this result to the case in which $f\colon X \to \mathbf{R}^n$?

3. Show that $f\colon X \to \mathbf{R}^m$ is uniformly continuous on X if, and only if, each coordinate function, $f_i(\cdot)\,[i = 1, \ldots, m]$, is uniformly continuous on X.

Note: In the next two exercises (the examples are from Moore [1972a]), you may take it as given that the function $g(y) = \sqrt{y}$ is continuous on $\mathbf{R}_+$.

4. Show that the following function is continuous on $X = \mathbf{R}^2_+$:

$$f(x) = (1/2) \cdot \Big((x_1 + x_2 + 1) + [(x_2 - 1)^2 + x_1 \cdot (x_1 + 2x_2 + 2)]^{1/2}\Big).$$

5. Show that the function f is continuous on X, where

$$X = \{\mathbf{x} \in \mathbf{R}^2_+ \mid x_1 + x_2 \geq 1\},$$

and f is defined on X by:

$$f(\mathbf{x}) = (1/2) \cdot \Big([x_1 + 3x_2] + [(x_1 + 3x_2)^2 - 8x_2]^{1/2}\Big).$$

6. [In this problem, we shall denote points in $\mathbf{R}^2$ by '(x, y)' rather that '(x_1, x_2).'] Consider the following functions, all of which are defined on $\mathbf{R}^2_+ \setminus \{\mathbf{0}\}$. In each case, determine whether or not $\lim_{(x,y)\to(0,0)} f(x, y)$ exists; and if it exists, find the limit.

a. $f(x,y) = \dfrac{x^2 - y^2}{x^2 + y^2}$, b. $f(x,y) = \dfrac{xy}{x^2 + (x-y)^2}$, c. $f(x,y) = \dfrac{x^2 + y^2}{|x| + |y|}$.

3.2 Continuity and Compactness

In this section we will begin by studying a different way of characterizing continuous functions. We will frequently find such a characterization to be useful in economic analysis, and it is of fundamental importance in generalizing the notion of continuous functions to abstract topological spaces, as we shall find in Chapter 7. The new characterization hinges around the notion of the inverse image of a set under a function, a definition which we have already presented (1.4.2), but which it will be convenient to repeat here.

3.17. Definition. Let $f\colon X \to \mathbf{R}^m$, where $X \subseteq \mathbf{R}^n$. Then, given any subset, $A \subseteq \mathbf{R}^m$, we define $f^{-1}(A)$, the **inverse image of A (under f)**, or **f-inverse of A**, by:

$$f^{-1}(A) = \{x \in X \mid f(x) \in A\}.$$

The following result deals with some purely set-theoretic properties of the inverse image which we shall often find useful in this and later chapters.

3.18. Proposition. *Let $f\colon X \to Y$. Then:*
1. *for any $A \subseteq Y, f^{-1}(Y \setminus A) = X \setminus f^{-1}(A)$;*
2. *for any family of subsets of $Y, \{B_a \mid a \in A\}$, we have:*

$$f^{-1}\Big(\bigcup_{a\in A} B_a\Big) = \bigcup_{a\in A} f^{-1}(B_a), \tag{3.10}$$

and

$$f^{-1}\Big(\bigcap_{a\in A} B_a\Big) = \bigcap_{a\in A} f^{-1}(B_a). \tag{3.11}$$

Proof. To prove part 1, let A be a subset of Y, and let $x \in X \setminus f^{-1}(A)$. Then $f(x) \notin A$, so that $f(x) \in Y \setminus A$, and thus:

$$x \in f^{-1}(Y \setminus A).$$

Conversely, suppose $x \in f^{-1}(Y \setminus A)$. Then, by definition of the inverse image, $f(x) \in Y \setminus A$, that is, $f(x) \notin A$, and thus:

$$x \in X \setminus f^{-1}(A).$$

To prove part 2, suppose $x \in \bigcup_{a\in A} f^{-1}(B_a)$. Then there exists $a' \in A$ such that $x \in f^{-1}(B_{a'})$. But then, $f(x) \in B_{a'}$, so that

$$f(x) \in \bigcup_{a\in A} B_a;$$

and thus:

$$x \in f^{-1}\Big(\bigcup_{a\in A} B_a\Big).$$

Conversely, suppose $x \in f^{-1}\left(\bigcup_{a \in A} B_a\right)$. Then:

$$(\exists \alpha \in A)\colon f(x) \in B_\alpha.$$

But then:

$$x \in f^{-1}(B_\alpha) \subseteq \bigcup_{a \in A} f^{-1}(B_a).$$

The proof of (3.11) will be left as an exercise. □

Notice that the above proposition can be interpreted as stating that the inverse-image operator preserves set-theoretic operations (complement, union, and intersection).

3.19. Lemma. *Let $f\colon X \to \mathbf{R}^n$, where X is a subset of $\mathbf{R}^m$. Then f is continuous on X if, and only if, for every open set U in $\mathbf{R}^n$, we have:*

$$f^{-1}(U) \textit{ is open relative to } X.$$

Proof. Suppose first that f is continuous on X, let U be an open subset of $\mathbf{R}^n$, and let $\mathbf{x}^*$ be an arbitrary element of $f^{-1}(U)$. Then $f(\mathbf{x}^*) \in U$; so that, since U is open (in $\mathbf{R}^n$), there exists $\epsilon > 0$ such that:

$$N[f(\mathbf{x}^*), \epsilon] \subseteq U. \tag{3.12}$$

However, since f is continuous on X, there exists $\delta > 0$ such that:

$$(\forall \mathbf{x} \in N(\mathbf{x}^*, \delta) \cap X)\colon \|f(\mathbf{x}) - f(\mathbf{x}^*)\| < \epsilon; \tag{3.13}$$

and we see from (3.12) and (3.13) that:

$$N(\mathbf{x}^*, \delta) \cap X \subseteq f^{-1}(U).$$

Since $\mathbf{x}^* \in f^{-1}(U)$ was arbitrary, it now follows from Theorem 1.72 that $f^{-1}(U)$ is open relative to X.

Now suppose that, for any open set, $U \subseteq \mathbf{R}^n$, $f^{-1}(U)$ is open relative to X; and let $\mathbf{x}^* \in X$ be arbitrary, and $\epsilon > 0$ be given. Then

$$V \equiv N[f(\mathbf{x}^*), \epsilon] \subseteq \mathbf{R}^n$$

is open, and therefore $f^{-1}(V)$ is open relative to X. But then, since $\mathbf{x}^* \in f^{-1}(V)$, it follows from Theorem 1.72 that there exists a positive number δ such that:

$$N(\mathbf{x}^*, \delta) \cap X \subseteq f^{-1}(V);$$

and it then follows from the definitions of V and $f^{-1}(V)$ that:

$$(\forall \mathbf{x} \in N(\mathbf{x}^*, \delta) \cap X)\colon \|f(\mathbf{x}) - f(\mathbf{x}^*)\| < \epsilon. \quad \square$$

3.20. Theorem. *Let $f\colon X \to \mathbf{R}^n$, where $X \subseteq \mathbf{R}^m$, and suppose $f(X) \subseteq Y \subseteq \mathbf{R}^n$. Then f is continuous on X if, and only if, for each subset of Y, U, which is open relative to Y, we have that $f^{-1}(U)$ is open relative to X.*

Proof. If $U_1 \subseteq Y$ is open relative to Y, there exists $U_2 \subseteq \mathbf{R}^n$, which is open in $\mathbf{R}^n$, and satisfies:

$$U_1 = Y \cap U_2.$$

Thus, using 3.18.2, we see that

$$f^{-1}(U_1) \equiv f^{-1}(Y \cap U_2) = f^{-1}(Y) \cap f^{-1}(U_2) = X \cap f^{-1}(U_2) = f^{-1}(U_2);$$

and our conclusion then follows from Lemma 3.19. □

3.21. Example. Suppose that a consumer's preference relation, P, is representable by a continuous utility function, $u\colon X \to \mathbf{R}$. Then, given any $\mathbf{x} \in X$, it follows from the definition of a utility function that:

$$P\mathbf{x} = \{\mathbf{y} \in X \mid \mathbf{y}P\mathbf{x}\} = \{\mathbf{y} \in X \mid u(\mathbf{y}) > u(\mathbf{x})\},$$

and

$$\mathbf{x}P = \{\mathbf{y} \in X \mid \mathbf{x}P\mathbf{y}\} = \{\mathbf{y} \in X \mid u(\mathbf{y}) < u(\mathbf{x})\}.$$

Thus it follows at once from Theorem 3.20 that for each $\mathbf{x} \in X$, the sets $P\mathbf{x}$ and $\mathbf{x}P$ must both be open relative to X. Consequently, we see that if P is representable by a continuous utility function, then P must be continuous, as defined in 1.78. □

While we have just characterized continuous functions as those functions having the property that the inverse image of every open set is open, continuous functions can equally well be categorized as those functions having the property that the inverse image of each closed set is closed. The proof of the following is an easy application of Theorem 3.20 and Proposition 3.18, and will be left as an exercise.

3.22. Theorem. (Corollary). *Let $f\colon X \to \mathbf{R}^m$, where X is a subset of $\mathbf{R}^n$, and let Y be such that $f(X) \subseteq Y \subseteq \mathbf{R}^m$. Then f is continuous on X if, and only if, for each set $C \subseteq Y$ which is closed relative to $Y, f^{-1}(C)$ is closed relative to X.*

Using the material just developed we can provide a particularly simple proof that the composition of two continuous functions must itself be continuous. However, to do so, we need to make use of the following lemma, the proof of which will be left as an exercise.

3.23. Lemma. *Let $f\colon X \to Y$ and $g\colon Y \to Z$, where X, Y, and Z are non-empty sets, and define the composite function, h, on X by:*

$$h(x) = g[f(x)] \quad \textit{for}\, x \in X.$$

Then, for any $W \subseteq Z$, we have:

$$h^{-1}(W) = f^{-1}[g^{-1}(W)].$$

3.24. Theorem. *Suppose $f\colon X \to \mathbf{R}^n$ and $g\colon Y \to \mathbf{R}^p$, where $X \subseteq \mathbf{R}^m$ and $f(X) \subseteq Y \subseteq \mathbf{R}^n$; and suppose that f is continuous on X, and that g is continuous on Y. Then the composite function, $h = g \circ f$, defined by:*

$$h(x) = g[f(x)] \quad \text{for } x \in X,$$

is continuous on X.

Proof. Let $U \subseteq \mathbf{R}^p$ be an arbitrary open set. Then, by Theorem 3.20, $g^{-1}(U)$ is open relative to Y. However, using Lemma 3.23 and Theorem 3.20 once again, we then see that:

$$h^{-1}(U) = f^{-1}[g^{-1}(U)]$$

is open relative to X; so that it follows from yet another use of Theorem 3.20 that h is continuous on X. □

In economic theory we are often interested in conditions under which we can be sure that a continuous real-valued function achieves a maximum value on some set. As you are probably already aware, a sufficient condition for the existence of such a maximum is that the set be compact. While the following definition is somewhat old-fashioned, it is equivalent to the more modern definition of compactness in the context of $\mathbf{R}^n$, and will provide us with a very simple proof that any continuous real-valued function achieves a maximum on a compact subset of its domain.

3.25. Definition. (*Compactness à la Bolzano-Weierstrass.*) A subset, A, of $\mathbf{R}^n$ is **compact** (or **sequentially compact**) iff, any sequence of points from A, $\langle \mathbf{x}_q \rangle$, has a subsequence which converges to a point in A.

In other words, A is compact if, and only if, given any sequence, $\langle \mathbf{x}_q \rangle$, of points from A, there exists an element, $\mathbf{x}^* \in A$, and a subsequence of $\langle \mathbf{x}_q \rangle$, $\langle \mathbf{x}_{q_i} \rangle$, such that:

$$\lim_{i \to \infty} \mathbf{x}_{q_i} = \mathbf{x}^*.$$

3.26. Theorem. *(Weierstrass). Let $f\colon X \to \mathbf{R}$, where X is a non-empty compact subset of $\mathbf{R}^n$, and suppose f is continuous on X. Then f takes on both a maximum and a minimum value on X; that is, there exist $\mathbf{y}, \mathbf{z} \in X$ such that:*

$$(\forall \mathbf{x} \in X)\colon f(\mathbf{y}) \leq f(\mathbf{x}) \leq f(\mathbf{z}).$$

Proof. We will only prove the existence of a maximum value; the proof that a minimum value exists can be done similarly, and will be left as an exercise.[4]

[4] This proof borrows heavily from Nikaido [1968].

Define

$$\alpha = \begin{cases} \sup f(X) & \text{if } f(X) \text{ is bounded,} \\ +\infty & \text{if } f(X) \text{ is unbounded.} \end{cases}$$

We wish to prove two things: (a) α must be finite, and (b) $f(\mathbf{x}^*) = \alpha$, for some $\mathbf{x}^* \in X$.

Now, whether or not α is finite, there exists a sequence of points from X, $\langle \mathbf{x}_q \rangle$, such that:

$$\lim_{q \to \infty} f(\mathbf{x}_q) = \alpha. \tag{3.14}$$

However, since X is compact, there exists a subsequence of $\langle \mathbf{x}_q \rangle$, $\langle \mathbf{x}_{q_i} \rangle$, and a point $\mathbf{x}^* \in X$, such that

$$\mathbf{x}_{q_i} \to \mathbf{x}^*. \tag{3.15}$$

But then it follows from (3.14) that

$$\lim_{i \to \infty} f(\mathbf{x}_{q_i}) = \alpha;$$

while by (3.15) and the fact that f is continuous on X,

$$\lim_{i \to \infty} f(\mathbf{x}_{q_i}) = f(\mathbf{x}^*).$$

Therefore, $f(\mathbf{x}^*) = \alpha$; which establishes both (a) and (b) from the previous paragraph. □

Our next result provides a very convenient characterization of the compact subsets of $\mathbf{R}^n$. Probably you are already familiar with its content.

3.27. Theorem. *A non-empty subset of* $\mathbf{R}^n$, X, *is compact if, and only if, it is both closed and bounded.*

Proof. Suppose first that X is compact. Then, since the function, f, defined on X by:

$$f(\mathbf{x}) = \|\mathbf{x}\|,$$

is continuous, it follows from Theorem 3.26 that $\|\mathbf{x}\|$ attains a maximum on X at, say $\bar{\mathbf{x}} \in X$. But then, defining $\epsilon = \|\bar{\mathbf{x}}\| + 1$, it follows easily that $X \subseteq N(\mathbf{0}, \epsilon)$; and thus that X is bounded.

Suppose now that $\mathbf{x}^* \in \overline{X}$. Then, as we have noted earlier, there exists a sequence, $\langle \mathbf{x}_q \rangle \subseteq X$, such that

$$\lim \mathbf{x}_q \to \mathbf{x}^*. \tag{3.16}$$

Moreover, since X is compact, it also follows that there exists a subsequence of $\langle \mathbf{x}_q \rangle$, $\langle \mathbf{x}_{q_i} \rangle$, and a point $\mathbf{x}^\dagger \in X$ such that:

$$\lim \mathbf{x}_{q_i} \to \mathbf{x}^\dagger. \tag{3.17}$$

However, since $\langle \mathbf{x}_{q_i} \rangle$ is a subsequence of $\langle \mathbf{x}_q \rangle$, it follows from (3.16) that $\mathbf{x}_{q_i} \to \mathbf{x}^*$; and we then have from (3.17) that $\mathbf{x}^* = \mathbf{x}^\dagger \in X$.

Suppose now that X is closed and bounded, and let $\langle \mathbf{x}_q \rangle$ be a sequence of points of X. Since $\langle \mathbf{x}_q \rangle$ is then bounded, it is an immediate consequence of Theorem 2.67 that there exists a subsequence of $\langle \mathbf{x}_q \rangle$, $\langle \mathbf{x}_{q_i} \rangle$, and a point $\mathbf{x}^* \in \mathbf{R}^n$, such that $\mathbf{x}_{q_i} \to \mathbf{x}^*$. However, since $\langle \mathbf{x}_{q_i} \rangle \subseteq X$, it then follows that $\mathbf{x}^* \in \overline{X}$; and hence, since X is closed, that $\mathbf{x}^* \in X$. □

It is important to notice that the closure involved in Theorem 3.27 is with respect to $\mathbf{R}^n$. We will return to this point shortly, but in the meantime, let's take a look at an economic application of some of these ideas.

3.28. Example. Let X be a consumption set (recall Example 1.27.1) which is closed and **bounded below**; that is, suppose X satisfies:

$$(\exists \mathbf{z} \in \mathbf{R}^n)(\forall \mathbf{x} \in X) \colon \mathbf{x} \geq \mathbf{z}.$$

Define

$$\Omega = \{(\mathbf{p}, w) \in \mathbf{R}^{n+1} \mid \mathbf{p} \gg \mathbf{0} \;\&\; (\exists \mathbf{x} \in X) \colon \mathbf{p} \cdot \mathbf{x} \leq w\},$$

and $b \colon \Omega \mapsto X$ by:[5]

$$b(\mathbf{p}, w) = \{\mathbf{x} \in X \mid \mathbf{p} \cdot \mathbf{x} \leq w\}.$$

Then:

$$(\forall (\mathbf{p}, w) \in \Omega) \colon b(\mathbf{p}, w) \text{ is compact and non-empty.} \tag{3.18}$$

Proof of equation (3.18). It is an immediate consequence of our definition of Ω that if $(\mathbf{p}, w) \in \Omega$, then $b(\mathbf{p}, w)$ is non-empty. To show that $b(\mathbf{p}, w)$ is compact, we begin by defining $\mathbf{y} \in \mathbf{R}^n$ by:

$$y_i = \frac{w - \sum_{\substack{j=1 \\ j \neq i}}^{n} p_j z_j}{p_i} \quad \text{for } i = 1, \ldots, n.$$

Now, let $b(\mathbf{p}, w)$, and suppose, by way of obtaining a contradiction, that for some $k \in \{1, \ldots, n\}$,

$$x_k > y_k. \tag{3.19}$$

Then we have:

$$\mathbf{p} \cdot \mathbf{x} = p_k x_k + \sum_{j \neq k} p_j x_j > p_k \cdot \left(\frac{w - \sum_{j \neq k} p_j z_j}{p_k} \right) + \sum_{j \neq k} p_j z_j = w,$$

[5] We use the notation '$b \colon \Omega \mapsto X$' to indicate that b is a correspondence with domain Ω and images which are subsets of X; in other words, $b \colon \Omega \to \mathcal{P}(X)$.

which contradicts the assumption that $\mathbf{x} \in b(\mathbf{p}, w)$. Thus we see that

$$b(\mathbf{p}, w) \subseteq X \cap C,$$

where C is the closed cube defined by:

$$C = \{\mathbf{x} \in \mathbf{R}^n \mid \mathbf{z} \leq \mathbf{x} \leq \mathbf{y}\}.$$

Therefore,[6] $b(\mathbf{p}, w)$ is bounded. Since it follows from Theorem 3.22 that the set

$$H = \{\mathbf{x} \in \mathbf{R}^n \mid \mathbf{p} \cdot \mathbf{x} \leq w\}$$

is closed,[7] and $b(\mathbf{p}, w) = X \cap H$, we also see that $b(\mathbf{p}, w)$ is closed (Theorem 1.62). Therefore, it follows from Theorem 3.27 that $b(\mathbf{p}, w)$ is compact. □

Now let G be a preference relation on X, and suppose that G is representable by a continuous real-valued utility function, $u\colon X \to \mathbf{R}$ (see Definition 1.31). Then it follows at once from the Weierstrass Theorem (Theorem 3.26) that for $(\mathbf{p}, w) \in \Omega$, the set

$$\begin{aligned} h(\mathbf{p}, w) &= \{\mathbf{x} \in b(\mathbf{p}, w) \mid (\forall \mathbf{y} \in b(\mathbf{p}, w))\colon \mathbf{x}G\mathbf{y}\} \\ &= \{\mathbf{x} \in b(\mathbf{p}, w) \mid (\forall \mathbf{y} \in b(\mathbf{p}, w))\colon u(\mathbf{x}) \geq u(\mathbf{y})\}, \end{aligned}$$

is non-empty. Thus we see that in this case, the consumer's demand correspondence, $h\colon \Omega \mapsto X$, is well-defined. In a later chapter, we will be considering generalizations of this result, as well as investigating the properties of $h(\cdot)$. □

In the previous section, we saw that a real-valued function may be continuous without being uniformly continuous. If, however, the domain of the function is compact, then the function is continuous if, and only if, it is uniformly continuous. This follows from Proposition 3.29, which is stated below. We will not attempt to prove 3.29 at this point, but we will prove a generalization of it in Chapter 8.

3.29. Proposition. *Suppose $f\colon X \to \mathbf{R}$, where X is a non-empty subset of $\mathbf{R}^n$. If a subset of X, Y, is compact, and if f is continuous on Y, then f is uniformly continuous on Y.*

Notice that Y can equal X in the above result. Less trivially, it should be noted that the set Y must be bounded and closed in $\mathbf{R}^n$; it may be closed in X (if X is not itself closed in $\mathbf{R}^n$) without being closed in $\mathbf{R}^n$. Thus, for example, if $X = \mathbf{R}_{++}$, the set $Y =]0, 1]$ is closed in X, and is bounded as well. However, Y is nonetheless *not* a compact set, since it is not closed in the underlying space ($\mathbf{R}$ in this case). Correspondingly, you can verify

[6]See Exercise 5 at the end of Section 1.4.
[7]Notice that $f(\mathbf{x}) \equiv \mathbf{p} \cdot \mathbf{x}$ is a continuous function.

that the function f defined in Example 3.15 of the previous section is *not* uniformly continuous on Y, *nor does it attain a maximum on* Y. Notice also that the sequence $\langle x_n \rangle$ defined by:

$$x_n = 1/n \quad \text{for } n = 1, 2, \dots ,$$

is a sequence of points from Y which converges to a point in $\mathbf{R}$, namely to 0; but 0 is nonetheless not an element of Y. This shows directly that Y is not compact, since we have shown that it does not satisfy Definition 3.25.

Exercises.

1. Suppose $f: X \to \mathbf{R}^n$, where X is a non-empty subset of $\mathbf{R}^m$, and let $\mathbf{a} \in \mathbf{R}^n$. Show that, if f is continuous, then the set Y defined by:

$$Y = \{\mathbf{x} \in X \mid f(\mathbf{x}) = \mathbf{a}\},$$

is closed in X.

2. Prove (a) Theorem 3.22, and (b) Lemma 3.23.

3. Show that the inner product function is **bi-continuous** on $\mathbf{R}^n$; that is, if $\mathbf{x}^*, \mathbf{y}^* \in \mathbf{R}^n$, and $\epsilon > 0$ is given, then there exists $\delta > 0$ such that, for all $\mathbf{x}, \mathbf{y} \in \mathbf{R}^n$,

$$\|\mathbf{x} - \mathbf{x}^*\| < \delta \;\&\; \|\mathbf{y} - \mathbf{y}^*\| < \delta \Rightarrow |\mathbf{x} \cdot \mathbf{y} - \mathbf{x}^* \cdot \mathbf{y}^*| < \epsilon.$$

4. Suppose $f: \mathbf{R}^m \to \mathbf{R}^n$, and define the set $S(r)$ by

$$S(r) = \{\mathbf{x} \in \mathbf{R}^m \mid \|\mathbf{x}\| = r\} \quad \text{for } r \in \mathbf{R}_+.$$

a. Show that if f is continuous, then, for each $r \in \mathbf{R}_+$,

$$g(r) \equiv \max\{\|f(\mathbf{x})\| \mid \mathbf{x} \in S(r)\},$$

exists.

b. Using the notation and definitions of part a of this problem, suppose that f is real-valued ($n = 1$) and linear. Can you find the value of $g(r)$, for each $r \in \mathbf{R}_+$? [Hint: Begin by considering the case in which $r = 1$, and think about using Proposition 1.82 and the Cauchy-Schwarz inequality.]

5. Suppose $f: \mathbf{R}_+ \to \mathbf{R}$ is continuous, and define $g: \mathbf{R}_+ \to \mathbf{R}$ by:

$$g(y) = \max\{f(x) \mid x \in [0, y]\} \quad \text{for } y \in \mathbf{R}_+.$$

Show that g is continuous on $\mathbf{R}_+$.

6. Prove that if $f: X \to \mathbf{R}^n$ and $g: X \to \mathbf{R}^n$, where X is a non-empty subset of $\mathbf{R}^m$, f and g are uniformly continuous on X, and if $\alpha \in \mathbf{R}$, then the functions (a) αf, and (b) $f + g$, are also uniformly continuous on X.

7. Show that if $f: X \to \mathbf{R}^n$, where X is a non-empty bounded subset of $\mathbf{R}^m$, and f is uniformly continuous on X, then f is bounded on X.

3.3 Semi-Continuous Functions

In key portions of microeconomic theory it is possible to replace the assumption that behavioral functions appearing in the theory are continuous with the weaker assumption that they are only semi-continuous. This not only allows us to gain generality in a mathematical sense, in many applications to economics it allows us to significantly extend the theory in a methodological or behavioral sense as well. We will be looking at some of these applications later on in this section. In the meantime, we define upper- and lower semi-continuity for real-valued functions as follows.

3.30. Definition. Let $f\colon X \to \mathbf{R}$, where X is a non-empty subset of $\mathbf{R}^n$. We shall say that f is **upper semi-continuous** (respectively, **lower semi-continuous) at** $\mathbf{x}^* \in X$ iff the following holds: for each $\epsilon > 0$, there exists $\delta > 0$ such that:

$$(\forall \mathbf{x} \in N(\mathbf{x}^*, \delta) \cap X)\colon f(\mathbf{x}) < f(\mathbf{x}^*) + \epsilon \text{ [respectively, } f(\mathbf{x}) > f(\mathbf{x}^*) - \epsilon\text{]}.$$

We shall say that f is **upper semi-continuous** (respectively, **lower semi-continous**) on X iff f is upper semi-continuous (respectively, lower semi-continuous) at each $\mathbf{x} \in X$.

Loosely speaking, a function is upper semi-continuous if it has no 'upward jumps at a point,' and is lower semi-continuous if it has no 'downward jumps at a point.' The phrases in quotation marks have to be interpreted a bit carefully, however, for the functions in Figures 3.1.a and 3.1.b, below, are upper (but not lower) semi-continuous, while the functions graphed in Figures 3.2.a and 3.2.b are lower (but not upper) semi-continuous. In particular, in each of Figures 3.1.a and 3.1.b, the function is upper, but not lower, semi-continuous at $\mathbf{x}^*$; while in each of Figures 3.2.a and 3.2.b, the function is lower, but not upper, semi-continuous at $\mathbf{x}^*$.

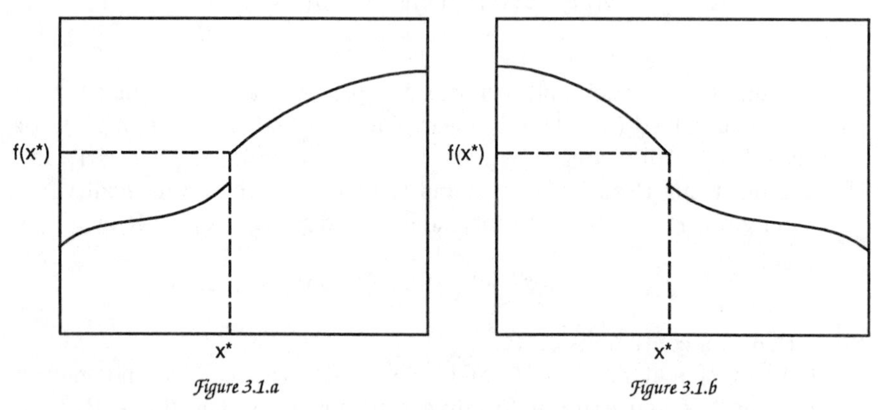

Figure 3.1: Upper Semi-Continuous Functions.

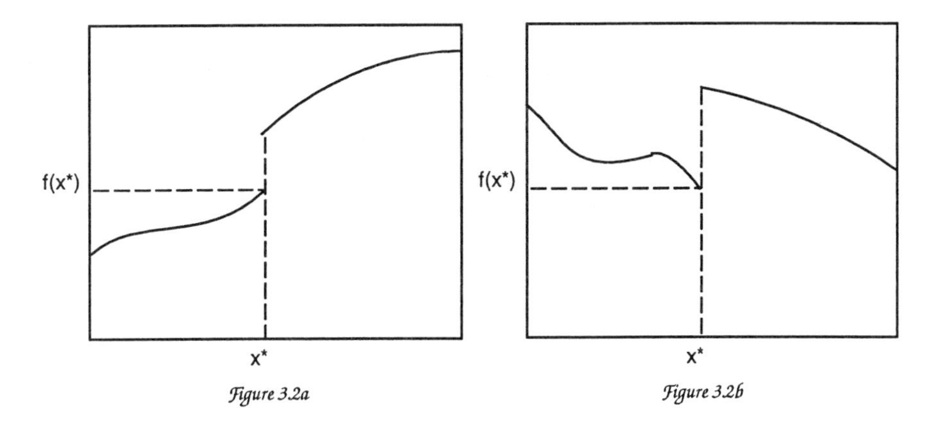

Figure 3.2: Lower Semi-Continuous Functions.

3.31. Examples/Exercises.

1. Let X be the unit interval in $\mathbf{R}$, that is, $X = [0, 1]$, and let f be defined on X by:

$$f(x) = \begin{cases} 0 & \text{for } 0 \leq x < 1, \text{ and} \\ 1 & \text{for } x = 1. \end{cases}$$

Show that f is upper, but not lower semi-continuous at $x = 1$.

2. A definition which one encounters frequently in the mathematical literature, but of which we shall make little use in this book is the following.

Let X be a subset of $\mathbf{R}$. A function $f\colon X \to \mathbf{R}$ is said to be **continuous from the left** (respectively, **right**) at a point $x^* \in X$, iff, for every $\epsilon > 0$, there exists $\delta > 0$ such that:

$$(\forall x \in N(x^*, \delta))\colon x < x^* \text{ [respectively, } x > x^*] \Rightarrow |f(x) - f(x^*)| < \epsilon.$$

Show that the function f defined on the unit interval by:

$$f(x) = \begin{cases} x & \text{for } 0 \leq x \leq 1/2, \text{ and} \\ x + 1 & \text{for } 1/2 < x \leq 1, \end{cases}$$

is lower semi-continuous and continuous from the left at $x = 1/2$, but is neither upper semi-continuous not continuous from the right at that point. On the other hand, the function g defined on X by:

$$g(x) = \begin{cases} x & \text{for } 0 \leq x \leq 1/2, \text{ and} \\ x - 1 & \text{for } 1/2 < x \leq 1, \end{cases}$$

is both upper semi-continuous and continuous from the left at $x = 1/2$, but is neither lower semi-continuous nor continuous from the right at that point.

3. It is common in the economic theory of the firm to assume that there exists a function, $\varphi\colon \mathbf{R}^n_+ \to \mathbf{R}_+$ (a **production function**) which describes

the firm's technology in the sense that, for each vector of input quantities, $\mathbf{v} \in \mathbf{R}^n_+, \varphi(\mathbf{v})$ is the maximum quantity of output the firm can produce, given these input quantities. It is also usual to assume that φ satisfies some sort of continuity condition. I would like to take a moment to consider what sort of continuity condition might be appropriate to use here, and why.

Generally speaking, continuity is *not* an empirically meaningful condition. This is because, if we are investigating an actual production process, the best we can do is to observe a *finite* collection of input vectors, $\{\mathbf{v}_1, \ldots, \mathbf{v}_T\}$, together with the associated set of output values, $\{x_1, \ldots, x_T\}$. If these observations are consistent with the hypothesis that there exists a function $\varphi\colon \mathbf{R}^n_+ \to \mathbf{R}_+$ such that:

$$x_t = \varphi(\mathbf{v}_t) \quad \text{for } t = 1, \ldots, T;$$

then it can be shown that *they are also consistent with the additional assumption that φ is continuous.* Because of this, one is tempted to conclude that continuity is an (empirically) 'free' assumption, and to think no more about it.

There is, however, a pitfall here, and it hinges upon the interpretation of the quantities involved. In economics we usually assume that the input and output quantities are both measured per unit of time, and thus are *flows*. If this is the case, then insofar as I can see, the additional assumption that φ is continuous is completely innocuous. However, there are situations in which we may want to interpret the quantity of output as a *stock*, rather than a flow. For example, in a given year, a farmer is interested primarily in the amount of corn he is able to harvest that year, which is a stock; a foundry may contract to produce a certain number of castings for a customer, which is not only a stock, but presumably is also necessarily an integer. In production processes in which output is naturally treated as a stock, it is very likely to be the case that only a finite, or at most a countable number of values for output are possible; in which case we say that the production function is **discrete-valued**. Here is where our difficulty regarding continuity comes to the fore, for while it is possible for a function $\varphi\colon \mathbf{R}^n_+ \to \mathbf{R}_+$ to be both upper semi-continuous and discrete-valued (as will be illustrated by our next example), in Chapter 8 we will see that *it is not possible for such a function to be both discrete-valued and continuous*! Consequently, generalizing our treatment of the theory of the firm by assuming that the production function is only upper semi-continuous, rather than continuous, is more than a mathematical nicety!

4. Define the function $N\colon \mathbf{R}_+ \to \mathbf{R}_+$ by (see Exercise 10 at the end of Section 1.2):

$$N(v) = \text{that unique integer, } n, \text{ satisfying } n \leq v < n + 1;$$

in other words, let $N(v)$ be the largest integer less than or equal to v. It is easy to show that $N(\cdot)$ is upper semi-continuous; in fact, if we let $a \in \mathbf{R}$ be any strictly positive constant, and define $f\colon \mathbf{R}_+ \to \mathbf{R}_+$ by:

$$f(v) = aN(v) \quad \text{for } v \in \mathbf{R}_+,$$

then it is easy to prove that f is upper semi-continuous. In Figure 3.3, below, I have graphed the function N. □

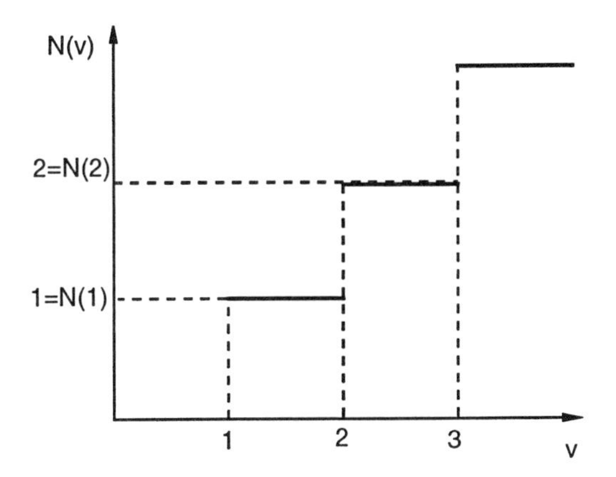

Figure 3.3: The 'Floor' Function.

Our first result characterizes upper semi-continuity of a function in a manner analogous to the way in which Theorem 3.20 characterized continuous functions. The three corollaries which follow the proof of Proposition 3.32 are nearly immediate consequences of that result, and their formal proofs will be left as exercises.

3.32. Proposition. *Let $f\colon X \to \mathbf{R}$, where X is a non-empty subset of $\mathbf{R}^n$. Then f is upper semi-continuous on X if, and only if, for each real number a, the set L_a given by:*

$$L_a = f^{-1}(]-\infty, a[) = \{\mathbf{x} \in X \mid f(\mathbf{x}) < a\}$$

is open relative to X.

Proof. Suppose first that f is upper semi-continuous, let $a \in \mathbf{R}$, and let $\mathbf{x}^* \in L_a$ be arbitrary. Then

$$f(\mathbf{x}^*) < a, \tag{3.20}$$

so that, if we define ϵ by:

$$\epsilon = [a - f(\mathbf{x}^*)]/2,$$

we see that $\epsilon > 0$. Since f is upper semi-continuous, there exists $\delta > 0$ such that:

$$(\forall \mathbf{x} \in N(\mathbf{x}^*, \delta) \cap X)\colon f(\mathbf{x}) < f(\mathbf{x}^*) + \epsilon = [a + f(\mathbf{x}^*)]/2. \tag{3.21}$$

However, by (3.20), we see that $[a + f(\mathbf{x}^*)]/2 < a$; and thus it follows from (3.21) that:

$$(\forall \mathbf{x} \in N(\mathbf{x}^*, \delta) \cap X)\colon f(\mathbf{x}) < a,$$

and thus:

$$N(\mathbf{x}^*, \delta) \cap X \subseteq L_a.$$

It follows that L_a is open relative to X.

Suppose now that, for each $a \in \mathbf{R}$, L_a is open relative to X, let $\mathbf{x}^* \in X$ be arbitrary, and let $\epsilon > 0$ be given. Defining

$$a = f(\mathbf{x}^*) + \epsilon, \tag{3.22}$$

we see that $\mathbf{x}^* \in L_a$; so that, since L_a is open relative to X, it follows that there exists $\delta > 0$ such that:

$$N(\mathbf{x}^*, \delta) \cap X \subseteq L_a.$$

From the definitions of L_a and a, we can then conclude that:

$$(\forall \mathbf{x} \in N(\mathbf{x}^*, \delta) \cap X)\colon f(\mathbf{x}) < a. \quad \square$$

3.33. Corollary. *Let $f\colon X \to \mathbf{R}$, where X is a nonempty subset of $\mathbf{R}^n$. Then f is upper semi-continuous on X if, and only if, for each $a \in \mathbf{R}$, the set $\overline{U}_a$ defined by:*

$$\overline{U}_a = f^{-1}([a, +\infty[) = \{\mathbf{x} \in X \mid f(\mathbf{x}) \geq a\},$$

is closed relative to X.

3.34. Corollary. *Let $f\colon X \to \mathbf{R}$, where X is a nonempty subset of $\mathbf{R}^n$. Then f is lower semi-continuous on X if, and only if, for each real number, a, the set U_a given by:*

$$U_a = f^{-1}(]a, +\infty[) = \{\mathbf{x} \in X \mid f(\mathbf{x}) > a\},$$

is open relative to X.

3.35. Corollary. *Let $f\colon X \to \mathbf{R}$, where X is a nonempty subset of $\mathbf{R}^n$. Then f is lower semi-continuous on X if, and only if, for each real number, a, the set*

$$\overline{L}_a = f^{-1}(] - \infty, a]) = \{\mathbf{x} \in X \mid f(\mathbf{x}) \leq a\},$$

is closed relative to X.

3.36. Example. We can generalize Example 3.21 as follows. Suppose once again that a consumer's preference relation, P, is representable by a utility function, $u\colon X \to \mathbf{R}$. Then it follows from the same basic considerations that were used in 3.21 that (using 3.32) if u is upper semi-continuous, then P must be upper semi-continuous, as defined in 1.78. Similarly, it follows from 3.34 that if u is lower semi-continuous, then P must be lower semi-continuous. These facts are, of course, the reason for the terminology introduced in Definition 1.78. □

In Section 5 we will look at more developments in economic theory which utilize upper and lower semi-continuous functions; in the meantime, let's look at some more of the properties of these functions. Upper and lower semi-continuity are preserved under most of the same arithmetic operations that preserve the continuity of functions. There are some exceptions to this general rule, however, so that it behooves us to be a bit careful in working with such functions. We begin by considering the sum of upper semi-continuous functions; the proof of the following result will be left as an exercise.

3.37. Proposition. *If $f\colon X \to \mathbf{R}$ and $g\colon X \to \mathbf{R}$ are both upper semi-continuous on $X \subseteq \mathbf{R}^n$, then the function $h = f + g$ is upper semi-continuous on X.*

The sum of an upper semi-continuous and a lower semi-continuous function is not necessarily either lower or upper semi-continuous, as is shown by the following example. Notice that this example also demonstrates that the difference of upper semi-continuous functions is not necessarily upper semi-continuous, and the same is true of lower semi-continuous functions.

3.38. Example. Let f and g be defined on $X \equiv [0,1] \subseteq \mathbf{R}$ by:

$$f(x) = \begin{cases} 1 & \text{for } 0 \leq x \leq 1/3, \\ 0 & \text{for } 1/3 < x \leq 1, \end{cases}$$

and

$$g(x) = \begin{cases} 0 & \text{for } 0 \leq x \leq 2/3, \\ 1 & \text{for } 2/3 < x \leq 1, \end{cases}$$

respectively. Then f is upper semi-continuous on X and g is lower semi-continuous on X, but $h = f + g$ is neither upper nor lower semi-continuous on X. □

3.39. Proposition. *Let X be a subset of $\mathbf{R}^n$, and suppose $f_i\colon X \to \mathbf{R}$ is upper semi-continuous at $\mathbf{x}^* \in X$ for $i = 1, \ldots, n$. Then the function $f\colon X \to \mathbf{R}$ defined by:*

$$f(\mathbf{x}) = \max_i f_i(\mathbf{x}) \quad \textit{for}\, \mathbf{x} \in X,$$

is upper semi-continuous at $\mathbf{x}^*$.

Proof. Let $\epsilon > 0$ be given. Since each f_i is upper semi-continuous at $\mathbf{x}^*$, it follows that for each i there exists $\delta_i > 0$ such that:

$$(\forall \mathbf{x} \in N(\mathbf{x}^*, \delta_i) \cap X) \colon f_i(\mathbf{x}) < f_i(\mathbf{x}^*) + \epsilon. \tag{3.23}$$

Define $\delta = \min\{\delta_1, \ldots, \delta_n\}$, and let $\mathbf{x} \in N(\mathbf{x}^*, \delta) \cap X$ be arbitrary. Letting $k \in \{1, \ldots, n\}$ be such that

$$f_k(\mathbf{x}) \geq f_i(\mathbf{x}) \quad \text{for } i = 1, \ldots, n,$$

we have by (3.23) and the definition of f, that:

$$f(\mathbf{x}) = f_k(\mathbf{x}) < f_k(\mathbf{x}^*) + \epsilon \leq f(\mathbf{x}^*) + \epsilon.$$

Since $\mathbf{x} \in N(\mathbf{x}^*, \delta) \cap X$ was arbitrary, it follows that f is upper semi-continuous at $\mathbf{x}^*$. □

3.40. Proposition. *Let X be a subset of $\mathbf{R}^n$, let A be a non-empty (index) set; and suppose that for each $a \in A$, $f_a \colon X \to \mathbf{R}$, and that, for some $\mathbf{x}^* \in X$, we have:*

1. *f_a is upper semi-continuous at $\mathbf{x}^*$, for each $a \in A$, and*
2. *$\inf_{a \in A} f_a(\mathbf{x}^*)$ is finite.*

Then the function $f \colon X \to \overline{\mathbf{R}}$ [where '$\overline{\mathbf{R}}$' denotes the extended real number system, $\overline{\mathbf{R}} \equiv \mathbf{R} \cup \{-\infty, +\infty\}$], defined by:

$$f(\mathbf{x}) = \inf_{a \in A} f_a(\mathbf{x}),$$

is upper semi-continuous at $\mathbf{x}^*$.

Proof. Let $\epsilon > 0$ be given. It follows from the definition of the infimum that there exists $b \in A$ such that:

$$f_b(\mathbf{x}^*) < f(\mathbf{x}^*) + \epsilon/2,$$

and, since f_b is upper semi-continuous at $\mathbf{x}^*$, there exists $\delta > 0$ such that:

$$(\forall \mathbf{x} \in N(\mathbf{x}^*, \delta) \cap X) \colon f_b(\mathbf{x}) < f_b(\mathbf{x}^*) + \epsilon/2.$$

But then we see that:

$$(\forall \mathbf{x} \in N(\mathbf{x}^*, \delta) \cap X) \colon f(\mathbf{x}) \leq f_b(\mathbf{x}) < f_b(\mathbf{x}^*) + \epsilon/2 < f(\mathbf{x}^*) + \epsilon. \quad \square$$

Notice that the above result allows the function f to be equal to $-\infty$ over part of its domain.[8] It should also be noted that, while 3.40 allows for an arbitrary family of functions over which the infimum is being taken, 3.39 may fail if there are even a countable number of functions involved; as is shown by the following example.

[8] In fact, the result remains correct even if we drop the condition that f is finite at $\mathbf{x}^*$; however, a proof of this last statement requires some new definitions regarding the continuity of extended real-valued functions, and is something which we shall not pursue further here.

3.41. Example. Let f_n be defined on $[0,1] \subseteq \mathbf{R}$ by:

$$f_n(x) = \begin{cases} 1 & \text{for } 0 \leq x \leq 1 - 1/n, \\ -nx + n & \text{for } 1 - 1/n < x \leq 1. \end{cases}$$

Obviously each f_n is upper semi-continuous (in fact, each is continuous) on X. However, defining g on X by:

$$g(x) = \sup_n f_n(x) \quad \text{for } x \in [0,1],$$

we have

$$g(x) = \begin{cases} 1 & \text{for } 0 \leq x < 1, \\ 0 & \text{for } x = 1. \end{cases}$$

Obviously g is not upper semi-continuous at $x = 1$. □

That the product of two upper semi-continuous functions is not necessarily upper semi-continuous is shown by the following example. If, however, both are positive, then their product *is* upper semi-continuous, as is established in 3.43, below.

3.42. Example. Let $X = \mathbf{R}_+$, and define f and g on X by:

$$f(x) = \begin{cases} -1/2 & \text{for } 0 \leq x \leq 1, \\ -1 & \text{for } x > 1, \end{cases}$$

and

$$g(x) = -1 \qquad \text{for } x \in X.$$

It is easy to show that both f and g are upper semi-continuous on X; however, it is equally easy to show that the product function, $h = f \cdot g$, is *not* upper semi-continuous at $x = 1$. □

3.43. Proposition. *Suppose $f\colon X \to \mathbf{R}_+$ and $g\colon X \to \mathbf{R}_+$ are both upper semi-continuous at a point $\mathbf{x}^* \in X$, where X is a subset of $\mathbf{R}^n$. Then the product function $h\colon X \to \mathbf{R}_+$, defined by:*

$$h(\mathbf{x}) = f(\mathbf{x}) \cdot g(\mathbf{x}) \quad \textit{for}\, \mathbf{x} \in X,$$

is also upper semi-continuous at $\mathbf{x}^*$.

Proof. Let $\epsilon > 0$ be given. Defining:

$$\eta = \min\{1, \epsilon/2[f(\mathbf{x}^*) + 1]\}, \tag{3.24}$$

we note that it follows from the upper semi-continuity of g at $\mathbf{x}^*$, that there exists $\delta_1 > 0$ such that:

$$(\forall \mathbf{x} \in N(\mathbf{x}^*, \delta_1) \cap X)\colon g(\mathbf{x}) < g(\mathbf{x}^*) + \eta. \tag{3.25}$$

Similarly, from the upper semi-continuity of f, there exists $\delta_2 > 0$ such that:

$$(\forall \mathbf{x} \in N(\mathbf{x}^*, \delta_2) \cap X)\colon f(\mathbf{x}) < f(\mathbf{x}^*) + \epsilon/2[g(\mathbf{x}^*) + 1]. \tag{3.26}$$

Defining $\delta = \min\{\delta_1, \delta_2\}$, we see that if $\mathbf{x} \in N(\mathbf{x}^*, \delta) \cap X)$, then it follows from (3.26) and the fact that $g(\mathbf{x})$ is always nonnegative that:

$$g(\mathbf{x}) \cdot [f(\mathbf{x}) - f(\mathbf{x}^*)] \leq g(\mathbf{x}) \cdot \left(\frac{\epsilon}{2[g(\mathbf{x}^*) + 1]}\right); \tag{3.27}$$

while by (3.24) and (3.25):

$$g(\mathbf{x}) \cdot \left(\frac{\epsilon}{2[g(\mathbf{x}^*) + 1]}\right) < [g(\mathbf{x}^*) + 1] \cdot \left(\frac{\epsilon}{2[g(\mathbf{x}^*) + 1]}\right) = \frac{\epsilon}{2}. \tag{3.28}$$

Similarly, by (3.24), (3.25), and the fact that $f(\mathbf{x}^*) \geq 0$:

$$f(\mathbf{x}^*) \cdot [g(\mathbf{x}) - g(\mathbf{x}^*)] \leq \left(\frac{f(\mathbf{x}^*)}{f(\mathbf{x}^*) + 1}\right) \cdot \left(\frac{\epsilon}{2}\right) < \frac{\epsilon}{2}. \tag{3.29}$$

Consequently, using (3.27)–(3.29), we see that if $\mathbf{x} \in N(\mathbf{x}^*, \delta) \cap X$, then:

$$\begin{aligned} h(\mathbf{x}) - h(\mathbf{x}^*) &= f(\mathbf{x}) \cdot g(\mathbf{x}) - f(\mathbf{x}^*) \cdot g(\mathbf{x}^*) \\ &= f(\mathbf{x}) \cdot g(\mathbf{x}) - f(\mathbf{x}^*) \cdot g(\mathbf{x}) + f(\mathbf{x}^*) \cdot g(\mathbf{x}) - f(\mathbf{x}^*) \cdot g(\mathbf{x}^*) \\ &= g(\mathbf{x}) \cdot [f(\mathbf{x}) - f(\mathbf{x}^*)] + f(\mathbf{x}^*) \cdot [g(\mathbf{x}) - g(\mathbf{x}^*)] < \epsilon/2 + \epsilon/2 = \epsilon; \end{aligned}$$

and thus:

$$h(\mathbf{x}) < h(\mathbf{x}^*) + \epsilon.$$

We conclude, therefore, that h is upper semi-continuous at $\mathbf{x}^*$. □

We know that the composition of two continuous functions is itself continuous. Unfortunately, the composition of two upper semi-continuous functions is not necessarily upper semi-continuous, as is shown by the following example. On the other hand, Proposition 3.45 provides a sufficient condition for the composition of two upper semi-continuous functions to be upper semi-continuous.

3.44. Example. Let $X = Y = [0, 1] \subseteq \mathbf{R}$, and let $f\colon X \to Y$ and $g\colon X \to Y$ be defined by:

$$f(x) = \begin{cases} 0 & \text{for } 0 \leq x < 1/2, \\ 1 & \text{for } 1/2 \leq x \leq 1, \end{cases}$$

and

$$g(y) = -y \quad \text{for } y \in Y,$$

respectively. Then f is upper semi-continuous, and g is continuous, but you should have no difficulty in showing that $h = g \circ f$ is *not* upper semi-continuous at $x = 1/2$. □

3.45. Proposition. *Suppose* $f\colon X \to Y$ *and* $g\colon Y \to \mathbf{R}$, *where* $X \subseteq \mathbf{R}^n$ *and* $Y \subseteq \mathbf{R}$, *and suppose* g *is non-decreasing on* Y. *If* f *is upper semi-continuous at* $\mathbf{x}^* \in X$, *and* g *is upper semi-continuous at* $f(\mathbf{x}^*) \in Y$, *then the composition function,* h, *defined on* X *by:*

$$h(\mathbf{x}) = g[f(\mathbf{x})],$$

is upper semi-continuous at $\mathbf{x}^*$.

Proof. Let $\epsilon > 0$ be given. Since g is non-decreasing and upper semi-continuous at $y^* \equiv f(\mathbf{x}^*)$, there exists $\eta > 0$ such that:

$$(\forall y \in Y)\colon [y < y^* + \eta] \Rightarrow [g(y) < g(y^*) + \epsilon]; \tag{3.30}$$

and, since f is upper semi-continuous at $\mathbf{x}^*$, there exists $\delta > 0$ such that:

$$(\forall \mathbf{x} \in N(\mathbf{x}^*, \delta) \cap X)\colon f(\mathbf{x}) < f(\mathbf{x}^*) + \eta. \tag{3.31}$$

Combining (3.30) and (3.31) with the definition of y^*, it then follows that:

$$(\forall \mathbf{x} \in N(\mathbf{x}^*, \delta) \cap X)\colon h(\mathbf{x}) = g[f(\mathbf{x})] < g[f(\mathbf{x}^*)] + \epsilon = h(\mathbf{x}^*) + \epsilon;$$

and we see that h is upper semi-continuous at $\mathbf{x}^*$. □

Exercises.

1. Prove that, for any positive real number, α, the function

$$f(v) = \alpha \cdot N(v),$$

defined in Example 3.33.4, is upper semi-continuous on $\mathbf{R}_+$.

2. Define the function n^* on $\mathbf{R}_+$ by:

$$n^*(y) = \text{that unique integer, n, satisfying: } n - 1 < y \leq n.$$

Show that, for any positive real number, β, the function g defined on $\mathbf{R}_+$ by:

$$g(y) = \beta \cdot n^*(y) \quad \text{for } y \in \mathbf{R}_+,$$

is lower, but not upper semi-continuous on $\mathbf{R}_+$. At what points in $\mathbf{R}_+$ is g not upper semi-continuous?

3. Show that if $f\colon X \to \mathbf{R}$ is upper semi-continuous at a point $\mathbf{x}^* \in X$, then there exists $\delta > 0$ such that f is bounded on $N(\mathbf{x}^*, \delta) \cap X$.

4. Define the function $f\colon [0,1] \to \{0,1\}$ by:

$$f(x) = \begin{cases} 1 & \text{if } x \text{ is a rational number, and} \\ 0 & \text{if } x \text{ is an irrational number.} \end{cases}$$

Are there points in the domain of f at which it is upper semi-continuous? Which ones? Are there points at which f is lower semi-continuous? Are there any points in the domain at which f is both upper and lower semi-continuous?

5. Prove the following (compare 3.45). Suppose $f: X \to Y$ and $g: Y \to \mathbf{R}$, where $X \subseteq \mathbf{R}^n$ and $Y \subseteq \mathbf{R}$, and suppose g is non-decreasing. If f is lower semi-continuous at $\mathbf{x} \in X$, and g is lower semi-continuous at $f(\mathbf{x}^*) \in Y$, then the composition function, $h = g \circ f$, is lower semi-continuous at $\mathbf{x}^*$.

6. It will probably come as no surprise to you that each of Propositions 3.39, 3.40, and 3.43 has an analogue for lower semi-continuous functions. It is a worthwhile exercise to state and prove all of these analogous results.

3.4 Limits Inferior and Superior of a Function *

In this section, we shall briefly discuss some more technical aspects of upper and lower semi-continuity; more specifically, the relationship between semi-continuity and some limit concepts which we shall define shortly. Those of you who have not previously studied semi-continuous functions may wish to simply skim through this section to get a general idea of what it is about for now, and use it for reference later.

We begin by defining $\overline{\mathbf{R}}$, the **extended real number system**, by:

$$\overline{\mathbf{R}} = \mathbf{R} \cup \{-\infty, +\infty\};$$

that is, we simply append to the real number system the two entities plus and minus infinity. One can define an arithmetic for the extended reals in a fairly obvious way; although be forewarned that there doesn't seem to be any appropriate definition of the sum of $-\infty$ and $+\infty$. However, our only concern at the moment is with the order properties of the extended reals, which are defined in an even more obvious way; namely $+\infty$ is greater than anything else in $\overline{\mathbf{R}}$, $-\infty$ is smaller than anything else, and we retain the usual ordering on $\mathbf{R}$. With this ordering on $\overline{\mathbf{R}}$, every non-empty subset of $\overline{\mathbf{R}}$ has both a supremum and an infimum in $\overline{\mathbf{R}}$. We use these ideas in the following definitions.

Suppose $f: X \to \mathbf{R}$, where X is a subset of $\mathbf{R}^n$, and let X' be the set of limit points of X (Definition 1.69). Then we can define the function $g: X' \times \mathbf{R}_{++} \to \overline{\mathbf{R}}$ by:

$$g(\mathbf{x}, \delta) = \sup_{\mathbf{y} \in N'(\mathbf{x}, \delta) \cap X} f(\mathbf{y}) \equiv \sup\{f(\mathbf{y}) \mid \mathbf{y} \in N'(\mathbf{x}, \delta) \cap X\}.$$

Now, for any $(\mathbf{x}, \delta) \in X' \times \mathbf{R}_{++}$, $g(\mathbf{x}, \delta)$ is well-defined, as an element of the extended real numbers. Moreover, we obviously have, for each $(\mathbf{x}, \delta) \in X' \times \mathbf{R}_{++}$,

$$(\forall \mathbf{y} \in N'(\mathbf{x}, \delta) \cap X): f(\mathbf{y}) \leq g(\mathbf{x}, \delta). \tag{3.32}$$

Given any $\mathbf{x} \in X'$, we can also consider:

$$\lim_{\delta \to 0+} g(\mathbf{x}, \delta). \tag{3.33}$$

Moreover, it is easy to see that, for any $\mathbf{x} \in X'$, $g(\mathbf{x}, \cdot)$ is a decreasing function of δ. Therefore, the limit in (3.33) will always exist (in the extended reals), and will be given by:

$$\lim_{\delta \to 0+} g(\mathbf{x}, \delta) = \inf_{\delta \in \mathbf{R}_{++}} g(\mathbf{x}, \delta). \tag{3.34}$$

This limit is called the **limit superior (lim sup)** of f at $\mathbf{x}^*$, and is formally defined in the following.

3.46. Definitions. Let $f : X \to \mathbf{R}$, where X is a subset of $\mathbf{R}^n$, and let $\mathbf{x}^*$ be a limit point of X. We define the **limit superior (lim sup)** and **limit inferior (lim inf) of f at $\mathbf{x}^*$**, which we denote by:

$$\text{`}\limsup_{\mathbf{x} \to \mathbf{x}^*} f(\mathbf{x})\text{' and `}\liminf_{\mathbf{x} \to \mathbf{x}^*} f(\mathbf{x}),\text{'}$$

respectively, by:

$$\limsup_{\mathbf{x} \to \mathbf{x}^*} f(\mathbf{x}) = \inf_{\delta > 0} \left[\sup\{f(\mathbf{x}) \mid \mathbf{x} \in N'(\mathbf{x}^*, \delta) \cap X\} \right] = \lim_{\delta \to 0+} g(\mathbf{x}, \delta) \tag{3.35}$$

and

$$\liminf_{\mathbf{x} \to \mathbf{x}^*} f(\mathbf{x}) = \sup_{\delta > 0} \left[\inf\{f(\mathbf{x}) \mid \mathbf{x} \in N'(\mathbf{x}^*, \delta) \cap X\} \right]. \tag{3.36}$$

Similar considerations to those preceding the statement of Definition 3.46 show that the limit inferior defined in (3.36) is well-defined, and that it is the limit as $\delta \to 0^+$ of the infimum appearing with the brackets in (3.36). The following examples will, hopefully, make these notions a bit clearer.

3.47. Examples/Exercises.

1. Let $X = [0, 1]$, and define $f : X \to \mathbf{R}$ by:

$$f(x) = \begin{cases} 0 & \text{for } 0 \leq x < 1/3 \text{ or } 2/3 \leq x \leq 1, \\ 1 & \text{for } 1/3 \leq x < 2/3. \end{cases}$$

Here it is easy to show that

$$\limsup_{x \to 1/3} f(x) = 1 = f(1/3), \quad \liminf_{x \to 1/3} f(x) = 0,$$

while

$$\limsup_{x \to 2/3} f(x) = 1, \quad \liminf_{x \to 2/3} f(x) = 0 = f(2/3);$$

and that f is upper semi-continuous at $x = 1/3$, and lower semi-continuous at $x = 2/3$. We shall demonstrate shortly that this is illustrative of the general situation; in the meantime, however, consider a second (and probably familiar) example.

2. Define the function, f, on $X = [0, 1]$ by:

$$f(x) = \begin{cases} 0 & \text{if } x \text{ is a rational number,} \\ 1 & \text{if } x \text{ is an irrational number.} \end{cases}$$

Here it is easy to see that, for each $x^* \in X$:

$$\limsup_{x \to x^*} f(x) = 1 \quad \text{and} \quad \liminf_{x \to x^*} f(x) = 0.$$

The question which I would like you to consider is this: Is f upper semi-continuous at any points $x^* \in X$? Is f lower semi-continuous anywhere on X?

3. In all of the examples considered thus far, both $\liminf f(x)$ and $\liminf f(x)$ have been finite. However, we can show that $\limsup$ can be equal to either positive or negative infinity, as follows.

a. We first consider the function $f\colon \mathbf{R}_+ \to \mathbf{R}_+$ defined by:

$$f(x) = \begin{cases} 1/x & \text{for } x > 0, \\ 0 & \text{for } x = 0. \end{cases}$$

Here it should be apparent that $g(0, \delta) = +\infty$ for all $\delta > 0$, and hence:

$$\lim_{x \to 0} f(x) = \limsup_{x \to 0} f(x) = \lim_{\delta \to 0+} g(0, \delta) = +\infty.$$

b. Now define $f\colon \mathbf{R}_+ \to \mathbf{R}$ by:

$$f(x) = \begin{cases} \log x & \text{for } x > 0, \\ 0 & \text{for } x = 0. \end{cases}$$

Here you should have no difficulty in showing that $g(0, \delta) = \log \delta$ for all $\delta > 0$; and thus that:

$$\limsup_{x \to 0} f(x) = -\infty.$$

4. Given a function $f\colon X \to \mathbf{R}$, define $h\colon X' \times \mathbf{R}_{++} \to \overline{\mathbf{R}}$ by:

$$h(\mathbf{x}, \delta) = \inf\{f(\mathbf{y}) \mid \mathbf{y} \in N'(\mathbf{x}, \delta) \cap X\}.$$

An argument similar to that presented in 3.47.3, above, shows that:

$$\lim_{\delta \to 0+} h(\mathbf{x}, \delta) = \liminf_{\mathbf{y} \to \mathbf{x}} f(\mathbf{y}).$$

It is also easy to show that, for any $(\mathbf{x}, \delta) \in X' \times \mathbf{R}_{++}$,

$$h(\mathbf{x}, \delta) \leq g(\mathbf{x}, \delta).$$

It is also clear that $h(\mathbf{x}, \cdot)$ is non-decreasing in δ, for each $\mathbf{x} \in X'$, and thus it is now easy to show that, for all $\mathbf{x} \in X'$:

$$\liminf_{\mathbf{y}\to\mathbf{x}} f(\mathbf{y}) \leq \limsup_{\mathbf{y}\to\mathbf{x}} f(\mathbf{y}). \quad \square$$

3.48. Theorem. *Let $f\colon X \to \mathbf{R}$, where X is a subset of $\mathbf{R}^n$, and let $\mathbf{x}^* \in X$ be a limit point of X. Then f is upper semi-continuous at $\mathbf{x}^*$ if, and only if, we have:*

$$f(\mathbf{x}^*) \geq \limsup_{\mathbf{x}\to\mathbf{x}^*} f(\mathbf{x}). \tag{3.37}$$

Proof.

1. In order to show that the upper semi-continuity of f implies (3.37), we shall establish the contrapositive. Accordingly, suppose (3.37) is *not* satisfied. Then, defining

$$z = \limsup_{\mathbf{x}\to\mathbf{x}^*} f(\mathbf{x}), \tag{3.38}$$

there exists $\epsilon^* > 0$ such that:

$$f(\mathbf{x}^*) + 2\epsilon^* < z. \tag{3.39}$$

We now distinguish two cases.

a. $z = +\infty$. In this case, it follows at once from the definition of lim sup that every neighborhood of $\mathbf{x}^*$ contains a point $\mathbf{x}$ such that, say:

$$f(\mathbf{x}) > f(\mathbf{x}^*) + 100,$$

and thus it is clear that f is not upper semi-continuous at $\mathbf{x}^*$ in this case.

b. $z \in \mathbf{R}$. Here it follows from the properties of the supremum that, given an arbitrary $\delta > 0$, there exists $\mathbf{x}_\delta \in N(\mathbf{x}^*, \delta) \cap X$ such that:

$$f(\mathbf{x}_\delta) > \sup\{f(\mathbf{x}) \mid \mathbf{x} \in N'(\mathbf{x}^*, \delta) \cap X\} - \epsilon^* \geq z - \epsilon^*$$

(where the second inequality is from the definition of the lim sup); and thus from (3.39),

$$f(\mathbf{x}_\delta) > f(\mathbf{x}^*) + \epsilon^*. \tag{3.40}$$

Since δ was an arbitrary positive number, it follows that every neighborhood of $\mathbf{x}^*$ contains a point satisfying (3.40), and therefore f is not upper semi-continuous at $\mathbf{x}^*$.

2. Suppose f satisfies (3.37), and let $\epsilon > 0$ be given. We distinguish two cases, as follows.

a. $\limsup_{\mathbf{x}\to\mathbf{x}^*} f(\mathbf{x}) = -\infty$. Since

$$\limsup_{\mathbf{x}\to\mathbf{x}^*} f(\mathbf{x}) = \lim_{\delta\to 0+} g(\mathbf{x}^*, \delta),$$

we see that there exists $\delta' > 0$ such that, for all δ such that $0 < \delta \leq \delta'$, we have:

$$\limsup_{\mathbf{x}\to\mathbf{x}^*} f(\mathbf{x}) = \sup\{f(\mathbf{x}) \mid \mathbf{x} \in N'(\mathbf{x}^*, \delta) \cap X\} < f(\mathbf{x}^*) + \epsilon.$$

But then, in particular, we have:

$$(\forall \mathbf{x} \in N'(\mathbf{x}^*, \delta') \cap X)\colon f(\mathbf{x}) \leq g(\mathbf{x}^*, \delta') < f(\mathbf{x}^*) + \epsilon;$$

b. $z \equiv \limsup_{\mathbf{x}\to\mathbf{x}^*} f(\mathbf{x}) \in \mathbf{R}$.[9] By the definition of lim sup, we see that in this case, there exists $\delta^* > 0$ such that:

$$(\forall \delta \in]0, \delta^*])\colon g(\mathbf{x}^*, \delta) < z + \epsilon. \tag{3.41}$$

But then, using (3.37) and (3.41), we see that:

$$(\forall \mathbf{x} \in N'(\mathbf{x}^*, \delta^*) \cap X)\colon f(\mathbf{x}) < f(\mathbf{x}^*) + \epsilon;$$

and it follows that f is upper semi-continuous at $\mathbf{x}^*$. □

With the help of Theorem 3.48, we can usefully extend Weierstrass' Theorem, as follows.

3.49. Theorem. *Let $f\colon X \to \mathbf{R}$, where X is a compact subset of $\mathbf{R}^n$, and suppose f is upper semi-continuous on X. Then f attains a maximum on X; that is, there exists $\mathbf{x}^* \in X$ such that:*

$$(\forall \mathbf{x} \in X)\colon f(\mathbf{x}) \leq f(\mathbf{x}^*).$$

Proof.[10] Define $z = \sup\{f(\mathbf{x}) \mid \mathbf{x} \in X\}$. We need to prove that (a) z is finite, and (b) there exists $\mathbf{x}^* \in X$ such that $f(\mathbf{x}^*) = z$.

By definition of the supremum, there exists a sequence, $\langle \mathbf{x}_q \rangle \subseteq X$, such that:

$$\lim_{q\to\infty} f(\mathbf{x}_q) = z; \tag{3.42}$$

and, since X is compact, $\langle \mathbf{x}_q \rangle$ contains a convergent subsequence, which for simplicity in notation we shall also denote by '$\langle \mathbf{x}_q \rangle$.' Letting

$$\mathbf{x}^* \equiv \lim_{q\to\infty} \mathbf{x}_q, \tag{3.43}$$

we note that (using Proposition 3.48) it follows from (3.43) and the upper semi-continuity of f that

$$\lim_{q\to\infty} f(\mathbf{x}_q) \leq \limsup_{\mathbf{x}\to\mathbf{x}^*} f(\mathbf{x}) \leq f(\mathbf{x}^*).$$

---

[9] Notice that it follows from (3.37) that z is either finite or equal to $-\infty$.

[10] You may find it useful to compare the argument which follows to the proof of Theorem 3.26.

It then follows from (3.42) and the definition of z that $z = f(\mathbf{x}^*)$. □

Exercises. (Note: in exercises 1–4, suppose $f\colon X \to \mathbf{R}$, with $X \subseteq \mathbf{R}^n$.)

1. If $\mathbf{x}^* \in X'$ is such that there exists some $\delta' > 0$ such that f is bounded on $N'(\mathbf{x}^*, \delta') \cap X$, then both $\limsup_{\mathbf{x}\to\mathbf{x}^*} f(\mathbf{x})$ and $\liminf_{\mathbf{x}\to\mathbf{x}^*} f(\mathbf{x})$ are finite.

2. Show that, for all $\mathbf{x} \in X'$,

$$\liminf_{\mathbf{y}\to\mathbf{x}} f(\mathbf{y}) \leq \limsup_{\mathbf{y}\to\mathbf{x}} f(\mathbf{y}).$$

3. Show that f is lower semi-continuous at $\mathbf{x}^* \in X'$ if, and only if:

$$f(\mathbf{x}^*) \leq \liminf_{\mathbf{x}\to\mathbf{x}^*} f(\mathbf{x}).$$

4. Show that, with g and h defined on $X' \times \mathbf{R}_{++}$ as in this section, we have, for any $(\mathbf{x}^*, \delta^*) \in X' \times \mathbf{R}_{++}$:

a. $g(\mathbf{x}^*, \delta^*)$ is finite if, and only if, f is bounded above on $N(\mathbf{x}^*, \delta^*)$, and

b. $h(\mathbf{x}^*, \delta^*)$ is finite if, and only if, f is bounded below on $N(\mathbf{x}^*, \delta^*)$.

5. State and prove an analogue of Theorem 3.49 for the case of lower semi-continuous functions.

3.5 Transformation Functions

In this and the next section we will look at some applications of the material we have been studying to the theory of cost and production functions. We will also develop some new mathematical material which will be useful in our study of cost and production, as well as in other areas of economics. In this section, our primary concern will be with the 'new mathematical material' to which I just referred, and which centers around the notion of a transformation function, defined as follows.

3.50. Definition. A function $F\colon \mathbf{R}_+ \to \mathbf{R}_+$ is said to be a **transformation function** iff:[11]

1. $F(0) = 0, F(v^*) > 0$, for some $v^* \in \mathbf{R}_{++}$, and

F is:

2. non-decreasing, and
3. upper semi-continuous on $\mathbf{R}_+$.

The function (or class of functions) F, defined in Example 3.31.4 of Section 3 is a transformation function. The following provides a second example of a transformation function.

[11] The notion of a transformation function was introduced into economic analysis by S Jacobsen [1968]; and was further developed by R. Shephard [1970, pp. 23–30]. The discussion here of transformation and input requirement functions, 'pseudo-inverses,' etc., is, I believe, a slight refinement of their treatment.

3.51. Example. Let α and β be positive constants, with $\beta \geq 1$, and define $F\colon \mathbf{R}_+ \to \mathbf{R}_+$ by:

$$F(v) = \begin{cases} 0 & \text{for } 0 \leq v < \beta, \text{ and} \\ \alpha \log v & \text{for } v \geq \beta. \end{cases}$$

It is easy to establish that F is a transformation function. Viewed as a production function, F indicates that some minimal positive input quantity must be available before any positive output is obtained. □

In Example 3.31.3 of Section 3, we discussed the rationale for the assumption that a production function is upper semi-continuous. In fact, in the remainder of this chapter, we will maintain the following convention.

3.52. Definition. A function $\varphi\colon \mathbf{R}^n_+ \to \mathbf{R}^m_+$ will be said to be a **production function** iff φ satisfies the following three conditions:

1. $\varphi(\mathbf{0}) = \mathbf{0}$, and for some $\mathbf{v}^* \in \mathbf{R}^n_+ \setminus \{\mathbf{0}\}, \varphi(\mathbf{v}^*) > 0$,

and φ is:

2. non-decreasing, and
3. upper semi-continuous on $\mathbf{R}^n_+$.

We then define the **extended producible set corresponding to** φ, X_0, by:

$$X_0 = \{\mathbf{x} \in \mathbf{R}^m_+ \mid (\exists \mathbf{v} \in \mathbf{R}^n_+)\colon \varphi(\mathbf{v}) \geq \mathbf{x} \geq \mathbf{0}\}; \tag{3.44}$$

the **producible set corresponding to** φ, X, by:

$$X = \{\mathbf{x} \in \mathbf{R}^m_+ \setminus \{\mathbf{0}\} \mid (\exists \mathbf{v} \in \mathbf{R}^n_+)\colon \varphi(\mathbf{v}) \geq \mathbf{x} > \mathbf{0}\} = X_0 \setminus \{\mathbf{0}\}; \tag{3.45}$$

and the (competitive) **cost function for** $\varphi, c : \mathbf{R}^n_+ \times X_0 \to \mathbf{R}_+$, by:

$$c(\mathbf{w}, \mathbf{x}) = \inf\{\mathbf{w} \cdot \mathbf{v} \mid \mathbf{v} \in \mathbf{R}^n_+ \;\&\; \varphi(\mathbf{v}) \geq \mathbf{x}\} \quad \text{for } (\mathbf{w}, \mathbf{x}) \in \mathbf{R}^n_+ \times X_0. \tag{3.46}$$

While it will be convenient for us to use the above definition in this section, it should be noted that in the case where $m \geq 2$ (that is, where more than one commodity is being produced), we will generally use correspondences, rather than functions, to characterize the technological possibilities available; and, of course, the other side of the coin is that except in this section we will normally consider only real-valued production functions. Moreover, while the 'producible set' terminology and notation introduced above will seem a bit more cumbersome than necessary in this section, in the next section and in Chapter 6 we will be primarily interested in the producible set, X, rather than X_0.

Notice that the cost function, $c(\mathbf{w}, \mathbf{x})$, is well defined (and finite), for each $(\mathbf{w}, \mathbf{x}) \in \mathbf{R}^n_+ \times X_0$; since the set of numbers over which we are taking the infimum in equation (3.46), above, is a set of nonnegative numbers, and thus is bounded below. In the next section, we will discuss the reason for using the infimum, rather than a minimum, in this definition.

Clearly a transformation function can be thought of as a production function for the very special case in which there is only one output and one (variable) input. While the principal reasons for our interest in transformation functions center around other applications which we shall take up later, it will be useful for the present discussion to concentrate upon the one variable input (or 'total product function') interpretation of a transformation function.

In line with this interpretation, consider for a moment the problem of deriving a cost function from a transformation function, F. In the special case in which F is one-to-one, this derivation is simplicity itself, at least conceptually: given any desired production level, $x \in F(\mathbf{R}_+)$, there exists a unique v such that

$$x = F(v)$$

[and, of course, $v = F^{-1}(x)$], and the (variable) cost of producing x is given by:

$$c(w, x) = wv,$$

where $w \in \mathbf{R}_+$ is the price per unit of v. More generally, if we define

$$X_0 = F(\mathbf{R}_+),$$

then the variable cost function, $c\colon \mathbf{R}_+ \times X_0 \to \mathbf{R}_+$, is given by:

$$c(w, x) = wF^{-1}(x) \quad \text{for } (w, x) \in \mathbf{R}_+ \times X_0.$$

3.53. Example. Suppose $F\colon \mathbf{R}_+ \to \mathbf{R}_+$ is defined by

$$F(v) = av^{1/2},$$

where a is a positive constant. In this case, $X_0 = \mathbf{R}_+$, and F^{-1} is given by:

$$F^{-1}(x) = [x/a]^2,$$

so that $c(w, x)$ is given by:

$$c(w, x) = (\alpha w)x^2 \quad \text{for } (w, x) \in \mathbf{R}^2_+,$$

where we define $\alpha = (1/a)^2$. □

For an arbitrary transformation function the derivation of the corresponding cost function may not be quite so simple as in the above example, for two reasons. First, since a transformation function is not necessarily one-to-one, it may not possess a conventional inverse. Secondly, the definition of the producible set which we are using means that we may not have $X_0 \subseteq F(\mathbf{R}_+)$.[12] Clearly, a more general notion of an inverse function

[12]We are allowing for a certain amount of 'free disposal' in our definitions. A more complete discussion of the reasons for, and effects of this assumption, is included in the next section.

is needed here. In order to set the stage for the introduction of such a generalization, we begin by considering the following result. It provides an operational characterization of inverse functions which will be of considerable use to us in our study of cost and production functions.

3.54. Proposition. *Suppose X and Y are non-empty sets, and that f and g are functions such that $f\colon X \to Y$, $g\colon Y \to X$, and f and g satisfy:*
1. $(\forall x \in X)\colon g[f(x)] = x,$
and
2. $(\forall y \in Y)\colon f[g(y)] = y.$
Then both f and g are one-to-one and onto (and, of course, $g = f^{-1}$ and $f = g^{-1}$).

Proof. To prove that f is onto Y, let $\bar{y} \in Y$ be arbitrary. Then by hypothesis 2,

$$f[g(\bar{y})] = \bar{y},$$

so that $\bar{y} \in f(X)$. Hence, f is onto Y.

To prove that f is one-to-one, suppose that for some $x, x' \in X$, we have:

$$y^* = f(x) = f(x').$$

Then, by hypothesis 1,

$$x = g[f(x)] = g(y^*) = g[f(x')] = x';$$

and it follows that f is one-to-one.

The proof that g is 1-1 and onto is symmetric, and will be left as an exercise. □

The following simply generalizes the conditions in the hypotheses of 3.54 to obtain our definition of a pseudo-inverse.

3.55. Definition. Let $F\colon V \to \mathbf{R}^m_+$, where V is a non-empty subset of $\mathbf{R}^n$, and define X_0 by:

$$X_0 = \{\mathbf{x} \in \mathbf{R}^m_+ \mid (\exists \mathbf{v} \in V)\colon F(\mathbf{v}) \geq \mathbf{x}\}. \tag{3.47}$$

We shall say that a function $f\colon X_0 \to V$ is a **pseudo-inverse of F** iff:
1. f is non-decreasing, and
2. f and F satisfy:

$$(\forall \mathbf{v} \in V)\colon f[F(\mathbf{v})] \leq \mathbf{v}, \tag{3.48}$$

and

$$(\forall \mathbf{x} \in X_0)\colon F[f(\mathbf{x})] \geq \mathbf{x}. \tag{3.49}$$

Notice that if a transformation function, F, has a conventional inverse, then F^{-1} is also a pseudo-inverse of F. Moreover, it can be shown that the pseudo-inverse of a transformation function plays the same role in the definition of the cost function corresponding to a transformation function, F, as does the conventional inverse in the case where F is one-to-one.[13] Rather than concentrating our attention upon pseudo-inverses, however, we will develop the theory of a closely related, but somewhat more economically meaningful concept, the input requirement function dual to a production function. While we have already made use of input requirement functions in Sections 2.4 and 2.5, we have not as yet attempted to develop a notion of an input requirement function dual to a production function. Informally, however, it is simply the function whose value at an output, x, is the minimum quantity of input which can be used to produce x. We can formally define such a function for the case where n inputs are used to produce m outputs as follows.[14]

3.56. Definition. If $\varphi\colon \mathbf{R}^n_+ \to \mathbf{R}^m_+$ is a production function, and

$$X = \{\mathbf{x} \in \mathbf{R}^m_+ \mid (\exists \mathbf{v} \in \mathbf{R}^n_+) \colon \varphi(\mathbf{v}) \geq \mathbf{x} > \mathbf{0}\},$$

is the corresponding producible set, we shall say that a function, f is the **input requirement function dual to φ** iff $f\colon X \to \mathbf{R}^n_+$, and satisfies, for all $\mathbf{x} \in X$:

$$\varphi[f(\mathbf{x})] \geq \mathbf{x}, \tag{3.50}$$

and

$$(\forall \mathbf{v} \in \mathbf{R}^n_+) \colon \varphi(\mathbf{v}) \geq \mathbf{x} \Rightarrow \mathbf{v} \geq f(\mathbf{x}). \tag{3.51}$$

Obviously if f is the input requirement function dual to a production function, φ, then for $(\mathbf{w}, \mathbf{x}) \in \mathbf{R}^n_+ \times X$, the cost function for φ will be given by:

$$c(\mathbf{w}, \mathbf{x}) = \mathbf{w} \cdot f(\mathbf{x}). \tag{3.52}$$

Notice that we are not considering the possibility of defining our input requirement function at zero output. Essentially, there are two reasons for this. The first reason is simply that it is mathematically more convenient not to do so. In applications it will be convenient not to consider zero output, because we will want the function f to be *positive* at all points of its domain;

[13]This will, in fact follow from Proposition 3.57.

[14]As was mentioned earlier, when $m > 1$ outputs are being produced, we shall generally use mathematical concepts other than functions to describe the technology. However, the basic idea of a vector-valued production function is clear enough [just rather too specialized, as we shall discover in Chapter 6], and will be useful in our present discussion.

and notice that if φ is a production function and f is the input requirement function dual to φ, then we will have:

$$(\forall \mathbf{x} \in X): f(\mathbf{x}) > \mathbf{0}. \tag{3.53}$$

Secondly, it will turn out that if we were to define the input requirement function at zero output, then condition 3.52.1 (and consistency of our definitions) will require that we set $f(\mathbf{0}) = \mathbf{0}$ for the dual input requirement function. From an economic point of view this is all right, as long as we are interpreting the inputs being used as *variable* inputs, and as long as:

$$\lim_{\mathbf{x} \to 0} f(\mathbf{x}) = \mathbf{0}. \tag{3.54}$$

If, however, the limit in (3.54) is not zero (and our assumptions do not imply that it is), then we may wish to consider the product of the input price vector, $\mathbf{w}$, with the limiting value as fixed cost. In fact, the expression for the cost function given by (3.52), above, can be interpreted as being either total cost or variable cost, depending upon the behavior of the input requirement function as it approaches zero output. Actually defining the input requirement function at zero output, however, requires us to commit ourselves as to whether we are dealing with total or variable costs. Thus we gain a considerable amount of flexibility by the simple expedient of not requiring zero output to be in the domain of f. Consistently with these considerations, *we will also modify the definition just given by restricting the domain of the cost function for* φ *to* $\mathbf{R}^n_+ \times X$.

While it would be nice if all production functions possessed a dual input requirement function, such is, unfortunately, not the case. However, if a production function, φ, posesses a pseudo-inverse, then the restriction of same to the producible set is the input requirement function dual to φ; as we shall now establish.

3.57. Proposition. *If* $\varphi: \mathbf{R}^n_+ \to \mathbf{R}^m_+$ *is a production function, with corresponding producible and extended producible sets,* X *and* X_0*, respectively, and if* $g: X_0 \to \mathbf{R}^n_+$ *is the pseudo-inverse of* φ*, then the restriction of* g *to* X *is the input requirement function dual to* φ*.*

Proof. Let $f: X \to \mathbf{R}^n_+$ be the restriction of g to X. It follows immediately from equation (3.49) of Definition 3.55 that f satisfies the first condition of Definition 3.56 [equation (3.50)]. To prove that f satisfies the second condition in this latter definition, let $\mathbf{x}^* \in X$ be arbitrary, and suppose $\mathbf{v} \in \mathbf{R}^n_+$ is such that $\varphi(\mathbf{v}) \geq \mathbf{x}^*$. Then it follows that $\varphi(\mathbf{v}) \in X$, and, since f is non-decreasing:

$$f[\varphi(\mathbf{v})] \geq f(\mathbf{x}^*). \tag{3.55}$$

However, from the definition of a pseudo-inverse:

$$\mathbf{v} \geq f[\varphi(\mathbf{v})]; \tag{3.56}$$

and, from (3.55) and (3.56) it then follows that:

$$\mathbf{v} \geq f(\mathbf{x}^*). \quad \square$$

While the above result establishes that the pseudo inverse of a production function (where one exists) is, essentially, the input requirement function dual to the production function; the latter function may exist in cases where no pseudo-inverse exists, in principle. In any case, we will concentrate our attention upon input requirement functions, rather than pseudo-inverses, in the remainder of this section. On the other hand, we will nonetheless often find the above result useful; as is illustrated in the following examples.

3.58. Examples/Exercises.

1. Consider the transformation function defined in Example 3.31.4 of Section 3, that is:

$$F(v) = \alpha N(v) \quad \text{for } v \in \mathbf{R}_+,$$

where α is a positive constant. In this case, the producible set, X, is given by

$$X = \{x \in \mathbf{R}_{++} \mid (\exists v \in \mathbf{R}_+) : F(v) \geq x\} = \mathbf{R}_{++};$$

and, letting $f \colon X \to \mathbf{R}_{++}$ be given by:

$$f(x) = n^*(x/\alpha),$$

where $n^*(\cdot)$ is the 'ceiling function' defined in Exercise 2 at the end of Section 3.3, we have, for $v \in \mathbf{R}_+$:

$$f[F(v)] = n^*\Big([\alpha N(v)]/\alpha\Big) = n^*[N(v)] = N(v) \leq v. \tag{3.57}$$

On the other hand, for $x \in X = \mathbf{R}_{++}$:

$$F[f(x)] = \alpha N[n^*(x/\alpha)] = \alpha n^*(x/\alpha).$$

However, since

$$n^*(x/\alpha) - 1 < x/\alpha \leq n^*(x/\alpha),$$

we see that

$$x \leq \alpha n^*(x/\alpha);$$

and therefore:

$$F[f(x)] \geq x. \tag{3.58}$$

Combining (3.57) and (3.58), we see that it follows from Proposition 3.57 that f is the input requirement function dual to F.

2. Define the function $\varphi \colon \mathbf{R}^n_+ \to \mathbf{R}_+$ by

$$\varphi(\mathbf{v}) = A \prod\nolimits_{i=1}^{n} v_i^{\alpha_i},$$

where A is a positive constant, $\alpha_i > 0$, for $i = 1, \ldots, n$, and

$$\sum_{i=1}^{n} \alpha_i = 1.$$

Now define the function $g\colon \mathbf{R}_+ \to \mathbf{R}^n_+$ by:

$$g(x) = (x/A, x/A, \ldots, x/A).$$

Is g the input requirement function dual to φ? Why or why not? If your answer is that g is not the dual to φ, can you show that no dual input requirement function exists in this case?

3. Consider the function $\varphi\colon \mathbf{R}^n_+ \to \mathbf{R}^m_+$ defined by:

$$\varphi(\mathbf{v}) = (\beta_1\sigma(\mathbf{v}), \ldots, \beta_m\sigma(\mathbf{v})) = \sigma(\mathbf{v})\mathbf{b},$$

where $\beta_j > 0$ for $j = 1, \ldots, m$,

$$\mathbf{b} \stackrel{\text{def}}{=} (\beta_1, \ldots, \beta_m),$$

$\sigma\colon \mathbf{R}^n_+ \to \mathbf{R}_+$ is defined by:

$$\sigma(\mathbf{v}) = \min\{v_1/\alpha_1, \ldots, v_n/\alpha_n\},$$

and $\alpha_i \in \mathbf{R}_{++}$, for $i = 1, \ldots, n$.

You should have no trouble in verifying the fact that φ is a production function, with corresponding producible set, X, given by:

$$X = \{\mathbf{x} \in \mathbf{R}^m_+ \mid (\exists \mathbf{v} \in \mathbf{R}^n_+)\colon \varphi(\mathbf{v}) \geq \mathbf{x} > \mathbf{0}\} = \mathbf{R}^m_+.$$

Now, define $\tau\colon \mathbf{R}^m_+ \to \mathbf{R}_+$ by

$$\tau(\mathbf{x}) = \max\{x_1/\beta_1, \ldots, x_m/\beta_m\},$$

and $g\colon \mathbf{R}^m_+ \to \mathbf{R}^n_+$ by:

$$g(\mathbf{x}) = (\tau(\mathbf{x})\alpha_1, \ldots, \tau(\mathbf{x})\alpha_n) \equiv \tau(\mathbf{x})\mathbf{a},$$

where the β_j's ($j = 1, \ldots, m$) and α_i's ($i = 1, \ldots, n$) are from the definition of φ, and the vector $\mathbf{a}$ is defined by

$$\mathbf{a} = (\alpha_1, \ldots, \alpha_n).$$

Now, if $\mathbf{v} \in \mathbf{R}^n_+$,

$$g[\varphi(\mathbf{v})] = \tau[\varphi(\mathbf{v})]\mathbf{a},$$

and

$$\tau[\varphi(\mathbf{v})] = \tau[\sigma(\mathbf{v})\mathbf{b}] = \max\{\sigma(\mathbf{v})\beta_1/\beta_1, \ldots, \sigma(\mathbf{v})\beta_m/\beta_m\} = \sigma(\mathbf{v});$$

so that

$$g[\varphi(\mathbf{v})] = \sigma(\mathbf{v})\mathbf{a}.$$

But then, from the definition of σ, we see that for an arbitrary $i \in \{1, \dots, n\}$:

$$\alpha_i \sigma(\mathbf{v}) \leq \alpha_i \cdot (v_i/\alpha_i) = v_i;$$

and therefore,

$$g[\varphi(\mathbf{v})] = \sigma(\mathbf{v})\mathbf{a} \leq \mathbf{v}. \tag{3.59}$$

Similarly, if $\mathbf{x} \in \mathbf{R}^m_+$,

$$\varphi[g(\mathbf{x})] = \sigma[g(\mathbf{x})]\mathbf{b},$$

and, since

$$g(\mathbf{x}) = \tau(\mathbf{x})\mathbf{a};$$

we see that

$$\sigma[g(\mathbf{x})] = \min\{\tau(\mathbf{x})\alpha_1/\alpha_1, \dots, \tau(\mathbf{x})\alpha_n/\alpha_n\} = \tau(\mathbf{x}).$$

However, for $j \in \{1, \dots, m\}$,

$$\tau(\mathbf{x})\beta_j = \left(\max_k \frac{x_k}{\beta_k}\right) \cdot \beta_j \geq (x_j/\beta_j) \cdot \beta_j = x_j;$$

and thus we see that

$$\varphi[g(\mathbf{x})] = \sigma[\tau(\mathbf{x})\mathbf{a}]\mathbf{b} \geq \mathbf{x}. \tag{3.60}$$

Since g is non-decreasing, it follows from (3.59) and (3.60) that g is the pseudo-inverse of φ; and it also follows that the restriction of g to X is the input requirement function dual to φ. □

While a production function may not possess a dual input requirement function, we will presently demonstrate that in the special case of most interest in this section, that in which there is only one input (and in particular, where the production function is actually a transformation function), the dual function always exists. In the meantime, we note in our next two results that (1) there is *at most* one input requirement function dual to a given production function, and (2) a given input requirement function can be dual to no more than one production function.

3.59. Proposition. *Suppose φ: $\mathbf{R}^n_+ \to \mathbf{R}^m_+$, let X be the corresponding producible set, and suppose both f: $X \to V$ and f^*: $X \to V$ are input requirement functions dual to φ. Then $f \equiv f^*$.*

Proof. Suppose f and f^* are both dual to φ, and let $\mathbf{x}' \in X$ be arbitrary. Then, using the fact that f^* satisfies equation (3.50), and defining $\mathbf{v}' = f^*(\mathbf{x}')$, we have:

$$\varphi(\mathbf{v}') \equiv \varphi[f^*(\mathbf{x}')] \geq \mathbf{x}'.$$

Thus, since f is dual to φ, we must have:

$$f^*(\mathbf{x}') \equiv \mathbf{v}' \geq f(\mathbf{x}').$$

A symmetric argument establishes that:

$$f(\mathbf{x}') \geq f^*(\mathbf{x}'). \quad \square$$

3.60. Proposition. *Suppose $\varphi_i\colon \mathbf{R}^n_+ \to \mathbf{R}^m_+$ is a production function, for $i = 1, 2$, and that f is the dual input requirement function for both φ_1 and φ_2. Then $\varphi_1 \equiv \varphi_2$.*

Proof. Let $\mathbf{v}^* \in \mathbf{R}^n_+$ be arbitrary. If $\varphi_1(\mathbf{v}^*) = \mathbf{0}$, then we obviously have:

$$\varphi_2(\mathbf{v}^*) \geq \varphi_1(\mathbf{v}^*).$$

Suppose, therefore, that $\varphi_1(\mathbf{v}^*) > \mathbf{0}$, and define:

$$\mathbf{x}^* = \varphi_1(\mathbf{v}^*).$$

Then, since f is dual to φ_1, we must have:

$$\mathbf{v}^* \geq f(\mathbf{x}^*);$$

and thus, since φ_2 is non-decreasing:

$$\varphi_2(\mathbf{v}^*) \geq \varphi_2[f(\mathbf{x}^*)]. \tag{3.61}$$

On the other hand, since f is dual to φ_2, we also have:

$$\varphi_2[f(\mathbf{x}^*)] \geq \mathbf{x}^* \equiv \varphi_1(\mathbf{v}^*);$$

and combining this with (3.61), we have:

$$\varphi_2(\mathbf{v}^*) \geq \varphi_1(\mathbf{v}^*).$$

A symmetric argument establishes that:

$$\varphi_1(\mathbf{v}^*) \geq \varphi_2(\mathbf{v}^*). \quad \square$$

As was mentioned earlier, in the special case in which there is only one variable input, we can show that a dual input requirement function will *always* exist, as follows.[15] Given a production function, $\varphi\colon \mathbf{R}_+ \to \mathbf{R}_+^m$, and its associated producible set, X, consider the set $V(\mathbf{x})$ defined for $\mathbf{x} \in X$ by:

$$V(\mathbf{x}) = \{v \in \mathbf{R}_+ \mid \varphi(v) \geq \mathbf{x}\}. \tag{3.62}$$

Since φ is upper semi-continuous, it follows at once from Exercise 7 at the end of Section 3.3 that $V(\mathbf{x})$ is closed in $\mathbf{R}_+$, which in turn is closed in $\mathbf{R}$. Furthermore, for $\mathbf{x}^* \in X$, it follows from the definition of the producible set that there exists $\hat{v} \in \mathbf{R}_+$, such that $\varphi(\hat{v}) \geq \mathbf{x}^*$; and thus we see that

$$V^* \stackrel{\text{def}}{=} V(\mathbf{x}^*) \cap [0, \hat{v}]$$

is a non-empty, compact set. Thus V^* contains a minimal element, v^*, and obviously $v^* = \min V(x^*)$ as well. Since $x \in X_0$ was arbitrary, it follows that we can define the function $f\colon X_0 \to \mathbf{R}_+$ by

$$f(x) = \min V(x) \quad \text{for } x \in X \tag{3.63}$$

[that is, the function in equation (3.63) is well-defined]; and obviously f is the input requirement function dual to φ. We will make use of the facts just established in proving the following result.

3.61. Theorem. *Suppose $F\colon \mathbf{R}_+ \to \mathbf{R}_+$ is a transformation function, and let X be the producible set for F. Then there exists an input requirement function, $f\colon X \to \mathbf{R}_+$ which is dual to F. Moreover, f satisfies the following conditions:*

1. *X is a non-degenerate sub-interval of $\mathbf{R}_+$ and $0 \in \overline{X}$,*
2. *$(\forall x \in X)\colon f(x) > 0$,*
3. *f is non-decreasing,*
4. *f is lower semi-continuous,*
5. *either f is unbounded on X, or there exists $\beta \in \mathbf{R}_{++}$ such that $X =]0, \beta]$.*

Proof. We have already shown that the function f defined in equation (3.63), above, is the input requirement function dual to F (and is well-defined). The proof that f satisfies properties 1-3 will be left as an exercise.

4. To prove that f is lower semi-continuous on X, it suffices to show that, for all $\alpha \in \mathbf{R}$, the set $X(\alpha)$ defined by

$$X(\alpha) = \{x \in X \mid f(x) \leq \alpha\}$$

[15] The idea of a technology in which only one variable input is used to produce many different commodities simultaneously may seem hopelessly unrealistic at first glance. However, there are a number of situations in which it may be useful to think of things in this way. For example, in an arid country, arable soil and labor to till the soil may be available in super-abundance, so that it might be quite reasonable to treat irrigation water as the only (scarce) input in agricultural production. As another example, in the theory of public goods, it is common to consider a situation in which a single private good [possibly tax revenues (?)] can be used as an input in the production of several different public goods.

is closed in X (Corollary 3.35). One case is easy: if $\alpha < 0$, then $X(\alpha) = \emptyset$, which is obviously closed in X. To prove closure for the remaining case, let $\alpha \geq 0$, and let $x^* \in X$ and $\langle x_q \rangle$ be a sequence satisfying:

$$x_q \in X(\alpha) \quad \text{for } q = 1, 2, \dots, \text{ and } x_q \to x^*.$$

Since $x_q \in X(\alpha)$ for each q, we have from the definition of $X(\alpha)$ that:

$$\alpha \geq f(x_q) \equiv \min\{v \in \mathbf{R}_+ \mid F(v) \geq x_q\} \quad \text{for } q = 1, 2, \dots.$$

Consequently, since $\alpha \in \mathbf{R}_+$ and F is non-decreasing,

$$F(\alpha) \geq x_q \quad \text{for } q = 1, 2, \dots.$$

But then it follows at once that

$$x^* = \lim_{q \to \infty} x_q \leq F(\alpha),$$

and thus that

$$f(x^*) = \min\{v \in \mathbf{R}_+ \mid F(v) \geq x^*\} \leq \alpha.$$

Therefore, $x^* \in X(\alpha)$, and it follows that $X(\alpha)$ is closed in X.

5. Suppose f is bounded on X. Then there exists $\bar{v} \in \mathbf{R}_+$ such that

$$(\forall x \in X): f(x) \leq \bar{v}.$$

However, from equation (3.63), we then have

$$(\forall x \in X): \min\{v \in \mathbf{R}_+ \mid F(v) \geq x\} \leq \bar{v};$$

and thus, since F is non-decreasing,

$$(\forall x \in X): F(\bar{v}) \geq x. \tag{3.64}$$

But then, defining $\beta = F(\bar{v})$, it follows from (3.64) and the definition of X that

$$X =]0, \beta]. \quad \square$$

Now let's take the properties which we established in the above result, and use them as the definition of the *class* of functions which we shall call (one-input, one-output) input requirement functions, as follows.

3.62. Definition. Let X be a non-degenerate sub-interval of $\mathbf{R}_{++}$, satisfying $0 \in \overline{X}$, and let $f: X \to \mathbf{R}_{++}$. We shall say that f is an **input requirement function** iff f and X satisfy:

1. f is non-decreasing on X,
2. f is lower semi-continuous on X,
3. either f is unbounded on X, or there exists $b \in \mathbf{R}_{++}$ such that $X =]0, b]$.

If we were to dream up 'reasonable' requirements for an input requirement function, then we would surely include the assumption that it is nondecreasing; as to why we probably do not wish to require that such functions be strictly increasing, take a look back at the discussion in Example 3.31.3 of Section 3 and 3.58.1. The rationale for the other two conditions may be less clear (although we will consider a part of this rationale later), but remember that we have already shown that the input requirement functions dual to transformation functions do satisfy *all* of the conditions of Definition 3.62. You may also find it useful to consider the following examples.

3.63. Examples.

1. The function f used in Example 3.58.1, above:

$$f(x) = \beta n^*(x),$$

where $n^*(\cdot)$ is the 'ceiling function,' is an input requirement function. In this case, $X = \mathbf{R}_{++}$, and f is unbounded on X.

2. Let $\alpha > 0$, and define f on $X \equiv]0, \alpha[$ by:

$$f(x) = \frac{x}{\alpha - x}.$$

It is easy to verify the fact that f is an input requirement fuction. Notice that, while X is an open interval in this case, f is unbounded on X.

3. Recall the transformation function introduced in Example 3.51:

$$F(v) = \begin{cases} 0 & \text{for } 0 \leq v < \beta, \text{and} \\ \alpha \log v & \text{for } v \geq \beta, \end{cases}$$

where $\alpha > 0$ and $\beta \geq 1$. Show that the input requirement function dual to F is given by:

$$f(x) = \begin{cases} \beta & \text{for } 0 < x \leq \alpha \log \beta, \\ \exp(x/\alpha) & \text{for } x > \alpha \log \beta, \end{cases}$$

and $X = \mathbf{R}_{++}$ in this case [probably the easiest way to proceed here is to show that f is the pseudo-inverse of F on X]. □

As we showed in Example 3.58.1, the fact that the input requirement function dual to a transformation function is the restriction to X of the pseudo-inverse of that function provides a convenient means for establishing that a function, f, believed to be the input requirement function dual to a transformation function, F, is indeed the dual function. On the other hand, the formula in equation (3.63) often provides a method for finding the dual input requirement function of a function, F, when we do not know in advance what form it may take.

From Theorem 3.61 and Proposition 3.59 we know that, corresponding to every transformation function, f, there is a unique input requirement function dual to F. As it happens, every input requirement function is dual to a unique transformation function, as is shown by the following.

3.64. Theorem. *Suppose $f: X \to \mathbf{R}_{++}$ is an input requirement function, where X is a non-degerate sub-interval of $\mathbf{R}_{++}$ such that $0 \in \overline{X}$, and define F on $\mathbf{R}_+$ by:*

$$F(v) = \begin{cases} 0 & \text{for } P(v) = \emptyset, \text{ and} \\ \max P(v) & \text{for } P(v) \neq \emptyset, \end{cases} \tag{3.65}$$

where we define:

$$P(v) = \{x \in X \mid f(x) \leq v\} \quad \text{for } v \in \mathbf{R}_+.$$

Then:

1. *F is a transformation function.*
2. *f is the input requirement function dual to F, and*
3. *F is the only transformation function to which f is dual.*

Proof.

1. In order to prove that F is well-defined by the formula (3.65), let $v^* \in \mathbf{R}_+$, and suppose that:

$$P(v^*) = \{x \in X \mid f(x) \leq v^*\},$$

is non-empty. Then, since f is lower semi-continuous, we see that $P(v^*)$ is closed in X. In order to show that F is well-defined, we need to show that $P(v^*)$ has a maximal element, and to do this, we distinguish two cases.

a. $(\forall x \in X)\colon f(x) \leq v^*$. In this case, f is bounded, and thus, from the definition of an input requirement function, it follows that X is of the form

$$X =]0, \beta],$$

for some $\beta > 0$. Obviously in this case,

$$F(v^*) \equiv \max\{x \in X \mid f(x) \leq v\} = \beta.$$

b. $(\exists x' \in X)\colon f(x') > v^*$. In this case we see that, since f is non-decreasing,

$$P(v^*) \subseteq [0, x'].$$

Thus it is clear that $P(v^*)$ is bounded, as well as closed, and consequently it contains a maximal element, x^*. It then follows, of course, that

$$F(v^*) = x^* = \max P(v^*).$$

It is obvious that F satisfies $F(0) = 0$, and that F is non-decreasing. To show that F is upper semi-continuous, we recall that from 3.33 it suffices to show that, for each $x^* \in \mathbf{R}$, the set $V(x^*)$ defined by

$$V(x^*) = \{v \in \mathbf{R}_+ \mid F(v) \geq x^*\}$$

is closed. To do this, we distinguish three cases.

a. $x^* \leq 0$. In this case, $V(x^*) = \mathbf{R}_+$, which is a closed set.

b. $x^* > 0$ and $(\exists x' \in X)\colon x' \geq x^*$. In this case, we note that, since X is an interval and $x^* > 0$, we have $x^* \in X$. Now suppose $\langle v_q \rangle \subseteq V(x^*)$ is a sequence such that

$$v_q \to v^*,$$

for some v^*. Since $\mathbf{R}_+$ is closed, it is apparent that $v^* \in \mathbf{R}_+$, and, from the definitions of F and $V(x^*)$, we see that

$$\max\{x \in X \mid f(x) \leq v_q\} \geq x^* \quad \text{for } q = 1, 2, \ldots .$$

But then we see that

$$v_q \geq f(x^*) \quad \text{for } q = 1, 2, \ldots ,$$

and it then follows that

$$v^* \geq f(x^*).$$

However, this means that x^* is an element of the set $P(v^*) \equiv \{x \in X \mid f(x) \leq v^*\}$, and from the definition of F we then have

$$F(v^*) \geq x^*;$$

so that $v^* \in V(x^*)$. Consequently we see that $V(x^*)$ is a closed set in this case.

c. $x^* > 0$ and $(\forall x \in X)\colon x^* > x$. In this case, it follows from the definition of F that, for all $v \in \mathbf{R}_+$,

$$F(v) = \max\{x \in X \mid f(x) \leq v\} < x^*.$$

Therefore, in this case we see that $V(x^*) = \emptyset$; which is, of course, a closed set.

Upon review, we see that in each possible case, $V(x^*)$ is a closed set; and it follows that F is upper semi-continuous.

2. To prove that f is the input requirement function dual to F, let $x^* \in X$ be arbitrary. Then

$$x^* \in \{x \in X \mid f(x) \leq f(x^*)\},$$

and thus

$$F[f(x^*)] = \max\{x \in X \mid f(x) \leq f(x^*)\} \geq x^*.$$

Moreover, if $\hat{v} \in \mathbf{R}_+$ is such that:

$$F(\hat{v}) \geq x^*,$$

then, since $x^* > 0$, we see that $P(\hat{v}) \neq \emptyset$, and:

$$F(\hat{v}) = \max\{x \in X \mid f(x) \leq \hat{v}\} \stackrel{\text{def}}{=} \hat{x} \geq x^*.$$

Since f is non-decreasing, it then follows that:

$$f(x^*) \leq f(\hat{x}) \leq \hat{v}.$$

From the argument of the above paragraph, we see that f is the input requirement function dual to F. By Proposition 3.59, f is the *only* input requirement function dual to F; and by Proposition 3.60, F is the only transformation function to which f is dual. □

It follows from Theorems 3.61 and 3.64 that there is a one-to-one correspondence between the family of transformation functions and the family of input requirement functions: given any transformation function there exists a unique dual input requirement function, and conversely. One of the beauties of this fact is that, if we are given a transformation function F, we are justified in referring to *the* input requirement function dual to F, even if we cannot exhibit it. We may, of course, be able to use the formula in equation (3.63) to find the dual input requirement function, f, but it will sometimes be impractical or effectively impossible to do so. Nonetheless, Theorem 3.61 assures us that f (the input requirement function dual to F) exists, and states the qualitative properties f will satisfy as well. Similarly, we can confidently refer to *the* transformation function dual to a given input requirement function f, even if we cannot exhibit said function. Thus, for example, let a, b, and c be positive constants satisfying

$$3ac \geq b^2, \tag{3.66}$$

and define f on $X = \mathbf{R}_{++}$ by:

$$f(x) = ax - bx^2 + cx^3. \tag{3.67}$$

It can be shown that, given (3.66), f is an input requirement function (Exercise). Thus we are justified in referring to the transformation function dual to f even though in this case it is very difficult indeed to *find* said transformation function.

If we have a transformation function, F, which describes a production process, then by Theorem 3.61, the input requirement function dual to F, f, is given by:

$$f(x) = \min\{v \in \mathbf{R}_+ \mid F(v) \geq x\}. \tag{3.68}$$

Therefore, if the price of the input, v, is positive, then for each $x \in X$, the quantity of the input which a cost-minimizing production unit would use to produce x is given by $f(x)$. Consequently, if 'w' denotes the price of v, we can express the unit's cost function as:

$$c(w,x) = w \cdot f(x) \quad \text{for } (w,x) \in \mathbf{R}_+ \times X. \tag{3.69}$$

It is then an easy consequence of Theorems 3.61 and 3.64 that there is a one-to-one correspondence between cost functions of the form (3.69) and production (transformation) functions, $F(\cdot)$; that is, if a firm's technology can be described by a transformation function, F, then (as we have just seen) its cost function will be of the form (3.69), where f is the (unique) input requirement function dual to F.[16] Conversely, if a firm has a cost function of the form (3.69), where $f(\cdot)$ is an input requirement function by our definition, then there is exactly one transformation function which could describe the firm's technology, and that is the transformation function dual to f. In Chapter 6 we shall find that this one-to-one correspondence between production relationships and cost functions holds in a much more general context.

Exercises.

1. Let F be given by

$$F(v) = a\log(v+1) \quad \text{for } v \in \mathbf{R}_+,$$

where a is a positive constant. Show that the input requirement function dual to F is given by

$$f(x) = \exp(x/a) - 1 \quad \text{for } x \in X = \mathbf{R}_{++}.$$

2. Let F be given by

$$F(v) = \frac{av}{1+v},$$

where a is a positive constant. Show that here the producible set is given by $X =]0,a[$, and the input requirement function dual to F, f, is given by:

$$f(x) = \frac{x}{a-x} \quad \text{for } x \in X$$

(compare Example 3.63.2).

3. Let F be given by:

$$F(v) = Av^a,$$

where A and a are positive constants. Show that $X = \mathbf{R}_{++}$, and that the dual input requirement function, f, is given by:

$$f(x) = \left(\frac{x}{A}\right)^{1/a}.$$

[16] If it is a price-taker in the market for its input, as we have been assuming to be the case throughout this section.

4. Let F be given

$$F(v) = \begin{cases} av - bv^2 & \text{for } 0 \leq v \leq a/2b, \\ a^2/4b & \text{for } v > a/2b, \end{cases} \tag{3.70}$$

where a and b are both positive constants. Show that in this case, $X =]0, a^2/4b]$, and that the dual input requirement function, f, is given by:

$$f(x) = \frac{a - (a^2 - 4bx)^{1/2}}{2b}.$$

5. Consider the function $F : \mathbf{R}_+ \to \mathbf{R}_+$ defined by

$$F(v) = \frac{aN(v)}{1 + N(v)},$$

where $N(\cdot)$ is the 'largest integer' function considered previously, and $a \in \mathbf{R}_{++}$. Show that F is a transformation function, and find the input requirement function dual to F. [Hint: see Example 3.31.4, Exercise 2, above, and Exercise 6, below.]

6. Show that if F and G are transformation functions, then the composition function, H, defined on $\mathbf{R}_+$ by

$$H(v) = G[F(v)],$$

is a transformation function as well. Show also that the input requirement function dual to H is the function h, given by:

$$h(x) = f[g(x)],$$

where f and g are the input requirement functions dual to F and G, respectively.

3.6 Production and Cost Functions

Before continuing our study of cost and production in the single-output case, we should pause momentarily to establish some conventions to which we shall adhere throughout the remainder of this book. First of all, we shall hereafter refer to the production unit involved as a 'firm.' It should be recognized, however, that in this context the word 'firm' is being used as an undefined primitive term in the theory. More pragmatically, we do not mean to confine our discussion to cases in which the production unit is a firm by any particular conventional definition; it is simply the entity which controls the production process.

Secondly, it should be noted that there are two disposability assumptions which we have implicitly made in our treatment thus far, and we should make them explicit at this point. In the first place, we are assuming throughout that production functions are non-decreasing (Definition 3.54.2). This assumption would appear to be fairly innocuous if we treat input quantitites as *amounts available per unit of time*, although less so if we interpret the quantities as available stocks. We will return to this point shortly; first, however, let's take note of the fact that in our definition of the producible set, X, corresponding to a production function, we have also slipped in another type of disposability assumption. The issue is this: if we say that the function $\varphi\colon \mathbf{R}^n_+ \to \mathbf{R}_+$ delineates the production possibilities available to the firm, what output quantities might we then observe the firm to produce? If we denote the set of possible outputs (excluding zero output) by 'X,' then what we might call the 'literalistic' interpretation of this assumption would be that

$$X = \varphi(\mathbf{R}^n_+) \setminus \{0\}; \tag{3.71}$$

that is, X is the set of all output quantities, x, such that there exists an input quantity, $\mathbf{v}$, such that

$$x = \varphi(\mathbf{v}) > 0.$$

In fact, we rather took this position implicitly ourselves when we argued in Section 3 that it made sense to assume that the production function was only upper semi-continuous, and not continuous. We argued there that making the weaker assumption allowed us to take either a 'stock' of a 'flow' interpretation of the quantities with which we are dealing; a distinction which is only important if we take (3.71), above, to be our definition of the producible set. On the other hand, in Definition 3.52, we defined the producible set, X, for φ, by

$$X = \{x \in \mathbf{R}_+ \mid (\exists \mathbf{v} \in \mathbf{R}^n_+)\colon \varphi(\mathbf{v}) \geq x > 0\}. \tag{3.72}$$

While we shall use (3.72), rather than (3.71), as our definition of the producible set, this does not constitute as much of an about-face as it appears. Our primary interest in the theory of the firm is in terms of analysis of the market behavior of such an entity; and the firm's market behavior will be the same whether we take (3.71) or (3.72) as our definition of the producible set, as long as the firm's revenue function is increasing in the quantity of output. We can see this as follows.

Suppose a firm has revenue and output expenditure functions $\rho\colon \mathbf{R}_+ \to \mathbf{R}_+$, and $\omega\colon \mathbf{R}^n_+ \to \mathbf{R}_+$, respectively, that $\rho(\cdot)$ is increasing, and that x^* is in X, but that there exists no $\mathbf{v}^* \in \mathbf{R}^n_+$ such that $x^* = \varphi(\mathbf{v}^*)$. Our definition of the producible set, as given in (3.72), above, implies, of course, that x^* can nonetheless be produced. However, let $\mathbf{v}^*$ be any vector of inputs satisfying

$$0 < x^* \leq \varphi(\mathbf{v}^*).$$

Then it follows from our assumption regarding x^* that

$$0 < x^* < \varphi(\mathbf{v}^*), \tag{3.73}$$

and thus, defining

$$\bar{x} = \varphi(\mathbf{v}^*),$$

we see that we will have

$$\rho(x^*) - \omega(\mathbf{v}^*) < \rho(\bar{x}) - \omega(\mathbf{v}^*).$$

Therefore, a profit-maximizing firm would never choose the input-output pair $(\mathbf{v}^*, x^*)$ in this case. In fact, the argument requires only that the firm's revenue function be increasing in output quantity to guarantee that it makes no difference whether we take the firm's producible set to be given by (3.71) or (3.72). As it happens, (3.72) is more convenient to work with, and consequently, this is what we shall use.

Essentially the same considerations as were just presented can be used to justify our assumption that production functions are non-decreasing. We will not attempt to formally develop this argument here, but it is a worthwhile exercise to consider which input quantities would be chosen by a profit-maximizing (or cost-minimizing) firm having the production function defined in Example 3.58.1.

The following result, sets forth the basic properties of (competitive) cost functions for the one-output case.

3.65. Theorem. *If* $\varphi\colon \mathbf{R}^n_+ \to \mathbf{R}_+$ *is a production function, and* $c\colon \mathbf{R}^n_+ \times X \to \mathbf{R}_+$ *is the cost function for* φ*, then* $c(\cdot)$ *satisfies the following:*

for each $\mathbf{w} \in \mathbf{R}^n_+, c(\mathbf{w}, \cdot)$ *is:*

1.a. non-decreasing in x*, and*

1.b. if $\mathbf{w} \in \mathbf{R}^n_{++}$*, then* $c(\mathbf{w}, \cdot)$ *is lower semi-continuous in* x *on* X*;*

and

for each $x \in X, c(\cdot, x)$ *is:*

2.a. positively homogeneous of degree one,

2.b. increasing,

2.c. concave, and

2.d. continuous in $\mathbf{w}$ *on* $\mathbf{R}^n_+$*.*

The proof of property 1.a of Theorem 3.65 is very easy, and will be left as an exercise. The properties of the cost function as a function of $\mathbf{w}$, for fixed output, as set out in 2.a–2.d of the above result are very easy implications of some results on 'support functions' which we will be studying in Chapter 6, and we shall defer the proof of these conclusions until that time. This leaves 1.b of the conclusion, which we shall prove shortly. Before we do, however, let's consider some further aspects of the cost function.

In the previous section, we defined the cost function for φ by:

$$c(\mathbf{w}, x) = \inf\{\mathbf{w} \cdot \mathbf{v} \mid \mathbf{v} \in \mathbf{R}^n_+ \ \& \ \varphi(\mathbf{v}) \geq x\} \quad \text{for } (\mathbf{w}, x) \in \mathbf{R}^n_+ \times X \quad (3.74)$$

This may well have struck you as being a bit strange, and it is in fact quite common to define the cost function as:

$$c(\mathbf{w}, x) = \min\{\mathbf{w} \cdot \mathbf{v} \mid \mathbf{v} \in \mathbf{R}^n_+ \ \& \ \varphi(\mathbf{v}) \geq x\} \quad \text{for } (\mathbf{w}, x) \in \mathbf{R}^n_+ \times X \quad (3.75)$$

There is, however, a very good reason for our not using (3.75) as our definition of the cost function; it may not be defined for all $\mathbf{w} \in \mathbf{R}^n_+$. Moreover, this may be the case even if the production function is continuous, and not just upper semi-continuous, as is shown by the following example.

3.66. Example. Let $\varphi\colon \mathbf{R}^2_+ \to \mathbf{R}_+$ be defined by

$$\varphi(v_1, v_2) = A v_1^{a_1} v_2^{a_2} \quad \text{for } \mathbf{v} = (v_1, v_2) \in \mathbf{R}^2_+,$$

where $A > 0$ and $a_i > 0$, for $i = 1, 2$. In this case the producible set, X, is given by

$$X = \mathbf{R}_{++};$$

but if $\bar{x} \in \mathbf{R}_{++}$, and $\bar{\mathbf{w}}$ is any element of $\mathbf{R}^2_+$ of the form

$$\bar{\mathbf{w}} = (0, \bar{w}_2), \text{ with } \bar{w}_2 > 0;$$

then, as you should have no trouble in verifying, there is no $\bar{\mathbf{v}} \in \mathbf{R}^2_+$ satisfying

$$\varphi(\bar{\mathbf{v}}) \geq \bar{x} \text{ and } \bar{\mathbf{w}} \cdot \bar{\mathbf{v}} \leq \bar{\mathbf{w}} \cdot \mathbf{v} \text{ for all } \mathbf{v} \in \mathbf{R}^2_+ \text{ such that } \varphi(\mathbf{v}) \geq \bar{x}.$$

On the other hand, the infimum in (3.74) is well-defined, and we have in this case:

$$\inf\{\bar{\mathbf{w}} \cdot \mathbf{v} \mid \mathbf{v} \in \mathbf{R}^2_+ \ \& \ \varphi(\mathbf{v}) \geq \bar{x}\} = 0.$$

In other words, while the function in (3.75) is not defined for this production function and these values for $\bar{x}$ and $\bar{\mathbf{w}}$, the function in (3.74) is well-defined in this situation. □

To make a long story short, Example 3.66 shows that there are some technical difficulties in using equation (3.75) as our definition of the cost function. On the other hand, as is noted in the following, these difficulties occur only on the boundaries of $\mathbf{R}^n_+$.

3.67. Proposition. *Let $\varphi\colon \mathbf{R}^n_+ \to \mathbf{R}_+$ be a production function, let $x^* \in X$, and let $\mathbf{w}^* \in \mathbf{R}^n_{++}$. Then*

$$c^*(\mathbf{w}^*, x^*) \equiv \min\{\mathbf{w}^* \cdot \mathbf{v} \mid \mathbf{v} \in \mathbf{R}^n_+ \;\&\; \varphi(\mathbf{v}) \geq x^*\} \tag{3.76}$$

is well-defined; that is, there exists $\mathbf{v}^* \in \mathbf{R}^n_+$ such that $\varphi(\mathbf{v}^*) \geq x^*$ and

$$(\forall \mathbf{v} \in \mathbf{R}^n_+)\colon \varphi(\mathbf{v}) \geq x^* \Rightarrow \mathbf{w}^* \cdot \mathbf{v} \geq \mathbf{w}^* \cdot \mathbf{v}^*.$$

Proof. Let $\mathbf{w}^* \in \mathbf{R}^n_{++}$, and let $x^* \in X$. It follows from the definition of the producible set, X, that there exists $v^* \in \mathbf{R}_+$ such that

$$\varphi(\mathbf{v}^*) \geq x^*. \tag{3.77}$$

Consider the set V defined by

$$V = \{\mathbf{v} \in \mathbf{R}^n_+ \mid \mathbf{w}^* \cdot \mathbf{v} \leq \mathbf{w}^* \cdot \mathbf{v}^*\}.$$

Since $\mathbf{w}^* \gg \mathbf{0}$, it can be shown that V is bounded; in exactly the same way, in fact, that we showed the budget set in Example 3.28 to be bounded. Furthermore, from the continuity of the inner product function and Theorem 3.22, we see that V is closed; so that it follows, therefore, that V is a compact set.

Now consider the set $V(x^*)$ defined by

$$V(x^*) = \{\mathbf{v} \in \mathbf{R}^n_+ \mid \varphi(\mathbf{v}) \geq x^*\}.$$

From equation (3.77) we see that $V(x^*)$ is non-empty, and it follows from 3.33 that it is a closed set. Therefore, the set V^* defined by:

$$V^* = V \cap V(x^*),$$

is compact; and, by (3.77) and the definition of V, it is also non-empty. Consequently, there exists an element $\bar{\mathbf{v}} \in V^*$ satisfying

$$(\forall \mathbf{v} \in V^*)\colon \mathbf{w}^* \cdot \mathbf{v} \geq \mathbf{w}^* \cdot \bar{\mathbf{v}};$$

and from the definitions of V^* and $V(x^*)$, it then follows that:

$$\mathbf{w}^* \cdot \bar{\mathbf{v}} = \min\{\mathbf{w}^* \cdot \mathbf{v} \mid \mathbf{v} \in \mathbf{R}^n_+ \;\&\; \varphi(\mathbf{v}) \geq x^*\}. \quad \square$$

Returning to the question of why it is that we use (3.74) as our definition of the cost function; notice that, with $c^*(\cdot)$ defined as in (3.75), above, we have, for any $x^* \in X$:

$$(\forall \mathbf{w} \in \mathbf{R}^n_{++})\colon c(\mathbf{w}, x^*) = c^*(\mathbf{w}, x^*). \tag{3.78}$$

Furthermore, if $\mathbf{w}^*$ is an element of the boundary of $\mathbf{R}^n_+$, then clearly there exists a sequence $\langle \mathbf{w}^q \rangle$ from $\mathbf{R}^n_{++}$ such that:

$$\mathbf{w}^q \rightarrow \mathbf{w}^*. \tag{3.79}$$

From equations (3.78) and (3.79), and the continuity of $c(\cdot, x^*)$, it then follows that:[17]

$$c(\mathbf{w}^*, x^*) = \lim_{q \to \infty} c^*(\mathbf{w}^q, x^*); \tag{3.80}$$

and thus we can see that the use of (3.74) as our definition of the cost function for $\varphi(\cdot)$ amounts to simply extending the domain of $c^*(\cdot)$ is a very natural and convenient way.

We can now return to complete our proof of Theorem 3.65.

Proof of 1.b of Theorem 3.65. Let $\mathbf{w}^* \in \mathbf{R}^n_{++}$ be arbitrary. Then we note that it suffices to prove (see Corollary 3.35) that for each real number, a, the set

$$X(a) \stackrel{\text{def}}{=} \{x \in X \mid c(\mathbf{w}^*, x) \leq a\},$$

is closed in X. In order to do this, we distinguish two cases.

1. $a < 0$. Here we obviously have $X(a) = \emptyset$, which is closed in X.

2. $a \geq 0$. In this case, let $\langle x_q \rangle$ be a sequence from $X(a)$ and x^* be an element of X such that

$$x_q \rightarrow x^*. \tag{3.81}$$

We wish to prove that it must then be the case that $x^* \in X(a)$ as well.

Now, since $\langle x_q \rangle$ is a sequence from $X(a)$, we have

$$c(\mathbf{w}^*, x_q) \leq a \quad \text{for } q = 1, 2, \ldots ; \tag{3.82}$$

and it follows from Proposition 3.67 that for each q, there exists $\mathbf{v}_q \in \mathbf{R}^n_+$ satisfying

$$\varphi(\mathbf{v}_q) \geq x_q \text{ and } \mathbf{w}^* \cdot \mathbf{v} = c(\mathbf{w}^*, x_q) \quad \text{for } q = 1, 2, \ldots ; \tag{3.83}$$

Moreover, if we define the set H by

$$H = \{\mathbf{v} \in \mathbf{R}^n_+ \mid \mathbf{w}^* \cdot \mathbf{v} \leq a\};$$

it can be shown that H is a compact set.[18] Moreover, it follows from (3.82) and (3.83) that $\mathbf{v}_q \in H$, for each q; and the Bolzano-Weierstrass Theorem

[17] Notice that (3.80) will follow for any sequence of vectors from $\mathbf{R}^n_{++}$ which satisfies (3.79).

[18] This can be done in almost exactly the same way in which we showed that $b(\mathbf{p}, w)$ was compact in Example 3.28.

then implies that we can assume, without loss of generality, that there exists $\mathbf{v}^* \in H$ such that

$$\mathbf{v}_q \to \mathbf{v}^* \text{ and } \mathbf{w}^* \cdot \mathbf{v}^* \leq a. \tag{3.84}$$

Now let $\epsilon \in \mathbf{R}_{++}$ be arbitrary. Then, since φ is upper semi-continuous at $\mathbf{v}^*$, there exists $\delta > 0$ such that:

$$(\forall \mathbf{v} \in N(\mathbf{v}^*, \delta) \cap \mathbf{R}^n_+) : \varphi(\mathbf{v}) < \varphi(\mathbf{v}^*) + \epsilon. \tag{3.85}$$

However, since $\mathbf{v}_q \to \mathbf{v}^*$, it also follows that there exists q^* such that for all $q \geq q^*$,

$$\mathbf{v}_q \in N(\mathbf{v}^*, \delta) \cap \mathbf{R}^n_+. \tag{3.86}$$

Thus, from equations (3.83)–(3.86), it follows that for all $q \geq q^*$:

$$x_q \leq \varphi(\mathbf{v}_q) < \varphi(\mathbf{v}^*) + \epsilon.$$

But it then follows from (3.81) that

$$x^* \leq \varphi(\mathbf{v}^*) + \epsilon;$$

and, since ϵ was arbitrary, we then see that

$$x^* \leq \varphi(\mathbf{v}^*). \tag{3.87}$$

From (3.84) and (3.87) we then have

$$c(\mathbf{w}^*, x^*) \leq \mathbf{w}^* \cdot \mathbf{v}^* \leq a;$$

and we then conclude that $x^* \in X(a)$, and thus that $X(a)$ is closed in X. □

Somewhat surprisingly, conclusion 1.b of Theorem 3.65 does not necessarily hold if $\mathbf{w}^* \notin \mathbf{R}^n_{++}$. In fact, in the following example, we show that there exists a *continuous* production function for which the function $c^*(\cdot)$ defined in equation (3.75), above, is well-defined over all of $\mathbf{R}^n_+ \times X$; but where we can also find $\mathbf{w}^* \in \mathbf{R}^n_+$ and $x^* \in X$ such that $c^*(\mathbf{w}^*, \cdot)$ is not lower semi-continuous in x at x^*.[19]

3.68. Example. In order to simplify the typing in this example, we shall denote the quantities of the two inputs by 'x' and 'y,' and the quantity of output by 'z.' Our production function φ: $\mathbf{R}^2_+ \to \mathbf{R}_+$ is then defined by:

$$\varphi(x, y) = \begin{cases} x + y & \text{for } x + y < 2, \\ 4y/[2 + y - x] & \text{for } 2 \leq x + y, x \leq y \,\&\, 0 \leq x < 2, \\ 4x/[2 + x - y] & \text{for } 2 \leq x + y, y < x \,\&\, 0 \leq y < 2, \\ 2\min\{x, y\} & \text{for } \min\{x, y\} \geq 2. \end{cases}$$

[19] If you have worked through Exercise 5, at the end of the previous section, you have already seen an example in which the cost function is not *upper* semi-continuous in x, for any non-zero value of $\mathbf{w}$.

It can be shown that φ is a continuous function;[20] and, since it is also non-decreasing, positive at all strictly positive points (x, y), and satisfies:

$$\varphi(0,0) = 0,$$

it is a production function, by our definition.[21]

As indicated earlier, we can show that, for each nonnegative $\mathbf{w} \in \mathbf{R}^2_+$ and each $z \in \mathbf{R}_+$, there exists an (x, y) which minimizes the cost of producing z. However, for the moment we will only concern ourselves with what happens when $\mathbf{w}^*$ is given by

$$\mathbf{w}^* = (1,0).$$

You should have no difficulty in showing that the values of x and y given by the functions:

$$x = \xi(z) = \begin{cases} 0 & \text{for } 0 \leq z < 4, \text{and} \\ z/2 & \text{for } 4 \leq z; \end{cases}$$

and

$$y = \psi(z) = \begin{cases} z & \text{for } 0 \leq z < 2, \\ 2z/(4-z) & \text{for } 2 \leq z < 4, \text{and} \\ z/2 & \text{for } 4 \leq z; \end{cases}$$

respectively, minimize the cost of producing z, given the vector of input prices $\mathbf{w}^* = (1, 0)$. Consequently, we have:

$$c(\mathbf{w}^*, z) = c^*(\mathbf{w}^*, z) = \begin{cases} 0 & \text{for } 0 \leq z < 4, \text{and} \\ z/2 & \text{for } 4 \leq z. \end{cases}$$

Obviously, $c(\mathbf{w}^*, \cdot)$ is *not* lower semi-continuous in z at $z = 4$. □

Ronald Shephard [1953] introduced the notion of homotheticity into the economics literature, and it has proved to be a tremendously useful concept. We shall conclude this section by developing some of the theory of homothetic production relationships for the single-output case.

3.69. Definition. We shall say that a function, $\varphi\colon \mathbf{R}^n_+ \to \mathbf{R}_+$ is **homothetic** iff there exists a transformation function, $F\colon \mathbf{R}_+ \to \mathbf{R}_+$, and a function $\sigma\colon \mathbf{R}^n_+ \to \mathbf{R}_+$ satisfying:

1. $(\forall \mathbf{v} \in \mathbf{R}^n_+)\colon \varphi(\mathbf{v}) = F[\sigma(\mathbf{v})]$, and
2. σ is positively homogeneous of degree one.

If σ is also a production function, we will say that φ is a **homothetic production function**.

[20] You will be asked to prove this as an exercise in Chapter 8.

[21] It can also be shown that φ is concave, but this is not our concern at the moment.

Since the function σ defined by

$$\sigma(\mathbf{v}) = 0 \quad \text{for all } \mathbf{v} \in \mathbf{R}^n_+,$$

is positively homogeneous of degree one, it follows at once that a function may be homothetic without being a production function. However, it can be shown that if σ is a production function (and F is a transformation function), then $\varphi = F \circ \sigma$ is also a production function. I will leave the proof of this fact as an exercise.

Notice also that if a function, φ, is positively homogeneous of degree $k > 0$, then it is homothetic; a fact which we can establish as follows. If we define the functions $\sigma \colon \mathbf{R}^n_+ \to \mathbf{R}_+$ and $F \colon \mathbf{R}_+ \to \mathbf{R}_+$ by:

$$\sigma(\mathbf{v}) = [\varphi(\mathbf{v})]^{1/k} \quad \text{for } \mathbf{v} \in \mathbf{R}^n_+,$$

and

$$F(y) = y^k \quad \text{for } y \in \mathbf{R}_+,$$

respectively, then σ is positively homogeneous of degree one, F is a transformation function, and, for each $\mathbf{v}$:

$$\varphi(\mathbf{v}) = F[\sigma(\mathbf{v})].$$

If φ is a production function with producible set, X, we define the **input-requirement correspondence for** φ by:

$$V(x) = \{\mathbf{v} \in \mathbf{R}^n_+ \mid \varphi(\mathbf{v}) \geq x\} \quad \text{for } x \in X;$$

that is, for $x \in X$, we define the **input-requirement set for** x by:

$$V(x) = \{\mathbf{v} \in \mathbf{R}^n_+ \mid \varphi(\mathbf{v}) \geq x\}.$$

We can characterize the input-requirement sets for a homothetic production function in a very useful alternative manner, as is shown in the following result.

3.70. Theorem. *If $\varphi \colon \mathbf{R}^n_+ \to \mathbf{R}_+$ is a homothetic production function, so that*

$$\varphi(\mathbf{v}) = F[\sigma(\mathbf{v})] \quad \textit{for } \mathbf{v} \in \mathbf{R}^n_+,$$

where F is a transformation function, and $\sigma \colon \mathbf{R}^n_+ \to \mathbf{R}_+$ is a production function which is positively homogeneous of degree one, then the input-requirement correspondence for φ can be written as:

$$V(x) = \{\mathbf{v} \in \mathbf{R}^n_+ \mid \sigma(\mathbf{v}) \geq f(x)\} \quad \textit{for } x \in X, \tag{3.88}$$

where f is the input requirement function dual to F. In fact, if we define the set V by:

$$V = \{\mathbf{v} \in \mathbf{R}^n_+ \mid \sigma(\mathbf{v}) \geq 1\}, \tag{3.89}$$

then

$$(\forall x \in X)\colon V(x) = f(x)V. \tag{3.90}$$

Proof.
1. Let $x^* \in X$ be arbitrary, and let $\mathbf{v}^* \in V(x^*)$, where

$$V(x^*) = \{\mathbf{v} \in \mathbf{R}^n_+ \mid F[\sigma(\mathbf{v})] \geq x^*\}.$$

Then we have

$$F[\sigma(\mathbf{v}^*)] \geq x^*, \tag{3.91}$$

and thus, since f is the input requirement function dual to F:

$$f(x^*) \leq \sigma(\mathbf{v}^*). \tag{3.92}$$

Conversely, if $\bar{\mathbf{v}}$ is such that

$$\sigma(\bar{\mathbf{v}}) \geq f(x^*);$$

then, using the fact that F is non-decreasing, and that f is the input-requirement function dual to F, we have:

$$F[\sigma(\bar{\mathbf{v}})] \geq F[f(x^*)] \geq x^*, \tag{3.93}$$

and thus we see that $\bar{\mathbf{v}} \in V(x^*)$.

From (3.92) and (3.93), we see that

$$V(x^*) = \{\mathbf{v} \in \mathbf{R}^n_+ \mid \sigma(\mathbf{v}) \geq f(x^*)\};$$

and, since $x^* \in X_0$ was arbitrary, we then conclude that (3.88) holds.

2. To prove that (3.90) holds, with V defined as in (3.89), let $x^\dagger \in X$ be arbitrary, and let $\mathbf{v}^* \in V(x^\dagger)$. Then by definition of $V(x^\dagger)$ and part 1 of our proof, we have

$$\sigma(\mathbf{v}^*) \geq f(x^\dagger); \tag{3.94}$$

so that, defining

$$\mathbf{v}^\dagger = [1/f(x^\dagger)]\mathbf{v}^*;$$

we have, using (3.94) and the homogeneity of σ:

$$\sigma(\mathbf{v}^\dagger) = \sigma\left([1/f(x^\dagger)]\mathbf{v}^*\right) = [1/f(x^\dagger)]\sigma(\mathbf{v}^*) \geq 1.$$

Therefore, $\mathbf{v}^\dagger \in V$, and, since

$$\mathbf{v}^* = f(x^\dagger)\mathbf{v}^\dagger,$$

it follows that $\mathbf{v}^* \in f(x^\dagger)V$, and therefore that

$$V(x^\dagger) \subseteq f(x^\dagger)V. \tag{3.95}$$

Conversely, let $\mathbf{v}' \in f(x^\dagger)V$. Then there exists $\mathbf{v} \in V$ such that

$$\mathbf{v}' = f(x^\dagger)\mathbf{v} \text{ and } \sigma(\mathbf{v}) \geq 1, \tag{3.96}$$

and using (3.96) and the homogeneity of σ, we have:

$$\sigma(\mathbf{v}') = \sigma[f(x^\dagger)\mathbf{v}] = f(x^\dagger)\sigma(\mathbf{v}) \geq f(x^\dagger).$$

Hence,

$$\mathbf{v}' \in V(x^\dagger),$$

and we conclude that

$$f(x^\dagger)V \subseteq V(x^\dagger). \tag{3.97}$$

Combining (3.95) and (3.97), we obtain equation (3.90). □

The cost function associated with a homothetic production function takes a particularly simple and convenient form, as is shown in the following result.

3.71. Theorem. *If $\varphi(\mathbf{v}) = F[\sigma(\mathbf{v})]$ is a homothetic production function, then the corresponding cost function is given by:*

$$c(\mathbf{w}, x) = \gamma(\mathbf{w}) \cdot f(x) \quad \textit{for } (\mathbf{w}, x) \in \mathbf{R}^n_+ \times X, \tag{3.98}$$

where

1. *f is the input requirement function dual to F, and*
2. *the function $\gamma\colon \mathbf{R}^n_+ \to \mathbf{R}_+$ is defined by:*

$$\gamma(\mathbf{w}) = \inf\{\mathbf{w} \cdot \mathbf{v} \mid \sigma(\mathbf{v}) \geq 1\} \quad \textit{for } \mathbf{w} \in \mathbf{R}^n_+.$$

Proof. Let $(\mathbf{w}, x) \in \mathbf{R}^n_+ \times X$. We have by Theorem 3.70 that

$$V(x) = f(x)V,$$

where

$$V = \{\mathbf{v} \in \mathbf{R}^n_+ \mid \sigma(\mathbf{v}) \geq 1\}.$$

Hence,

$$c(\mathbf{w}, x) = \inf\{\mathbf{w} \cdot \mathbf{v} \mid \mathbf{v} \in V(x)\} = \inf\{\mathbf{w} \cdot \mathbf{v} \mid \mathbf{v} \in f(x)V\}$$

$$= f(x) \cdot \{\mathbf{w} \cdot \mathbf{v} \mid \mathbf{v} \in V\} = f(x) \cdot \gamma(\mathbf{w}). \quad \square$$

Given a cost function, $c(\cdot)$, we often wish to consider the function $C(\cdot)$ defined on X by:

$$C(x) = c(\mathbf{w}^*, x) \quad \text{for } x \in X,$$

where $\mathbf{w}^* \in \mathbf{R}^n_+$ is a fixed vector of input prices. From Theorem 3.71 we see that in the homothetic case, C will be given by:

$$C(x) = \gamma^* f(x) \quad \text{where } \gamma^* \equiv \gamma(\mathbf{w}^*); \tag{3.99}$$

and notice that if we were to fix $\mathbf{w}$ at a different value, our new $C(\cdot)$ function [giving cost as a function of x, for a different value of $\mathbf{w}$] would simply be a different scalar multiple of $f(\cdot)$. Consequently, it is apparent that in the homothetic case, the form of the $C(\cdot)$ function, for a fixed value of $\mathbf{w}$, is determined solely by the input requirement function, f (and thus by the transformation function to which f is dual).

Another important aspect of this result, at least from the standpoint of applied microeconomics, is that in the homothetic case we can use the formula (3.98) to obtain a whole family of cost functions very quickly, once we know one of them. The idea is this: suppose $\sigma \colon \mathbf{R}^n_+ \to \mathbf{R}_+$ is a production function which is positively homogeneous of degree one, and suppose we have obtained the function $\gamma(\cdot)$, where

$$\gamma(\mathbf{w}) = \inf\{\mathbf{w} \cdot \mathbf{v} \mid \sigma(\mathbf{v}) \geq 1\} \quad \text{for } \mathbf{w} \in \mathbf{R}^n_+. \tag{3.100}$$

[*Notice that $\gamma(\cdot)$ will be the average cost function for $\sigma(\cdot)$.*[22]] For each transformation function, $F \colon \mathbf{R}_+ \to \mathbf{R}_+$, we obtain a homothetic production function by defining

$$\varphi(\mathbf{v}) = F[\sigma(\mathbf{v})];$$

and it follows from Theorem 3.71 that the cost function for φ is simply the product of the function γ obtained from (3.100) and the input requirement function dual to F. This makes it practical to try several different function forms for $\varphi(\cdot)$ in doing econometric estimation work with cost data; and notice also that one can select the appropriate form for the input requirement function, $f(\cdot)$, by looking at a plot of cost against output, for a particular

[22] See Exercises 1 and 7 at the end of this section.

(fixed) value of w; a procedure which will be particularly natural if one is working with cross-section data.

Exercises.

1. Show that if $\varphi\colon \mathbf{R}^n_+ \to \mathbf{R}_+$ is a production function which is positively homogeneous of degree $k > 0$, then the corresponding cost function is given by:

$$c(\mathbf{w}, x) = \gamma(\mathbf{w}) \cdot x^{1/k},$$

where the function γ is defined by:

$$\begin{aligned}\gamma(\mathbf{w}) &= \inf\{\mathbf{w}\cdot\mathbf{v} \mid \mathbf{v} \in \mathbf{R}^n_+ \;\&\; \varphi(\mathbf{v})^{1/k} \geq 1\} \\ &= \inf\{\mathbf{w}\cdot\mathbf{v} \mid \mathbf{v} \in \mathbf{R}^n_+ \;\&\; \varphi(\mathbf{v}) \geq 1\}.\end{aligned}$$

2. Show that if $\sigma(\cdot)$ takes the Cobb-Douglas form:

$$\sigma(\mathbf{v}) = A\prod\nolimits_{i=1}^{n} v_i^{\alpha_i},$$

where A is a positive constant, $\alpha_i \geq 0$ for $i = 1, \ldots, n$, and

$$\sum\nolimits_{i=1}^{n} \alpha_i = 1,$$

then the corresponding cost function is given by:

$$c(\mathbf{w}, x) = \left[B\prod\nolimits_{i=1}^{n} w_i^{\alpha_i}\right] \cdot x \quad \text{for } (\mathbf{w}, x) \in \mathbf{R}^n_+ \times X,$$

where

$$B = 1/\left[A\prod\nolimits_{i=1}^{n} (\alpha_i)^{\alpha_i}\right].$$

3. Show that if $\sigma(\cdot)$ takes the CES[23] form:

$$\sigma(\mathbf{v}) = \left[\sum\nolimits_{i=1}^{n} \alpha_i v_i^{-\rho}\right]^{-1/\rho}$$

where $-1 \leq \rho < 0$, or $0 < \rho$ and $\alpha_i \geq 0$, for $i = 1, \ldots, n$,

then the corresponding cost function is given by:

$$c(\mathbf{w}, x) = x \cdot \left[\sum\nolimits_{i=1}^{n} (\alpha_i w_i^{\rho})^{1/(1+\rho)}\right]^{(1+\rho)/\rho} \quad \text{for } (\mathbf{w}, x) \in \mathbf{R}^n_+ \times X.$$

4. Consider the production function, $\varphi\colon \mathbf{R}^n_+ \to \mathbf{R}_+$ defined by

$$\varphi(\mathbf{v}) = \alpha \cdot \sigma(\mathbf{v})/[1 + \sigma(\mathbf{v})],$$

[23] Constant Elasticity of Substitution.

where α is a positive constant. What is the cost function for φ, if (a) σ is the Cobb-Douglas function used in Exercise 2, above? (b) σ is the CES function used in Exercise 3, above?

5. Consider the production function, $\varphi\colon \mathbf{R}^n_+ \to \mathbf{R}_+$, defined by

$$\varphi(\mathbf{v}) = \alpha \cdot N[\sigma(\mathbf{v})],$$

where α is a positive constant, and $N(\cdot)$ is the 'floor function' used in Example 3.60.1. What is the cost function for φ, if (a) σ is the Cobb-Douglas function used in Exercise 2, above? (b) σ is the CES function used in Exercise 3, above?

6. Consider the function $c\colon \mathbf{R}^n_+ \times X \to \mathbf{R}_+$ defined by

$$c(\mathbf{w}, x) = \left[ax - bx^2 + cx^3\right] \cdot \left[\sum_{i=1}^{n} (\alpha_i w_i^{\rho})^{1/(1+\rho)}\right]^{(1+\rho)/\rho}$$

$$\text{for } (\mathbf{w}, x) \in \mathbf{R}^n_+ \times X;$$

where a, b, and c are positive constants, with

$$3ac \geq b^2,$$

$\alpha_i \geq 0$ for $i = 1, \ldots, n$; and either $-1 \leq \rho < 0$ or $0 < \rho$. Is $c(\cdot)$ the cost function corresponding to some production function? How do you know?

7. Show, without using Theorem 3.71, that if $\sigma(\cdot)$ is a production function which is positively homogeneous of degree one, then the function $\gamma : \mathbf{R}^n_+ \to \mathbf{R}_+$ defined by:

$$\gamma(\mathbf{w}) = \inf\{\mathbf{w} \cdot \mathbf{v} \mid \sigma(\mathbf{v}) \geq 1\} \quad \text{for } \mathbf{w} \in \mathbf{R}^n_+,$$

is the average cost function for $\sigma(\cdot)$.

3.7 Sequences of Functions and Limit Functions

We often have occasion to deal with functions which are defined as the limit of a sequence of functions. Thus, for example, in Chapter 2 we considered (Example 2.46), the infinite series, $f(x)$, defined by:

$$f(x) = \sum_{m=0}^{\infty} x^m/m!; \tag{3.101}$$

which we found to converge, for all values of $x \in \mathbf{R}$. We can view this example in a slightly different, though logically equivalent way, as follows. Define the function f_q on $\mathbf{R}$ as the qth partial sum of the series; that is,

$$f_q(x) = \sum_{m=0}^{q} x^m/m! \quad \text{for } x \in \mathbf{R}. \tag{3.102}$$

Since we know that, for each $x \in \mathbf{R}$, the sequence of partial sums of the series converges, we can see that it makes sense to say that the sequence of functions, $\langle f_q \rangle$, converges to the function f, defined in (3.101).[24] The question is, how do we formally define what we mean by such convergence? One way of defining convergence is given by the following.

3.72. Definition. Suppose $f_q\colon X \to \mathbf{R}^m$, for $q = 1, 2, \ldots$, where X is a non-empty subset of $\mathbf{R}^n$. We shall say that $\langle f_q \rangle$ **converges pointwise**, or that $\langle f_q \rangle$ is **pointwise convergent**, iff, for each $\mathbf{x}^* \in X$, the limit

$$\lim_{q \to \infty} f_q(\mathbf{x}^*), \tag{3.103}$$

exists (and is finite); in other words, $\langle f_q \rangle$ is **pointwise convergent** iff, for each $\mathbf{x}^* \in X$, the sequence $\langle f_q(\mathbf{x}^*) \rangle$ is convergent. If $f\colon X \to \mathbf{R}^m$, we then say that $\langle f_q \rangle$ **converges pointwise to f** iff, for each $\mathbf{x} \in X$, we have:

$$\lim_{q \to \infty} f_q(\mathbf{x}) = f(\mathbf{x}). \tag{3.104}$$

In this latter case, we shall refer to f as the **pointwise limit function** of the sequence $\langle f_q \rangle$.

You can easily verify the fact that the sequence of functions defined in (3.102), above, converges pointwise to the function f defined in equation (3.101), above.[25] Of course, whenever $\langle f_q \rangle$ is a sequence of functions which is pointwise convergent, we can define a function, $f\colon X \to \mathbf{R}^m$, by:

$$f(\mathbf{x}) = \lim_{q \to \infty} f_q(\mathbf{x}) \quad \text{for } \mathbf{x} \in X;$$

and we will then find that, by definition, $\langle f_q \rangle$ converges pointwise to f. As it turns out, however, this is not a very satisfactory way of defining the convergence of a sequence of functions; and shortly we will consider a better definition. Before doing so, however, and before investigating the issue of what is wrong with the present definition, let's take a look at some more examples of pointwise convergence.

3.73. Examples.

1. Let $\varphi\colon \mathbf{R} \to \mathbf{R}$ be any function, and define the sequence of functions, $\langle f_q \rangle$, by:

$$f_q(x) = (1/q) \cdot \varphi(x) \quad \text{for } x \in \mathbf{R}.$$

In this case, it should be clear that the sequence $\langle f_q \rangle$ is pointwise convergent, whatever the function φ looks like. Moreover, if we define $f\colon \mathbf{R} \to \mathbf{R}$ by:

$$f(x) = 0 \quad \text{for } x \in \mathbf{R},$$

---

[24] In this case, as we noted earlier, the function f is better known as $f(x) = \exp(x)$.

[25] That is, it converges pointwise to the function $\exp(x)$.

then $\langle f_q \rangle$ converges pointwise to f.

2. Let $X =]-1, 1]$, and define the sequence $\langle f_n \rangle$ on X by:

$$f_n(x) = x^n \quad \text{for } x \in X.$$

Here you should have no trouble in showing that $\langle f_n \rangle$ converges pointwise to the function f defined on X by:

$$f(x) = \begin{cases} 0 & \text{for } -1 < x < 1, \text{ and} \\ 1 & \text{for } x = 1. \end{cases}$$

3. Consider the sequence of functions,

$$f_n(x) = \sum\nolimits_{m=0}^{n} \frac{x^2}{(1+x^2)^m} \quad \text{for } x \in \mathbf{R}, \text{ and } n = 0, 1, \ldots .$$

We will show that $\langle f_n \rangle$ converges pointwise to the function f, defined by:

$$f(x) = \begin{cases} 0 & \text{for } x = 0, \text{ and} \\ 1 + x^2 & \text{for } x \neq 0. \end{cases}$$

We shall establish convergence for the case in which $x \neq 0$ (we'll leave the case in which $x = 0$ as an exercise). We begin by noting that, for $x \neq 0$:

$$f_n(x) - \left(\frac{1}{1+x^2}\right) \cdot f_n(x) = x^2 - \frac{x^2}{(1+x^2)^{n+1}};$$

and thus:

$$\left(\frac{x^2}{1+x^2}\right) \cdot f_n(x) = x^2 \cdot \left(1 - \frac{1}{(1+x^2)^{n+1}}\right),$$

so that:

$$f_n(x) = 1 + x^2 - \left(\frac{1}{1+x^2}\right)^n .$$

Thus we see that, for each $x \in \mathbf{R} \setminus \{0\}$,

$$f_n(x) \to 1 + x^2,$$

and notice that we have also proved that:

$$\sum\nolimits_{m=0}^{\infty} \frac{x^2}{(1+x^2)^m} = \begin{cases} 0 & \text{if } x = 0, \text{ and} \\ 1 + x^2 & \text{for } x \neq 0. \end{cases}$$

4. Consider the sequence $\langle f_n \rangle$ defined on $\mathbf{R}$ by:

$$f_n(x) = x^{2n}/(1 + x^{2n}) \quad \text{for } x \in \mathbf{R}, \text{ and } n = 1, 2, \ldots .$$

Show that $\langle f_n \rangle$ converges pointwise to the function, f, defined by:

$$f(x) = \begin{cases} 0 & \text{for } |x| < 1, \\ 1/2 & \text{for } |x| = 1, \text{ and} \\ 1 & \text{for } |x| > 1. \end{cases} \quad \square$$

In the first of the preceding examples, we looked at a case in which f, the pointwise limit function, is continuous (whatever the properties of φ). However, in each of the next three examples the pointwise limit function was not continuous, despite the fact that each of the functions f_n was continuous. In fact, in the fourth of the examples, the pointwise limit function was not only discontinuous, it was neither upper, nor lower semi-continuous. The alternative definition of convergence, which we present next, does not allow these inconsistencies.

3.74. Definition. Suppose $f_q\colon X \to \mathbf{R}^m$ for $q = 1, 2, \ldots$, where X is a non-empty subset of $\mathbf{R}^n$. We shall say that $\langle f_q \rangle$ **converges to a function** f (or that $\langle f_q \rangle$ **converges uniformly to** f, or that $\langle f_q \rangle$ is **uniformly convergent**),[26] and we write '$f_q \to f$,' if, and only if:

a. $f\colon X \to \mathbf{R}^m$, and

b. for each $\epsilon > 0$, there exists a positive integer, p, such that for all $q \geq p$, we have:

$$(\forall \mathbf{x} \in X)\colon \|f_q(\mathbf{x}) - f(\mathbf{x})\| < \epsilon.$$

Notice the difference between this and Definition 3.72. In the former case, the sequence $\langle f_q \rangle$ is said to converge pointwise to f iff, for each $\epsilon > 0$, and for each $\mathbf{x}^* \in X$, there exists p (which can depend upon both ϵ and $\mathbf{x}^*$) such that, for all $q \geq p$,

$$\|f_q(\mathbf{x}^*) - f(\mathbf{x}^*)\| < \epsilon.$$

On the other hand, in Definition 3.74, p can depend only upon ϵ, and not upon $\mathbf{x}^*$. Thus, consider the following example.

3.75. Example. Let f_q be defined on $\mathbf{R}_+$ by:

$$f_q(x) = \begin{cases} 1/qx & \text{for } x > 0, \text{ and} \\ 0 & \text{for } x = 0. \end{cases}$$

Here it follows as a special case of Example 3.73.1, above, that $\langle f_q \rangle$ is pointwise convergent to the function f defined on $\mathbf{R}_+$ by:

$$f(x) = 0 \quad \text{for } x \in \mathbf{R}_+.$$

However, $\langle f_q \rangle$ does not converge (uniformly) to f. To see this, let ϵ be equal to, say, one, and let q be any positive integer. Then, given any x satisfying

$$0 < x < 1/2q,$$

we will have:

$$|f_q(x) - f(x)| = f_q(x) = 1/qx > 2.$$

Consequently, we see that $\langle f_q \rangle$ does not converge (uniformly) to f □

[26] Insofar as I am aware, the original terminology for this property is that '$\langle f_n \rangle$ is uniformly convergent;' and this is the terminology which you will probably find to be used in most of the mathematics texts which you encounter.

We will examine the differences in the meaning of Definitions 3.72 and 3.74 in more detail shortly; but first let's examine some properties which, if satisfied by all of the f_q, where $f_q \to f$, will also be satisfied by f.

3.76. Proposition. *Suppose $f_q\colon X \to \mathbf{R}^m$ is bounded, for $q = 1, 2, \ldots$, where X is a non-empty subset of $\mathbf{R}^n$, and that $f_q \to f$. Then f is bounded as well.*

Proof. Let $\langle a_q \rangle$ be a sequence of real numbers satisfying:

$$(\forall \mathbf{x} \in X)\colon \|f_q(\mathbf{x})\| < a_q \quad \text{for } q = 1, 2, \ldots.$$

Since $f_q \to f$, there exists a positive integer, p, such that, for all $q \geq p$, we have:

$$(\forall \mathbf{x} \in X)\colon \|f_q(\mathbf{x}) - f(\mathbf{x})\| < 1. \tag{3.105}$$

But then, using (3.105) we see that, for any $\mathbf{x}^* \in X$, we will have:

$$\begin{aligned}\|f(\mathbf{x}^*)\| &= \|f(\mathbf{x}^*) - f_p(\mathbf{x}^*) + f_p(\mathbf{x}^*)\| \\ &\leq \|f(\mathbf{x}^*) - f_p(\mathbf{x}^*)\| + \|f_p(\mathbf{x}^*)\| < 1 + a_p.\end{aligned}$$

Thus we see that f is bounded on X □

3.77. Proposition. *Suppose $f_q\colon X \to \mathbf{R}^m$ is continuous, for $q = 1, 2, \ldots$, where X is a non-empty subset of $\mathbf{R}^n$, and that $f_q \to f$. Then f is continuous as well.*

Proof. Let $\mathbf{x}^*$ be an arbitrary element of X, and let $\epsilon > 0$ be given. Since $f_q \to f$, there exists a positive integer, p, such that, for all $q \geq p$,

$$(\forall \mathbf{x} \in X)\colon \|f_q(\mathbf{x}) - f(\mathbf{x})\| < \epsilon/3. \tag{3.106}$$

Furthermore, since f_p is a continuous function, there exists $\delta > 0$ such that

$$(\forall \mathbf{x} \in N(\mathbf{x}^*, \delta) \cap X)\colon \|f_p(\mathbf{x}) - f_p(\mathbf{x}^*)\| < \epsilon/3. \tag{3.107}$$

But then, utilizing (3.106) and (3.107), we see that if $\mathbf{x} \in N(\mathbf{x}^*, \delta) \cap X$, we will have:

$$\begin{aligned}\|f(\mathbf{x}) - f(\mathbf{x}^*)\| \leq{}& \|f(\mathbf{x}) - f_p(\mathbf{x})\| \\ &+ \|f_p(\mathbf{x}) - f_p(\mathbf{x}^*)\| + \|f_p(\mathbf{x}^*) - f(\mathbf{x}^*)\| < \epsilon.\end{aligned}$$ □

Returning to our discussion of the difference between convergence and pointwise convergence, we begin by noting that we could have equivalently defined (uniform) convergence by the condition in the following result; the proof of which I shall leave as an exercise.

3.78. Proposition. *Suppose $f_q\colon X \to \mathbf{R}^m$ for $q = 1, 2, \ldots$, where X is a non-empty subset of $\mathbf{R}^n$, and that $f\colon X \to \mathbf{R}^m$. Then $f_q \to f$ if, and only if, for every $\epsilon > 0$, there exists a positive integer, p, such that for all $q \geq p$, we have:*

$$\sup_{\mathbf{x} \in X} \|f_q(\mathbf{x}) - f(\mathbf{x})\| < \epsilon. \tag{3.108}$$

Suppose now that $f_q \to f$, and, for the sake of convenience, that $m = 1$; that is, that:

$$f_q\colon X \to \mathbf{R}, \quad \text{for } q = 1, 2, \ldots, \text{ and } f\colon X \to \mathbf{R}.$$

In this case, if $\epsilon > 0$ is given, it follows from (3.108) that there exists a positive integer, p, such that, for all $q \geq p$, and for all $\mathbf{x} \in X$:

$$f(\mathbf{x}) - \epsilon < f_q(\mathbf{x}) < f(\mathbf{x}) + \epsilon. \tag{3.109}$$

Thus, if the graph of f is as shown in Figure 3.4, below, condition (3.109) is equivalent to the statement that, for all $q \geq p$, the graph of f_q must lie within the area bounded by $\mathcal{G}_f + \epsilon$ and $\mathcal{G}_f - \epsilon$. Consequently, we can see that it makes sense in this case to say that, for all $q \geq p$, f_q must lie within an ϵ-neighborhood of f.[27]

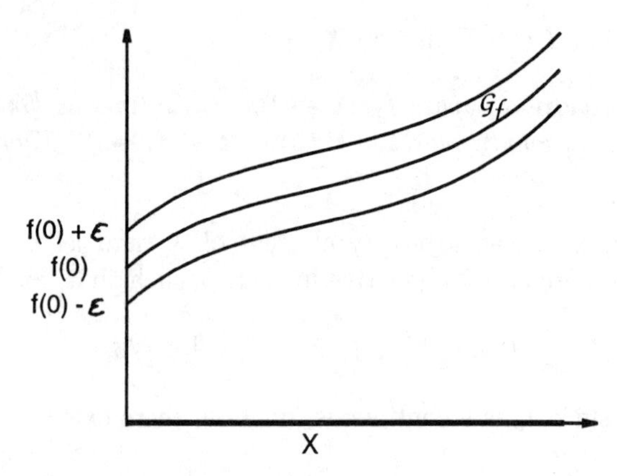

Figure 3.4: The 'Epsilon Neighborhood' of a Function.

We have already shown that, if $f_q \to f$, and each f_q is continuous, then so is f. It turns out that it is also true that, if each f_q is upper or lower semi-continuous, then f will be as well.

3.79. Theorem. *Suppose $f_q\colon X \to \mathbf{R}$ is upper (respectively, lower) semi-continuous, for $q = 1, 2, \ldots$, where X is a non-empty subset of $\mathbf{R}^n$, and $f_q \to f$. Then f is upper (respectively, lower) semi-continuous as well.*

[27] We will define these terms a bit more exactly in Chapter 7.

Proof. We shall prove this result for the case in which each f_q is upper semi-continuous. A similar argument establishes the result for the case in which f_q is lower semi-continuous.

Let $\mathbf{x}^* \in X$, and let $\epsilon > 0$ be given. Since $f_q \to f$, there exists a positive integer, p, such that, for all $q \geq p$, and for all $\mathbf{x} \in X$, we will have:

$$|f_q(\mathbf{x}) - f(\mathbf{x})| = |f(\mathbf{x}) - f_q(\mathbf{x})| < \epsilon/3;$$

and thus

$$-\epsilon/3 < f_q(\mathbf{x}) - f(\mathbf{x}) < \epsilon/3, \tag{3.110}$$

and:

$$-\epsilon/3 < f(\mathbf{x}) - f_q(\mathbf{x}) < \epsilon/3. \tag{3.111}$$

Since f_p is upper semi-continuous at $\mathbf{x}^*$, there also exists $\delta > 0$ such that, for all $\mathbf{x} \in N(\mathbf{x}^*, \delta) \cap X$, we will have:

$$f_p(\mathbf{x}) - f_p(\mathbf{x}^*) < \epsilon/3. \tag{3.112}$$

Combining (3.110)–(3.112), we see that if $\mathbf{x} \in N(\mathbf{x}^*, \delta) \cap X$:

$$f(\mathbf{x}) - f(\mathbf{x}^*) = f(\mathbf{x}) - f_p(\mathbf{x}) + f_p(\mathbf{x}) - f_p(\mathbf{x}^*) + f_p(\mathbf{x}^*) - f(\mathbf{x}^*) < \epsilon. \quad \square$$

We conclude this section by establishing a very useful criterion for (uniform) convergence. You will probably immediately recognize the fact that it is derived from the Cauchy criterion.

3.80. Proposition. *Suppose $f_q \colon X \to \mathbf{R}^m$, for $q = 1, 2, \ldots$, where X is a non-empty subset of $\mathbf{R}^n$. Then there exists a function, f, such that $f_q \to f$ if, and only if, the following condition is satisfied: for each $\epsilon > 0$, there exists a positive integer, p, such that for all $q, q' \geq p$:*

$$\sup_{\mathbf{x} \in X} \|f_q(\mathbf{x}) - f_{q'}(\mathbf{x})\| < \epsilon. \tag{3.113}$$

Proof. Suppose $f_q \to f$, and let $\epsilon > 0$ be given. By 3.78 there exists a positive integer, p, such that, for all $q \geq p$,

$$\sup_{\mathbf{x} \in X} \|f_q(\mathbf{x}) - f(\mathbf{x})\| < \epsilon/3. \tag{3.114}$$

Consequently, if $q, q' \geq p$, and $\mathbf{x}^*$ is an arbitrary element of X, we will have:

$$\|f_q(\mathbf{x}^*) - f_{q'}(\mathbf{x}^*)\| \leq \|f_q(\mathbf{x}^*) - f(\mathbf{x}^*)\| + \|f(\mathbf{x}^*) - f_{q'}(\mathbf{x}^*)\| < 2\epsilon/3;$$

and it follows at once that

$$\sup_{\mathbf{x} \in X} \|f_q(\mathbf{x}) - f_{q'}(\mathbf{x})\| \leq 2\epsilon/3 < \epsilon.$$

Suppose now that (3.113) holds, and let $\mathbf{x}^*$ be an arbitrary element of X. If we consider the sequence $\langle f_q(\mathbf{x}^*)\rangle$, it is clear that this is a Cauchy sequence; and thus, by Theorem 2.66, there exists an element $f(\mathbf{x}^*) \in \mathbf{R}^m$ such that $f_q(\mathbf{x}^*) \to f(\mathbf{x}^*)$. Since $\mathbf{x}^*$ was an arbitrary element of X, it is then clear that we can define a function, $f\colon X \to \mathbf{R}^m$, by:

$$f(\mathbf{x}) = \lim_{q\to\infty} f_q(\mathbf{x}) \quad \text{for } \mathbf{x} \in X;$$

and, by definition, f_q converges *pointwise* to f. What we now need to prove is that f_q converges (uniformly) to f.

Accordingly, let $\epsilon > 0$ be given. By condition (3.113), there exists a positive integer, p, such that for all $q, q' \geq p$,

$$\sup_{\mathbf{x}\in X} \|f_q(\mathbf{x}) - f_{q'}(\mathbf{x})\| < \epsilon/2. \tag{3.115}$$

Now, let $\mathbf{x}^* \in X$ be arbitrary, and let $q \geq p$. We wish to prove that

$$\|f_q(\mathbf{x}^*) - f(\mathbf{x}^*)\| < \epsilon. \tag{3.116}$$

To establish (3.116), we first recall that, since f_q converges pointwise to f, there exists p^* such that, for all $n \geq p^*$,

$$\|f_n(\mathbf{x}^*) - f(\mathbf{x}^*)\| < \epsilon/2, \tag{3.117}$$

and we define $p' = \max\{p, p^*\}$, where p is from (3.115). But then it follows from (3.115) and (3.117) that, for $q \geq p$,

$$\|f_q(\mathbf{x}^*) - f(\mathbf{x}^*)\| \leq \|f_q(\mathbf{x}^*) - f_{p'}(\mathbf{x}^*)\| + \|f_{p'}(\mathbf{x}^*) - f(\mathbf{x}^*)\| < \epsilon;$$

which verifies (3.116). Since $\mathbf{x}^*$ was arbitrary, it now follows that $f_q \to f$. □

Exercises.

1. Prove Proposition 3.78. (This is essentially a very easy exercise, except that there is one little trap into which you may fall while doing the necessity portion of this proof. See Exercise 2, below.)

2. Let $X = \mathbf{R}_+$, and define f_q on X by:

$$f_q(x) = x/[q(1+x)] \quad \text{for } q = 1, 2, \ldots;$$

and do the following.

a. Show that $f_q \to f$, where f is the constant function:

$$f(x) = 0 \quad \text{for } x \in \mathbf{R}_+.$$

b. Show that if we take $p = 4$, then for all $q \geq p$, and all $x \in \mathbf{R}_+$:

$$|f_q(x) - 0| = x/[q(1+x)] = f_q(x) < 1/4.$$

c. Show that:

$$\sup_{x \in X} f_4(x) = 1/4.$$

3. Show that if $f_q \colon X \to \mathbf{R}^m$ is a sequence of bounded functions, and $f_q \to f$, then $\langle f_q \rangle$ is **uniformly bounded**; that is, there exists a real number, α, such that for all q and for all $x \in X$:

$$\|f_q(\mathbf{x})\| \leq \alpha.$$

4. Show that the sequence $\langle f_q \rangle$ defined on $X = [0, 1]$ by:

$$f_q(x) = 1/(1 + qx) \quad \text{for } q = 1, 2, \ldots,$$

is pointwise convergent on X. Is $\langle f_q \rangle$ uniformly convergent on X? Explain.

5. Show that the sequence $\langle f_q \rangle$ defined on $X = \mathbf{R}_+$ by:

$$f_q(x) = x/\exp(qx) \quad \text{for } q = 1, 2, \ldots,$$

is (uniformly) convergent on X.

Chapter 4

Linear Spaces

4.1 Introduction

Formally, a linear space is a collection of things (which we will call 'vectors' or 'points'), together with two operations, vector addition and scalar multiplication. The latter operation (scalar multiplication) is defined for each possible combination, or pair, consisting of a vector from the linear space and an element (which we will generally call a 'scalar') from a given *algebraic field*. In this text, the only fields with which we will be dealing in defining a linear space are the number systems $\mathbf{Q}$ (the rational numbers), $\mathbf{R}$ (the real numbers), and $\mathbf{C}$ (the complex numbers); and most often with $\mathbf{R}$, for that matter. Nevertheless, it will be worth our while to present a formal definition of a field, and to use it in our definition of a linear space, because this will enable us to state and prove results which apply to linear spaces defined over any of the three fields listed above, rather than having to state and prove results separately for each of the three cases.

4.1. Definition. A **field** $\mathcal{F} = (\mathbf{F}, +, \cdot)$ consists of a set of elements, $\mathbf{F}$, called scalars or numbers, together with two binary operations,[1] $+$ and $\cdot$, which we will call 'addition' and 'multiplication,' respectively, and which satisfy the following axioms:

1 (**Axioms for addition**). The binary operation addition satisfies, for each $\alpha, \beta, \gamma \in \mathbf{F}$:

a. Associativity: $\alpha + (\beta + \gamma) = (\alpha + \beta) + \gamma$.

b. Commutativity: $\alpha + \beta = \beta + \alpha$.

c. Existence of an additive identity: there exists a scalar, $0 \in \mathbf{F}$, satisfying, for each $\alpha \in \mathbf{F}$:

$$\alpha + 0 = 0 + \alpha = \alpha.$$

[1]Recall that a **binary operation** on a set X is a function, $f \colon X \times X \to X$. Thus a binary operation maps each pair of points from X into another point in X.

d. Existence of an additive inverse: for each $\alpha \in \mathbf{F}$, there exists an element, $-\alpha \in \mathbf{F}$, called the **negative of α**, satisfying:

$$\alpha + (-\alpha) = (-\alpha) + \alpha = 0.$$

2. **(Axioms for multiplication)**. The binary operation multiplication satisfies, for each $\alpha, \beta, \gamma \in \mathbf{F}$:

a. Associativity: $\alpha \cdot (\beta \cdot \gamma) = (\alpha \cdot \beta) \cdot \gamma$.

b. Commutativity: $\alpha \cdot \beta = \beta \cdot \alpha$.

c. Existence of a multiplicative identity: there exists a scalar, $1 \in \mathbf{F}$, satisfying, $1 \neq 0$, and, for each $\alpha \in \mathbf{F}$:

$$\alpha \cdot 1 = 1 \cdot \alpha = \alpha.$$

d. Existence of a multiplicative inverse: for each $\alpha \in \mathbf{F}$ such that $\alpha \neq 0$, there exists an element, $\alpha^{-1} \in \mathbf{F}$, called the **inverse of α**, satisfying:

$$\alpha \cdot \alpha^{-1} = \alpha^{-1} \cdot \alpha = 1.$$

3. **(Distributive Law)**. For each $\alpha, \beta, \gamma \in \mathbf{F}$, we have:

$$\alpha \cdot (\beta + \gamma) = \alpha \cdot \beta + \alpha \cdot \gamma = (\beta + \gamma) \cdot \alpha.$$

In dealing with fields, we will generally abuse the notation just introduced by denoting the field by, for example, '$\mathbf{F}$,' instead of '$\mathcal{F} = (\mathbf{F}, +, \cdot)$;' in other words, we will generally just use the underlying set of scalars in referring to the field, instead of using the ordered triple notation just introduced. The student who has not previously encountered the formal definition of a field should be sure to verify the fact that each of the three systems, $\mathbf{Q}$, $\mathbf{R}$, and $\mathbf{C}$, is a field by the above definition (with addition and multiplication defined in the usual way, and with the usual meanings attached to '0' and '1'). With this definition of a field, we can now define a linear space as follows.

4.2. Definition. We shall say that $\mathcal{L} = (L, \oplus, \otimes)$ is a **linear space over the field F** iff $\mathcal{L}$ satisfies:

1. **(Closure)**. L is a nonempty set (the elements of which we shall call '**vectors**,' or, sometimes, '**points**'), $\oplus$ is a binary operation on L,[2] and $\otimes\colon \mathbf{F} \times L \to L$ [and, given $\alpha \in \mathbf{F}$, and $x \in L$, we use the notation '$\alpha \otimes x$' to denote the value of $\otimes$ for the pair (α, x)].

2. **(Axioms for vector addition)**. $\oplus$ satisfies, for each $x, y, z \in L$:

a. Associativity: $x \oplus (y \oplus z) = (x \oplus y) \oplus z$.

b. Commutativity: $x \oplus y = y \oplus x$.

c. Existence of an additive identity: there exists a vector $\mathbf{0} \in L$

[2]That is, for each $x, y \in L$, $x \oplus y$ is defined, and is an element of L.

(called the **null vector**, or the **origin**, or sometimes simply **zero**) satisfying (for each $x \in L$):

$$x \oplus \mathbf{0} = \mathbf{0} \oplus x = x.$$

d. Existence of an additive inverse: for each $x \in L$, there exists an element $-x \in L$ (called the **negative of x** or **minus x**) satisfying:

$$x \oplus (-x) = (-x) \oplus x = \mathbf{0}.$$

3. (**Axioms for scalar multiplication**). $\otimes$ satisfies, for each $\alpha, \beta \in \mathbf{F}$, and each $x \in L$:
 a. $(\alpha \cdot \beta) \otimes x = \alpha \otimes (\beta \otimes x)$.
 b. $1 \otimes x = x$ (where '1' denotes the multiplicative identity for $\mathbf{F}$).
4. (**Distributive Laws**). For each $\alpha, \beta \in \mathbf{F}$, and each $x, y \in L$:
 a. $(\alpha + \beta) \otimes x = (\alpha \otimes x) \oplus (\beta \otimes x)$.
 b. $\alpha \otimes (x \oplus y) = (\alpha \otimes x) \oplus (\alpha \otimes y)$.

Notice that if $\mathbf{F}$ is a field, then $\mathbf{F}$ is a linear space over $\mathbf{F}$ (with $\oplus = +$, and $\otimes = \cdot$); and thus a linear space is a generalization of a field. Our next two examples are sufficiently important that I will present them in the form of theorems, although I will leave the first proof as an exercise.

4.3. Theorem. *Let* $\mathbf{F}$ *be a field, and let '*$\mathbf{F}^n$*' denote its n-fold cartesian product; that is, let* $\mathbf{F}^n$ *be the set of all n-tuples,* $\mathbf{a} = (\alpha_1, \ldots, \alpha_n)$ *such that:*

$$\alpha_i \in \mathbf{F} \quad \textit{for}\, i = 1, \ldots, n;$$

and define $\oplus$ *and* $\otimes$ *on* $\mathbf{F}^n \times \mathbf{F}^n$ *and* $\mathbf{F} \times \mathbf{F}^n$*, respectively, by:*

$$\mathbf{a} \oplus \mathbf{b} = (\alpha_1 + \beta_1, \alpha_2 + \beta_2, \ldots, \alpha_n + \beta_n)$$
$$\textit{for}\, \mathbf{a} = (\alpha_1, \ldots, \alpha_n) \textit{ and } \mathbf{b} = (\beta_1, \ldots, \beta_n),$$

and

$$\alpha \otimes \mathbf{b} = (\alpha \cdot \beta_1, \ldots, \alpha \cdot \beta_n) \quad \textit{for}\, \alpha \in \mathbf{F} \textit{ and } \mathbf{b} \in \mathbf{F}^n.$$

Then $\mathfrak{L} = (\mathbf{F}^n, \oplus, \otimes)$ *is a linear space over* $\mathbf{F}$.

Notice that, as a special case of the above result, it follows that the familiar linear space $\mathbf{R}^n$ is a linear space over $\mathbf{R}$. An even more general class of linear spaces (which actually includes, in a sense, the class developed in 4.3) is introduced in the following result.

4.4. Theorem. *Let* X *be any nonempty set, let* $\mathfrak{L} = (L, \oplus, \otimes)$ *be a linear space over a field,* $\mathbf{F}$*, and define* $\mathbf{S}$ *by:*

$$\mathbf{S} = \{f \mid f\colon X \to L\};$$

that is, $\mathbf{S}$ *is the set of all functions mapping* X *into* L *(this set is often denoted by '*L^X*'). If we define* $+$ *and* $*$ *on* $\mathbf{S} \times \mathbf{S}$ *and* $\mathbf{F} \times \mathbf{S}$ *by:*

$$(f + g)(x) = f(x) \oplus g(x) \quad \text{for each } x \in X; \text{ and } f, g, \in \mathbf{S}, \tag{4.1}$$

and

$$(\alpha * f)(x) = \alpha \otimes f(x) \quad \text{for each } \alpha \in \mathbf{F}, f \in \mathbf{S}, \text{ and } x \in X, \tag{4.2}$$

respectively, then $\mathcal{S} = (\mathbf{S}, +, *)$ *is a linear space over* $\mathbf{F}$.

Proof. We will present the outline of a proof, details will be left as an exercise. We first show that the functions set out in equations (4.1) and (4.2), above, are well-defined. If we begin by considering the first definition, we note that, for any f and g in S, if $x \in X$, then $f(x)$ and $g(x)$ are both elements of L. Therefore, $f(x) \oplus g(x)$ is (defined and is) a vector in L. Consequently, it follows that the function $f + g$ set forth in (4.1) is a mapping from X into L, and is therefore an element of S. A similar argument establishes that the function $\alpha * f$ defined in (4.2) is an element of S.

It is then easy to verify the fact that $\mathcal{S} = (S, +, *)$ is a linear space over $\mathbf{F}$, with additive identity (zero vector) θ given by:

$$\theta(x) = \mathbf{0} \quad \text{for each } x \in X,$$

where '$\mathbf{0}$' denotes the additive identity for L. The fact that $\mathcal{S}$ satisfies the axioms for vector addition and scalar multiplication, as well as the distributive laws, follows readily from the fact that they are satisfied by the underlying space, $\mathcal{L} = (L, \oplus, \otimes)$. □

Notice that $\mathbf{R}^n$ can be regarded as the collection of all mappings from the first n positive integers, $N_n \stackrel{\text{def}}{=} \{1, \ldots, n\}$, into $\mathbf{R}$. Similarly, $\mathbf{C}^n$ can be regarded as the collection of all mappings from N_n into the complex numbers, $\mathbf{C}$. In fact, by extending this reasoning, you should be able to show that Theorem 4.3 is actually a special case of Theorem 4.4. It also follows immediately from Theorem 4.4 that the collection of all sequences of real numbers, which we will sometimes denote by $\mathbf{R}^\infty$, is a linear space.

Before proceeding further in our study of linear spaces, I would like to simplify our notation somewhat. In the first place, from here on we will revert to the more familiar notation, $+$, to denote vector addition; thus we will denote the sum of two elements of L, x and y by '$x + y$' rather than '$x \oplus y$.' The reason that the original notation was introduced was to allow us to carefully distinguish between vector summation on L and scalar summation on $\mathbf{F}$ in the statement of the axioms for a linear space. In actually working with linear spaces, however, it will be clear from the context whether we are dealing with summation on L or summation on $\mathbf{F}$

when we use the symbol '+;' particularly so since we will hereafter generally use bold-face type in indicating elements of L, denoting these elements by '$\mathbf{x}$,' '$\mathbf{y}$,' '$\mathbf{z}$,' etc. Similarly, we will hereafter denote the product of a scalar $a \in F$ and a vector $\mathbf{x} \in L$ by simple juxtaposition, '$a\mathbf{x}$,' rather that '$a \otimes \mathbf{x}$.' Finally, we will use the same simplification we have used in dealing with fields in that we will refer to the underlying set L as the linear space, rather than using the notation $\mathcal{L} = (L, \oplus, \otimes)$. Just to make sure that there are no misunderstandings, let's re-state the definition of a linear space using the notational simplifications just introduced. Hopefully the reader will then see the advantages of both the notation used in the original statement, and of that used in the simplification.

4.2′. Definition. We shall say that L is a **linear space over the field F** iff L is a nonempty set (the elements of which we shall call '**vectors**,' or, sometimes, '**points**), for which two operations, vector addition and scalar multiplication, are defined, and which satisfy the following conditions:

1. (**Axioms for vector addition**). The operation of vector addition satisfies, for each $\mathbf{x}, \mathbf{y}, \mathbf{z} \in L$:

   a. Closure: $\mathbf{x} + \mathbf{y}$ is defined and is an element of L.

   b. Associativity: $\mathbf{x} + (\mathbf{y} + \mathbf{z}) = (\mathbf{x} + \mathbf{y}) + \mathbf{z}$.

   c. Commutativity: $\mathbf{x} + \mathbf{y} = \mathbf{y} + \mathbf{x}$.

   d. Existence of an additive identity: there exists a vector $\mathbf{0} \in L$ (called the **null vector**, or the **origin**, or sometimes simply **zero**) satisfying (for each $\mathbf{x} \in L$):

$$\mathbf{x} + \mathbf{0} = \mathbf{0} + \mathbf{x} = \mathbf{x}.$$

   e. Existence of an additive inverse: for each $\mathbf{x} \in L$, there exists an element $-\mathbf{x} \in L$ (called the **negative of x** or **minus x**) satisfying:

$$\mathbf{x} + (-\mathbf{x}) = (-\mathbf{x}) + \mathbf{x} = \mathbf{0}.$$

2. (**Axioms for scalar multiplication**). The operation of scalar multiplication satisfies, for each $a, b \in \mathbf{F}$, and each $\mathbf{x} \in L$:

   a. Closure: $a\mathbf{x}$ is defined, and is an element of L.

   b. Associativity: $(a \cdot b)\mathbf{x} = a(b\mathbf{x})$.

   c. $1\mathbf{x} = \mathbf{x}$ (where '1' denotes the multiplicative identity for $\mathbf{F}$).

3. (**Distributive Laws**). For each $a, b \in \mathbf{F}$, and each $\mathbf{x}, \mathbf{y} \in L$:

   a. $(a + b)\mathbf{x} = a\mathbf{x} + b\mathbf{x}$.

   b. $a(\mathbf{x} + \mathbf{y}) = a\mathbf{x} + a\mathbf{y}$.

Our use of bold-face type to denote vectors will enable us, where convenient, to use normal Roman letters to denote scalars. Thus, for example, we will generally use the generic notation:

$$\mathbf{x} = (x_1, \dots, x_n), \mathbf{y} = (y_1, \dots, y_n), \mathbf{z} = (z_1, \dots, z_n),$$

etc., to denote elements of $\mathbf{R}^n$.

We conclude this section by proving some very basic facts about linear spaces. In each of these results, we will suppose that L is a linear space over some field, $\mathbf{F}$; and *in each reference to an Axiom in a proof, we will be referring to the axioms in Definition* 4.2′.

4.5. Proposition. *The additive identity,* $\mathbf{0}$*, for* L *is unique; in fact if* $\mathbf{z}^* \in L$ *is such that there exists* $\mathbf{x}^* \in L$ *satisfying:*

$$\mathbf{z}^* + \mathbf{x}^* = \mathbf{x}^*, \tag{4.3}$$

then $\mathbf{z}^* = \mathbf{0}$.

Proof. Suppose $\mathbf{x}^*$ and $\mathbf{z}^*$ satisfy (4.3). Then using Axioms 1.d, 1.e, 1.b, our hypothesis [Equation (4.3)], and Axiom 1.e in turn, we have:

$$\begin{aligned}\mathbf{z}^* = \mathbf{z}^* + \mathbf{0} = \mathbf{z}^* + [\mathbf{x}^* + (-\mathbf{x}^*)]\\ = (\mathbf{z}^* + \mathbf{x}^*) + (-\mathbf{x}^*) = \mathbf{x}^* + (-\mathbf{x}^*) = \mathbf{0}.\end{aligned} \quad \square$$

4.6. Proposition. *For any* $\mathbf{x} \in L$*, we have* $0\mathbf{x} = \mathbf{0}$ *(where '0' denotes the additive identity in* $\mathbf{F}$*).*

Proof. Using Axioms 2.c, 3.a, and 2.c in turn, we have:

$$0\mathbf{x} + \mathbf{x} = 0\mathbf{x} + 1\mathbf{x} = (0 + 1)\mathbf{x} = 1\mathbf{x} = \mathbf{x}.$$

It then follows at once from Proposition 4.5 that $0\mathbf{x} = \mathbf{0}$. $\square$

4.7. Proposition. *For any* $\mathbf{x} \in L$*, the additive inverse of* $\mathbf{x}$*,* $-\mathbf{x}$*, is unique; that is, if:*

$$\mathbf{x} + \mathbf{y} = \mathbf{x} + \mathbf{z} = \mathbf{0}, \tag{4.4}$$

then:

$$\mathbf{y} = \mathbf{z} = -\mathbf{x}.$$

Proof. Suppose $\mathbf{y}$ and $\mathbf{z}$ satisfy (4.4). Then, using Axiom 1.d, (4.4), Axioms 1.b and 1.c, (4.4), and Axiom 1.d in turn, we have:

$$\mathbf{y} = \mathbf{y} + \mathbf{0} = \mathbf{y} + (\mathbf{x} + \mathbf{z}) = (\mathbf{y} + \mathbf{x}) + \mathbf{z} = (\mathbf{x} + \mathbf{y}) + \mathbf{z} = \mathbf{0} + \mathbf{z} = \mathbf{z}. \quad \square$$

4.8. Proposition. *For all* $a \in \mathbf{F}$*, and all* $\mathbf{x} \in L$*, we have:*

$$-(a\mathbf{x}) = (-a)\mathbf{x} = [(-1) \cdot a]\mathbf{x} = a[(-1)\mathbf{x}] = a(-\mathbf{x}) \tag{4.5}$$

[in particular, $-\mathbf{x} = (-1)\mathbf{x}$*].*

Proof. Using Axiom 3.a and Proposition 4.6 in turn, we have:

$$a\mathbf{x} + (-a)\mathbf{x} = (a - a)\mathbf{x} = 0\mathbf{x} = \mathbf{0}.$$

Thus it follows from Proposition 4.7 that:

$$-(a\mathbf{x}) = (-a)\mathbf{x}, \tag{4.6}$$

which verifies the first equality in (4.5). The second equality in (4.5) follows from (4.6) [applied to $\mathbf{F}$], the third equality follows from Axiom 2.b and the commutativity of multiplication in $\mathbf{F}$, while the last equality follows from (4.6) [with $a = 1$]. □

Exercises.

1. Does the set of integers, $I = \{0, 1, -1, 2, -2, \dots\}$, together with the usual definitions of addition and multiplication, constitute a field? Explain why or why not.

2. Show that the additive inverse relationship is symmetric; that is, if $\mathbf{x} = -\mathbf{y}$, then $\mathbf{y} = -\mathbf{x}$.

3. Prove the following laws of cancellation for a linear space, L: If $\mathbf{x}, \mathbf{y}, \mathbf{z} \in L$ are such that

$$\mathbf{x} + \mathbf{y} = \mathbf{x} + \mathbf{z},$$

or if, for some scalar, $a \neq 0$,

$$a\mathbf{y} = a\mathbf{z},$$

then $\mathbf{y} = \mathbf{z}$.

4. If $\mathbf{x}, \mathbf{y} \in L$, show that there exists $\mathbf{z} \in L$ such that $\mathbf{x} + \mathbf{z} = \mathbf{y}$. Is $\mathbf{z}$ unique?

5. If $\mathbf{x}, \mathbf{y} \in L$, is it necessarily the case that there exists a scalar, a, such that $a\mathbf{x} = \mathbf{y}$? (Compare with exercise 4, above.) Justify your answer.

6. Show that the set $F = \{0, 1, 2, 3, 4\}$, together with addition and multiplication modulo 5, constitute a field; where we define these two operations in the following way.

Addition modulo five: for $x, y \in F$, which we shall denote by '$x \oplus y$,' is defined as the remainder of $(x+y)$ after dividing by 5; that is, for $x, y \in F$, there exist a unique nonnegative integer, p, and a unique $r \in F$, such that:

$$x + y = 5p + r,$$

and we define:

$$x \oplus y = r.$$

Thus, for example,

$$3 \oplus 4 = 2 \text{ (since } 3 + 4 = 7 = 5 \cdot 1 + 2\text{)}.$$

Similarly, we define **multiplication modulo 5**, $x \otimes y$, as that unique $r \in F$ such that

$$xy = 5p + r,$$

for a (unique) positive integer, p. Thus, for example, $4 \otimes 4 = 1$.

7. Is the set $f = \{0\}$, together with addition and multiplication modulo 1, a field? Why or why not? How about the set $F = \{0, 1\}$, together with addition and multiplication modulo 2? How about $F = \{0, 1, 2\}$, together with addition and multiplication modulo 3? How about $F = \{0, 1, 2, 3\}$, together with addition and multiplication modulo 4?

4.2 Linear Combinations and Subspaces

In this section we will assume, unless specifically stated otherwise, that L is a linear space over some arbitrary field, $\mathbf{F}$.

4.9. Definition. If X is a subset of L and $\mathbf{z}$ is an element of L, we shall say that **z is a linear combination of X** iff there exist $n \in N, \mathbf{x}_1, \ldots, \mathbf{x}_n \in X$, and $a_1, \ldots, a_n \in \mathbf{F}$ satisfying:

$$\mathbf{z} = \sum_{i=1}^{n} a_i \mathbf{x}_i. \tag{4.7}$$

If X and Y are nonempty subsets of L, we shall say that **Y depends upon X** (or that **X generates Y**) iff each element of Y is a linear combination of X.

While I will not provide a formal proof of this, it follows from Axioms 1.a and 1.b of 4.2′ that an expression like the right-hand-side of (4.7) is well-defined and unambiguous, and we will hereafter use such expressions freely and without further comment. Moreover, I will leave it as an easy exercise for the reader to show that if $\mathbf{x}$ and $\mathbf{y}$ are linear combinations of X, and a and b are scalars, then the vector $a\mathbf{x} + b\mathbf{y}$ is also a linear combination of X.

The terminology 'Y depends upon X' makes it sound as if X is somehow a bigger set that Y; however, this is not necessarily the case. For example, if we define $\mathbf{e}_i \in \mathbf{R}^n$ by:

$$\mathbf{e}_i = (\delta_{i1}, \delta_{i2}, \ldots, \delta_{in}) \quad \text{for } i = 1, \ldots, n; \tag{4.8}$$

where δ_{ij} is the Kronecker delta function:

$$\delta_{ij} = \begin{cases} 1 & \text{if } i = j \\ 0 & \text{if } i \neq j, \end{cases}$$

we see that if $\mathbf{x} = (x_1, \ldots, x_n)$ is an arbitrary element of $\mathbf{R}^n$, we have:

$$\mathbf{x} = \sum_{i=1}^{n} x_i \mathbf{e}_i.$$

Thus we see that $\mathbf{R}^n$ depends upon (or is generated by) $\{\mathbf{e}_1, \ldots, \mathbf{e}_n\}$.

4.10. Definition. We shall say that a non-empty subset, X, of L, is a **(linear) subspace (of L)** iff, for each pair of vectors, $\mathbf{x}$ and $\mathbf{y}$ from S, and each pair of scalars, a and b from $\mathbf{F}$, we have:

$$a\mathbf{x} + b\mathbf{y} \in S.$$

If X contains a non-null element, $\mathbf{x} \neq \mathbf{0}$, and $S \neq L$, then S will be said to be a **proper subspace (of L)**.

4.11. Examples/Exercises.

1. Notice that L itself is a subspace (of L), as is $S^* \equiv \{\mathbf{0}\}$.
2. Show that every subspace of L must contain the origin in L.
3. It follows from the definition of a subspace, via an easy induction argument, that if S is a linear subspace, and $\mathbf{z}$ is a linear combination of S, then $\mathbf{z} \in S$.
4. Show that if, for each $a \in A, S_a$ is a subspace of L, then:

$$S \stackrel{\text{def}}{=} \bigcap_{a \in A} S_a,$$

is also a subspace of L.

5. Show that if S is a subspace of L, then S is a linear space over $\mathbf{F}$, with vector addition and scalar multiplication defined as in L. This is, of course, the reason for the terminology 'subspace.'
6. Show that in $\mathbf{R}^3$ the set S defined by:

$$S = \{\mathbf{x} \in \mathbf{R}^3 \mid x_3 = 0\},$$

is a subspace of $\mathbf{R}^3$.

7. Show that if $\mathbf{p}$ is a non-zero vector in $\mathbf{R}^n$ then the set H defined by:

$$H = \{\mathbf{x} \in \mathbf{R}^n \mid \mathbf{p} \cdot \mathbf{x} = 0\},$$

is a subspace of $\mathbf{R}^n$. It is an example of a hyperplane, a kind of set we will be studying in detail later on. Notice also that the set S defined in Example 6, just above, is another example of this kind. □

Before proceeding further with our general study of linear spaces, it may be well to spend a little time taking a first look at several examples of linear spaces which we will use frequently in later chapters.

4.12. Examples. Subspaces of $\mathbf{R}^\infty$. We noted earlier that $\mathbf{R}^\infty$, the collection of all (infinite) sequences of real numbers, was a linear space. In our present discussion, we will use the generic notation, '$\mathbf{x} = (x_1, x_2, \ldots), \mathbf{y} = (y_1, y_2, \ldots)$,' etc., to denote elements of $\mathbf{R}^\infty$. Recall that we define vector addition and scalar multiplication on $\mathbf{R}^\infty$ by:

$$\mathbf{x} + \mathbf{y} = (x_1 + y_1, x_2 + y_2, \ldots) \text{ and } a\mathbf{x} = (a\mathbf{x}_1, a\mathbf{x}_2, \ldots),$$

respectively, for $\mathbf{x}, \mathbf{y} \in \mathbf{R}^\infty$, and $a \in \mathbf{R}$.

1. The space of bounded sequences, m. Recall that a sequence, $\mathbf{x} \in \mathbf{R}^\infty$, is said to be bounded iff there exists a real number, $a \geq 0$, such that:

$$|x_n| \leq a \quad \text{for } n = 1, 2, \ldots .$$

Letting 'm' denote the set of all bounded sequences of real numbers, it is easy to prove that, for all $\mathbf{x}, \mathbf{y} \in m$, and all $a \in \mathbf{R}$,

$$\mathbf{x} + \mathbf{y} \in m \quad \& \quad a\mathbf{x} \in m.$$

Therefore we see that m is a subspace of $\mathbf{R}^\infty$ (and, as indicated above, we will refer to m as the **space of bounded sequences**.)

2. The space of convergent sequences, c. If we define c by:

$$c = \{\mathbf{x} \in \mathbf{R}^\infty \mid (\exists a \in \mathbf{R}) : x_n \to a\},$$

it follows at once from Theorem 2.9 that c is a subspace of $\mathbf{R}^\infty$. As a matter of fact, it also follows from Proposition 2.8 that c is a subset of m, and thus can also be thought of as a subspace of m (remember that a subspace is a linear space in its own right).

3. The space of null sequences, c_0. If we define c_0 by:

$$c_0 = \{\mathbf{x} \in \mathbf{R}^\infty \mid x_n \to 0\},$$

it follows at once from Theorem 2.9 that c_0 is a subspace of $\mathbf{R}^\infty$ (and, of course, it is a subspace of c and m as well).

4. The space ℓ_2. We define the set ℓ_2 by:

$$\ell_2 = \{\mathbf{x} \in \mathbf{R}^\infty \mid \sum_{n=1}^{\infty} |x_n|^2 < \infty\}; \tag{4.9}$$

that is, ℓ_2 is the set of all sequences such that the sequence of partial sums,

$$s_n = \sum_{i=1}^{n} |x_i|^2 \quad \text{for } n = 1, 2, \ldots ,$$

converges (see Section 3 of Chapter 2).[3] It is an immediate consequence of 2.37.1 that ℓ_2 is a subset of c_0.

To show that ℓ_2 is a subspace, we begin by noting that for any real numbers, a and b, we have

$$\begin{aligned} 2(a^2 + b^2) - (a + b)^2 &= 2(a^2 + b^2) - (a^2 + 2ab + b^2) \\ &= a^2 - 2ab + b^2 = (a - b)^2 \geq 0, \end{aligned}$$

[3]The use of the absolute value, $|x_i|^2$, rather than x_i^2, is unnecessary here, since we are dealing with sequences of real numbers. However, in actuality, ℓ_2 is normally defined as the set of all sequences of complex numbers satisfying the inequality in (4.9); and, by using absolute values throughout this discussion, all of our arguments are thereby correct for sequences of complex numbers, as well as for sequences of real numbers.

so that

$$(a+b)^2 \leq 2(a^2+b^2),$$

and it follows that for all n:

$$\sum_{i=1}^{n} |x_i + y_i|^2 \leq 2\left(\sum_{i=1}^{n} |x_i|^2 + \sum_{i=1}^{n} |y_i|^2\right). \tag{4.10}$$

Now, if $\mathbf{x}, \mathbf{y} \in \ell_2$, and we define the sequences of partial sums $\mathbf{s}$ and $\mathbf{t}$ by:

$$s_n = \sum_{i=1}^{n} |x_i|^2 \text{ and } t_n = \sum_{i=1}^{n} |y_i|^2 \quad \text{for } n = 1, 2, \dots,$$

it follows from the definition of ℓ_2 that there exist real numbers, α and β, such that $s_n \to \alpha$ and $t_n \to \beta$. Moreover, since $\mathbf{s}$ and $\mathbf{t}$ are both increasing sequences of nonnegative terms, we can also see that:

$$s_n \leq \alpha \text{ and } t_n \leq \beta \quad \text{for } n = 1, 2, \dots.$$

It then follows from (4.10) that if $\mathbf{x}, \mathbf{y} \in \ell_2$, and $a, b \in \mathbf{R}$, we have:

$$\begin{aligned}\sum_{i=1}^{n} |ax_i + by_i|^2 &\leq 2\left(\sum_{i=1}^{n} |ax_i|^2 + \sum_{i=1}^{n} |by_i|^2\right) \\ &\leq 2(a^2\alpha + b^2\beta) \quad \text{for } n = 1, 2, \dots,\end{aligned}$$

and thus it follows from 2.42 that $a\mathbf{x} + b\mathbf{y} \in \ell_2$. Therefore, ℓ_2 is a subspace of $\mathbf{R}^\infty$ (and of c_0, c, and m as well).

5. Show that the set $\boldsymbol{f}$ defined by:

$$\boldsymbol{f} = \{\mathbf{x} \in \mathbf{R}^\infty \mid (\exists m \in N)(\forall n \geq m)\colon x_n = 0\}$$

is a subspace of $\mathbf{R}^\infty$. This set can be described as the collection of all sequences for which all but a finite number of terms are equal to zero. It is easy to show that $\boldsymbol{f}$ is a subspace of ℓ_2.

6. Choice over time. It is often natural to suppose that an economic agent chooses a sequence of quantities of a commodity, x_t, for $t = 1, \dots, T$; which sequence we can clearly identify with a point $\mathbf{x}$ in $\mathbf{R}^T$. In analyzing such a choice, and the consequences thereof, it is obviously very convenient that the collection of all such sequences is a linear space (and, of course, a very familiar one as well).

On the other hand, for many economists there are difficulties with the notion of an arbitrary finite planning horizon; that is, the choice of a finite integer, T, represents a point of concern. There are at least three reasons why this might be a problem. First, the specification of a particular value of T is at best arbitrary; and, since the conclusions of the analysis may be sensitive to the *particular* value of T chosen (although it also may not be), this may mean that the analysis is effectively much less general than it appears to be. Secondly, unless one makes arbitrary and sometimes mildly strange assumptions about the connection between the agent's goals and the desired value of x_T, one often finds apparently reasonable assumptions yielding patently fallacious conclusions in this context. For example, if x_t represents a firm's derived demand for the stock of a capital good, the assumption of profit maximization as the underlying goal will typically result in the firm's choosing a capital stock level which decreases to zero as one approaches T.

Finally, the existence of the finite horizon, T, means that in general the Tth period is going to have to be treated differently than the other time periods; there is an asymmetry which enters the analysis which is at best inconvenient. Consequently, the life of the analyst is very often made a great deal simpler by the supposition that the agent plans indefinitely far into the future (in effect, treats T as if it were equal to $+\infty$); and thus, if the assumption doesn't appear to be in flagrant conflict with empirical observations of the economic phenomenon one is attempting to analyze (and an infinite planning horizon may be a reasonable assumption for corporations, or, indeed, for certain choices made by a family), it may be desirable to suppose that the agent makes a choice of an element, $\mathbf{x}$, from $\mathbf{R}^\infty$, or from m. In either case, however, the set of all conceivable choices (the 'commodity space') is still a linear space; and, in many ways these spaces are as easy to deal with as if we were working in $\mathbf{R}^T$. □

4.13. Example. The Spaces $\mathfrak{F}_{[a,b]}$and $C_{[a,b]}$. It is an immediate consequence of Theorem 4.4 that, given any real numbers a and b such that $a < b$, the set $\mathfrak{F}_{[a,b]}$ defined by

$$\mathfrak{F}_{[a,b]} = \{f \colon [a,b] \to \mathbf{R}\},$$

is a real linear space. Furthermore, it follows at once from Theorem 3.10 that the subset, $C_{[a,b]}$, consisting of all elements of $\mathfrak{F}_{[a,b]}$ which are *continuous* on $[a,b]$, is a subspace. The space $C_{[a,b]}$ is sufficiently important in analysis that the notation we have used here is quite standard. Moreover, in the special case in which $a = 0$ and $b = 1$, the space is generally denoted simply by 'C,' rather than '$C_{[0,1]}$.'

Once again these linear spaces are of interest in certain parts of economic analysis. It has been argued, for example, that one cannot sensibly model pure competition without assuming that there are a continuum of economic agents. The point is this: in most economic analysis, the set, A, of agents is taken to be of the form

$$A = \{1, \ldots, m\}, \tag{4.11}$$

where m is some positive integer. However, no matter how large we choose m, each agent will have some influence on the total market. Traditionally, economic theorists have claimed that this is all right; that as long as this influence is 'sufficiently small,' it is perfectly reasonable to suppose that the agents *act as if* they have no influence at all on prices in the market. More recently, however, many economic theorists have felt that this argument is not adequate, and have argued that it is only by assuming that there is a continuum of agents that one obtains a satisfactory model of a 'purely competitive' market.

Whichever side of this issue one chooses, the structure of the analysis changes a great deal less than one might at first suppose. In general equilibrium analysis, we have [since the work of Arrow and Debreu in the mid-1950's] represented allocations available to the ith agent as vectors $\mathbf{x}_i \in \mathbf{R}^n$, for $i = 1, \ldots, m$. This has meant that we could consider **allocations** (for the economy) as vectors

$$\mathbf{x} = (\mathbf{x}_1, \ldots, \mathbf{x}_m) \in \mathbf{R}^{mn},$$

(or as $m \times n$ matrices $[x_{ij}]$), where

$$\mathbf{x}_i = (x_{i1}, \ldots, x_{in}) \in \mathbf{R}^n,$$

is the commodity bundle available to the ith agent, and 'x_{ij}' represents the amount of the jth commodity available to the ith agent, for $j = 1, \ldots, n$. However, after studying the examples in this section, you should have no difficulty in seeing that we could equally well describe such an allocation by a function $f\colon A \to \mathbf{R}^n$, where A is the set defined in (4.11), and $\mathbf{x}_i = f(i)$ represents the commodity bundle available to the ith agent. If we now suppose that there is a continuum of agents in the economy; representing the set of agents, as, say:

$$A = [0, 1], \tag{4.12}$$

then we can see that allocations can now be considered to be functions $\mathbf{f}\colon A \to \mathbf{R}^n$. The convenient thing about this is that the set of all such allocations,

$$\mathcal{F}^n \stackrel{\text{def}}{=} \{\mathbf{f}\colon [0, 1] \to \mathbf{R}^n\},$$

is (by Theorem 4.4) also a real linear space. In fact, it is a linear space very much like $\mathcal{F}_{[0,1]}$, and we will consider its properties more formally in Chapter 10. □

Our next definition will probably look a bit peculiar at first reading, but be patient; we will explain the definition in some detail after our initial statement.

4.14. Definition. If X is a non-empty subset of L, we define the **subspace spanned by X**, denoted by '$Sp(X)$,' as the smallest (in the sense of set inclusion) linear subspace of L containing X.

The idea of the preceding definition is the following. Define the set $\mathcal{S}$ by

$$\mathcal{S} = \{S \subseteq L \mid S \text{ is a subspace of } L \;\&\; X \subseteq S\}.$$

Clearly $\mathcal{S}$ contains L itself, and thus is non-empty. Therefore, if we define the set T by

$$T = \bigcap_{S \in \mathcal{S}} S,$$

it follows from 4.11.4 that T is a subspace of L; and it is an immediate consequence of the definition of $\mathcal{S}$ that $X \subseteq T$. Moreover, notice that if S^* is any subspace containing X, then $S^* \in \mathcal{S}$, and therefore $T \subseteq S^*$. It is in this sense that $T = Sp(X)$ is the *smallest* linear subspace of L which contains X.

In Definition 4.15, we have defined the subspace spanned by X as a Kuratowski closure operation. While it is the first example which we have encountered of such a closure operation, it is but the first of many; such things as the convex hull and the conical closure of a subset of $\mathbf{R}^n$, the transitive closure of a binary relation, and the topology generated by a family of sets, can all be viewed as examples of Kuratowski closure operations. While I will not attempt a formal definition of a Kuratowski closure operation at this time, the basic idea is as follows. Suppose we are interested in the 'smallest' subset of a space which both contains a given set, X, and satisfies a certain property, $\mathbf{P}$; that is, suppose we seek a set, Y, which contains X, satisfies property $\mathbf{P}$, and which is also such that if Z is any other subset of the space satisfying these two conditions, then $Y \subseteq Z$. Consider the family, $\mathcal{C}$, of subsets of the given space, T, defined by

$$\mathcal{C} = \{C \subseteq T \mid X \subseteq C \;\&\; C \text{ satisfies property } \mathbf{P}\}.$$

If we can establish two things:

1. the family $\mathcal{C}$ is non-empty; that is, there exists at least one subset of T which both contains X and satisfies property $\mathbf{P}$, and
2. the intersection of an arbitrary family of subsets satisfying $\mathbf{P}$ also satisfies $\mathbf{P}$;

then the desired set Y can be defined as:

$$Y = \bigcap_{C \in \mathcal{C}} C.$$

It was, of course, exactly these two conditions which we checked in justifying our definition of the linear subspace spanned by a set.

We will study a second example of a Kuratowski closure operation in Section 6 of this chapter. In the meantime, we will consider a second characterization of $Sp(X)$, as follows.

4.15. Proposition. *If X is a nonempty subset of L, then $Sp(X)$ is the collection of all linear combinations of X.*

Proof. Let 'X^*' denote the collection of all linear combinations of X. We wish to prove that $X^* = Sp(X)$. Accordingly, we note first that it follows immediately from 4.11.3 that any subspace containing X must also contain X^*. Therefore, in particular,

$$X^* \subseteq Sp(X).$$

To prove the converse, we recall that if $\mathbf{x}$ and $\mathbf{y}$ are linear combinations of X (and therefore are elements of X^*), and a and b are scalars from $\mathbf{F}$, then $a\mathbf{x} + b\mathbf{y}$ is again a linear combination of X (and thus is an element of X^*). Consequently, we see that X^* is a linear subspace; and, since X^* obviously contains X, and $Sp(X)$ is the *smallest* subspace containing X, it then follows that:

$$Sp(X) \subseteq X^*. \quad \square$$

It follows from Proposition 4.15 that we can think of $Sp(X)$ in either of two ways: (a) as the smallest linear subspace containing X, or (b) as the collection of all vectors which are obtainable as linear combinations of vectors from X. Both representations will prove to be useful to us in our work with linear spaces.

4.16. Definition. If X is a finite subset of L, say,

$$X = \{\mathbf{x}_1, \ldots, \mathbf{x}_m\},$$

we shall say that X is **linearly independent** iff, for every set of m scalars from $\mathbf{F}$, $\{a_1, \ldots, a_m\}$, we have:

$$\sum_{i=1}^{m} a_i\mathbf{x}_i = \mathbf{0} \Rightarrow a_1 = a_2 \cdots = a_m = 0;$$

otherwise X is said to be **linearly dependent**.[4] If X is an arbitrary (possibly infinite) subset of L, we shall say that X is **linearly independent** iff every finite subset of X is linearly independent.

4.17. Examples.

1. Consider once again the vectors $\mathbf{e}_i (i = 1, \ldots, n)$ in $\mathbf{R}^n$ which were defined in Equation (4.8) of this section. If $a_i, i = 1, \ldots, n$, is any set of n real numbers, we have:

$$\sum_{i=1}^{n} a_i\mathbf{e}_i = (a_1, a_2, \ldots, a_n);$$

[4]Notice that we assume here that the $\mathbf{x}_i$'s are all distinct points; a convention we shall follow whenver we write '$X = \{\mathbf{x}_1, \ldots, \mathbf{x}_n\}$.'

so that it follows at once that if

$$\sum\nolimits_{i=1}^{n} a_i \mathbf{e}_i = \mathbf{0},$$

then:

$$a_1 = a_2 = \cdots = a_n = 0.$$

Consequently, we see that $\{\mathbf{e}_1, \ldots, \mathbf{e}_n\}$ is a linearly independent subset of $\mathbf{R}^n$.

2. Expanding upon the previous example, define $\mathbf{e}_i \in \mathbf{R}^\infty$, for $i = 1, 2, \ldots$, by:

$$\mathbf{e}_i = (\delta_{i1}, \delta_{i2}, \ldots, \delta_{ij}, \ldots) \quad \text{for } i = 1, 2, \ldots,$$

where once again, δ_{ij} is the Kronecker delta, for $i, j = 1, 2, \ldots$. The same basic argument as was used in the previous example shows that $\{\mathbf{e}_i \mid i \in N\}$ is a linearly independent subset of $\mathbf{R}^\infty$. Moreover, notice that this set is contained in the subspace $\boldsymbol{f}$ defined in 4.12.5, above. Therefore we have shown that $\boldsymbol{f}$ contains a (countably) infinite subset which is linearly independent. □

Notice that if X is a linearly independent subset of L, then we must have $\mathbf{0} \notin X$. Moreover, if X is linearly independent, then it must not contain any $\mathbf{x}$ which is a linear combination of points in X; for if $\mathbf{x}, \mathbf{x}_1, \ldots, \mathbf{x}_m \in L$, and $a_1, \ldots, a_m \in \mathbf{F}$ are such that

$$\mathbf{x} = \sum\nolimits_{i=1}^{m} a_i \mathbf{x}_i,$$

then

$$\mathbf{x} - \sum\nolimits_{i=1}^{m} a_i \mathbf{x}_i = \mathbf{0};$$

and we see that $\{\mathbf{x}, \mathbf{x}_1, \ldots, \mathbf{x}_m\}$ is a finite subset of X which is linearly dependent.

Actually, our primary concern in this chapter will be with the linear independence, or lack thereof, of finite subsets of L; and with linear spaces of finite dimension as well, for that matter. I will remind you of what is meant by finite dimension shortly, but first consider the following.

4.18. Theorem. *If X is a linearly independent subset of L, and $\mathbf{x} \in Sp(X)$ is non-null, then there exists a unique (finite) subset of X, $\{\mathbf{x}_1, \ldots, \mathbf{x}_m\}$, and a unique set of scalars, $\{a_1, \ldots, a_m\}$, satisfying:*

$$\sum\nolimits_{i=1}^{m} a_i \mathbf{x}_i = \mathbf{x} \text{ and } a_i \neq 0 \quad \text{for } i = 1, \ldots, m. \tag{4.13}$$

Proof. It follows at once from the assumption that $\mathbf{x} \in Sp(X)$ that there exists a finite subset of $X, \{\mathbf{x}_1, \ldots, \mathbf{x}_k\}$, and a set of scalars, $\{a_1, \ldots, a_k\}$, satisfying:

$$\sum_{i=1}^{k} a_i\mathbf{x}_i = \mathbf{x};$$

and if some of the a_i's are zero, we can eliminate the corresponding $\mathbf{x}_i$'s to obtain sets $\{\mathbf{x}_1, \ldots, \mathbf{x}_m\}$ and $\{a_1, \ldots, a_m\}$ satisfying (4.13).[5]

Now suppose, by way of obtaining a contradiction, that there exist sets $\{\mathbf{x}_1, \ldots, \mathbf{x}_m\}$ and $\{\mathbf{y}_1, \ldots, \mathbf{y}_n\}$ in X, and sets of non-zero scalars, $\{a_1, \ldots, a_m\}$ and $\{b_1, \ldots, b_n\}$ in $\mathbf{F}$, such that:

$$\{\mathbf{x}_1, \ldots, \mathbf{x}_m\} \neq \{\mathbf{y}_1, \ldots, \mathbf{y}_n\}, \text{ or } \{a_1, \ldots, a_m\} \neq \{b_1, \ldots, b_n\},$$

and

$$\sum_{i=1}^{m} a_i\mathbf{x}_i = \mathbf{x} = \sum_{j=1}^{n} b_j\mathbf{y}_j; \tag{4.14}$$

and we may suppose, for the sake of convenience (and without loss of generality), that $m \leq n$. From (4.14) we have:

$$\sum_{i=1}^{m} a_i\mathbf{x}_i - \sum_{j=1}^{n} b_j\mathbf{y}_j = \mathbf{0}. \tag{4.15}$$

We can complete our proof by considering four possible cases [although most readers will probably feel that equation (4.15) has already established the basic contradiction!]

Case 1. $\{\mathbf{x}_1, \ldots, \mathbf{x}_m\} \cap \{\mathbf{y}_1, \ldots, \mathbf{y}_n\} = \emptyset$. Here it follows immediately from (4.15) that $\{\mathbf{x}_1, \ldots, \mathbf{x}_m, \mathbf{y}_1, \ldots, \mathbf{y}_n\}$ is a finite linearly dependent subset of X, which contradicts our assumption.

I will leave Case 2:

$$\{\mathbf{x}_1, \ldots, \mathbf{x}_m\} = \{\mathbf{y}_1, \ldots, \mathbf{y}_n\},$$

and Case 3:

$$\{\mathbf{x}_1, \ldots, \mathbf{x}_m\} \subseteq \{\mathbf{y}_1, \ldots, \mathbf{y}_n\}, \text{ but } \{\mathbf{x}_1, \ldots, \mathbf{x}_m\} \neq \{\mathbf{y}_1, \ldots, \mathbf{y}_n\},$$

as exercises.

Case 4. $\{\mathbf{x}_1, \ldots, \mathbf{x}_m\} \neq \{\mathbf{y}_1, \ldots, \mathbf{y}_n\}, \{\mathbf{x}_1, \ldots, \mathbf{x}_m\} \cap \{\mathbf{y}_1, \ldots, \mathbf{y}_n\} \neq \emptyset$, and $\{\mathbf{x}_1, \ldots, \mathbf{x}_m\} \setminus \{\mathbf{y}_1, \ldots, \mathbf{y}_n\} \neq \emptyset$.[6] Here we may suppose, re-labeling if necessary, that there exist $n_1, n_2 \in N$ such that

$$n_1 + n_2 = n, n \geq m > n_1,$$

[5]Notice that at least one of the $a_i's$ in the original set of scalars must be non-zero, since we have assumed that $\mathbf{x}$ is non-null.

[6]Since we are supposing throughout that $n \geq m$, we must also have $\{\mathbf{y}_1, \ldots, \mathbf{y}_n\} \setminus \{\mathbf{x}_1, \ldots, \mathbf{x}_m\} \neq \emptyset$ in this case.

and

$$\mathbf{y}_j = \mathbf{x}_j \quad \text{for } j = 1, \ldots, n_1,$$

while

$$\mathbf{y}_j \notin \{\mathbf{x}_1, \ldots, \mathbf{x}_m\}, \quad \text{for } j = n_1 + 1, \ldots n_1 + n_2.$$

Letting $p = m + n_2$, we then define $\{\mathbf{z}_1, \ldots, \mathbf{z}_p\} \subseteq X$, and $\{c_1, \ldots c_p\} \subseteq \mathbf{F}$ by:

$$\mathbf{z}_i = \begin{cases} \mathbf{x}_i & \text{for } i = 1, \ldots, m, \text{ and} \\ \mathbf{y}_{i+n_1-m} & \text{for } i = m+1, \ldots, m+n_2; \end{cases}$$

and

$$c_i = \begin{cases} a_i - b_i & \text{for } i = 1, \ldots, n_1, \\ a_i & \text{for } i = n_1 + 1, \ldots, m, \text{ and} \\ -b_{i-m+n_1} & \text{for } i = m+1, \ldots, m+n_2. \end{cases}$$

Using the above definitions and equation (4.15) in turn, we then have:

$$\sum_{i=1}^{p} c_i \mathbf{z}_i = \sum_{i=1}^{m} a_i \mathbf{x}_i - \sum_{j=1}^{n} b_j \mathbf{y}_j = \mathbf{0},$$

which, since not all the c_i's are zero,[7] contradicts our assumption that X is a linearly independent set. □

4.19. Definition. If S is a subspace of L, we shall say that S **has (finite) dimension $n \in N$**, and write 'dim$(S) = n$,' iff

a. there exists a subset of S, $\{\mathbf{x}_1, \ldots, \mathbf{x}_n\}$, which is linearly independent, and

b. any subset of S containing more than n elements is linearly dependent.

If, for each $n \in N$, S has a linearly independent subset containing n elements, then we shall say that S is **infinite-dimensional**.

Notice that it follows from our discussion in Example 4.17.1 that $\mathbf{R}^n$ has dimension of at least n, since we there defined a linearly independent subset of $\mathbf{R}^n$ containing n elements. In fact, as I am sure you are already aware, $\mathbf{R}^n$ has dimension n; as will follow immediately from 4.17.1 and Theorem 4.22, below. Before proving this theorem, however, let's first consider the following.

4.20. Proposition. *If S is a linear subspace of L having dimension r, and $X = \{\mathbf{x}_1, \ldots, \mathbf{x}_r\}$ is a linearly independent subset of S, then $S = Sp(X)$; in fact, if $\mathbf{x}$ is an element of S, then there exist unique scalars, $a_1, \ldots, a_r \in \mathbf{F}$ such that:*

$$\mathbf{x} = \sum_{i=1}^{r} a_i \mathbf{x}_i. \tag{4.16}$$

[7] Recall that none of the a_i's equals zero, nor does $b_j = 0$ for any j.

Proof. It follows immediately from the definition of the space spanned by X, and the fact that S is a linear subspace containing X, that $Sp(X) \subseteq S$. To prove the converse, let $\mathbf{x}$ be an arbitrary element of S. Then, since the dimension of S is r, we see that it must be the case that $\{\mathbf{x}, \mathbf{x}_1, \ldots, \mathbf{x}_r\}$ is linearly dependent. Therefore, there exist scalars, $a_0, a_1, \ldots, a_r$, not all of which are zero, such that:

$$a_0\mathbf{x} + \sum_{i=1}^{r} a_i\mathbf{x}_i = \mathbf{0}. \tag{4.17}$$

Moreover, it must then be the case that $a_0 \neq 0$; for otherwise (4.17) would contradict the assumed linear independence of X. Therefore, defining

$$b_i = -a_i/a_0 \quad \text{for } i = 1, 2, \ldots, r,$$

we have from (4.17) that:

$$\mathbf{x} = \sum_{i=1}^{r} b_i\mathbf{x}_i.$$

Thus we see that $\mathbf{x} \in Sp(X)$; and, since $\mathbf{x}$ was an arbitrary element of S, it follows that $S \subseteq Sp(X)$.

Having shown that $S = Sp(X)$ it now follows immediately from Theorem 4.18 that, given any $\mathbf{x} \in S$, there exist unique scalars, $a_1, \ldots, a_r$ satisfying (4.16). □

If we compare the conclusions of Theorem 4.18 and Proposition 4.20, there is something which may be mildly confusing, at least at first glance. Suppose $X = \{\mathbf{x}_1, \ldots, \mathbf{x}_r\}$ is a finite set of linearly independent vectors, let $\mathbf{x}$ be an arbitrary element of $Sp(X)$, and suppose that there exist scalars, $a_1, \ldots, a_p \in \mathbf{F}$, where $1 \leq p < r$, such that:

$$\mathbf{x} = \sum_{i=1}^{p} a_i\mathbf{x}_i.$$

If this is the case, then obviously we can define $a_{p+1} = 0$, and equally well write:

$$\mathbf{x} = \sum_{i=1}^{p+1} a_i\mathbf{x}_i.$$

The question is, does this contradict 4.18, or does it contradict 4.20, or what? Since I am sure that this issue will not puzzle you for very long (if at all), I will leave you to resolve it.

4.21. Example. Defining

$$S = \{\mathbf{x} \in \mathbf{R}^3 \mid \mathbf{e} \cdot \mathbf{x} = \mathbf{0}\} \quad \text{where } \mathbf{e} = (1, 1, 1),$$
$$\mathbf{x}_1 = (1, -1, 0) \;\text{ and }\; \mathbf{x}_2 = (1, 1, -2);$$

show that:

1. $S = Sp(\{\mathbf{x}_1, \mathbf{x}_2\})$, and
2. if $\mathbf{y} = (y_1, y_2, y_3)$ is an arbitrary element of S, then:

$$\mathbf{y} = a_1\mathbf{x}_1 + a_2\mathbf{x}_2,$$

where

$$a_1 = y_1 + y_3/2 \text{ and } a_2 = -y_3/2. \quad \square$$

Our next result is the key to a great many of the relationships involving linear independence, linear combinations, and the solutions to linear equations, among other things. The proof is a bit messy, but if you bear with it, you will discover that the argument is conceptually quite simple.

4.22. Theorem. *If $X = \{\mathbf{x}_1, \ldots, \mathbf{x}_m\}$ depends upon $Y = \{\mathbf{y}_1, \ldots, \mathbf{y}_n\}$, and $m > n$ then X is linearly dependent.*

Proof. We note first that if $\mathbf{x}_i = \mathbf{0}$ for some i, then it follows at once that X is linearly dependent. Therefore we will suppose that $\mathbf{0} \notin X$. To avoid some clumsiness in notation, we will distinguish two cases at the outset.

1. Suppose first that $n = 1$. Then $m \geq 2$, and, since X depends upon Y, there exist a_1 and a_2 such that:

$$\mathbf{x}_i = a_i\mathbf{y} \quad \text{for } i = 1, 2. \tag{4.18}$$

Furthermore, since $\mathbf{0} \notin X$, we must have $a_i \neq 0$ for $i = 1, 2$. But from (4.18) we have:

$$a_2\mathbf{x}_1 - a_1\mathbf{x}_2 = a_2(a_1\mathbf{y}) - a_1(a_2\mathbf{y}) = \mathbf{0};$$

and thus we see that $\{\mathbf{x}_1, \mathbf{x}_2\}$ (and therefore X) is linearly dependent.

2. Suppose now that $n \geq 2$. Then there exist $a_{ij}(i = 1, \ldots, m; j = 1, \ldots, n)$ such that:

$$\mathbf{x}_i = \sum_{j=1}^{n} a_{ij}\mathbf{y}_j \quad \text{for } i = 1, \ldots, m. \tag{4.19}$$

Now, since $\mathbf{x}_1 \neq \mathbf{0}$, at least one of $a_{11}, \ldots, a_{1n}$ is non-zero; and we may suppose, re-labeling if necessary, that $a_{11} \neq 0$. But then we have from (4.19) that:

$$\mathbf{y}_1 = (1/a_{11})\mathbf{x}_1 - \sum_{j=2}^{n}(a_{1j}/a_{11})\mathbf{y}_j. \tag{4.20}$$

Substituting (4.20) into (4.19), we then obtain:

$$\mathbf{x}_2 = (a_{21}/a_{11})\mathbf{x}_1 + \sum_{j=2}^{n}\left(\frac{a_{11}a_{2j} - a_{21}a_{1j}}{a_{11}}\right)\mathbf{y}_j. \tag{4.21}$$

At this point we can distinguish two cases, as follows.

a. $a_{11}a_{2j} - a_{21}a_{1j} = 0$ for $j = 2, \ldots, n$. Here we have from (4.21) that:

$$\mathbf{x}_2 = (a_{21}/a_{11})\mathbf{x}_1;$$

so it follows that $\{\mathbf{x}_1, \mathbf{x}_2\}$ is a linearly dependent set, and thus that X is linearly dependent as well.

b. $a_{11}a_{2j} - a_{21}a_{1j} \neq 0$, for some $j \in \{2, \ldots, n\}$. Here we may suppose, re-labeling if necessary, that

$$a_{11}a_{22} - a_{21}a_{12} \neq 0.$$

We can then solve (4.21) for $\mathbf{y}_2$ to obtain:

$$\mathbf{y}_2 = (a_{11}a_{22} - a_{21}a_{12})^{-1} \cdot \left[-a_{21}\mathbf{x}_1 + a_{11}\mathbf{x}_2 + \sum\nolimits_{j=3}^{n}(a_{21}a_{1j} - a_{11}a_{2j})\mathbf{y}_j\right]$$

This solution for $\mathbf{y}_2$ can now be substituted into (4.20), so that both $\mathbf{y}_1$ and $\mathbf{y}_2$ can be expressed as linear combinations of $\{\mathbf{x}_1, \mathbf{x}_2, \mathbf{y}_3, \ldots, \mathbf{y}_n\}$.

Proceeding in this way, we will, at the initiation of the qth step [after having just completed the $(q-1)^{st}$ step], have shown that $\{\mathbf{x}_1, \ldots, \mathbf{x}_{q-1}\}$ (and thus X) is linearly dependent, or we will have obtained each of $\mathbf{y}_1, \ldots, \mathbf{y}_{q-1}$ as a linear combination of $\mathbf{x}_1, \ldots, \mathbf{x}_{q-1}, \mathbf{y}_q, \ldots, \mathbf{y}_n$:

$$\mathbf{y}_k = \sum\nolimits_{i=1}^{q-1} b_{ki}\mathbf{x}_i + \sum\nolimits_{j=q}^{n} b_{kj}\mathbf{y}_j \quad \text{for } k = 1, \ldots, q-1. \tag{4.22}$$

Equations (4.22) can then be substituted into the qth equation in (4.19) to obtain an equation of the form:

$$\mathbf{x}_q = \sum\nolimits_{i=1}^{q-1} c_i\mathbf{x}_i + \sum\nolimits_{j=q}^{n} c_j\mathbf{y}_j. \tag{4.23}$$

At this point, we can again distinguish two cases.

a. $c_j = 0$ for $j = q, \ldots, n$. In this case we have from (4.23) that:

$$\mathbf{x}_q = \sum\nolimits_{i=1}^{q-1} c_i\mathbf{x}_i;$$

from which it follows at once that $\{\mathbf{x}_1, \ldots, \mathbf{x}_q\}$, and thus X, is linearly dependent.

b. $c_j \neq 0$ for some $j \in \{q, \ldots, n\}$. Here we can suppose, re-labeling if necessary, that $c_q \neq 0$, and obtain from (4.23):

$$\mathbf{y}_q = \sum\nolimits_{i=1}^{q-1}(-c_i/c_q)\mathbf{x}_i + (1/c_q)\mathbf{x}_q - \sum\nolimits_{j=q+1}^{n}(c_j/c_q)\mathbf{y}_j.$$

Clearly this solution for $\mathbf{y}_q$ can be substituted into the expressions for $\mathbf{y}_1, \ldots, \mathbf{y}_{q-1}$ given in (4.22) to obtain these vectors as linear combinations of $\{\mathbf{x}_1, \ldots, \mathbf{x}_q, \mathbf{y}_{q+1}, \ldots, \mathbf{y}_n\}$ as well.

Proceeding in this fashion, after the nth step we will either have shown some subset of X containing n or fewer elements to be linearly dependent, or we will have obtained scalars $b_{ij}(i,j = 1,\ldots,n)$ satisfying:

$$\mathbf{y}_j = \sum_{j=1}^{n} b_{ij}\mathbf{x}_j \quad \text{for } i = 1,\ldots,n. \tag{4.24}$$

But then we see that it follows from (4.19) and (4.24) that we can obtain $\mathbf{x}_{n+1}$ (recall that $m > n$) as a linear combination of $\mathbf{x}_1,\ldots,\mathbf{x}_n$; from which it follows that $\{\mathbf{x}_1,\ldots,\mathbf{x}_n,\mathbf{x}_{n+1}\}$, and thus X is linearly dependent. □

4.23. Definition. If S is a subspace of L, then a subset of S, X, is a **basis for S** iff (a) X is linearly independent, and (b) $S = Sp(X)$.

4.24. Proposition. *If S is a linear space having dimension n ($\dim S = n$), then every basis for S contains exactly n elements. Conversely, if T is a linearly independent subset of S containing n elements, then T is a basis for S.*

Proof. Let $X = \{\mathbf{x}_1,\ldots,\mathbf{x}_m\}$ be a basis for S. It follows from the definition of dimension that $m \leq n$, and that S contains a linearly independent set, $Y = \{\mathbf{y}_1,\ldots,\mathbf{y}_n\}$, containing n elements. However, since X is a basis for S, and Y is a subset of S, it also follows that Y depends upon X. Consequently, we have from Theorem 4.22 that $n \leq m$; and we conclude that $m = n$.

The converse statement: "If T is a linearly independent subset of S containing n elements, then T is a basis for S," is simply a re-statement of Proposition 4.20. □

4.25. Definition. If Y is a non-empty subset of L, we shall say that $X = \{\mathbf{x}_1,\ldots,\mathbf{x}_m\}$ is a **maximal linearly independent subset of Y** iff (a) X is a linearly independent subset of Y, and (b) any subset of Y containing more than m elements is linearly dependent.

It is possible that a subset of L may contain no finite maximal linearly independent subset. In particular, if Y is an infinite-dimensional linear subspace, then Y will contain no finite subset satisfying the above definition. Furthermore, notice that if $Y = \{\mathbf{0}\}$, then Y contains no maximal linearly independent subset. On the other hand, if Y contains a non-null element and is contained in some finite-dimensional subspace, then Y does contain a maximal linearly independent subset; as is shown in the following.

4.26. Proposition. *Suppose Y is a subset of L, and that $\dim(S) = m \geq 1$, where we define $S = Sp(Y)$. Then Y contains a maximal linearly independent subset, $X = \{\mathbf{x}_1,\ldots,\mathbf{x}_m\}$, containing m elements, and X is a basis for S. Conversely, if Y contains a maximal linearly independent subset, X, containing m elements, then $\dim(S) = m$, and X is a basis for S.*

Proof. Suppose that $\dim(S) = m \geq 1$. Then Y must contain at least one non-null element, $\mathbf{x}_1$; for if not, then $Y = \{\mathbf{0}\}$, and $Sp(Y) = Y$, which contradicts the assumption that $\dim[Sp(Y)] \geq 1$. There are now two possibilities, either: (a) every element of Y is a scalar multiple of $\mathbf{x}_1$, or (b) there exists an element, $\mathbf{x}_2$ which is not a scalar multiple of $\mathbf{x}_1$. In the latter case, $\{\mathbf{x}_1, \mathbf{x}_2\}$ is obviously a linearly independent set, and there again two possibilities, either: (a) every element of Y is a linear combination of $\{\mathbf{x}_1, \mathbf{x}_2\}$, or (b) there exists an element $\mathbf{x}_3 \in Y$ such that $\mathbf{x}_3$ is *not* a linear combination of $\{\mathbf{x}_1, \mathbf{x}_2\}$. Continuing in this way, we obtain at the kth step a linearly independent set, $\{\mathbf{x}_1, \dots, \mathbf{x}_k\}$, and there are basically the same two possibilities: (a) every element of Y is a linear combination of $\{\mathbf{x}_1, \dots, \mathbf{x}_k\}$, or (b) there exists a element $\mathbf{x}_{k+1} \in Y$ which is not a linear combination of $\{\mathbf{x}_1, \dots, \mathbf{x}_k\}$. In the latter case, it is easy to see that $\{\mathbf{x}_1, \dots, \mathbf{x}_k, \mathbf{x}_{k+1}\}$ is a linearly independent set, and we can again distinguish two possibilities, and so on. Notice, however, that it follows at once from Theorem 4.23 and the assumption that $\dim[Sp(Y)] = m$ that this process must terminate in m steps or less, because Y cannot contain a linearly independent subset consisting of more than m vectors.

Suppose, then, that we have found a linearly independent subset, $X = \{\mathbf{x}_1, \dots, \mathbf{x}_n\}$, such that $n \leq m$, and each element of Y is a linear combination of X, and define $T = Sp(X)$. Then we see that $Y \subseteq T$, and, since T is a linear subspace, it follows immediately from the definition of $Sp(Y)$ that

$$S \stackrel{\text{def}}{=} Sp(Y) \subseteq T.$$

On the other hand, since X is a subset of Y, and S is a linear subspace containing Y, we also have $T \subseteq S$. Therefore $T = S$, and it then follows immediately that $m = n$ and that X is a basis for S.

To establish the converse statement, notice that if $X = \{\mathbf{x}_1, \dots, \mathbf{x}_m\}$ is a maximal linearly independent subset of Y, then it is clear that every element of Y is a linear combination of X. The argument of the preceding paragraph then establishes that $\dim[Sp(Y)] = m$, and that X is a basis for $Sp(Y)$. □

Exercises.

1. Prove that S is a subspace if, and only if, S satisfies:
   a. $(\forall a \in \mathbf{F})(\forall \mathbf{x} \in S)\colon a\mathbf{x} \in S$, and
   b. $(\forall \mathbf{x}, \mathbf{y} \in S)\colon \mathbf{x} + \mathbf{y} \in S$.
2. Prove that m is a subspace of $\mathbf{R}^\infty$ (see 4.13.1).
3. Prove that E is a subspace of ℓ_2 (see 4.13.5).
4. Determine whether the following are linearly independent subsets of $\mathbf{R}^4$.
   a. $(1, -2, 3, -1), (5, 2, -1, 4), (-7, -10, 11, -11)$.
   b. $(10, 1, 4, -2), (-2, 3, 8, -1), (6, -1, 20, -3)$.

5. Consider the following three elements of C (see Example 4.14):

$$f_1(x) = -1 + x, f_2(x) = 2 - 3x, f_3(x) = -x.$$

Is the set $\{f_1, f_2, f_3\}$ linearly independent? Justify your answer.

4.3 Linear Transformations and Functionals

We will assume throught this section, unless specifically stated otherwise, that L_i is a linear space over some arbitrary field, $\mathbf{F}$, for $i = 1, 2$.

4.27. Definition. We shall say that a function $f\colon L_1 \to L_2$ is **linear** iff f satisfies:

$$(\forall a, b \in \mathbf{F})(\forall \mathbf{x}, \mathbf{y} \in L_1)\colon f(a\mathbf{x} + b\mathbf{y}) = af(\mathbf{x}) + bf(\mathbf{y}).$$

Linear functions are often called **linear transformations**, rather than functions; and if $L_2 = \mathbf{R}$, a linear function is usually called a **linear functional**. In the special case in which $f(\mathbf{x}) = \mathbf{0}$, for all $\mathbf{x} \in L_1$, we will say that f is **trivial**; otherwise we will say that f is **non-trivial**. You have probably encountered these definitions before; in any case, you have certainly dealt with linear functions and functionals before, as you will be reminded by the first two of the following examples.

4.28. Examples/Exercises.

1. Let $\mathbf{p}$ be a fixed non-zero vector in $\mathbf{R}^n$, and define $f\colon \mathbf{R}^n \to \mathbf{R}$ by:

$$f(\mathbf{x}) = \mathbf{p} \cdot \mathbf{x} \quad \text{for } \mathbf{x} \in \mathbf{R}^n.$$

Then f is a linear function; more specifically, it is a linear functional.

2. Let A be an $m \times n$ (real) matrix, and define $f\colon \mathbf{R}^n \to \mathbf{R}^m$ by:

$$f(\mathbf{x}) = A\mathbf{x} \quad \text{for } \mathbf{x} \in \mathbf{R}^n.$$

Show that f is a linear function.

3. Consider the space C defined in 4.13, above, and define $\varphi\colon C \to \mathbf{R}$ by:

$$\varphi(f) = \int_0^1 f(t)\,dt.$$

Prove that φ is a linear functional.

4. Let $\mathbf{x}^*$ be a fixed non-zero vector, and define $S = Sp(\{\mathbf{x}^*\})$ and $f\colon S \to \mathbf{F}$ by:

$$f(\mathbf{x}) = a,$$

where $a \in \mathbf{F}$ is such that $\mathbf{x} = a\mathbf{x}^*$ (see Exercise 2 at the end of the previous section). Prove that f is a linear function. □

The definition of a linear function can be re-stated verbally as: 'a function is linear if (and only if) it maps a linear combination of two points in the domain into the same linear combination of the images of those two points in the range space.' It turns out that a very simple induction argument will suffice to show that this same relationship holds for any finite linear combination of points in the domain. The formal statement of this relationship is set forth in the following, the proof of which will be left as an exercise.

4.29. Proposition. *If $f\colon L_1 \to L_2$ is a linear function, $\mathbf{x}_1, \ldots, \mathbf{x}_m$ are elements of L_1, and $a_1, \ldots, a_m$ are elements of $\mathbf{F}$, then:*

$$f\left(\sum_{i=1}^{m} a_i\mathbf{x}_i\right) = \sum_{i=1}^{m} a_i f(\mathbf{x}_i).$$

Notice that if $f\colon L_1 \to L_2$ is a linear function, and S_i is a linear subspace of L_i, for $i = 1, 2$, then $\mathbf{0} \in f(S_1)$ [in particular, $f(\mathbf{0}) = \mathbf{0}$], and $\mathbf{0} \in f^{-1}(S_2)$. In fact, both of these subsets are actually subspaces, as is shown by the following result.

4.30. Proposition. *Suppose $f\colon L_1 \to L_2$ is a linear function, and that S_1 and S_2 are subspaces of L_1 and L_2, respectively. Then $f(S_1)$ is a subspace of L_2, and $f^{-1}(S_2)$ is a subspace of L_1.*

Proof. Let $\mathbf{y}, \mathbf{y}^* \in f(S_1)$, and let $a, b \in \mathbf{F}$. By definition of $f(S_1)$, there exist $\mathbf{x}, \mathbf{x}^* \in S_1$ such that:

$$\mathbf{y} = f(\mathbf{x}) \ \&\ \mathbf{y}^* = f(\mathbf{x}^*);$$

and, since S_1 is a subspace, we have:

$$a\mathbf{x} + b\mathbf{x}^* \in S_1.$$

But then, since f is linear, we see that:

$$f(a\mathbf{x} + b\mathbf{x}^*) = af(\mathbf{x}) + bf(\mathbf{x}^*) = a\mathbf{y} + b\mathbf{y}^* \in f(S_1);$$

and it follows that $f(S_1)$ is a subspace of L_2.

Now, let $\mathbf{x}_1, \mathbf{x}_2 \in f^{-1}(S_2)$, and let $a, b \in \mathbf{F}$. By definition of $f^{-1}(S_2)$, we see that:

$$\mathbf{y}_i \stackrel{\text{def}}{=} f(\mathbf{x}_i) \in S_2 \quad \text{for } i = 1, 2;$$

and, since S_2 is a subspace, we have:

$$a\mathbf{y}_1 + b\mathbf{y}_2 \in S_2.$$

But then we see that:

$$f(a\mathbf{x}_1 + b\mathbf{x}_2) = af(\mathbf{x}_1) + bf(\mathbf{x}_2) = a\mathbf{y}_1 + b\mathbf{y}_2 \in S_2,$$

and it follows that

$$a\mathbf{x}_1 + b\mathbf{x}_2 \in f^{-1}(S_2). \quad \square$$

4.31. Definition. If $f: L_1 \to L_2$ is a linear function, then the set $f^{-1}(\{\mathbf{0}\})$ is called the **kernel** of the function, or transformation, and is denoted by '$\mathbf{k}_f$.'

Of course it follows from 4.30 that the kernel of a linear transformation, $f: L_1 \to L_2$, is a subspace of L_1. We get a very convenient necessary and sufficient condition for f to be one-to-one, which is based upon $\mathbf{k}_f$, as follows.[8]

4.32. Proposition. *If $f: L_1 \to L_2$ is a linear function, then f is one-to-one if, and only if, $\mathbf{k}_f = \{\mathbf{0}\}$.*

Proof. Suppose first that $\mathbf{k}_f = \{\mathbf{0}\}$, and let $\mathbf{x}, \mathbf{x}^* \in L_1$ be such that $f(\mathbf{x}) = f(\mathbf{x}^*)$. Then, since f is linear,

$$f(\mathbf{x} - \mathbf{x}^*) = f(\mathbf{x}) - f(\mathbf{x}^*) = \mathbf{0},$$

and we see that $\mathbf{x} - \mathbf{x}^* \in \mathbf{k}_f$. Therefore, since $\mathbf{k}_f = \{\mathbf{0}\}$, it follows that $\mathbf{x} = \mathbf{x}^*$, and we conclude that f is one-to-one.

Now suppose that $\mathbf{k}_f \neq \{\mathbf{0}\}$, so that there exists a vector, $\mathbf{x}^*$, satisfying:

$$\mathbf{x}^* \neq \mathbf{0} \;\&\; f(\mathbf{x}^*) = \mathbf{0}.$$

But then we have:

$$f(\mathbf{x}^*) = \mathbf{0} = f(\mathbf{0}) \;\&\; \mathbf{x}^* \neq \mathbf{0},$$

and thus f is not one-to-one. Consequently, if f is one-to-one, then $\mathbf{k}_f = \{\mathbf{0}\}$. □

In Section 5 of the previous chapter, we proved a result, Proposition 3.55, which characterized inverse functions. We will put it to work in the proof of the following.

4.33. Proposition. *Suppose $f: L_1 \to L_2$ is both linear and one-to-one, and define $S = f(L_1)$. Then the function $g: S \to L_1$ defined by:*

$$g(\mathbf{y}) = \textit{that } \mathbf{x} \in L_1 \textit{ such that } f(\mathbf{x}) = \mathbf{y},$$

(that is, $g = f^{-1}$) is linear, one-to-one, and onto L_1.

Proof. It should be obvious that g is well-defined[9] and onto L_1. To prove that g is one-to-one, let $\mathbf{y}, \mathbf{y}^* \in S$ be such that:

$$\mathbf{x} \stackrel{\text{def}}{=} g(\mathbf{y}) = g(\mathbf{y}^*).$$

[8] At some risk of confusion, we have used the same symbol, '$\mathbf{0}$,' to denote the origin in L_2 in Definition 4.31 and the origin in L_1 in Proposition 4.32.

[9] That is, for each $\mathbf{y} \in f(L_1)$, there is exactly one $\mathbf{x} \in L_1$ such that $\mathbf{x} = g(\mathbf{y})$.

Then, by definition of g, we see that:

$$\mathbf{y} = f(\mathbf{x}) = \mathbf{y}^*,$$

and thus $\mathbf{y} = \mathbf{y}^*$.

To prove that g is linear, let $\mathbf{y}_1, \mathbf{y}_2 \in S$, and let $a, b \in \mathbf{F}$. Then, using, in turn, Proposition 3.55, the linearity of f, and Proposition 3.55 a second time, we obtain:

$$\begin{aligned} g(a\mathbf{y}_1 + b\mathbf{y}_2) &= g\Big(af\big[g(\mathbf{y}_1)\big] + bf\big[g(\mathbf{y}_2)\big]\Big) \\ &= g\Big(f\big[ag(\mathbf{y}_1) + bg(\mathbf{y}_2)\big]\Big) = ag(\mathbf{y}_1) + bg(\mathbf{y}_2). \quad \square \end{aligned}$$

4.34. Definition. A linear function, $f \colon L_1 \to L_2$, which is also one-to-one and onto is said to be an **isomorphism**; and if such a function exists, L_1 is said to be **isomorphic to L_2**.

Notice that if L_1 is isomorphic to L_2, so that there exists a linear and one-to-one function, f, mapping L_1 onto L_2, then it follows at once from 4.33 that f^{-1} is an isomorphism from L_2 to L_1. Consequently, if L_1 is isomorphic to L_2, then L_2 is also isomorphic to L_1; and if we know that either of these conditions hold, we will generally simply say that L_1 and L_2 are **isomorphic**.

The following result will be very useful in our further discussion of isomorphisms. Its proof will be left as an exercise.

4.35. Proposition. *If $f \colon L_1 \to L_2$, and $g \colon L_2 \to L_3$ are both isomorphisms, then the composition function, $h = g \circ f$, is also an isomorphism.*

4.36. Example. Isomorphism as an Equivalence Relation. For purposes of our present discussion, let us denote the collection of all linear spaces over a particular field, $\mathbf{F}$ (to be held fixed in our discussion), by '$\mathfrak{L}$;' and define the relation '$\approx$' on $\mathfrak{L}$ by:

$$L_1 \approx L_2 \iff L_1 \text{ is isomorphic to } L_2.$$

It is easy to see that the relation $\approx$ is reflexive, since if $L \in \mathfrak{L}$, it is apparent that the identity mapping,

$$\mathbf{i}(\mathbf{x}) = \mathbf{x} \quad \text{for } \mathbf{x} \in L,$$

is an isomorphism from L onto itself. The discussion immediately following Definition 4.35 established that $\approx$ is symmetric, that is, for all $L_1, L_2 \in \mathfrak{L}$::

$$L_1 \approx L_2 \Leftrightarrow L_2 \approx L_1.$$

Furthermore, Proposition 4.35 implies that $\approx$ is transitive; for if $L_i \in \mathfrak{L}$ for $i = 1, 2, 3$, and we have $L_1 \approx L_2$ and $L_2 \approx L_3$, then it follows at once from

4.35 that $L_1 \approx L_3$; that is, that L_1 is isomorphic to L_3. Thus we see that the relation $\approx$ is an **equivalence relation** on $\mathfrak{L}$.

Suppose now that the relation $\simeq$ is an equivalence relation on a set X. We then define for each $x \in X$, the **equivalence class determined by x**, usually denoted by '$[x]$,' by:

$$[x] = \{y \in X \mid y \simeq x\}.$$

It can be shown (the exercise is not difficult) that two equivalence classes, say $[x]$ and $[y]$, either have no elements in common or are identical; that is,

$$(\forall x, y \in X)\colon [x] \cap [y] \neq \emptyset \Rightarrow [x] = [y]$$

(and, of course, $[x] = [y] \Leftrightarrow x \simeq y$). Thus the collection of equivalence classes is a partition of the set X. Moreover, one of the reasons for an interest in this particular partition is that, for some purposes, it makes essentially no difference which of the elements of a particular equivalence class is chosen; from some point of view we can treat all the elements of a given equivalence class as if they were identical, or 'identical for all practical purposes.' We frequently use this kind of idea in economics, the most familiar example of which arises in the theory of consumer choice. Suppose a consumer's preference relation, G, is a weak ordering of a consumption set, X. Then, as we saw in 1.26.2, the symmetric part of G, I (the consumer's indifference relation) is an equivalence relation on X. In this context, the equivalence class, $[x]$, for a particular $x \in X$, is simply the 'indifference curve' through x, and we learn in intermediate microtheory that, under these assumptions, different indifference curves never intersect. In this context, it is from the point of view of the consumer's preferences that two elements from a given equivalence class (or two points on the same indifference curve) are the same; from the standpoint of the consumer's tastes and/or desires, it makes no difference which of the two commodity bundles it has.

Returning to the equivalence relation with which we began our discussion, $\approx$ ('is isomorphic to'), on $\mathfrak{L}$, the sense in which two elements of a given equivalence class are the same is from the standpoint of linear algebra; or, to stretch a point a bit, from the standpoint of algebra in general. Suppose we consider the implications of this statement in a bit more detail.

If L_1 and L_2 are from the same equivalence class in $\mathfrak{L}$, then there exists an isomorphism, $f\colon L_1 \to L_2$. Now, since f is both one-to-one and onto, this means first of all that L_1 and L_2 have the same number of elements; to each element, $\mathbf{x}$, of L_1, there corresponds *exactly one* element of L_2, namely $\mathbf{y} = f(\mathbf{x})$. However, there is more to the concept of a linear space than just the underlying sets involved; and in this connection, we can also identify $\mathbf{x}$ with $f(\mathbf{x})$ from the standpoint of algebraic operations (recall also that L_1 and L_2 are defined over the same field), which we can see as follows. Suppose we use the notation '$\oplus_i$' to denote vector addition on L_i, for $i = 1, 2$. If $\mathbf{x}$

and $\mathbf{y}$ are elements of L_1, then by applying the binary operation, $\oplus_1$, to $\mathbf{x}$ and $\mathbf{y}$, we obtain a third element, $\mathbf{z}$, of L_1 (that is, $\mathbf{z} = \mathbf{x} \oplus_1 \mathbf{y}$). If we consider what happens to $f(\mathbf{x})$ and $f(\mathbf{y})$ when we apply the operation $\oplus_2$ in L_2, we note that, since f is linear,

$$f(\mathbf{x}) \oplus_2 f(\mathbf{y}) = f(\mathbf{x} \oplus_1 \mathbf{y}) = f(\mathbf{z});$$

that is, by adding the two points in L_2 with which we have identified $\mathbf{x}$ and $\mathbf{y}$, we obtain exactly the point with which we identify their sum, $\mathbf{z}$. Similarly, if a is a scalar in $\mathbf{F}$, then

$$f(a\mathbf{x}) = af(\mathbf{x}).$$

It is from this point of view that the two spaces can be considered essentially the same from an algebraic point of view; in particular, *notice that any and all implications which can be deduced from linear space assumptions regarding a particular linear space, L, apply equally in any other linear space, L^*, which is an element of [L].*

With this discussion in mind, we can perhaps gain a better understanding of our next result. □

4.37. Theorem. *If L is any real linear space of finite dimension, n, then L is isomorphic to* $\mathbf{R}^n$.

Proof. Since L is of dimension n, L has a basis, $X = \{\mathbf{x}_1, \ldots, \mathbf{x}_n\}$, containing n elements; and, since X is a basis for L, we know that for each $\mathbf{x} \in L$, there exists a unique n-tuple of real numbers, $\{a_1, \ldots, a_n\}$, such that:

$$\mathbf{x} = \sum_{i=1}^{n} a_i \mathbf{x}_i. \tag{4.25}$$

Because of this, we can represent each element of L in terms of a sum of the form of the right-hand-side of (4.25), and define $f \colon L \to \mathbf{R}^n$ by:

$$f\left(\sum_{i=1}^{n} a_i \mathbf{x}_i\right) = (a_1, a_2, \ldots, a_n) \in \mathbf{R}^n \quad \text{for } \mathbf{x} = \sum_{i=1}^{n} a_i \mathbf{x}_i \in L. \tag{4.26}$$

With f defined as in (4.26), it should be apparent that f is one-to-one and onto $\mathbf{R}^n$. To show that f is linear, suppose that:

$$\mathbf{x} = \sum_{i=1}^{n} a_i \mathbf{x}_i \text{ and } \mathbf{y} = \sum_{i=1}^{n} b_i \mathbf{x}_i,$$

are elements of L, and that $\alpha, \beta \in \mathbf{R}$. Then we see that:

$$\alpha\mathbf{x} + \beta\mathbf{y} = \alpha\Big(\sum_{i=1}^{n} a_i \mathbf{x}_i\Big) + \beta\Big(\sum_{i=1}^{n} b_i \mathbf{x}_i\Big) = \sum_{i=1}^{n} (\alpha a_i + \beta b_i)\mathbf{x}_i;$$

and thus,

$$\begin{aligned} f(\alpha\mathbf{x} + \beta\mathbf{y}) &= (\alpha a_1 + \beta b_1, \ldots, \alpha a_n + \beta b_n) \\ &= \alpha(a_1, \ldots, a_n) + \beta(b_1, \ldots, b_n) = \alpha f(\mathbf{x}) + \beta f(\mathbf{y}). \end{aligned}$$ □

The isomorphism just established means that, for most purposes, we can discover the properties of any n-dimensional real linear space by studying $\mathbf{R}^n$. In this connection, you may find it worthwhile to review Section 1.7 from the perspective of the discussion of this section. The isomorphism which we made use of in the proof of 4.37 will be sufficiently useful as to merit a formal definition, as follows.

4.38. Definition. If $Y = \{\mathbf{y}_1, \ldots, \mathbf{y}_n\}$ is a linearly independent subset of a real linear space, L, and we define $S = Sp(Y)$, then the linear function $f: S \to \mathbf{R}^n$ defined by:

$$f\Big(\sum_{i=1}^n a_i \mathbf{y}_i\Big) = (a_1, \ldots, a_n),$$

will be called **the isomorphism from S to $\mathbf{R}^n$ determined by Y**.

It should be fairly apparent that an argument essentially identical to that used to prove 4.37 can be used to establish the fact that if L is an n-dimensional linear space over $\mathbf{C}$, then L is isomorphic to $\mathbf{C}^n$; in fact the same basic argument shows that if L is an n-dimensional linear space over an arbitrary field, $\mathbf{F}$, then L is isomorphic to $\mathbf{F}^n$.

Exercises.

1. Let A be an $m \times n$ (real) matrix, and define $f: \mathbf{R}^n \to \mathbf{R}^m$ by:

$$f(\mathbf{x}) = A\mathbf{x} \quad \text{for } \mathbf{x} \in \mathbf{R}^n.$$

Show that f is a linear function. Suppose now that $\mathbf{b} \in \mathbf{R}^m$ is a (fixed) vector. Is the function $g: \mathbf{R}^n \to \mathbf{R}^m$ defined by:

$$g(\mathbf{x}) = A\mathbf{x} + \mathbf{b} \quad \text{for } \mathbf{x} \in \mathbf{R}^n,$$

also a linear function?

2. Show that the composition of two linear functions is also linear.

3. Show that if $f_i: L_1 \to L_2$ is linear, where L_2 is a real linear space, and $a_i \in \mathbf{R}$, for $i = 1, \ldots, n$, then the function f defined on L_1 by:

$$f(\mathbf{x}) = \sum_{i=1}^n a_i f_i(\mathbf{x}) \quad \text{for } \mathbf{x} \in L_1,$$

is linear.

4. Prove (a) Proposition 4.29, and (b) Proposition 4.35.

5. Let X be any non-empty set, and let L be a linear space over the field $\mathbf{F}$. Show that if we define $S = L^X$ as in Theorem 4.4, and let x^* be an arbitrary fixed element of X, then the function $\varphi: S \to L$ defined by:

$$\varphi(f) = f(x^*),$$

is a linear function on S.

4.4 Normed Linear Spaces

In Section 1.4, we briefly introduced the idea of a norm for $\mathbf{R}^n$, and considered the euclidean norm for $\mathbf{R}^n$ in some detail. In this section, we will consider the generalization of some of these ideas to an arbitrary linear space. However, *we will specifically assume that the field over which our linear space, L, is defined is one of the three fields,* $\mathbf{Q}$, $\mathbf{R}$, or $\mathbf{C}$.

4.39. Definition. We shall say that a function $\rho\colon L \to \mathbf{R}_+$ is a **norm (for L)** iff, for all $\mathbf{x}, \mathbf{y} \in L$, and $a \in \mathbf{F}$:

1. $\rho(\mathbf{x}) = 0 \iff \mathbf{x} = \mathbf{0}$.
2. $\rho(\mathbf{x}+\mathbf{y}) \leq \rho(\mathbf{x}) + \rho(\mathbf{y})$.
3. $\rho(a\mathbf{x}) = |a| \cdot \rho(\mathbf{x})$.

A linear space equipped with a norm is called a **normed linear space**.

Strictly speaking, a normed linear space involves a third idea, the topology induced by the norm. We will postpone most of our considerations regarding this aspect of a normed linear space until Chapter 7. However, in this section we will consider some of the basic properties of open sets in normed linear spaces.

4.40. Examples/Exercises.

1. In $\mathbf{R}^n$, the usual euclidean norm,

$$\|\mathbf{x}\| = (\mathbf{x} \cdot \mathbf{x})^{1/2} = \left[\sum\nolimits_{i=1}^n x_i^2\right]^{1/2},$$

is a norm by the above definition; and thus $\mathbf{R}^n$ is a normed linear space.

2. On the space m (the space of bounded real sequences), define ρ by:

$$\rho(\mathbf{x}) = \sup_n |x_n|.$$

Prove that this is a norm for m.

3. On the space $C_{[a,b]}$, define ρ by:

$$\rho(f) = \max_{x\in[a,b]} |f(x)|.$$

Show that ρ is a norm for $C_{[a,b]}$. □

In Section 4 of Chapter 1, we discussed the way in which the euclidean metric could be defined from the euclidean norm. As it turns out, one can conveniently define a metric from the norm on any normed linear space, as we will see shortly. First we must formally define what we mean by a metric, however.

4.41. Definition. We shall say that a function, $d\colon L \times L \to \mathbf{R}_+$ is a **metric for L** iff, for all $\mathbf{x}, \mathbf{y}, \mathbf{z} \in L$, we have:

1. $d(\mathbf{x}, \mathbf{y}) = 0 \iff \mathbf{x} = \mathbf{y}$.
2. $d(\mathbf{x}, \mathbf{y}) = d(\mathbf{y}, \mathbf{x})$.
3. $d(\mathbf{x}, \mathbf{z}) \leq d(\mathbf{x}, \mathbf{y}) + d(\mathbf{y}, \mathbf{z})$.

If $d(\cdot)$ is a metric on L, then $d(\mathbf{x}, \mathbf{y})$ provides a measure of the 'distance between $\mathbf{x}$ and $\mathbf{y}$.' Notice that the properties listed in the above definition coincide with our intuitive notions of a measure of distance, and are satisfied by the euclidean distance function on $\mathbf{R}^n$.[10] If L is a normed linear space, then a metric for L can be defined from the norm in exactly the same way as was done in Section 1.4; as is shown in our next result. The proof is easy, and will be left as an exercise.

4.42. Theorem. *If L is a normed linear space, with norm ρ, and if we define $d\colon L \times L \to \mathbf{R}_+$ by:*

$$d(\mathbf{x}, \mathbf{y}) = \rho(\mathbf{x} - \mathbf{y}),$$

then d is a metric for L. Furthermore, d satisfies the following two additional conditions: for all $\mathbf{x}, \mathbf{y}, \mathbf{z} \in L$, and all $a \in \mathbf{F}$, we have:

1. **(homogeneity)**: $d(a\mathbf{x}, a\mathbf{y}) = |a| \cdot d(\mathbf{x}, \mathbf{y})$, *and*

2. **(translation invariance)**: $d(\mathbf{x} + \mathbf{z}, \mathbf{y} + \mathbf{z}) = d(\mathbf{x}, \mathbf{y})$.

In dealing with normed linear spaces, we will often use the metric d defined in 4.42 without explicit comment. For example, suppose L is a linear space with norm $\rho, \mathbf{x}^*$ is an element of L, and ϵ is a strictly positive real number. We then define the **neighborhood of $\mathbf{x}^*$ with radius ϵ** by:

$$N(\mathbf{x}^*, \epsilon) = \{\mathbf{y} \in L \mid \rho(\mathbf{x}^* - \mathbf{y}) < \epsilon\};$$

in other words, we will define such a neighborhood in exactly the same general way in which such neighborhoods were defined for euclidean spaces. Of course, it follows from 4.42 that $N(\mathbf{x}^*, \epsilon)$ can be thought of as the set of all points in L whose 'distance' (in terms of the metric defined in 4.42) from $\mathbf{x}^*$ is less than ϵ. Notice the similarity in the definitions to follow to those which were presented for euclidean space in Chapters 1 and 3.

4.43. Definitions. If L is a linear space with norm ρ, and X is a subset of L, we shall say that:

1. $\mathbf{x}^* \in X$ is an **interior point of X** iff there exists $\epsilon > 0$ such that $N(\mathbf{x}^*, \epsilon) \subseteq X$. The collection of all interior points of X will be called the **interior of X**, and denoted by '$int(X)$.'

2. X is **open**, or is an **open set**, iff $X = int(X)$.[11]

3. $\mathbf{x}^* \in L$ is a **point of closure of X** iff, for each $\epsilon > 0$,

$$N(\mathbf{x}^*, \epsilon) \cap X \neq \emptyset.$$

The collection of all points of closure of X will be called the **closure of X**, and will be denoted by '$\overline{X}$.'

4. X is **closed**, or is a **closed set**, iff its complement, $L \setminus X$, is open.

---

[10] Compare Theorem 1.49 with Definition 4.41.

[11] By convention we define $int(\emptyset) = \emptyset$, and thus the empty set, $\emptyset$, is an open set.

As was the case in $\mathbf{R}^n$, it can be shown that X is closed if, and only if, it is equal to its closure (which is always a closed set); and $\overline{X}$ can equivalently be defined as the smallest closed subset of L which contains X. We will prove these facts as special cases of more general results which will be studied in Chapter 7. In the meantime, let's take a look at a couple of additional definitions.

4.44. Definitions. Suppose $\langle \mathbf{x}_n \rangle$ is a sequence of points from L, where L is a normed linear space with norm ρ. We shall say that $\langle \mathbf{x}_n \rangle$:

1. **converges to $\mathbf{x}^* \in L$**, and write '$\mathbf{x}_n \to \mathbf{x}^*$,' or

$$\text{`}\lim_{n\to\infty} \mathbf{x}_n = \mathbf{x}^*,\text{'}$$

iff, for each $\epsilon > 0$, there exists an integer, m, such that for all $n \geq m$, $\rho(\mathbf{x}_n - \mathbf{x}^*) < \epsilon$.

2. is a **Cauchy sequence** iff, given any $\epsilon > 0$, there exists an integer, m, such that for all $p, q \geq m$:

$$\rho(\mathbf{x}_p - \mathbf{x}_q) < \epsilon.$$

As was the case when we were dealing with sequences from $\mathbf{R}^n$, the first of the preceding definitions can be paraphrased as '$\langle \mathbf{x}_n \rangle$ converges to $\mathbf{x}^*$ iff each ϵ-neighborhood of $\mathbf{x}^*$ contains all but a finite number of terms of the sequence.'

It is easy to show that if $\langle \mathbf{x}_n \rangle$ converges, then it is a Cauchy sequence. On the other hand, in an arbitrary normed linear space, a Cauchy sequence may not be convergent. However, if it is the case that every Cauchy sequence in a linear space, L, converges to a point in L, we will say that L is **complete**. In particular, a complete normed linear space is called a **Banach space**. We will not worry especially about the issue of completeness at the moment, deferring our general consideration of this property until Chapter 10; on the other hand, notice that it is an immediate consequence of Theorem 2.66 that $\mathbf{R}^n$ is complete (and thus is a Banach space). We will defer further consideration of Banach spaces *per se* until Chapter 10; although in the next section we will define a particular type of Banach space, and begin the study of its properties.

We can define continuous functions from one normed linear space into another in a very natural way, as follows.

4.45. Definitions. Suppose $f\colon L_1 \to L_2$, where L_i is a normed linear space with norm ρ_i, for $i = 1, 2$. We shall say that f is:

1. **continuous at $\mathbf{x}^* \in L_1$** iff, given any $\epsilon > 0$, there exists $\delta > 0$, such that:

$$(\forall \mathbf{y} \in N(\mathbf{x}^*, \delta))\colon \rho_2[f(\mathbf{y}) - f(\mathbf{x}^*)] < \epsilon;$$

and we shall say that f is **continuous on a subset, $X \subseteq L_1$**, iff f is continuous at each $\mathbf{x} \in X$. If f is continuous on L_1, we shall simply say that f is **continuous.**

2. **uniformly continuous on a set $X \subseteq L_1$** iff, for each $\epsilon > 0$, there exists $\delta > 0$ such that:

$$(\forall \mathbf{x}, \mathbf{y} \in X)\colon \rho_1(\mathbf{x} - \mathbf{y}) < \delta \Rightarrow \rho_2[f(\mathbf{y}) - f(\mathbf{x})] < \epsilon.$$

If f is uniformly continuous on L_1, we shall simply say that f is **uniformly continuous.**

Notice the similarity between the above definitions and those we formulated for euclidean space [Definitions 3.1 and 3.16]; and that the distinction between continuity and uniform continuity is essentially the same as between the corresponding concepts for euclidean space. In the following we consider two very simple examples of continuous functions on spaces quite different from $\mathbf{R}^n$.

4.46. Examples.

1. Let L be the space $C_{[a,b]}$ which was defined in 4.13, and let $x^* \in [a, b]$ be fixed. We then define $\varphi\colon L \to \mathbf{R}$ by:

$$\varphi(f) = f(x^*).$$

Now, let $\epsilon > 0$ be given. If $f, g \in L$ are such that $d(f, g) < \epsilon$, then we have:

$$\begin{aligned} |\varphi(f) - \varphi(g)| &= |f(x^*) - g(x^*)| \equiv |(f - g)(x^*)| \\ &\leq \max_{x \in [a,b]} |(f - g)(x)| \equiv \rho(f - g) \equiv d(f, g) < \epsilon. \end{aligned}$$

Consequently, we see that φ is continuous; in fact, is uniformly continuous on L.

2. Let L be any real normed linear space, with norm $\|\cdot\|$, let A be a non-empty subset of L, and define the function $\delta(\mathbf{x}, A)$ on L by:

$$\delta(\mathbf{x}, A) = \inf_{\mathbf{y} \in A} \|\mathbf{x} - \mathbf{y}\| \quad \text{for } \mathbf{x} \in L.$$

Then $\delta(\cdot)$ is uniformly continuous on L, which we can show as follows.

Let $\epsilon > 0$ be given, and let $\mathbf{x}, \mathbf{x}' \in L$ be such that $\|\mathbf{x} - \mathbf{x}'\| < \epsilon/2$. By definition of an infimum, there exists $\mathbf{y}^* \in A$ such that:

$$\|\mathbf{x} - \mathbf{y}^*\| < \delta(\mathbf{x}, A) + \epsilon/2,$$

and thus we see that:

$$\delta(\mathbf{x}', A) \leq \|\mathbf{x}' - \mathbf{y}^*\| \leq \|\mathbf{x}' - \mathbf{x}\| + \|\mathbf{x} - \mathbf{y}^*\| < \epsilon/2 + \delta(\mathbf{x}, A) + \epsilon/2,$$

and therefore:

$$\delta(\mathbf{x}', A) - \delta(\mathbf{x}, A) < \epsilon.$$

A symmetric argument establishes that:

$$\delta(\mathbf{x}, A) - \delta(\mathbf{x}', A) < \epsilon,$$

and the uniform continuity of $\delta(\cdot)$ follows □

For the most part, we will have to postpone our study of continuity of functions between normed linear spaces for a later chapter; however, the following result is both relatively simple to prove and quite useful (compare with Theorems 3.20 and 3.21).

4.47. Theorem. *Suppose $f\colon L_1 \to L_2$, where L_i is a normed linear space with norm ρ_i, for $i = 1, 2$. Then the following conditions are mutually equivalent.*

1. *The function f is continuous.*
2. *For each open subset, U, of L_2, $f^{-1}(U)$ is open in L_1.*
3. *For each closed subset, C, of L_2, $f^{-1}(C)$ is closed in L_1.*

Proof. Suppose first that f is continuous, let U be an open subset of L_2, and let $\mathbf{x}^* \in f^{-1}(U)$. Then $f(\mathbf{x}^*) \in U$, and, since U is open, there exists $\epsilon > 0$ such that:

$$N[f(\mathbf{x}^*), \epsilon] \subseteq U. \tag{4.27}$$

However, since f is continuous, there exists $\delta > 0$ such that:

$$(\forall \mathbf{x} \in N(\mathbf{x}^*, \delta))\colon \rho_2[f(\mathbf{x}) - f(\mathbf{x}^*)] < \epsilon;$$

and it then follows from (4.27) that $N(\mathbf{x}^*, \delta) \subseteq f^{-1}(U)$. Consequently we see that $f^{-1}(U)$ is open in L_1.

Now suppose that f satisfies condition 2, and let C be a closed subset of L_2. Then $L_2 \setminus C$ is open, and thus, by hypothesis, $f^{-1}(L_2 \setminus C)$ is open in L_1. But then:

$$L_1 \setminus f^{-1}(L_2 \setminus C)$$

is closed in L_1, and we have from Proposition 3.18 that:

$$f^{-1}(C) = L_1 \setminus f^{-1}(L_2 \setminus C);$$

so it follows that condition 3 is satisfied.

Now suppose f satisfies condition 3, and let $\mathbf{x}^* \in L_1$ be arbitrary and $\epsilon > 0$ be given. Since $N[f(\mathbf{x}^*), \epsilon]$ is open, the set $L_2 \setminus N[f(\mathbf{x}^*), \epsilon]$ is closed in L_2. Therefore, by hypothesis,

$$f^{-1}\Big(L_2 \setminus N[f(\mathbf{x}^*), \epsilon]\Big)$$

is closed in L_1. But it then follows that:

$$L_1 \setminus f^{-1}\Big(L_2 \setminus N[f(\mathbf{x}^*), \epsilon]\Big)$$

is open in L_1; and by Proposition 3.18, we have:

$$L_1 \setminus f^{-1}\Big(L_2 \setminus N[f(\mathbf{x}^*), \epsilon]\Big) = f^{-1}\Big(N[f(\mathbf{x}^*), \epsilon]\Big).$$

Since this last set obviously contains $\mathbf{x}^*$, and is an open set, we see that there exists $\delta > 0$ such that:

$$N(\mathbf{x}^*, \delta) \subseteq f^{-1}\Big(N[f(\mathbf{x}^*), \epsilon]\Big);$$

and thus:

$$\big(\forall \mathbf{x} \in N(\mathbf{x}^*, \delta)\big): \rho_2[f(\mathbf{x}) - f(\mathbf{x}^*)] < \epsilon,$$

and it follows that f is continuous at $\mathbf{x}^*$. □

Linear functions are particularly convenient to deal with. As a first example of why this is the case, consider the following result.

4.48. Theorem. *Suppose $f: L_1 \to L_2$ is a linear function, where L_i ($i = 1, 2$) is a normed linear space. Then f is uniformly continuous on L_1 if, and only if, f is continuous at a single point of L_1.*

Proof. Let '$|\cdot|$' and '$\|\cdot\|$' denote the norms for L_1 and L_2, respectively; suppose f is continuous at a point $\mathbf{x}^* \in L_1$, and let $\epsilon > 0$ be given. Then, since f is continuous at $\mathbf{x}^*$, there exists $\delta > 0$ such that, for all $\mathbf{x} \in L$:

$$|\mathbf{x} - \mathbf{x}^*| < \delta \Rightarrow \|f(\mathbf{x}) - f(\mathbf{x}^*)\| < \epsilon.$$

Now let $\mathbf{x}', \mathbf{x}'' \in L$ be such that $|\mathbf{x}' - \mathbf{x}''| < \delta$, and define:

$$\mathbf{x} = \mathbf{x}' + \mathbf{x}^* - \mathbf{x}''.$$

Then we have:

$$|\mathbf{x} - \mathbf{x}^*| = |\mathbf{x}' + \mathbf{x}^* - \mathbf{x}'' - \mathbf{x}^*| = |\mathbf{x}' - \mathbf{x}''| < \delta,$$

so that:

$$\|f(\mathbf{x}) - f(\mathbf{x}^*)\| < \epsilon. \tag{4.28}$$

However, since f is linear:

$$f(\mathbf{x}) = f(\mathbf{x}' + \mathbf{x}^* - \mathbf{x}'') = f(\mathbf{x}') + f(\mathbf{x}^*) - f(\mathbf{x}''), \tag{4.29}$$

and thus from (4.28) and (4.29), we see that:

$$\|f(\mathbf{x}') - f(\mathbf{x}'')\| < \epsilon;$$

and therefore f is uniformly continuous.

The converse is, of course, immediate. □

You may have noticed that we have not considered the possibility of defining a linear function on a subset of a linear space; even though it would be perfectly sensible to consider linear functions whose domain is a subspace, X, say, of a linear space, L_1.[12] We did not do so, in fact, because if X is a linear subspace, then it can be considered to be a perfectly respectable linear space in its own right; and thus we do not need to separately consider linear functions whose domains are linear subspaces of some underlying larger linear space. Similarly, if X is a linear subspace of a linear space, L, upon which are defined a norm, ρ, and corresponding metric, d, then ρ and d are also a norm and metric, respectively, for X. Consequently, X itself can be considered to be a normed linear space. Therefore, it is also true that we do not need to separately consider linear functions whose domain is a linear subspace of some underlying normed linear space. Having said all of this, we have, hopefully, prepared the way for the following result.

4.49. Theorem. *Suppose $f\colon L_1 \to L_2$ is a linear function, where L_i ($i = 1, 2$) is a normed linear space; let '$|\cdot|$' and '$\|\cdot\|$' denote the norms for L_1 and L_2, respectively; and define S as the unit sphere in L_1, that is,*

$$S = \{\mathbf{x} \in L_1 \mid |\mathbf{x}| \leq 1\}.$$

Then f is uniformly continuous on L_1 if, and only if, it is bounded on S.

Proof.
1. Suppose f is bounded on S, and suppose $\alpha \in \mathbf{R}_{++}$ is such that:

$$(\forall \mathbf{x} \in S)\colon \|f(\mathbf{x})\| < \alpha.$$

Let $\epsilon > 0$ be given, and define δ by:

$$\delta = \epsilon/\alpha.$$

Then we note if $\mathbf{y} \in L_1$ is such that $0 < |\mathbf{y}| < \delta$, then $(1/|\mathbf{y}|)\mathbf{y} \in S$; so that:

$$\left\| f\left[(1/|\mathbf{y}|)\mathbf{y}\right] \right\| < \alpha,$$

and thus from the linearity of f,

$$\left\| f\left[(1/|\mathbf{y}|)\mathbf{y}\right] \right\| = \|f(\mathbf{y})\|/|\mathbf{y}| < \alpha,$$

so that:

$$\|f(\mathbf{y}) - f(\mathbf{0})\| = \|f(\mathbf{y})\| < \alpha \cdot |\mathbf{y}| < \alpha \cdot \delta = \epsilon.$$

[12] It wouldn't make a whole lot of sense to try to consider linear functions on domains which aren't linear subspaces, however; since it is only in this case that $a\mathbf{x} + b\mathbf{y}$ is an element of the set whenever, $\mathbf{x}$ and $\mathbf{y}$ are.

Therefore, f is continuous at $\mathbf{0}$, and it follows from 4.48 that f is uniformly continuous on L_1.

2. Now suppose that f is not bounded on S. We will show that for any $\delta > 0$, there exists $\mathbf{y} \in L_1$ such that $|\mathbf{y}| < \delta$, and $\|f(\mathbf{y})\| \geq 1$; from which it follows that f is not continuous at $\mathbf{0}$, and thus is not uniformly continuous on L_1. Accordingly, let $\delta > 0$ be arbitrary, except that we will suppose that $\delta < 1$ as well (which we can obviously do without loss of generality). Then, since f is unbounded on S, there exists $\mathbf{y}^* \in S$ such that $\|f(\mathbf{y}^*)\| > 2/\delta$. But then, defining $\mathbf{y}'$ by:

$$\mathbf{y}' = (\delta/2)\mathbf{y}^*,$$

we see that

$$|\mathbf{y}'| = |(\delta/2)\mathbf{y}^*| = (\delta/2)|\mathbf{y}^*| \leq \delta/2 < \delta;$$

while

$$\|f(\mathbf{y}') - f(\mathbf{0})\| = \|f(\mathbf{y}') - \mathbf{0}\| = \|f[(\delta/2)\mathbf{y}^*]\| = (\delta/2)\|f(\mathbf{y}^*)\| > 1. \quad \square$$

Before leaving the general topic of normed linear spaces, let's take a look at another result in this context.

4.50. Proposition. *If $\mathbf{y}_1, \ldots, \mathbf{y}_n$ are linearly independent vectors in the real normed linear space, L, then there exists a number $\eta > 0$ such that, for all $\mathbf{a} \in \mathbf{R}^n$,*

$$\sum_{i=1}^n |a_i| \leq \eta \cdot \left\| \sum_{i=1}^n a_i \mathbf{y}_i \right\|; \tag{4.30}$$

where '$\|\cdot\|$' denotes the norm on L.

Proof. Suppose that there exists no $\eta > 0$ satisfying (4.30), define the norm $|\cdot|$ on $\mathbf{R}^n$ by:

$$|\mathbf{a}| = \sum_{i=1}^n |a_i|,$$

and let:

$$B = \{\mathbf{x} \in \mathbf{R}^n \mid |\mathbf{x}| = 1\}.$$

Then we must have:

$$\mu \stackrel{\text{def}}{=} \inf_{\mathbf{a} \in B} \left\| \sum_{i=1}^n a_i \mathbf{y}_i \right\| = 0; \tag{4.31}$$

for otherwise it is easy to show that if we define $\eta = 1/\mu$, then this value for η satisfies the condition in (4.30)

Now, given (4.31), it follows that there exists a sequence $\langle \mathbf{a}_k \rangle \subseteq B$ such that, defining:

$$\mathbf{x}_k = \sum_{i=1}^{n} a_{ki}\mathbf{y}_i \quad \text{for } k = 1, 2, \dots,$$

we have:

$$\|\mathbf{x}_k\| = \left\| \sum_{i=1}^{n} a_{ki}\mathbf{y}_i \right\| \leq 1/k. \tag{4.32}$$

Since B is a compact set, it follows from the Bolzano-Weierstrass Theorem (Theorem 2.67) that we can assume, without loss of generality, that there exists $\mathbf{a}^* \in B$ such that $\mathbf{a}_k \to \mathbf{a}^*$; and defining

$$\mathbf{x}^* = \sum_{i=1}^{n} a_i^* \mathbf{y}_i,$$

we have:

$$\begin{aligned} \|\mathbf{x}_k - \mathbf{x}^*\| &= \left\| \sum_{i=1}^{n} a_{ki}\mathbf{y}_i - \sum_{i=1}^{n} a_i^* \mathbf{y}_i \right\| \\ &\leq \sum_{i=1}^{n} |a_{ki} - a_i^*| \cdot \|\mathbf{y}_i\| \leq \left(\max_i \|\mathbf{y}_i\| \right) \cdot \sum_{i=1}^{n} |a_{ki} - a_i^*|. \end{aligned} \tag{4.33}$$

Since $\mathbf{a}_k \to \mathbf{a}^*$, it follows readily from (4.33) that $\mathbf{x}_k \to \mathbf{x}^*$. However, from (4.32) it also follows that $\mathbf{x}_k \to \mathbf{0}$, so that $\mathbf{x}^* = \mathbf{0}$; which then implies:

$$\mathbf{x}^* = \sum_{i=1}^{n} a_i^* \mathbf{y}_i = 0.$$

But this is impossible, since $\mathbf{a}^* \neq \mathbf{0}$ and $\{\mathbf{y}_1, \dots, \mathbf{y}_n\}$ is a linearly independent set. □

We know from our earlier discussions that there is an isomorphism between any n-dimensional subset, S, of L, and $\mathbf{R}^n$; which can be defined as follows (see Definition 4.38). Let $\{\mathbf{y}_1, \dots, \mathbf{y}_n\}$ be a basis for S. Then the function $f\colon S \to \mathbf{R}^n$ defined by:

$$f\Big(\sum_{i=1}^{n} a_i \mathbf{y}_i\Big) = (a_1, \dots, a_n),$$

is an isomorphism between S and $\mathbf{R}^n$ (Theorem 4.37). Suppose then that $X \subseteq S$ is bounded, say:

$$(\forall \mathbf{x} \in X)\colon \|\mathbf{x}\| < \alpha.$$

If $\mathbf{a} \in f(X)$, so that $\mathbf{a} = f\big(\sum_{i=1}^{n} a_i \mathbf{y}_i\big)$, with:

$$\mathbf{x} \stackrel{\text{def}}{=} \sum_{i=1}^{n} a_i \mathbf{y}_i,$$

such that $\|\mathbf{x}\| < \alpha$, then it follows from Theorem 4.50 that:

$$\sum_{i=1}^{n} |a_i| \leq \|\mathbf{x}\| < \alpha.$$

In particular, if:

$$X = \{\mathbf{x} \in S \mid \|\mathbf{x}\| \leq 1\},$$

then $f(X)$ is bounded, and it follows from Theorem 4.49 that f is (uniformly) continuous on S.

Conversely, notice that the inverse of f is the function $g\colon \mathbf{R}^n \to S$ defined by:

$$g(\mathbf{a}) = \sum_{i=1}^{n} a_i \mathbf{y}_i.$$

Moreover, if $\mathbf{a} \in \mathbf{R}^n$ is such that $\sum_{i=1}^{n} |a_i| \leq 1$, it follows from the triangle inequality that:

$$\begin{aligned}\|g(\mathbf{a})\| = \left\|\sum_{i=1}^{n} a_i \mathbf{y}_i\right\| &\leq \sum_{i=1}^{n} |a_i| \cdot \|\mathbf{y}_i\| \\ &\leq \left(\max_i \|\mathbf{y}_i\|\right) \cdot \sum_{i=1}^{n} |a_i| \leq \left(\max_i \|\mathbf{y}_i\|\right).\end{aligned}$$

Consequently, it follows from Theorem 4.49 that the inverse mapping g is also continuous. The significance of this statement will become clearer in the course of our considerations in Chapter 7.

Exercises.

1. Prove Theorem 4.42.

2. Show that if a sequence, $\langle \mathbf{x}_n \rangle \subseteq L$ converges, then it is a Cauchy sequence.

3. If L_i is a normed linear space with norm ρ_i, for $i = 1, 2$, show that a function $f\colon L_1 \to L_2$ is continuous at $\mathbf{x}^* \in L_1$ if, and only if, given any $\langle \mathbf{x}_n \rangle \subseteq L_1$ such that $\mathbf{x}_n \to \mathbf{x}^*$, we have $f(\mathbf{x}_n) \to f(\mathbf{x}^*)$.

4. Let A be an nonempty subset of a real normed linear space, L, and define $\delta(\mathbf{x}, A)$ on L as in Example 4.46.2. Show that:

$$\overline{A} = \{\mathbf{x} \in L \mid \delta(\mathbf{x}, A) = 0\}.$$

5. Show that if L is a normed linear space, then:

a. L and $\emptyset$ are open subsets of L.

b. if $\mathcal{U}$ is a family of open subsets of L, then:

$$O \stackrel{\text{def}}{=} \bigcup_{U \in \mathcal{U}} U,$$

is an open set.

c. if U_1 and U_2 are open subsets of L, then $U \equiv U_1 \cap U_2$ is also an open subset of L.

6. Define the function ρ on $\mathbf{R}^n$ by:

$$\rho(\mathbf{x}) = \begin{cases} |x_1| & \text{if } x_1 \neq 0, \text{ and} \\ |x_2| & \text{if } x_1 = 0. \end{cases}$$

Is ρ a norm for $\mathbf{R}^3$? Is ρ a norm for $\mathbf{R}^2$?

4.5 Inner Product Spaces

In this section we will begin the study of a particulary interesting and useful sort of linear space. We begin with the following.

4.51. Definition. If L is a real linear space, we shall say that a function $f: L \times L \to \mathbf{R}$ is an **inner product (for L)** iff, writing '$\mathbf{x} \cdot \mathbf{y}$' in place of '$f(\mathbf{x}, \mathbf{y})$,' we have, for all $\mathbf{x}, \mathbf{y}, \mathbf{z} \in L$, and all $a \in \mathbf{R}$:

1. $\mathbf{x} \cdot \mathbf{x} \geq 0$, and $[\mathbf{x} \cdot \mathbf{x} = 0 \iff \mathbf{x} = \mathbf{0}]$,
2. $\mathbf{x} \cdot \mathbf{y} = \mathbf{y} \cdot \mathbf{x}$,
3. $(a\mathbf{x}) \cdot \mathbf{y} = a(\mathbf{x} \cdot \mathbf{y})$, and
4. $\mathbf{x} \cdot (\mathbf{y} + \mathbf{z}) = \mathbf{x} \cdot \mathbf{y} + \mathbf{x} \cdot \mathbf{z}$.

4.52. Definitions. A real linear space, L, equipped with an inner product, is called an **inner product space**. If an inner product space is also complete, it is called a **Hilbert space**.[13]

In the remainder of this section, we will always take L to be an inner product space; however, we will not be assuming that L is necessarily complete. In the next chapter, we will begin our formal study of Hilbert spaces *per se*.

Inner product spaces are of particular interest from the standpoint of economics, because in the context of such a space, L, we can generally define price and quantity concepts exactly as is done in $\mathbf{R}^n$, even if L is infinite-dimensional! We will elaborate upon this point a bit in the next examples, and at greater length in Chapter 10.

4.53. Examples/Exercises.

1. In $\mathbf{R}^n$, the familiar product defined by:

$$\mathbf{x} \cdot \mathbf{y} = \sum_{i=1}^{n} x_i y_i \quad \text{for } \mathbf{x}, \mathbf{y} \in \mathbf{R}^n,$$

satisfies properties 1–4 of Definition 4.51 (see Proposition 1.43), and thus is an inner product.

2. If we define a product on ℓ_2 by:

$$\mathbf{x} \cdot \mathbf{y} = \sum_{n=1}^{\infty} x_n y_n,$$

then we can show that this is an inner product, as follows.

Let $\mathbf{x}, \mathbf{y} \in \ell_2$ be arbitrary, let $p \in N$, and consider the vectors $\mathbf{x}^*$ and $\mathbf{y}^* \in \mathbf{R}^p$ defined from $\mathbf{x}$ and $\mathbf{y}$, respectively, by:

$$\mathbf{x}^* = (|x_1|, \ldots |x_p|) \text{ and } \mathbf{y}^* = (|y_1|, \ldots |y_p|).$$

[13] An inner product space is sometimes called a **Euclidean space**, and the term 'inner product space' is sometimes reserved for a linear space over the field **C**. Such a space is also sometimes called a **pre-Hilbert space**. In a mathematics text, the usual definition of a Hilbert space would allow the field over which the space is defined to be either **C** or **R**. In this text, however, we will only be concerned with real Hilbert spaces.

Then we have:

$$(\mathbf{x}^* - \mathbf{y}^*) \cdot (\mathbf{x}^* - \mathbf{y}^*) \geq 0,$$

and thus we obtain:

$$\sum_{n=1}^{p} |x_n y_n| \leq (1/2) \cdot \left(\sum_{n=1}^{p} x_n^2 + \sum_{n=1}^{p} y_n^2 \right);$$

from which it follows that the series $\sum_{n=1}^{\infty} x_n y_n$, is absolutely convergent. Propositions 2.40 and 2.42 then imply that the series converges, and thus that our product is well-defined. It is then easy to show that Properties 1–4 of Definition 4.51 are satisfied.

We can make use of the inner product function to define value concepts in ℓ_2 exactly as is done in $\mathbf{R}^n$, as follows. Suppose we interpret $\mathbf{x} \in \ell_2$ to be a commodity bundle,

$$\mathbf{x} = (x_1, x_2, \ldots, x_t, \ldots),$$

where, say, the quantity x_t denotes the quantity of a commodity available at time t, and that:

$$\mathbf{p} = (p_1, p_2, \ldots, p_t, \ldots)$$

is a vector of (current) prices of the commodity, with the quantity p_t being the (discounted) present price of a unit of the commodity for delivery in the tth period. Then the inner product, $\mathbf{p} \cdot \mathbf{x}$, is the present value of the commodity bundle, $\mathbf{x}$, given the price vector $\mathbf{p}$. We will elaborate upon this idea in later chapters; in the meantime, notice that this value function is as easy to deal with, in principle, as is the inner product function in $\mathbf{R}^n$. □

For an inner product space, L, we define the function $\|\cdot\|$ on L by:

$$\|\mathbf{x}\| = (\mathbf{x} \cdot \mathbf{x})^{1/2}. \tag{4.34}$$

We will use a generalization of our old friend, the Cauchy-Schwarz Inequality, to prove that the function defined in (4.34) is a norm for L. The proof of the second result is easy (following along the lines of the proof of Theorem 1.47), and will be left as an exercise.

4.54. Theorem. (Cauchy-Schwarz Inequality) *If L is an inner product space, then defining $\|\cdot\|$ on L as in (4.34), above, we have for all $\mathbf{x}, \mathbf{y} \in L$:*

$$|\mathbf{x} \cdot \mathbf{y}| \leq \|\mathbf{x}\| \cdot \|\mathbf{y}\|.$$

Proof. Let $\mathbf{x}, \mathbf{y} \in L$, and $\lambda \in \mathbf{R}$ be arbitrary. Then, using the properties of the inner product:

$$0 \leq (\lambda\mathbf{x} + \mathbf{y}) \cdot (\lambda\mathbf{x} + \mathbf{y}) = (\mathbf{x} \cdot \mathbf{x})\lambda^2 + (2\mathbf{x} \cdot \mathbf{y})\lambda + \mathbf{y} \cdot \mathbf{y}.$$

Therefore it follows from Lemma 1.44 that:

$$(\mathbf{x} \cdot \mathbf{x}) \cdot (\mathbf{y} \cdot \mathbf{y}) \geq (\mathbf{x} \cdot \mathbf{y})^2.$$

Taking the square root of both sides, we obtain the desired inequality. □

4.55. Proposition. *The function* $\|\cdot\|$ *defined on* L *in (4.34), above, is a norm for* L.

Since we have used the inner product to define the norm, it will probably come as no great surprise that the inner product function satisfies special continuity properties.

4.56. Proposition. *The inner product function is* **bi-continuous** *on an inner product space,* L*; that is, given any points* $\mathbf{x}^*, \mathbf{y}^* \in L$*, and any* $\epsilon > 0$*, there exists* $\delta > 0$ *such that for all* $\mathbf{x}, \mathbf{y} \in L$ *satisfying:*

$$\max\{\|\mathbf{x} - \mathbf{x}^*\|, \|\mathbf{y} - \mathbf{y}^*\|\} < \delta \Rightarrow |\mathbf{x} \cdot \mathbf{y} - \mathbf{x}^* \cdot \mathbf{y}^*| < \epsilon.$$

Proof. Given $\epsilon > 0$, let δ be defined by:

$$\delta = \min\{1, \epsilon/2(1 + \|\mathbf{x}^*\|), \epsilon/2(1 + \|\mathbf{y}^*\|)\}.$$

Then if $\mathbf{x}, \mathbf{y} \in L$ are such that:

$$\max\{\|\mathbf{x} - \mathbf{x}^*\|, \|\mathbf{y} - \mathbf{y}^*\|\} < \delta,$$

we note first that:

$$\|\mathbf{x}\| \leq \|\mathbf{x} - \mathbf{x}^*\| + \|\mathbf{x}^*\| < \|\mathbf{x}^*\| + 1;$$

so that:

$$\begin{aligned} &|\mathbf{x} \cdot \mathbf{y} - \mathbf{x}^* \cdot \mathbf{y}^*| \\ &\quad = |\mathbf{x} \cdot \mathbf{y} - \mathbf{x} \cdot \mathbf{y}^* + \mathbf{x} \cdot \mathbf{y}^* - \mathbf{x}^* \cdot \mathbf{y}^*| \leq |\mathbf{x} \cdot (\mathbf{y} - \mathbf{y}^*)| + |(\mathbf{x} - \mathbf{x}^*) \cdot \mathbf{y}^*| \\ &\quad\quad \leq \|\mathbf{x}\| \cdot \|\mathbf{y} - \mathbf{y}^*\| + \|\mathbf{y}^*\| \cdot \|\mathbf{x} - \mathbf{x}^*\| < \epsilon; \end{aligned}$$

where we have used the Cauchy-Schwarz inequality in obtaining the next-to-the-last inequality. □

Much of the geometry of an inner product space follows the constructs of Euclidean geometry; for example, we have the following, which is easily proved by using the inner product definition of the norm.

4.57. Proposition. **(Parallelogram Law)** *For any* $\mathbf{x}, \mathbf{y} \in L$*, we have:*

$$\|\mathbf{x} + \mathbf{y}\|^2 + \|\mathbf{x} - \mathbf{y}\|^2 = 2\|\mathbf{x}\|^2 + 2\|\mathbf{y}\|^2.$$

4.58. Definition. In an inner product space, we say that $\mathbf{x}, \mathbf{y} \in L$ are **orthogonal** iff $\mathbf{x} \cdot \mathbf{y} = 0$; in which case we shall also say that **$\mathbf{x}$ is orthogonal to $\mathbf{y}$**, and write '$\mathbf{x} \perp \mathbf{y}$.'

Notice that $\mathbf{0}$ is orthogonal to every $\mathbf{x} \in L$; and that $\mathbf{x}$ is orthogonal to $\mathbf{y}$ if, and only if, $\mathbf{y}$ is orthogonal to $\mathbf{x}$. Recall that for non-zero vectors $\mathbf{x}$ and $\mathbf{y}$ in $\mathbf{R}^2$ or $\mathbf{R}^3$, $\mathbf{x} \cdot \mathbf{y} = 0$ if, and only if, $\mathbf{0x}$ and $\mathbf{0y}$ [the directed line segments obtained by connecting $\mathbf{x}$ and $\mathbf{y}$, respectively, with the origin] form a ninety-degree angle. These considerations should make our next result look quite familiar.

4.59. Proposition. **(Pythagorean theorem)** *For any* $\mathbf{x}, \mathbf{y} \in L$ *such that* $\mathbf{x} \cdot \mathbf{y} = 0$, *we have:*

$$\|\mathbf{x} + \mathbf{y}\|^2 = \|\mathbf{x}\|^2 + \|\mathbf{y}\|^2.$$

Proof. Making use of the definition of the norm and the properties of the inner product, we have:

$$\begin{aligned}\|\mathbf{x} + \mathbf{y}\|^2 &= (\mathbf{x} + \mathbf{y}) \cdot (\mathbf{x} + \mathbf{y}) = \mathbf{x} \cdot \mathbf{x} + 2\mathbf{x} \cdot \mathbf{y} + \mathbf{y} \cdot \mathbf{y} \\ &= \mathbf{x} \cdot \mathbf{x} + \mathbf{y} \cdot \mathbf{y} = \|\mathbf{x}\|^2 + \|\mathbf{y}\|^2. \quad \square\end{aligned}$$

4.60. Definitions. A set of vectors, $X = \{\mathbf{x}_a \mid a \in A\}$, in L is said to be an **orthogonal system** iff:

1. each $\mathbf{x}_a$ is non-null, and
2. for each $a, b \in A$ such that $a \neq b$, we have:

$$\mathbf{x}_a \cdot \mathbf{x}_b = 0.$$

If, in addition to the preceding properties, X satisfies:

$$(\forall a \in A)\colon \mathbf{x}_a \cdot \mathbf{x}_a = 1,$$

then X is said to be an **orthonormal system**.

The set of unit coordinate vectors, $\mathbf{e}_i$ $(i = 1, \ldots, n)$ is, of course, an orthonormal system in $\mathbf{R}^n$; as is the set $\{\mathbf{e}_i \mid i \in N\}$ which was defined for ℓ_2 in 4.17.2.

4.61. Proposition. *If* $X = \{\mathbf{x}_a \mid a \in A\}$ *is an orthogonal system, then* X *is a linearly independent set.*

Proof. Suppose $\{\mathbf{x}_1, \ldots, \mathbf{x}_m\}$ is a subset of X, and that $\{a_1, \ldots, a_m\}$ is a set of real numbers satisfying

$$\sum_{i=1}^m a_i \mathbf{x}_i = \mathbf{0},$$

and let $h \in \{1, \ldots, m\}$ be arbitrary. Then, using the properties of the inner product and the fact that X is an orthogonal system, we have:

$$0 = \mathbf{x}_h \cdot \mathbf{0} = \mathbf{x}_h \cdot \left(\sum_{i=1}^m a_i \mathbf{x}_i\right) = \sum_{i=1}^m a_i (\mathbf{x}_h \cdot \mathbf{x}_i) = a_h (\mathbf{x}_h \cdot \mathbf{x}_h);$$

which, using 4.51.1, implies that $a_h = 0$. $\square$

4.62. Definition. If S is a linear subspace of L, a subset, X, of L which is an orthogonal (respectively, orthonormal) system is called an **orthogonal** (respectively, **orthonormal**) **basis for S** iff $Sp(X) = S$.

The sets $\mathbf{e}_i$ $(i = 1, \ldots, n)$ and $\{\mathbf{e}_i \mid i \in N\}$ discussed earlier in this section are orthonormal bases for $\mathbf{R}^n$ and f (see Example 4.12.5), respectively.

4.63. Definition. If S is a linear subspace of L, we define $S^\perp$, the **orthogonal complement of S**, by:

$$S^\perp = \{\mathbf{x} \in L \mid (\forall \mathbf{y} \in S) : \mathbf{x} \cdot \mathbf{y} = 0\}.$$

Notice that we always have $\mathbf{0} \in S^\perp$, for any subspace, S. In fact, we have the following; the proof of which will be left as an exercise.

4.64. Proposition. *If S is a linear subspace, then $S^\perp$ is a linear subspace of L as well, and*

$$S \cap S^\perp = \{\mathbf{0}\}.$$

The following theorem is an extremely useful result. However, while the proof is not really particularly difficult, at first reading you may want to assume that the dimension of S (and thus the number of elements of Y) is finite.

4.65. Theorem. *If Y is a basis for a subspace, S, of L, and Y consists of a finite or countable set of linearly independent vectors, $\mathbf{y}_1, \mathbf{y}_2, \ldots$, then there exists an orthonormal subset, $X = \{\mathbf{x}_1, \mathbf{x}_2, \ldots\}$, of S such that, for each n, $\{\mathbf{x}_1, \ldots, \mathbf{x}_n\}$ spans exactly the same subspace of S as does $\{\mathbf{y}_1, \ldots, \mathbf{y}_n\}$.*

Proof. We begin by defining

$$\mathbf{x}_1 = (1/\|\mathbf{y}_1\|)\mathbf{y}_1;$$

which, since $\mathbf{y}_1$ must be non-null, is a legitimate operation. Moreover, it follows immediately that $\mathbf{x}_1$ is a unit vector, and that $Sp\{\mathbf{x}_1\} = Sp\{\mathbf{y}_1\}$. Next, defining:

$$\mathbf{x}_2' = \mathbf{y}_2 - (\mathbf{x}_1 \cdot \mathbf{y}_2)\mathbf{x}_1;$$

we note that $\mathbf{x}_1 \cdot \mathbf{x}_2' = 0$, and that $\mathbf{x}_2'$ can be expressed as a linear combination of $\{\mathbf{y}_1, \mathbf{y}_2\}$. Since not all of the coefficients of the $\mathbf{y}_i$'s are zero in this linear combination, it follows that $\mathbf{x}_2' \neq \mathbf{0}$; so that we can define

$$\mathbf{x}_2 = (1/\|\mathbf{x}_2'\|)\mathbf{x}_2',$$

which is a unit vector orthogonal to $\mathbf{x}_1$. Since each $\mathbf{x}_i$ is a linear combination of $\{\mathbf{y}_1, \mathbf{y}_2\}$, it follows immediately that $\{\mathbf{x}_1, \mathbf{x}_2\} \subseteq Sp(\{\mathbf{y}_1, \mathbf{y}_2\})$. However, since by Proposition 4.61, $\{\mathbf{x}_1, \mathbf{x}_2\}$ is a linearly independent set, and since $Sp(\{\mathbf{y}_1, \mathbf{y}_2\})$ has dimension 2, it then follows that:

$$Sp(\{\mathbf{x}_1, \mathbf{x}_2\}) = Sp(\{\mathbf{y}_1, \mathbf{y}_2\}).$$

Now suppose we have obtained an orthonormal set, $\{\mathbf{x}_1, \ldots, \mathbf{x}_k\}$, such that

$$\begin{aligned}&\mathbf{x}_i \in Sp(\{\mathbf{y}_1, \ldots, \mathbf{y}_i\}) \text{ and}\\ &\qquad Sp(\{\mathbf{x}_1, \ldots, \mathbf{x}_i\}) = Sp(\{\mathbf{y}_1, \ldots, \mathbf{y}_i\}) \quad \text{for } i = 1, \ldots, k; \end{aligned} \tag{4.35}$$

and define:

$$\mathbf{x}'_{k+1} = \mathbf{y}_{k+1} - \sum\nolimits_{i=1}^{k} (\mathbf{x}_i \cdot \mathbf{y}_{k+1})\mathbf{x}_i. \tag{4.36}$$

Then, for $j \in \{1, \ldots, k\}$, we have:

$$\mathbf{x}_j \cdot \mathbf{x}'_{k+1} = \mathbf{x}_j \cdot \mathbf{y}_{k+1} - \sum\nolimits_{i=1}^{k} (\mathbf{x}_i \cdot \mathbf{y}_{k+1})\mathbf{x}_j \cdot \mathbf{x}_i = \mathbf{x}_j \cdot \mathbf{y}_{k+1} - \mathbf{x}_j \cdot \mathbf{y}_{k+1} = 0.$$

Thus, $\mathbf{x}'_{k+1}$ is orthogonal to each of the $\mathbf{x}_i$'s, and it follows from (4.35) and (4.36) that $\mathbf{x}'_{k+1}$ is a non-null linear combination of $\{\mathbf{y}_1, \ldots, \mathbf{y}_{k+1}\}$. Consequently, since $\{\mathbf{y}_1, \ldots, \mathbf{y}_{k+1}\}$ is a linearly independent set, we see that $\mathbf{x}'_{k+1}$ is non-zero. Thus we can define $\mathbf{x}_{k+1}$ by:

$$\mathbf{x}_{k+1} = (1/\|\mathbf{x}'_{k+1}\|)\mathbf{x}'_{k+1},$$

to obtain a unit vector; and, notice that $\mathbf{x}_{k+1}$ is orthogonal to each $\mathbf{x}_i$ ($i = 1, \ldots, k$), and is a linear combination of $\{\mathbf{y}_1, \ldots, \mathbf{y}_{k+1}\}$. Therefore, it follows that $\{\mathbf{x}_1, \ldots, \mathbf{x}_{k+1}\} \subseteq Sp(\{\mathbf{y}_1, \ldots, \mathbf{y}_{k+1}\})$; and, since $\{\mathbf{x}_1, \ldots, \mathbf{x}_{k+1}\}$ is linearly independent, and the dimension of the subspace is $k+1$, it follows that:

$$Sp(\{\mathbf{x}_1, \ldots, \mathbf{x}_{k+1}\}) = Sp(\{\mathbf{y}_1, \ldots, \mathbf{y}_{k+1}\}).$$

If Y is a finite set, then, proceeding in this way, we obtain the set X of the conclusion after a finite number of steps (and we will, of course, have the same number of elements in X as are in Y). If, on the other hand, Y is countably infinite, then it follows by induction that there exists a set X satisfying the statements in the conclusion of the result. □

Using the method developed in the proof of the preceding result (this method is called the 'Gram-Schmidt process,' after its original developers), we can obtain another useful result, which is almost a corollary of 4.65.

4.66. Theorem. *If $X = \{\mathbf{x}_1, \ldots, \mathbf{x}_m\}$ is an orthonormal subset of a finite-dimensional linear subspace, S, then we can extend X to an orthonormal basis for S.*

Proof. If $\dim(S) = m$, then $\{\mathbf{x}_1, \ldots, \mathbf{x}_m\}$ is a basis for S, and there is nothing to prove. Otherwise, there exists $\mathbf{y} \in S$ such that:

$$\mathbf{y} \notin Sp(\{\mathbf{x}_1, \ldots, \mathbf{x}_m\});$$

and, if we define (proceeding as in the proof of 4.65),

$$\mathbf{y}' = \mathbf{y} - \sum\nolimits_{i=1}^{m} (\mathbf{x}_i \cdot \mathbf{y})\mathbf{x}_i,$$

and

$$\mathbf{x}_{m+1} = (1/\|\mathbf{y}'\|)\mathbf{y}',$$

then we have obtained an orthonormal subset of S containing $m+1$ vectors. If, in fact, $\dim(S) > m+1$, then we can proceed in exactly the same way to obtain an orthonormal subset of S containing $m+2$ vectors. When our orthonormal subset, X, contains n vectors, where $n = \dim(S)$, then it follows from considerations similar to those of 4.65 that X is a basis for S; and, of course, the first m elements of X will be $\{\mathbf{x}_1, \ldots, \mathbf{x}_m\}$. □

The proof of our next result is quite easy, and will be left as an exercise.

4.67. Proposition. *If $X = \{\mathbf{x}_1, \ldots, \mathbf{x}_m\}$ is a basis for a linear subspace, S, then:*

$$S^{\perp} = \{\mathbf{y} \in L \mid \mathbf{y} \cdot \mathbf{x}_i = 0, \text{ for } i = 1, \ldots, m\}.$$

Our final result of this section is concerned with subspaces of $\mathbf{R}^n$, and is probably quite familiar to you already.

4.68. Proposition. *If S is a linear subspace of $\mathbf{R}^n$, then:*

$$\dim(S) + \dim(S^{\perp}) = n.$$

Proof. If $S = \{\mathbf{0}\}$, then $S^{\perp} = \mathbf{R}^n$, and the result is obvious. Otherwise, if $\dim(S) = m \geq 1$, it is an easy consequence of 4.65 that there exists an orthonormal basis for $S, X = \{\mathbf{x}_1, \ldots, \mathbf{x}_m\}$. Moreover, it also follows from Theorem 4.66 that X can be extended to an orthonormal basis for $\mathbf{R}^n$; which we can write as:

$$Z = \{\mathbf{x}_1, \ldots, \mathbf{x}_m, \mathbf{y}_1, \ldots, \mathbf{y}_q\}, \quad \text{where } q = n - m.$$

Clearly, if we define Y and T by:

$$Y = \{\mathbf{y}_1, \ldots, \mathbf{y}_q\}, \text{ and } T = Sp(Y),$$

respectively, then we must have:

$$T \subseteq S^{\perp}.$$

Moreover, if $\mathbf{x} \in S^{\perp}$, there exist real numbers, $a_1, \ldots, a_m, b_1, \ldots, b_q$, such that:

$$\mathbf{x} = \sum_{i=1}^{m} a_i \mathbf{x}_i + \sum_{j=1}^{q} b_j \mathbf{y}_j. \tag{4.37}$$

Now, let h be an arbitrary element of $\{1, \ldots, m\}$. Then, since $\mathbf{x} \in S^{\perp}$,

$$\begin{aligned} 0 = \mathbf{x}_h \cdot \mathbf{x} = \mathbf{x}_h \cdot \left(\sum_{i=1}^{m} a_i \mathbf{x}_i + \sum_{j=1}^{q} b_j \mathbf{y}_j\right) \\ = \sum_{i=1}^{m} a_i \mathbf{x}_h \cdot \mathbf{x}_i + \sum_{j=1}^{q} b_j \mathbf{x}_h \cdot \mathbf{y}_j = a_h \mathbf{x}_h \cdot \mathbf{x}_h = a_h. \end{aligned}$$

Consequently, we see that

$$\mathbf{x} = \sum_{j=1}^{q} b_j \mathbf{y}_j;$$

and it follows that $\mathbf{x} \in T \equiv Sp(Y)$, and thus that:

$$S^{\perp} \subseteq T.$$

From the reasoning of the above paragraph we have found that $S^{\perp} = Sp(Y)$; so that $\dim(S^{\perp}) = q = n - m$, and thus:

$$\dim(S) + \dim(S^{\perp}) = m + (n - m) = n. \quad \square$$

Exercises.

1. Show that if we define S in $\mathbf{R}^3$ by:

$$S = \{\mathbf{x} \in \mathbf{R}^3 \mid x_3 = 0\},$$

then $S^{\perp} = \{\mathbf{y} \in \mathbf{R}^3 \mid y_1 = y_2 = 0\}$.

2. Show that the set of vectors:

$$\{(1/\sqrt{2}, 1/\sqrt{2}, 0), (-1/\sqrt{2}, 1/\sqrt{2}, 0), (0, 0, -1)\}$$

is an orthonormal system in $\mathbf{R}^3$.

3. Prove (a) Proposition 4.55, (b) 4.57, (c) 4.64, and (d) 4.67.

4. Show that in $\mathbf{R}^n$, if $T = S^{\perp}$, then $S = T^{\perp}$; that is, if S is a linear subspace of $\mathbf{R}^n$, then $(S^{\perp})^{\perp} = S$. [Hint: study the proof of Proposition 4.68.]

5. Define $\otimes$ on m by:

$$\mathbf{x} \otimes \mathbf{y} = \sup_n x_n y_n \quad \text{for } \mathbf{x}, \mathbf{y} \in m.$$

Is $\otimes$ an inner product for m?

6. Define $\otimes$ on c by:

$$\mathbf{x} \otimes \mathbf{y} = (\lim_{n \to \infty} x_n) \cdot (\lim_{n \to \infty} y_n) \quad \text{for } \mathbf{x}, \mathbf{y} \in c.$$

Is $\otimes$ an inner product for c?

7. Show that if $f \colon \mathbf{R}^n \to \mathbf{R}^m$ is linear, then f is uniformly continuous on $\mathbf{R}^n$. [Hint: show that if $f(\mathbf{x})= \mathrm{A}\mathbf{x}$, then $\|f(\mathbf{x})\| \leq \|\mathbf{x}\| \cdot \left(\sum_{i=1}^{m} \|\mathbf{a}_i\|\right)$, where $\mathbf{a}_i$ is the ith row of A.]

8. Let L be an inner product space, let S be a subspace of L such that $\dim(S) = n$, and let $\{\mathbf{y}_1, \ldots, \mathbf{y}_n\}$ be an orthonormal basis for S. Show that if f is the isomorphism from S to $\mathbf{R}^n$ determined by $\{\mathbf{y}_1, \ldots, \mathbf{y}_n\}$ (Definition 4.38), then f preserves inner products; that is:

$$(\forall \mathbf{x}, \mathbf{y} \in S) \colon \mathbf{x} \cdot \mathbf{y} = f(\mathbf{x}) \cdot f(\mathbf{y}).$$

4.6 Product Spaces and Direct Sums

In this section we will study the product space determined by two or more linear spaces. In some sense you are already familiar with this basic idea; since $\mathbf{R}^n$ can be thought of as the product of $\mathbf{R}$ taken with itself n times.

4.69. Definition. Suppose $L_1, \ldots, L_m$ are linear spaces over the same field, $\mathbf{F}$. We define the **product space determined by the $\mathbf{L}_i$**, $L = \prod_{i=1}^m L_i$, by defining addition, $+$, and scalar multiplication on L by:

$$(\mathbf{x}_1, \ldots, \mathbf{x}_m) + (\mathbf{y}_1, \ldots, \mathbf{y}_m) = (\mathbf{x}_1 \oplus_1 \mathbf{y}_1, \ldots, \mathbf{x}_m \oplus_m \mathbf{y}_m)$$

and

$$a(\mathbf{x}_1, \ldots, \mathbf{x}_m) = (a\mathbf{x}_1, \ldots, a\mathbf{x}_m),$$

respectively, for $a \in \mathbf{F}$ and $\mathbf{x}_i, \mathbf{y}_i \in L_i$ for $i = 1, \ldots, m$; where '$\oplus_i$' denotes the addition operation on L_i, for $i = 1, \ldots, m$.

As noted in our next result, the product space L just defined is a linear space. Some of the elementary properties of such product spaces are then set out in the two propositions which follow. I will leave the proof of all three results as exercises.

4.70. Proposition. *The product space, L, defined in 4.69 is a linear space over $\mathbf{F}$.*

4.71. Proposition. *If L_i is a normed linear space over the field $\mathbf{F}$, and with norm ρ_i, for $i = 1, \ldots, m$, then the function ρ defined on the product space $L = \prod_{i=1}^m L_i$, by:*

$$\rho(\mathbf{x}_1, \ldots, \mathbf{x}_m) = \max\{\rho_1(\mathbf{x}_1), \ldots, \rho_m(\mathbf{x}_m)\},$$

is a norm for L. Moreover, the ϵ-neighborhood of a point $\mathbf{x} = (\mathbf{x}_1, \ldots, \mathbf{x}_m)$ is then given by:

$$N(\mathbf{x}, \epsilon) = \prod_{i=1}^m N_i(\mathbf{x}_i, \epsilon),$$

where $N_i(\mathbf{x}_i, \epsilon)$ is the epsilon-neighborhood of $\mathbf{x}_i$ in L_i; that is,

$$N_i(\mathbf{x}_i, \epsilon) = \{\mathbf{y} \in L_i \mid \rho_i(\mathbf{y} - \mathbf{x}_i) < \epsilon\} \quad \text{for } i = 1, \ldots, m.$$

4.72. Proposition. *Suppose $L_1, \ldots, L_m$ are linear spaces over the same field, $\mathbf{F}$, and that S_i is a linear subspace of L_i, for $i = 1, \ldots, m$. Then the product of the S_i's, $S = \prod_{i=1}^m S_i$, is a linear subspace of the product space.*

Until further notice, we will once again be maintaining the assumption that L is a linear space over some arbitrary field, $\mathbf{F}$.

4.73. Definition. If A and B are subsets of L, then we define $A + B$, the **sum of A and B**, by:

$$A + B = \{\mathbf{z} \in L \mid (\exists \mathbf{x} \in A, \mathbf{y} \in B) \colon \mathbf{z} = \mathbf{x} + \mathbf{y}\}.$$

In the special case in which one of the sets is a singleton, say $A = \{\mathbf{x}^*\}$, for some $\mathbf{x}^* \in L$, we will write '$\mathbf{x}^* + B$' in place of '$\{\mathbf{x}^*\} + B$,' and call $\mathbf{x}^* + B$ the **translate of B by x***.

4.74. Examples/Exercises.

1. In $\mathbf{R}^2$, let:

$$A = \{\mathbf{x} \in \mathbf{R}^2 \mid x_1 \geq 0 \;\&\; x_2 = 0\}$$

and

$$B = \{\mathbf{y} \in \mathbf{R}^2 \mid y_1 = 0 \;\&\; y_2 \geq 0\}.$$

Prove that:

$$A + B = \mathbf{R}^2_+.$$

2. Let $\epsilon > 0$, and define:

$$S = \{\mathbf{x} \in \mathbf{R}^n \mid \|\mathbf{x}\| < \epsilon\}.$$

Prove that, for any $\mathbf{x}^* \in \mathbf{R}^n$:

$$\mathbf{x}^* + S = N(\mathbf{x}^*, \epsilon).$$

3. Prove that if S is a linear subspace of L, and $\mathbf{x}^* \in S$, then $\mathbf{x}^* + S = S$; and, in fact:

$$S + S = S. \quad \square$$

4.75. Proposition. *If S_1 and S_2 are linear subspaces of L, then*

$$S \stackrel{\text{def}}{=} S_1 + S_2$$

is also a linear subspace of L.

Proof. Consider the product space $\mathbf{L} \equiv L \times L$. By Proposition 4.72, the set:

$$\mathbf{S} \stackrel{\text{def}}{=} S_1 \times S_2,$$

is a subspace of $\mathbf{L}$. Since it is apparent that vector addition can be considered to be a linear transformation from $\mathbf{L}$ to L, it then follows at once from Proposition 4.30 that $S \equiv f(\mathbf{S})$ is a linear subspace of L. $\square$

4.76. Definition. If S_1 and S_2 are subspaces of L, we say that **L is the direct sum of $\mathbf{S_1}$ and $\mathbf{S_2}$**, and write '$L = S_1 \oplus S_2$,' iff:

1. $L = S_1 + S_2$, and
2. $S_1 \cap S_2 = \{\mathbf{0}\}$.

4.77. Proposition. *If S is a linear subspace of $\mathbf{R}^n$, then $\mathbf{R}^n$ is the direct sum of S and $S^\perp$.*

Proof. Exercise. (Hint: look back at the proof of Proposition 4.68.) □

4.78. Examples/Exercises.

1. In $\mathbf{R}^3$, let:

$$S_1 = \{\mathbf{x} \in \mathbf{R}^3 \mid x_3 = 0\} \text{ and } S_2 = \{\mathbf{x} \in \mathbf{R}^3 \mid x_1 = x_2 = 0\}.$$

Show that $\mathbf{R}^3$ is the direct sum of S_1 and S_2.

2. Defining S_2 as in the previous example, let S_1 be defined by:

$$S_1 = \{\mathbf{x} \in \mathbf{R}^3 \mid x_2 = x_3\}.$$

Show that $\mathbf{R}^3$ is the direct sum of S_1 and S_2.

3. Let $X = \{\mathbf{x}_1, \ldots, \mathbf{x}_m\}$ be a linearly independent subset of L, let X_1 be a proper subset of X, and define $X_2 = X \setminus X_1, S = Sp(X)$, and $S_i = Sp(X_i)$, for $i = 1, 2$. Then $S = S_1 \oplus S_2$.

4. **Solutions of Linear Equations.** Suppose $\mathbf{A}$ is a real $m \times n$ matrix, and that $\mathbf{b} \in \mathbf{R}^m$. Then *exactly one* of the following holds: *either:* (a) there exists $\mathbf{x} \in \mathbf{R}^n$ such that:

$$\mathbf{A}\mathbf{x} = \mathbf{b},$$

or (b) there exists $\mathbf{y} \in \mathbf{R}^m$ satisfying:

$$\mathbf{y}'\mathbf{A} = \mathbf{0} \;\;\&\;\; \mathbf{b} \cdot \mathbf{y} = 1.$$

I will leave it as an exercise to prove that not both (a) and (b) can hold. To prove that one of these alternatives must hold, let '$\mathbf{a}_j$' denote the jth column of $\mathbf{A}$, for $j = 1, \ldots, n$, and let 'S' denote the subspace of $\mathbf{R}^m$ spanned by $\{\mathbf{a}_1, \ldots, \mathbf{a}_n\}$. Then, since $\mathbf{b} \in \mathbf{R}^m$, it follows from Proposition 4.77 that there exist unique vectors, $\mathbf{u} \in S$ and $\mathbf{v} \in S^\perp$ such that:

$$\mathbf{b} = \mathbf{u} + \mathbf{v}.$$

However, if (a) does not hold (that is, the equation $\mathbf{A}\mathbf{x} = \mathbf{b}$ has no solution), then it follows that $\mathbf{v} \neq \mathbf{0}$, and:

$$\mathbf{b} \cdot \mathbf{v} = (\mathbf{u} + \mathbf{v}) \cdot \mathbf{z} = \mathbf{u} \cdot \mathbf{v} + \mathbf{v} \cdot \mathbf{v} = \mathbf{v} \cdot \mathbf{v} > 0.$$

If we now define:

$$\mathbf{y} = (1/\mathbf{v} \cdot \mathbf{v})\mathbf{v},$$

we see that $\mathbf{y}$ satisfies condition (b). □

The previous examples illustrate at least two facts which are worth noting at this point. First of all, notice that in the first example, $S_1 = (S_2)^{\perp}$, and conversely; while in the second example, $S_1 \neq (S_2)^{\perp}$. Thus we see that an inner product space, L, may be equal to the direct sum of two subspaces which are not the orthogonal complements of one another. Secondly, the first two examples also show that L may be equal to the direct sum of S and T and of S and U as well, even though $U \neq T$.

4.79. Proposition. *Suppose L is the direct sum of S and T, and let $\mathbf{x} \in L$. Then there is exactly one $\mathbf{y} \in S$ and exactly one $\mathbf{z} \in T$ such that $\mathbf{x} = \mathbf{y} + \mathbf{z}$.*

Proof. Letting $\mathbf{x}$ be an arbitrary element of L, it follows immediately from the definition of direct sum that there exists $\mathbf{y} \in S$ and $\mathbf{z} \in T$ such that:

$$\mathbf{x} = \mathbf{y} + \mathbf{z}. \tag{4.38}$$

Suppose now that $\mathbf{y}^* \in S$ and $\mathbf{z}^* \in T$ are also such that:

$$\mathbf{x} = \mathbf{y}^* + \mathbf{z}^*. \tag{4.39}$$

Then from (4.38) and (4.39):

$$\mathbf{x} = \mathbf{y} + \mathbf{z} = \mathbf{y}^* + \mathbf{z}^*;$$

so that

$$\mathbf{y} - \mathbf{y}^* = \mathbf{z}^* - \mathbf{z}. \tag{4.40}$$

However, $\mathbf{y} - \mathbf{y}^* \in S$, and $\mathbf{z}^* - \mathbf{z} \in T$, so that we have from (4.40) that:

$$\mathbf{y} - \mathbf{y}^* \in S \cap T.$$

Since $S \cap T = \{\mathbf{0}\}$, it then follows that $\mathbf{y} = \mathbf{y}^*$ and $\mathbf{z} = \mathbf{z}^*$. □

You may have already noticed that there is a close relationship between the concept of a product space and that of a direct sum. In fact, if L is the direct sum of subspaces S_1 and S_2, then L is isomorphic to $S_1 \times S_2$. To verify this statement, define the function $f\colon L \to S_1 \times S_2$ by:

$$f(\mathbf{x}) = (\mathbf{y}, \mathbf{z}),$$

where $\mathbf{y} \in S_1$ and $\mathbf{z} \in S_2$ are that unique pair such that:

$$\mathbf{x} = \mathbf{y} + \mathbf{z}.$$

It is obvious that the function f is one-to-one and onto $S_1 \times S_2$. It is also easy to prove that f is linear, although I will leave the proof of this fact as an exercise.

4.80. Definition. A linear subspace, S, is said to have **codimension n** iff there exists a linear subspace, T, such that:

1. L is the direct sum of S and T, and
2. T has dimension n.

Notice that it follows from the above definition and the definition of direct sum that if S is a subspace of L having codimension n, then there exists $\mathbf{x}_1, \ldots, \mathbf{x}_n \in L$ such that, for every $\mathbf{x}$ in L, there exist a unique set of scalars, $a_1, \ldots, a_n$, and a unique $\mathbf{y} \in S$ such that:

$$\mathbf{x} = \sum_{i=1}^{n} a_i \mathbf{x}_i + \mathbf{y}.$$

Moreover, using Theorem 4.22 it can be shown that the codimension of a linear subspace is uniquely defined.

4.81. Definition. A subset, H, of L is said to be a **hyperplane** iff there exist a linear subspace of L, S, and a vector $\mathbf{x}^* \in L$ such that:

1. S has codimension one, and
2. $H = \mathbf{x}^* + S$.

4.82. Examples/Exercises.

1. Prove that if $H = \mathbf{x}^* + S$ is a hyperplane, then $\mathbf{x}^* \in H$; and that if $\mathbf{x}'$ is *any* vector in H, then $H = \mathbf{x}' + S$ as well.

2. Suppose H is a hyperplane, so that H is the translate of a linear subspace, S, having codimension one. Show that if $H \cap S \neq \emptyset$, then $H = S$.

3. When working in $\mathbf{R}^n$, we are used to saying that a subset, H, is a hyperplane if, and only if, there exists a non-zero vector, $\mathbf{p} \in \mathbf{R}^n$, and a real number, α, such that:

$$H = \{\mathbf{x} \in \mathbf{R}^n \mid \mathbf{p} \cdot \mathbf{x} = \alpha\}; \tag{4.41}$$

a characterization which may appear to be in conflict with Definition 4.81. However, suppose H satisfies (4.41), and define the set S by:

$$S = \{\mathbf{x} \in \mathbf{R}^n \mid \mathbf{p} \cdot \mathbf{x} = 0\}.$$

Then we note, first of all, that $S = [Sp\{\mathbf{p}\}]^{\perp}$; so that it follows from Proposition 4.64 that S is a linear space. Moreover, it follows from Proposition 4.77 that $\mathbf{R}^n$ is the direct sum of S and $Sp\{\mathbf{p}\}$; and therefore, since $Sp\{\mathbf{p}\}$ has dimension one, we see that S has codimension one.

Now define $\mathbf{x}^*$ by:

$$\mathbf{x}^* = (\alpha/\mathbf{p} \cdot \mathbf{p})\mathbf{p}.$$

Then, if $\mathbf{x} \in \mathbf{x}^* + S$, we note that there exists $\mathbf{y} \in S$ such that $\mathbf{x} = \mathbf{x}^* + \mathbf{y}$, and thus:

$$\mathbf{p} \cdot \mathbf{x} = \mathbf{p} \cdot (\mathbf{x}^* + \mathbf{y}) = \mathbf{p} \cdot \mathbf{x}^* + \mathbf{p} \cdot \mathbf{y} = \alpha + 0 = \alpha;$$

from which we conclude that:

$$\mathbf{x}^* + S \subseteq H.$$

Conversely, if $\mathbf{x} \in H$, then, since by Proposition 4.77 $\mathbf{R}^n$ is the direct sum of S and $Sp\{\mathbf{p}\}$, it follows that there exists $\mathbf{z} \in S$ and $a \in \mathbf{R}$ such that:

$$\mathbf{x} = a\mathbf{p} + \mathbf{z}. \tag{4.42}$$

However, since $\mathbf{x} \in H$, we have:

$$\alpha = \mathbf{p} \cdot \mathbf{x} = \mathbf{p} \cdot (a\mathbf{p} + \mathbf{z}) = a\mathbf{p} \cdot \mathbf{p} + \mathbf{p} \cdot \mathbf{z} = a\mathbf{p} \cdot \mathbf{p} + 0 = a\mathbf{p} \cdot \mathbf{p}.$$

Therefore, $a = \alpha/\mathbf{p} \cdot \mathbf{p}$, and it follows from (4.42) and the definition of $\mathbf{x}^*$ that $\mathbf{x} \in \mathbf{x}^* + S$. We can then conclude that:

$$H \subseteq \mathbf{x}^* + S. \quad \square$$

It follows from the example just discussed that the familiar definition of a hyperplane in $\mathbf{R}^n$ is consistent with Definition 4.81 in the sense that if H takes the form:

$$H = \{\mathbf{x} \in \mathbf{R}^n \mid \mathbf{p} \cdot \mathbf{x} = a\},$$

for some non-zero $\mathbf{p}$ and real number a, then H is a hyperplane by Definition 4.81. The converse is also true, a conclusion which is an immediate consequence of Theorem 4.86, below. Before considering this result, however, we will prove a result which generalizes the conclusion of our example to an arbitary real linear space.

4.83. Theorem. *Let $f: L \to \mathbf{R}$ be a non-trivial linear functional, where L is a real linear space, and let $a \in \mathbf{R}$. Then the set H defined by:*

$$H = \{\mathbf{x} \in L \mid f(\mathbf{x}) = a\},$$

is a hyperplane.

Proof. We begin by noting that, since f is non-trivial, there exists $\mathbf{y} \in L$ such that $f(\mathbf{y}) \neq 0$. If we then define $\mathbf{x}^*$ by:

$$\mathbf{x}^* = [1/f(\mathbf{y})]\mathbf{y}, \tag{4.43}$$

we see that:

$$f(\mathbf{x}^*) = f\big([1/f(\mathbf{y})]\mathbf{y}\big) = [1/f(\mathbf{y})]f(\mathbf{y}) = 1;$$

and therefore,

$$f(a\mathbf{x}^*) = af(\mathbf{x}^*) = a.$$

It follows that $a\mathbf{x}^* \in H$, and thus that $H \neq \emptyset$.

By Proposition 4.30, the set S defined by:

$$S = \mathbf{k}_f = \{\mathbf{x} \in L \mid f(\mathbf{x}) = 0\},$$

is a linear subspace of L; and defining the linear subspace P by:

$$P = Sp\{\mathbf{x}^*\},$$

where $\mathbf{x}^*$ is from (4.43), we note that it is obvious that:

$$\dim(P) = 1 \text{ and } S \cap P = \{\mathbf{0}\}.$$

Consequently, if we can show that L is the direct sum of S and P, and that:

$$H = a\mathbf{x}^* + S,$$

our result will follow.

In order to complete the argument that L is the direct sum of S and P, let $\mathbf{x} \in L$ be arbitrary, and define $b = f(\mathbf{x})$, and $\mathbf{z} = \mathbf{x} - b\mathbf{x}^*$. Then $f(\mathbf{z}) = 0$, so that $\mathbf{z} \in S$; and thus:

$$\mathbf{x} = b\mathbf{x}^* + \mathbf{z} \in P + S.$$

Since it is obvious that $P \cap S = \{\mathbf{0}\}$, it follows that L is the direct sum of P and S, and thus that S has codimension one.

Next we note that if $\mathbf{x}$ is of the form:

$$\mathbf{x} = a\mathbf{x}^* + \mathbf{y},$$

for some $\mathbf{y} \in S$, then:

$$f(\mathbf{x}) = af(\mathbf{x}^*) + f(\mathbf{y}) = a,$$

so that $\mathbf{x} \in H$. Therefore:

$$a\mathbf{x}^* + S \subseteq H.$$

Conversely, if $\mathbf{x} \in H$, and we define:

$$\mathbf{y}' = \mathbf{x} - a\mathbf{x}^*,$$

then:

$$f(\mathbf{y}') = f(\mathbf{x}) - af(\mathbf{x}^*) = a - a = 0,$$

so that $\mathbf{y}' \in S$. Since, with $\mathbf{y}'$ so defined,

$$\mathbf{x} = a\mathbf{x}^* + \mathbf{y}',$$

it follows that:

$$H \subseteq a\mathbf{x}^* + S. \quad \square$$

The following corollary is an almost immediate implication of 4.83. Its proof will be left as an exercise.

4.84. Corollary. *If f is any non-trivial linear functional on a real linear space, L, and $\mathbf{x}^*$ is any vector in L for which $f(\mathbf{x}^*) \neq 0$; then, given any $\mathbf{x} \in L$, there exists a unique $a \in \mathbf{R}$ and a unique $\mathbf{y} \in \mathbf{k}_f$ such that $\mathbf{x} = a\mathbf{x}^* + \mathbf{y}$.*

4.85. Proposition. *Suppose H is a hyperplane in a real linear space, L, so that H is the translate of a linear subspace, S, and suppose that $H \neq S$. Then, given any $\mathbf{x}^* \in H$, we have:*

1. *L is the direct sum of S and $Sp\{\mathbf{x}^*\}$, and*
2. *$H = \mathbf{x}^* + S$.*

Proof. Since S has codimension one, there exists a vector $\mathbf{z} \in L$ such that:

$$L = Sp\{\mathbf{z}\} \oplus S. \tag{4.44}$$

Furthermore, there exists $\mathbf{y}^* \in S$, and a scalar $a \in \mathbf{R}$ such that:

$$\mathbf{x}^* = a\mathbf{z} + \mathbf{y}^*. \tag{4.45}$$

Moreover, we must have $a \neq 0$, for otherwise it follows from (4.45) that $\mathbf{x}^* \in S$; in which case it would follow from 4.82.2 that $H = S$, contrary to our assumption. Thus we see that:

$$\mathbf{z} = (1/a)\mathbf{x}^* - (1/a)\mathbf{y}^*;$$

and it is easy to see (although I will leave the details as an exercise), that:

$$L = Sp\{\mathbf{x}^*\} \oplus S.$$

Part 2 of our conclusion is an immediate implication of 4.82.1. □

Our next result completes our characterization of hyperplanes in real linear spaces.

4.86. Theorem. *If L is a real linear space, and H is any hyperplane in L, then there exists a non-trivial linear functional, $f\colon L \to \mathbf{R}$, and a constant, $a \in \mathbf{R}$, such that:*

$$H = \{\mathbf{x} \in L \mid f(\mathbf{x}) = a\} = f^{-1}(\{a\}).$$

Proof. Since H is a hyperplane, there exists a linear space, S, and $\mathbf{x}^* \notin S$ such that:

$$(\forall \mathbf{x} \in L)(\exists \alpha(\mathbf{x}) \in \mathbf{R}, \mathbf{y} \in S)\colon \mathbf{x} = \alpha(\mathbf{x})\mathbf{x}^* + \mathbf{y}; \tag{4.46}$$

and, for some $a \in \mathbf{R}$:

$$H = a\mathbf{x}^* + S. \tag{4.47}$$

Consequently, we can define the functional f on L by:

$$f(\mathbf{x}) = \alpha(\mathbf{x}),$$

where, for $\mathbf{x} \in L$, $\alpha(\mathbf{x})$ is the unique real number satisfying (4.46).

Now suppose that $\mathbf{z}$ and $\mathbf{z}^*$ are vectors in L, and that a and b are real numbers. There exist unique vectors $\mathbf{y}$ and $\mathbf{y}^*$ in S such that:

$$\mathbf{z} = f(\mathbf{z})\mathbf{x}^* + \mathbf{y} \quad \text{and} \quad \mathbf{z}^* = f(\mathbf{z}^*)\mathbf{x}^* + \mathbf{y}^*. \tag{4.48}$$

But then we see from (4.48) that:

$$a\mathbf{z} + b\mathbf{z}^* = [af(\mathbf{z}) + bf(\mathbf{z}^*)]\mathbf{x}^* + a\mathbf{y} + b\mathbf{y}^*;$$

and, since $a\mathbf{y} + b\mathbf{y}^* \in S$, it then follows that:

$$f(a\mathbf{z} + b\mathbf{z}^*) = af(\mathbf{z}) + bf(\mathbf{z}^*).$$

Therefore, the function f is linear.

To prove that:

$$H = \{\mathbf{x} \in L \mid f(\mathbf{x}) = a\},$$

let $\mathbf{x} \in H$. Then from (4.47) we see that there exists some $\mathbf{y} \in S$ such that:

$$\mathbf{x} = a\mathbf{x}^* + \mathbf{y},$$

and thus by definition of $f, f(\mathbf{x}) = a$. Conversely, if $f(\mathbf{x}') = a$, it follows from the definition of f that there exists $\mathbf{y}' \in S$ such that:

$$\mathbf{x}' = a\mathbf{x}^* + \mathbf{y}';$$

and it follows from (4.47) that $\mathbf{x} \in H$. □

While we have chosen to represent the hyperplane H as:

$$H = \{\mathbf{x} \in L \mid f(\mathbf{x}) = a\}, \tag{4.49}$$

in the above result, it should be obvious that if H is not a linear subspace, then we could have chosen $a = 1$. On the other hand, if H *is* a linear subspace, then, in the notation of our proof, $H = S$, and it is only the value $a = 0$ which will satisfy (4.47) and (4.49).

4.87. Examples/Exercises.

1. Consider the space $C = \{f\colon [0,1] \to \mathbf{R} \mid f \text{ is continuous}\}$, and define the subset S by:

$$S = \{f \in C \mid f(0) = 0\}.$$

It is easily seen that S is a subspace of C, although I will leave the details of the argument as an exercise. Now consider the subset, T, of C defined by:

$$T = \{f \in C \mid f \text{ is constant on } [0,1]\}.$$

Once again you should have no difficulty in proving that T is a linear subspace of C, and clearly T has dimension one. We will now prove that $C = S \oplus T$, from which it will follow that S has codimension one; and is, therefore, a hyperplane.

To prove that $C = S \oplus T$, let f be an arbitrary element of C, and define the functions g and h on $[0,1]$ by:

$$g(x) = f(x) - f(0) \quad \text{and} \quad h(x) = f(0),$$

respectively. Then it is easy to see that $g \in S, h \in T$, and that $f = g + h$. Since it is also apparent that:

$$S \cap T = \{\mathbf{0}\},$$

it now follows that $C = S \oplus T$. In fact, it is readily seen that, given any $a \in \mathbf{R}$, if we define:

$$S_a = \{f \in C \mid f(0) = a\},$$

then S_a is a hyperplane.

Now consider the function, $\varphi \colon C \to \mathbf{R}$, defined by:

$$\varphi(f) = f(0) \quad \text{for } f \in C.$$

You should have no difficulty in proving that φ is linear, and thus is a linear functional; and notice that

$$S_a = \{f \in C \mid \varphi(f) = a\}.$$

2. Consider the space of convergent real-valued infinite sequences, $\mathbf{c}$, let $a \in \mathbf{R}$, and define:

$$S_a = \{\mathbf{x} \in \mathbf{c} \mid x_n \to a\}.$$

Using considerations similar to those developed in the previous example, you should easily be able to show that S_a is a hyperplane in $\mathbf{c}$. Moreover, if we define $f \colon \mathbf{c} \to \mathbf{R}$ by:

$$f(\mathbf{x}) = \lim_{n\to\infty} x_n \quad \text{for } \mathbf{x} \in \mathbf{c},$$

it is easy to see that f is a linear functional, and that:

$$S_a = \{\mathbf{x} \in \mathbf{c} \mid f(\mathbf{x}) = a\}. \quad \square$$

Exercises.

1. Prove Propositions: (a) 4.70, (b) 4.71, and (c) 4.72.
2. Show that, given any $a \in \mathbf{R}$, the subset of C defined by:

$$S_a = \left\{ f \in C \mid \int_0^1 f(t)\,dt = a \right\},$$

is a hyperplane in C. What is the linear subspace, S, of which S_a is the translate? Having found S, can you now find a one-dimensional subspace, T, such that $C = S \oplus T$?

3. Suppose L is an arbitrary linear space, and that S and T are linear subspaces of L such that $L = S \oplus T$. Show that, for each $\mathbf{x} \in L$, there exist unique vectors, $\pi_1(\mathbf{x}) \in S$ and $\pi_2(\mathbf{x}) \in T$, such that:

$$\mathbf{x} = \pi_1(\mathbf{x}) + \pi_2(\mathbf{x}).$$

Moreover, show that the two functions, π_1 and π_2, thus implicitly defined, are both linear.

4. Show that, if S is a linear subspace of $\mathbf{R}^n$, with $\dim(S) = k \geq 1$ (and $k < n$), and $f\colon S \to \mathbf{R}^m$ is a linear function, then there exists a unique $m \times n$ matrix, $\mathbf{B}$, satisfying:

a. $(\forall \mathbf{x} \in S)\colon f(\mathbf{x}) = \mathbf{B}\mathbf{x}$, and

b. $(\forall \mathbf{y} \in S^{\perp})\colon \mathbf{B}\mathbf{y} = \mathbf{0}$.

(Hint: defining $T = S^{\perp}$, let π_1 be the function defined in exercise 3, and consider the function $g\colon \mathbf{R}^n \to \mathbf{R}^m$ defined by $g(\mathbf{x}) = f[\pi_1(\mathbf{x})]$.)

5. Show that if S is a subspace of $\mathbf{R}^n$, then $(S^{\perp})^{\perp} = S$. (To prove that $(S^{\perp})^{\perp} \subseteq S$, try making use of 4.78.4.)

4.7 Affine Sets

An affine subset of a real linear space, L, can be regarded as a generalization of the notion of a linear subspace; on the other hand, we will also discover that a hyperplane is a special case of an affine set. Because of this latter relationship, there is a great deal of overlap between some of the results, and the proofs of the results, which we studied in the previous section in connection with hyperplanes, and some of the results concerning affine sets which we will be studying in this section. My feeling is that the results we are going to study here will be a great deal easier to understand because of the work which we did in the previous section; however, many of these results will generalize results of Section 6, and thus there is a certain amount of inefficiency in the way I have chosen to develop this material.

In Euclidean n-space, an affine set can be thought of as a set which contains every line generated by any two of its points. While we will once again be dealing with an arbitrary real linear space, L, and in such a space the idea of a line may have very little geometric meaning; we can, for purposes of the present discussion, define the line generated by two points in the space as follows. For $\mathbf{x}$ and $\mathbf{y}$ in L, define $L(\mathbf{x}, \mathbf{y})$, the **line generated by x and y**, as:

$$L(\mathbf{x}, \mathbf{y}) = \{\mathbf{z} \in L \mid (\exists \lambda \in \mathbf{R}) \colon \mathbf{z} = \lambda \mathbf{x} + (1 - \lambda)\mathbf{y}\}. \tag{4.50}$$

Notice that, for any $\mathbf{x}$ and $\mathbf{y}$ in L, $L(\mathbf{x}, \mathbf{y}) = L(\mathbf{y}, \mathbf{x})$.

4.88. Example. In $\mathbf{R}^2$ it follows from the 'two-point representation' of the line connecting $\mathbf{x} = (x_1, x_2)$ and $\mathbf{y} = (y_1, y_2)$, where $x_1 \neq y_1$, that $L(\mathbf{x}, \mathbf{y})$ is the set of all $\mathbf{z}$ in $\mathbf{R}^2$ satisfying the equation:

$$z_2 = \left(\frac{y_2 - x_2}{y_1 - x_1}\right) \cdot (z_1 - x_1) + x_2. \tag{4.51}$$

Show that this is equivalent to the definition just given; that is, show first that if $\mathbf{z} \in \mathbf{R}^2$ is of the form:

$$\mathbf{z} = \lambda \mathbf{y} + (1 - \lambda)\mathbf{x} = \mathbf{x} + \lambda(\mathbf{y} - \mathbf{x}),$$

then $\mathbf{z}$ satisfies equation (4.51); while if $\mathbf{z}$ satisfies equation (4.51) then there exists a real number λ satisfying the above equation. □

4.89. Definition. A non-empty subset, M, of L is said to be **affine** (or a **linear variety** or a **linear flat**) iff:

$$(\forall \mathbf{x}, \mathbf{y} \in M)(\forall \lambda \in \mathbf{R}) \colon \lambda \mathbf{x} + (1 - \lambda)\mathbf{y} \in M.$$

Since I do not intend to use the notation '$L(\mathbf{x}, \mathbf{y})$' outside of this section, I have not made use of it in the formal definition of an affine set. However, notice that a set M is affine iff, for each $\mathbf{x}$ and $\mathbf{y}$ in M, we have:

$$L(\mathbf{x}, \mathbf{y}) \subseteq M.$$

4.90. Examples/Exercises.

1. Any linear subspace of L (including $\{\mathbf{0}\}$ and L itself) is an affine set; in fact, an affine set is a generalization of the idea of a linear subspace.

2. Any hyperplane is an affine set; in fact, you can easily prove that if S is a linear subspace of L, and $\mathbf{x}^*$ is any element of L, then the set M defined by:

$$M = \mathbf{x}^* + S, \tag{4.52}$$

is an affine set. Later in this section, we will show that the converse is also true; that is, if M is an affine set, then there exists a linear subspace, S, and an element, $\mathbf{x}^*$, of L, such that (4.52) holds. In fact, it follows from this and our work with hyperplanes in the previous section that, in general, translations of linear subspaces are affine sets, and in the special case in which the linear subspace has codimension one, we call the affine set a hyperplane. Notice also that it follows as a special case of this example that any singleton subset of L is an affine set.

3. Suppose that $\mathbf{A}$ is an $m \times n$ matrix, and that $\mathbf{b}$ is an element of $\mathbf{R}^m$ such that the set M defined by:

$$M = \{\mathbf{x} \in \mathbf{R}^n \mid \mathbf{A}\mathbf{x} = \mathbf{b}\}, \tag{4.53}$$

is non-empty. Prove that M is an affine set.

We can look at this last example in a more geometric way as follows: let '$\mathbf{a}_{i\cdot}$' denote the ith row of $\mathbf{A}$, and define H_i by:

$$H_i = \{\mathbf{x} \in \mathbf{R}^n \mid a_{i\cdot} \cdot \mathbf{x} = b_i\} \quad \text{for } i = 1, \ldots, m.$$

Then the set M defined in (4.53) can equivalently be expressed as:

$$M = \bigcap\nolimits_{i=1}^{m} H_i;$$

that is, M is the intersection of the m hyperplanes H_i.

Yet another way of looking at M is as the inverse image of the vector $\mathbf{b}$; that is, if we let 'f' denote the linear transformation for which $\mathbf{A}$ is the matrix, then $M = f^{-1}(\mathbf{b})$. Thus, we have three ways of looking at a set M of this form: as the solution set of a (consistent) set of linear equations, as the intersection of m hyperplances, or as the inverse image of a vector in the range of a linear transformation. This last idea is generalized in our next example.

4. Let $f\colon L_1 \to L_2$ be a linear function, let $\mathbf{z} \in f(L_1)$, and define

$$M = f^{-1}(\mathbf{z}) = \{\mathbf{x} \in L_1 \mid f(\mathbf{x}) = \mathbf{z}\}.$$

Then M is an affine set. □

4.91. Proposition. *Suppose A and B are affine subsets of L, and that α and β are real numbers. Then the set:*

$$C \overset{def}{=} \alpha A + \beta B,$$

is an affine set. Moreover, if $\{M_a \mid a \in A\}$ is a family of affine subsets of L such that:

$$M \overset{def}{=} \bigcap_{a \in A} M_a,$$

is non-empty, then M is an affine set.

Proof. Exercise. □

Notice that it follows as a special case of this last proposition that if M is affine, and a is any real number, then the set aM is affine. Moreover, from this proposition and Example 4.90.2 it follows that a translate of an affine set is also affine.

4.92. Definition. If X is a subset of L and $\mathbf{z}$ is an element of L, we shall say that **z is an affine combination of X** iff there exist $n \in N, \mathbf{x}_1, \ldots, \mathbf{x}_n \in X$, and $a_1, \ldots, a_n \in \mathbf{R}$ satisfying:

$$\mathbf{z} = \sum_{i=1}^{n} a_i \mathbf{x}_i \text{ and } \sum_{i=1}^{n} a_i = 1.$$

One can then use a straightforward induction argument to prove the following result. I will leave the details as an exercise.

4.93. Proposition. *If M is an affine subset of L, and $\mathbf{x}$ is an affine combination of M, then $\mathbf{x}$ is an element of M; that is, M contains all its affine combinations.*

4.94. Proposition. *A subset, S, of L is a linear subspace if, and only if, S is affine and contains $\mathbf{0}$.*

Proof. If S is a linear subspace, then it is obvious that S is affine and contains the origin. Conversely, suppose the latter two conditions hold. If $\mathbf{x} \in S$ and $\theta \in \mathbf{R}$, then, since S is affine:

$$\theta\mathbf{x} = \theta\mathbf{x} + (1 - \theta)\mathbf{0} \in S. \tag{4.54}$$

Furthermore, if $\mathbf{x}$ and $\mathbf{y}$ are members of S, then:

$$(1/2)\mathbf{x} + (1/2)\mathbf{y} \in S;$$

and thus by (4.54):

$$2[(1/2)\mathbf{x} + (1/2)\mathbf{y}] = \mathbf{x} + \mathbf{y} \in S.$$

Therefore, S is a linear subspace. □

As noted earlier, if M is an affine set, then M is the translate of a linear subspace, and conversely; as we establish in the following result. For an illustration of the relationships being set out in Theorem 4.95, see Figure 4.1, on the next page.

4.95. Theorem. *Each affine subset of L is a translate of a unique linear subspace. In fact, if M is affine, then:*

$$S \stackrel{def}{=} M - M, \tag{4.55}$$

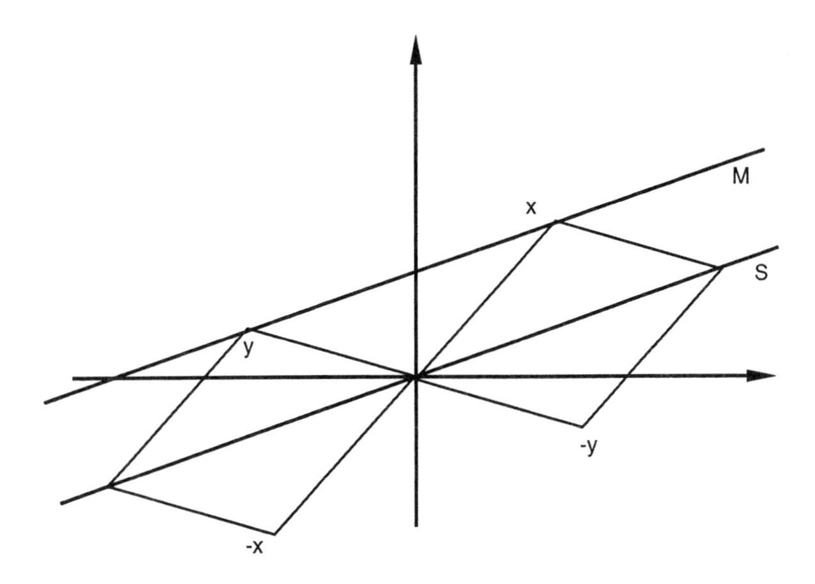

Figure 4.1: The Subspace Parallel to M.

is a linear subspace; and, if $\mathbf{x}^*$ *is any element of* M*, then:*

$$M = \mathbf{x}^* + S.$$

Furthermore, if $\mathbf{x} \in L$ *and* T *is a linear subspace such that* $M = \mathbf{x} + T$*, then* $\mathbf{x} \in M$ *and* $T = S$*.*

Proof.

1. We note first that it follows from Proposition 4.91 that the set S defined in equation (4.55) is an affine set; and, since S obviously contains the origin, it then follows from Proposition 4.94 that S is a linear subspace.

Now let $\mathbf{x}^*$ be an arbitrary element of M, and suppose that $\mathbf{x}$ is an element of $\mathbf{x}^* + S$. Then there exist $\mathbf{y}, \mathbf{z} \in M$ such that:

$$\mathbf{x} = \mathbf{x}^* + (\mathbf{y} - \mathbf{z});$$

and, since $\mathbf{x}^*, \mathbf{y}$, and $\mathbf{z}$ are all elements of M, we have:

$$2\mathbf{x}^* + (1-2)\mathbf{z} = 2\mathbf{x}^* - \mathbf{z} \in M,$$

and

$$2\mathbf{y} + (1-2)\mathbf{z} = 2\mathbf{y} - \mathbf{z} \in M.$$

But then it follows that:

$$(1/2)(2\mathbf{x}^* - \mathbf{z}) + (1 - 1/2)(2\mathbf{y} - \mathbf{z}) = \mathbf{x}^* + \mathbf{y} - \mathbf{z} = \mathbf{x},$$

is an element of M; and thus that:

$$\mathbf{x}^* + S \subseteq M.$$

Conversely, if $\mathbf{x} \in M$, then $\mathbf{x} - \mathbf{x}^* \in S$, and thus:

$$\mathbf{x} = \mathbf{x}^* + (\mathbf{x} - \mathbf{x}^*) \in \mathbf{x}^* + S;$$

so that we can also conclude that:

$$M \subseteq \mathbf{x}^* + S.$$

2. Now suppose that there exist $\mathbf{y}^* \in L$, and a linear subspace, T, such that we also have:

$$M = \mathbf{y}^* + T. \tag{4.56}$$

Then it follows immediately that $\mathbf{y}^* \in M$. To prove that $S = T$, we begin by noting that it follows from part 1 of our proof that:

$$M = \mathbf{y}^* + S. \tag{4.57}$$

Now suppose $\mathbf{z} \in T$. Then by (4.56):

$$\mathbf{y}^* + \mathbf{z} \in M,$$

and it follows from (4.57) that there exists $\mathbf{z}' \in S$ such that:

$$\mathbf{y}^* + \mathbf{z} = \mathbf{y}^* + \mathbf{z}',$$

and thus $\mathbf{z} = \mathbf{z}'$. Therefore $\mathbf{z} \in S$, and it follows that $T \subseteq S$. A similar argument, using (4.57) and then (4.56), establishes that $S \subseteq T$; and thus we conclude that $S = T$. □

Notice that if M is a linear subspace, and $\mathbf{x}^*$ is any point in M, then $M = \mathbf{x}^* + M$. In the general case, the subspace of which M is a translate can be represented in a somewhat simpler form than is given in 4.95, as is shown in the following; the proof of which will be left as an exercise.

4.96. Corollary. *If M is an affine subset of L, and $\mathbf{x}^*$ is any element of M, then we can write:*

$$M = \mathbf{x}^* + S,$$

where S is a linear subspace, and is given by:

$$S = M - \mathbf{x}^*.$$

4.97. Definition. We shall say that an affine set M **is of (finite) codimension n** iff M takes the form:

$$M = \mathbf{x}^* + S,$$

where S has codimension n.

Notice that the first part of the following result almost generalizes Theorem 4.83, while the second part does generalize Theorem 4.86.

4.98. Theorem. *If $f\colon L_1 \to L_2$ is a linear function, and $\mathbf{y}^*$ is an element of $f(L_1)$, then the set M defined by:*

$$M = \{\mathbf{x} \in L_1 \mid f(\mathbf{x}) = \mathbf{y}^*\}, \tag{4.58}$$

is an affine set. Conversely, if M is an affine set having codimension n, then there exists a linear function, $f\colon L \to \mathbf{R}^n$ mapping onto $\mathbf{R}^n$, and $\mathbf{a}^* \in \mathbf{R}^n$ such that:

$$M = \{\mathbf{x} \in L \mid f(\mathbf{x}) = \mathbf{a}^*\}, \tag{4.59}$$

and $\mathbf{k}_f = M - M$.

Proof. I will leave the proof that a set which can be expressed in the form (4.58) is affine as an exercise. To prove the converse, suppose M is of codimension n. Then there exists a linear subspace of L, S, having codimension n, and such that for any $\mathbf{x}^* \in M$, we have:

$$M = \mathbf{x}^* + S. \tag{4.60}$$

Moreover, there exists a linearly independent set, $X = \{\mathbf{x}_1, \dots, \mathbf{x}_n\}$, such that for any $\mathbf{x} \in L$, there is a unique $\mathbf{a} \in \mathbf{R}^n$ and $\mathbf{y} \in S$ satisfying:

$$\mathbf{x} = \sum_{i=1}^{n} a_i \mathbf{x}_i + \mathbf{y}. \tag{4.61}$$

Consequently, we can define a function $f\colon L \to \mathbf{R}^n$ by:

$$f(\mathbf{x}) = \mathbf{a},$$

where $\mathbf{a}$ is that element of $\mathbf{R}^n$ satisfying (4.61). It is easy to see that f is linear and onto $\mathbf{R}^n$, and obviously:

$$S = \mathbf{k}_f = \{\mathbf{x} \in L \mid f(\mathbf{x}) = \mathbf{0}\}.$$

To prove that M satisfies (4.59), define $\mathbf{a}^* = f(\mathbf{x}^*)$, where $\mathbf{x}^*$ is from (4.60). If $\mathbf{x} \in M$, there exists $\mathbf{y} \in S$ such that:

$$\mathbf{x} = \mathbf{x}^* + \mathbf{y};$$

and from the fact that $\mathbf{k}_f = S$, and the definition of $\mathbf{a}^*$,

$$f(\mathbf{x}) = f(\mathbf{x}^* + \mathbf{y}) = f(\mathbf{x}^*) + f(\mathbf{y}) = \mathbf{a}^* + \mathbf{0} = \mathbf{a}^*.$$

Conversely, if $\mathbf{x} \in L$ is such that $f(\mathbf{x}) = \mathbf{a}^*$, and we define $\mathbf{y}^* = \mathbf{x} - \mathbf{x}^*$, then we have:

$$f(\mathbf{y}^*) = f(\mathbf{x}) - f(\mathbf{x}^*) = \mathbf{a}^* - \mathbf{a}^* = \mathbf{0};$$

so that $\mathbf{y}^* \in S$. But then, since $\mathbf{x} = \mathbf{x}^* + \mathbf{y}^*$, it follows that:

$$\mathbf{x} \in \mathbf{x}^* + S = M. \quad \square$$

4.99. Example. Suppose M is an affine subset of $\mathbf{R}^m$, so that, by Theorem 4.95, $M = \mathbf{x}^* + S$, for some $\mathbf{x}^* \in M$ and S a linear subspace; and suppose $\dim(S) = n$. Then the codimension of M, which is equal to the codimension of S, is equal to $p \equiv m - n$; and by Theorem 4.98, there exists a linear function, $f \colon \mathbf{R}^m \to \mathbf{R}^p$, such that f maps onto $\mathbf{R}^p$, $\mathbf{k}_f = S$ and:

$$M = \{\mathbf{x} \in \mathbf{R}^m \mid f(\mathbf{x}) = \mathbf{b}\},$$

where $\mathbf{b} \equiv f(\mathbf{x}^*)$. But then, letting $\mathbf{A}$ be the matrix of the transformation f, we see that the rank of $\mathbf{A}$ is p, and:

$$M = \{\mathbf{x} \in \mathbf{R}^m \mid \mathbf{A}\mathbf{x} = \mathbf{b}\}, \tag{4.62}$$

while:

$$S = \{\mathbf{x} \in \mathbf{R}^m \mid \mathbf{A}\mathbf{x} = \mathbf{0}\}.$$

It is, of course, very easy to show that if M is of the form given in (4.62), then M is affine (as was already noted in 4.90.3). It is also an easy exercise, which I leave to you, to prove that if M can be written in this form, then the codimension of M is equal to the rank of $\mathbf{A}$. $\square$

4.100. Definition. If X is a non-empty subset of L, we define the **affine hull of X**, denoted by '*aff*(X),' as the smallest affine set containing X.

The definition just presented is our second encounter with a Kuratowski closure operation. Notice that, if X is non-empty, and we define:

$$\mathcal{A} = \{A \subseteq L \mid A \text{ is affine and } X \subseteq A\};$$

then $\mathcal{A}$ is non-empty, since $L \in \mathcal{A}$. Furthermore, if we define *aff*(X) by:

$$\mathit{aff}(X) = \bigcap_{A \in \mathcal{A}} A,$$

it follows from 4.91 and the definitions just given that $\mathit{aff}(X)$ is an affine set containing X. Furthermore, it is an immediate consequence of the definition of $\mathcal{A}$ that, if M is an affine set containing X, then $M \in \mathcal{A}$, and thus:

$$\mathit{aff}(X) \subseteq M.$$

Therefore, $\mathit{aff}(X)$ is the 'smallest' affine set containing X.

It should also be noted that, if X is an affine set, then $X = \mathit{aff}(X)$. We can see this by noting first that, for any subset of L (affine or not), we have:

$$X \subseteq \mathit{aff}(X);$$

while, if X is affine, then it is an affine set containing itself, and thus we also have:

$$\mathit{aff}(X) \subseteq X.$$

Our next result presents what we might call a 'constructive' characterization of $\mathit{aff}(X)$.

4.101. Proposition. *If X is a non-empty subset of L, then:*

$$\text{aff}(X) = \{\mathbf{x} \in L \mid \mathbf{x} \textit{ is an affine combination of } X\}. \tag{4.63}$$

Proof. For purposes of the present argument, denote the set on the right-hand-side of (4.63) by 'M.' It follows at once from Proposition 4.93 that M is contained in any affine set which contains X, and thus it is clear that:

$$M \subseteq \mathit{aff}(X). \tag{4.64}$$

Conversely, it is clear that any affine combination of two elements of M is again an affine combination of X, and therefore, M is an affine set. Since M also contains X, we see that:

$$\mathit{aff}(X) \subseteq M;$$

and, combining this with (4.64), our result follows. □

4.102. Definition. If M is an affine set, we define the **dimension of M**, denoted by 'dim(M),' by:

$$\dim(M) = \dim(M - M).$$

If X is a non-empty subset of L (not necessarily affine), we define the **dimension of X**, by:

$$\dim(X) = \dim[\mathit{aff}(X)];$$

that is, we define the dimension of X to be the dimension of the affine hull of X.

The preceding definition is both satisfactory for our purposes and will be useful in our later work; and notice that it agrees with 4.19 in the special case wherein X is a linear subspace. It should be mentioned, however, that it is not a very satisfactory notion of dimension for non-convex sets.

4.103. Proposition. *If $\mathbf{x}^*$ is any element of X, where X is a non-empty subset of L, and if we define $Y = X - \mathbf{x}^*$ and $S = Sp(Y)$; then we have:*

$$\text{aff}(X) = \mathbf{x}^* + S.$$

Proof. Clearly:

$$X \subseteq M \stackrel{\text{def}}{=} \mathbf{x}^* + S,$$

and, since M is affine, it then follows that:

$$\mathit{aff}(X) \subseteq M.$$

On the other hand, defining $A = \mathit{aff}(X)$ and $T = A - A$, we note that $Y \subseteq T$; and hence, since T is a linear subspace, it follows from the definition of $Sp(Y)$ that:

$$S = Sp(Y) \subseteq T.$$

Therefore,

$$M \equiv \mathbf{x}^* + S \subseteq \mathbf{x}^* + T = A \equiv \mathit{aff}(X). \quad \square$$

4.104. Definition. The vectors $\mathbf{x}_1, \ldots, \mathbf{x}_m$ are said to be **affinely dependent**, and the set $X = \{\mathbf{x}_1, \ldots, \mathbf{x}_m\}$ is said to be **affinely dependent** iff there exists a vector $\mathbf{a} \in \mathbf{R}^m$ such that $\mathbf{a} \neq \mathbf{0}$,

$$\sum_{i=1}^{m} a_i \mathbf{x}_i = \mathbf{0} \text{ and } \sum_{i=1}^{m} a_i = 0. \tag{4.65}$$

Otherwise [that is, if the only **a** which satisfies (4.65) is the zero vector in $\mathbf{R}^m$], X is said to be **affinely independent**.

The following sets forth the basic relationship between affine independence and linear independence. I will leave the proof as an exercise.

4.105. Proposition. *The set $X = \{\mathbf{x}_1, \ldots, \mathbf{x}_m\}$ is affinely independent if, and only if, for each $j \in \{1, \ldots, m\}$, the set X_j defined by:*

$$X_j = \{\mathbf{x}_i - \mathbf{x}_j \mid i \in \{1, \ldots, m\} \setminus \{j\}\},$$

is linearly independent.

4.106. Definition. Let Y be a non-empty subset of L. We shall say that $X = \{\mathbf{x}_1, \ldots, \mathbf{x}_m\}$ is a **maximal affinely independent subset of Y** iff (a) X is an affinely independent set, and (b) every subset of Y which contains more than m vectors is affinely dependent.

Any non-empty subset of finite dimension (even if it is a singleton set) contains a maximal affinely independent subset, as is noted in our next result. I will leave the proof as an exercise; it follows fairly easily from Propositions 4.103 and 4.105.

4.107. Proposition. *Suppose A is a nonempty subset of L. Then* $\dim(A) = m$, *where m is a positive integer, if, and only if, A contains a maximal affinely independent subset containing $m+1$ elements.*

4.108. Proposition. *Let A be a non-empty subset of L such that* $\dim(A) = m$, *and let $X = \{\mathbf{x}_1, \ldots, \mathbf{x}_{m+1}\}$ be any affinely independent subset of A containing $m+1$ vectors. Then, for each $\mathbf{x} \in$* aff(A) *there exists a unique $\mathbf{a} \in \mathbf{R}^{m+1}$ satisfying:*

$$\mathbf{x} = \sum\nolimits_{i=1}^{m+1} a_i \mathbf{x}_i \ \textit{and} \ \sum\nolimits_{i=1}^{m+1} a_i = 1. \tag{4.66}$$

Proof. Let $X = \{\mathbf{x}_1, \ldots, \mathbf{x}_{m+1}\}$ be an affinely independent subset,[14] and, let $\mathbf{x} \in \mathit{aff}(A)$ be arbitrary. Since X must be a maximal affinely independent subset of A, it then follows that there exists a nonnull $\mathbf{b} \in \mathbf{R}^{m+1}$ such that

$$\sum\nolimits_{i=1}^{m+1} b_i \mathbf{x}_i + b_{m+2}\mathbf{x} = \mathbf{0} \ \text{ and } \ \sum\nolimits_{i=1}^{m+2} b_i = 0. \tag{4.67}$$

Furthermore, since $X = \{\mathbf{x}_1, \ldots, \mathbf{x}_{m+1}\}$ is affinely independent, it is clear that we cannot have $b_{m+2} = 0$, so that we obtain from (4.67):

$$\mathbf{x} = \sum\nolimits_{i=1}^{m+1} (-b_i/b_{m+2})\mathbf{x}_i. \tag{4.68}$$

Furthermore, defining:

$$a_i = -b_i/b_{m+2} \quad \text{for } i = 1, \ldots, m+1, \tag{4.69}$$

it follows from the second equality in (4.67) that:

$$\sum\nolimits_{i=1}^{m+1} a_i = 1. \tag{4.70}$$

Now suppose that $\mathbf{a}^* \in \mathbf{R}^{m+1}$ is such that:

$$\mathbf{x} = \sum\nolimits_{i=1}^{m+1} a_i^* \mathbf{x}_i \ \text{ and } \ \sum\nolimits_{i=1}^{m+1} a_i^* = 1. \tag{4.71}$$

Then from (4.68)–(4.71), we have:

$$\mathbf{0} = \sum\nolimits_{i=1}^{m+1} a_i \mathbf{x}_i - \sum\nolimits_{i=1}^{m+1} a_i^* \mathbf{x}_i = \sum\nolimits_{i=1}^{m+1} (a_i - a_i^*)\mathbf{x}_i,$$

[14] Notice that it is an immediate consequence of Proposition 4.107 that such a set exists.

and:

$$\sum_{i=1}^{m+1}(a_i - a_i^*) = 1 - 1 = 0;$$

which, since X is affinely independent, implies:

$$a_i = a_i^* \quad \text{for } i = 1, \dots, m+1. \quad \square$$

The scalars, $a_1, \dots, a_{m+1}$ in the preceding proposition are called the 'barycentric coordinates' of $\mathbf{x}$. We define these more formally as follows.

4.109. Definition. Let $X = \{\mathbf{x}_1, \dots, \mathbf{x}_n\}$ be an affinely independent set, and let $\mathbf{x} \in \mathit{aff}\, X$. The unique scalars, $a_1, \dots, a_n$ such that:

$$\mathbf{x} = \sum_{i=1}^{n} a_i \mathbf{x}_i \text{ and } \sum_{i=1}^{n} a_i = 1,$$

are called the **barycentric coordinates of x with respect to X**.

Notice that, under the conditions of the above definition, and defining $M = \mathit{aff}\, X$, we will sometimes denote the barycentric coordinates of a point $\mathbf{x} \in M$ by:

$$\lambda(\mathbf{x}) = (\lambda_1(\mathbf{x}), \dots, \lambda_n(\mathbf{x})).$$

We have thus implicitly defined a mapping from M to $\mathbf{R}^n$, called the **barycentric coordinate mapping defined on M by X**. It is an example of an affine function, which we formally define as follows.

4.110. Definition. Suppose L_1 and L_2 are real linear spaces. A function $f \colon M \to L_2$, where M is an affine subset of L_1, is called **affine** iff it satisfies:

$$(\forall \mathbf{x}, \mathbf{y} \in M)(\forall \theta \in \mathbf{R}) \colon f[\theta\mathbf{x} + (1-\theta)\mathbf{y}] = \theta f(\mathbf{x}) + (1-\theta) f(\mathbf{y}).$$

As it turns out, an affine function is less different from a linear function than it might appear at first glance. The following result, which I will state without proof, sets out the basic relationship between the two kinds of mappings.

4.111. Proposition. *Suppose L_1 and L_2 are real linear spaces, and that $f \colon M \to L_2$ is an affine function, where M is an affine subset of L_1. Then, given any (fixed) $\mathbf{x}^* \in M$, the function g defined on $S \equiv M - M$ by:*

$$g(\mathbf{y}) = f(\mathbf{x}^* + \mathbf{y}) - f(\mathbf{x}^*) \quad \textit{for } \mathbf{y} \in S.$$

is linear. Conversely, if T is a linear subspace of L_1, $f \colon T \to L_2$ is a linear function, $\mathbf{x}'$ is an arbitrary element of L_1, and $\mathbf{a}$ is an arbitrary element of L_2, then the function F defined on $M \equiv \mathbf{x}' + T$ into L_2 by:

$$F(\mathbf{x}) = f(\mathbf{x} - \mathbf{x}') + \mathbf{a} \quad \textit{for } \mathbf{x} \in \mathbf{x}' + T,$$

is affine.

We will not make a great deal of use of the general notion of an affine function; however, the following two results (which we again state without proof) set out the basic properties of such functions.

4.112. Proposition. *Suppose L_1 and L_2 are real linear spaces, and that $f\colon M \to L_2$ is an affine function, where M is an affine subset of L_1. If A and B are non-empty affine subsets of M and $f(L_1)$, respectively, then $f(A)$ and $f^{-1}(B)$ are affine subsets of L_2 and M, respectively.*

4.113. Proposition. *Suppose that M and A are non-empty affine subsets of L_1 and L_2, respectively, that $f\colon M \to L_2$ and $g\colon A \to L_3$ are affine functions, and that $f(M) \subseteq A$. Then the composition function, $h\colon M \to L_3$ defined by:*

$$h(\mathbf{x}) = g[f(\mathbf{x})] \quad \textit{for}\, \mathbf{x} \in M,$$

is affine.

Exercises.

1. Show that if $\mathbf{x}_1, \ldots, \mathbf{x}_m$ are affinely independent, and $\mathbf{x}^*$ is an arbitrary element of L, the the set $\{\mathbf{x}_1 + \mathbf{x}^*, \ldots, \mathbf{x}_m + \mathbf{x}^*\}$ is also affinely independent. Show also that the preceding statement is no longer true if we substitute the word 'linearly' for the word 'affinely' wherever the latter appears.

2. Show that if $X = \{\mathbf{x}_1, \ldots, \mathbf{x}_{m+1}\}$ is an affinely independent subset of a real linear space, L, and we define $M = \mathit{aff}(X)$, then the barycentric coordinate mapping defined on M by X is an affine function

3. Show that if $f\colon M \to L_2$ is an affine function, where M is an affine subset of L_1, and $\mathbf{x}, \mathbf{x}_1, \ldots, \mathbf{x}_n \in M$ and $\mathbf{a} \in \mathbf{R}^n$ are such that:

$$\mathbf{x} = \sum_{i=1}^{n} a_i \mathbf{x}_i \ \text{ and } \ \sum_{i=1}^{n} a_i = 1,$$

then

$$f(\mathbf{x}) = \sum_{i=1}^{n} a_i f(\mathbf{x}_i).$$

4. Show that if $M_i \subseteq L_i$ is affine, for $i = 1, \ldots, n$, then:

$$M \stackrel{\text{def}}{=} \prod_{i=1}^{n} M_i,$$

is an affine subset of $L \equiv \prod_{i=1}^{n} L_i$.

5. Suppose $\mathbf{x}_1, \ldots, \mathbf{x}_m$ are vectors in L, a real linear space, and define the vectors $\mathbf{y}_i$ in $L \times \mathbf{R}$ by:

$$\mathbf{y}_i = (\mathbf{x}_i, 1) \quad \text{for}\, i = 1, \ldots, m.$$

Show that $\{\mathbf{x}_1, \ldots, \mathbf{x}_m\}$ is an affinely independent set iff $\{\mathbf{y}_1, \ldots, \mathbf{y}_m\}$ is a linearly independent set in $L \times \mathbf{R}$ (notice that $L \times \mathbf{R}$ is a real linear space).

6. Prove: Propositions (a) 4.93, and (b) 4.107, and (c) Corollary 4.96.

7. Show that if $f\colon M \to \mathbf{R}^m$ is an affine function, where M is an affine subset of $\mathbf{R}^n$, then f is uniformly continuous on M.

Chapter 5

Convex Sets and Functions

5.1 Convex Sets

Since the early 1950's, when such scholars as Arrow, Debreu, Gale, and Koopmans, among others, began showing the rest of the profession the usefulness of the theory of convex sets and functions in economic theory, a knowledge of this area of mathematics has become a very fundamental part of the knowledge required of a competent economic theorist. Since convexity is a concept which is so widely used in economic theory, you are probably already aware of many examples of its use therein. Accordingly, we shall dispense with our usual practice in this book, and make no attempt to incorporate into this chapter any examples of the use of convexity notions in economic analysis; however, Chapter 6 will be entirely devoted to topics from theoretical economics which make heavy use of convexity results.

Many of the more important results in this chapter will be concerned with sets in $\mathbf{R}^n$. However, most of the results which we will be developing are nearly as easily done for a general real linear space as they are for $\mathbf{R}^n$; and, where this is the case, it would seem wasteful not to develop the more general result. Consequently, *in this and the next section we shall always assume, unless specifically stated otherwise, that L is an arbitrary real linear space.* However, we shall also sometimes (explicitly) assume that L is a normed linear space as well; and, when we do, we shall denote the norm of $\mathbf{x}$ by '$\|\mathbf{x}\|$,' for $\mathbf{x} \in L$.

5.1. Definition. A subset, X, of L, will be said to be **convex** iff given any $\mathbf{x}$ and $\mathbf{y}$ in X, and any real number $\theta \in [0, 1]$, we have:

$$\theta\mathbf{x} + (1 - \theta)\mathbf{y} \in X.$$

It will probably be obvious to the reader that the notion of convexity can be viewed as a generalization of the idea of an affine set; and, correspondingly, any affine set (and thus, in particular, any linear subspace) is convex. Geometrically, a set is affine iff it contains the whole line generated by any two of its points, whereas a convex set need only contain the line *segments* connecting each pair of its points.

A decision which every author who intends to deal with convex sets needs to make at some point is whether or not he or she will call the empty set convex. There are precedents either way, but *in this book we will say that the empty set is convex.* This will require us to include the requirement that $X \neq \emptyset$ among the hypotheses of many of the results we will study, but it has seemed to me that this is preferable to the alternative.

One of the classic applications of the theory of convexity in mathematics is to the study of linear inequalities. Thus, for example, it will follow from results of this and the next section of this chapter (and it is easily proved directly) that if $f\colon L \to \mathbf{R}^n$ is affine, and $\mathbf{b}$ is an element of $\mathbf{R}^n$, then the subset, X, of L defined by:

$$X = \{\mathbf{x} \in L \mid f(\mathbf{x}) \geq \mathbf{b}\},$$

is convex. As a special case of this, if $\mathbf{A}$ is an $m \times n$ real matrix, and $\mathbf{b} \in \mathbf{R}^m$, then

$$\{\mathbf{x} \in \mathbf{R}^n \mid \mathbf{A}\mathbf{x} \geq \mathbf{b}\},$$

is a convex set. In particular, letting $\mathbf{A} = \mathbf{I}_n$, the $n \times n$ identity matrix, it follows from this latter fact that

$$\mathbf{R}^n_+ = \{\mathbf{x} \in \mathbf{R}^n \mid \mathbf{x} \geq \mathbf{0}\},$$

is convex.

In the following proposition, we develop a different sort of example of a convex set.

5.2. Proposition. *If L is a normed real linear space, $\mathbf{x}$ is an element of L, and ϵ is a positive real number, then the spherical neighborhood $N(\mathbf{x}, \epsilon)$, defined by*

$$N(\mathbf{x}, \epsilon) = \{\mathbf{y} \in L \mid \|\mathbf{y} - \mathbf{x}\| < \epsilon\},$$

is a convex set.

Proof. Let $\mathbf{y}, \mathbf{z} \in N(\mathbf{x}, \epsilon)$, and $\theta \in [0, 1]$. Then we have:

$$\begin{aligned} \|\theta\mathbf{y} + (1-\theta)\mathbf{z} - \mathbf{x}\| &= \|\theta(\mathbf{y} - \mathbf{x}) + (1-\theta)(\mathbf{z} - \mathbf{x})\| \\ &\leq \|\theta(\mathbf{y} - \mathbf{x})\| + \|(1-\theta)(\mathbf{z} - \mathbf{x})\| = \theta\|\mathbf{y} - \mathbf{x}\| + (1-\theta)\|\mathbf{z} - \mathbf{x})\| \\ &< \theta\epsilon + (1-\theta)\epsilon = \epsilon. \quad \square \end{aligned}$$

5.3. Definition. Let X be a nonempty subset of L. We shall say that $\mathbf{x} \in L$ is a **convex combination of X** iff there exists a (nonempty) finite subset of X, $\{\mathbf{x}_1, \ldots, \mathbf{x}_n\}$, and a nonnegative vector $\mathbf{a} \in \mathbf{R}^n_+$ such that:

$$\mathbf{x} = \sum\nolimits_{i=1}^{n} a_i\mathbf{x}_i \text{ and } \sum\nolimits_{i=1}^{n} a_i = 1.$$

Notice that if X is finite, say,

$$X = \{\mathbf{x}_1, \dots, \mathbf{x}_m\},$$

and $\mathbf{x}$ is a convex combination of X, then there exists a vector $\mathbf{b} \in \mathbf{R}^m_+$ such that:

$$\mathbf{x} = \sum_{i=1}^m b_i \mathbf{x}_i \text{ and } \sum_{i=1}^m b_i = 1;$$

in other words, if X is finite, we can express each convex combination of X as a sum involving all of the $\mathbf{x}_i$'s.

In the remainder of this section; in fact, in the remainder of this book, we shall be using the notation 'Δ_n' to denote the 'unit simplex' in $\mathbf{R}^n$; that is, we define Δ_n by:

$$\Delta_n = \left\{ \mathbf{a} \in \mathbf{R}^n_+ \mid \sum_{i=1}^n a_i = 1 \right\}.$$

We can use this notation to simplify, for example, the last part of Definition 5.3 to: "... $\{\mathbf{x}_1, \dots, \mathbf{x}_n\}$, and a vector $\mathbf{a} \in \Delta_n$ such that:

$$\mathbf{x} = \sum_{i=1}^n a_i \mathbf{x}_i."$$

As you probably already suspect, a set is convex if, and only if, it contains all of its convex combinations; a fact which follows from our next result.

5.4. Proposition. *If C is a convex subset of L, and $\mathbf{x} \in L$ is a convex combination of C, then $\mathbf{x} \in C$ (in other words, if C is convex, then it contains all of its convex combinations).*

Proof. If C is convex, it obviously contains all convex combinations involving either one or two of its elements. Suppose now that C contains any convex combination involving k (with $k \geq 2$) or fewer of its elements, and suppose

$$\mathbf{x} = \sum_{i=1}^{k+1} a_i \mathbf{x}_i, \text{ where } \mathbf{a} \in \Delta_{k+1} \text{ and } \{\mathbf{x}_1, \dots, \mathbf{x}_{k+1}\} \subseteq C.$$

If $a_{k+1} = 0$, then it follows at once from our induction hypothesis that $\mathbf{x} \in C$; and if $a_{k+1} = 1$, then it is obvious that $\mathbf{x} \in C$. Suppose, therefore, that $0 < a_{k+1} < 1$, and define the k-vector $\mathbf{b}$ by:

$$b_i = a_i/(1 - a_{k+1}) \quad \text{for } i = 1, \dots, k.$$

Then $\mathbf{b} \in \Delta_k$, so that it follows from our induction hypothesis that

$$\mathbf{y} \stackrel{\text{def}}{=} \sum_{i=1}^k b_i \mathbf{x}_i,$$

is an element of C; and thus, since C is convex, we then see that:

$$\mathbf{x} = (1 - a_{k+1})\mathbf{y} + a_{k+1}\mathbf{x}_{k+1},$$

is an element of C as well. □

The following three results are easily proved, and I will leave the proofs as exercises.

5.5. Proposition. *If $\mathcal{C} = \{C_\alpha \mid \alpha \in A\}$ is a family of convex subsets of L, then*

$$C \equiv \bigcap_{\alpha \in A} C_\alpha,$$

is a convex set.

5.6. Proposition. *If $C_i \subseteq L_i$ is convex, for $i = 1, \ldots, n$, then*

$$C \equiv \prod_{i=1}^{n} C_i,$$

is a convex subset of the product space, $L \equiv \prod_{i=1}^{n} L_i$.

5.7. Proposition. *If $f \colon M \to L_2$ is affine, where M is an affine subset of L_1, and C and D are convex subsets of M and L_2, respectively, then $f(C)$ is a convex subset of L_2, and $f^{-1}(D)$ is a convex subset of M.*

The following can be established by an easy induction argument. Details will be left as an exercise.

5.8. Proposition. *If C_i is convex and $a_i \in \mathbf{R}$, for $i = 1, \ldots, n$, then the set C defined by:*

$$C = \sum_{i=1}^{n} a_i C_i$$

is also convex.

5.9. Proposition. *If X is a convex subset of the normed linear space, L, then $\overline{X}$, the closure of X, is also convex.*

Proof. Suppose X is convex, let $\mathbf{x}, \mathbf{y} \in \overline{X}, \theta \in [0, 1]$, and let $\epsilon > 0$ be given. By definition of $\overline{X}$, there exist $\mathbf{x}'$ and $\mathbf{y}'$ in X such that

$$\|\mathbf{x} - \mathbf{x}'\| < \epsilon \text{ and } \|\mathbf{y} - \mathbf{y}'\| < \epsilon \tag{5.1}$$

However, since X is convex,

$$\theta \mathbf{x}' + (1 - \theta)\mathbf{y}' \in X,$$

and, using (5.1),

$$\begin{aligned} \|\theta \mathbf{x} + (1 - \theta)\mathbf{y} - [\theta \mathbf{x}' + (1 - \theta)\mathbf{y}']\| &\leq \theta \|\mathbf{x} - \mathbf{x}'\| + (1 - \theta)\|\mathbf{y} - \mathbf{y}'\| \\ &< \theta\epsilon + (1 - \theta)\epsilon = \epsilon; \end{aligned}$$

from which it follows that

$$\theta\mathbf{x} + (1-\theta)\mathbf{y} \in \overline{X}. \quad \square$$

Given our previous work with Kuratowski closure operations, the student should now have no difficulty in verifying the fact that the following is a proper and meaningful definition.

5.10. Definition. If X is a subset of L, we define the **convex hull of X**, denoted by '$\langle X \rangle$,' to be the smallest convex subset of L which contains X.

The following then provides a more operational definition of $\langle X \rangle$.

5.11. Proposition. *If X is a non-empty subset of L, then the convex hull of X is the collection of all convex combinations of X.*

Proof. Letting 'C' denote the collection of all convex combinations of X, we note that it follows immediately from 5.4 that any convex set containing X must contain C. Consequently, it follows that

$$C \subseteq \langle X \rangle.$$

To prove that C contains $\langle X \rangle$, let $\mathbf{x}$ and $\mathbf{y}$ be elements of C, let $\theta \in [0,1]$, and define:

$$\mathbf{z} = \theta\mathbf{x} + (1-\theta)\mathbf{y}.$$

From the definition of C we see that there exist subsets of X,

$$\{\mathbf{x}_1, \ldots, \mathbf{x}_m\} \text{ and } \{\mathbf{y}_1, \ldots, \mathbf{y}_n\},$$

and vectors

$$\mathbf{a} \in \Delta_m \text{ and } \mathbf{b} \in \Delta_n,$$

such that

$$\mathbf{x} = \sum_{i=1}^{m} a_i\mathbf{x}_i \text{ and } \mathbf{y} = \sum_{j=1}^{n} b_j\mathbf{y}_j.$$

But then, defining the $(m+n)$-vector $\mathbf{c}$ by:

$$c_k = \begin{cases} \theta a_k & \text{for } k = 1, \ldots, m, \\ (1-\theta)b_{k-m} & \text{for } k = m+1, \ldots, m+n, \end{cases}$$

we have

$$c_k \geq 0 \quad \text{for } k = 1, \ldots, m+n,$$

while

$$\sum_{k=1}^{m+n} c_k = \sum_{i=1}^{m} \theta a_i + \sum_{j=1}^{n} (1-\theta) b_j$$
$$= \theta \sum_{i=1}^{m} a_i + (1-\theta) \sum_{j=1}^{n} b_j = 1.$$

Moreover,

$$\mathbf{z} \stackrel{\text{def}}{=} \theta \mathbf{x} + (1-\theta)\mathbf{y} = \theta \sum_{i=1}^{m} a_i \mathbf{x}_i + (1-\theta) \sum_{j=1}^{n} b_j \mathbf{y}_j$$
$$= \sum_{i=1}^{m} c_i \mathbf{x}_i + \sum_{j=1}^{n} c_{j+m} \mathbf{y}_j.$$

Therefore we see that $\mathbf{z}$ is a convex combination of X, and it follows that $\mathbf{z} \in C$; and thus that C is convex. However, C obviously contains X; and, since $\langle X \rangle$ is the *smallest* convex set containing X, it then follows that

$$\langle X \rangle \subseteq C. \quad \square$$

It is worth our while to pause at this point to reflect upon what we have established in the result just proved. Let X be a non-empty subset of a real linear space, L, and define the sequence of sets $\langle X_n \rangle$ by:

$$\begin{aligned} &X_1 = X, \\ &X_2 = \{\mathbf{x} \in L \mid (\exists \mathbf{x}_1, \mathbf{x}_2 \in X, \theta \in [0,1]) \colon \theta \mathbf{x}_1 + (1-\theta)\mathbf{x}_2\}, \\ &\ldots \\ &X_n = \left\{\mathbf{x} \in L \mid (\exists \mathbf{x}_1, \ldots, \mathbf{x}_n \in X, \mathbf{a} \in \Delta_n) \colon \mathbf{x} = \sum_{i=1}^{n} a_i \mathbf{x}_i\right\}, \\ &\ldots; \end{aligned} \tag{5.2}$$

in other words, X_n is the set of all points in L which can be written as a convex combination of n (or fewer) elements of X. It is obvious from the definitions and Proposition 5.4 that:

$$X_n \subseteq X_{n+1} \subseteq \langle X \rangle \quad \text{for } n = 1, 2, \ldots; \tag{5.3}$$

however, in Proposition 5.11 we have shown that:

$$\langle X \rangle = \bigcup_{n=1}^{\infty} X_n.$$

While I introduced Proposition 5.11 by stating that it " ... provides a more operational definition of $\langle X \rangle$," it is clear that it does not necessarily provide a very practical method for finding the convex hull of a set. However, in special cases, we can sharpen this result considerably. Thus, in Figure 5.1, on the next page, $X_1 = X$ is the set of three (affinely independent) points $\{\mathbf{x}, \mathbf{y}, \mathbf{z}\}$. It is then easily seen that X_2 is the perimeter of the triangle formed by the three points, while X_3 is the whole triangle (interior and perimeter).

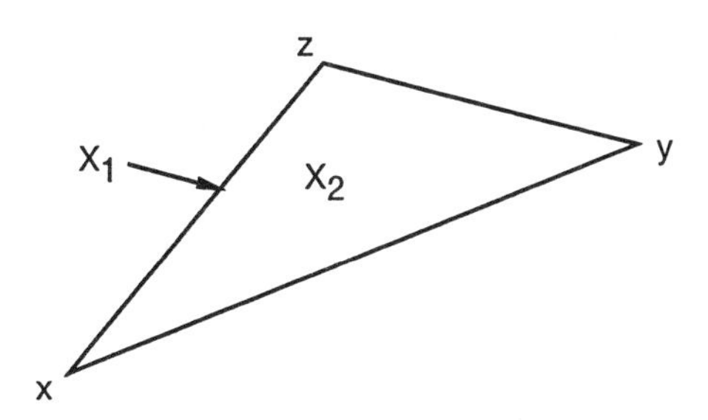

Figure 5.1: The Convex Hull.

Since X_3 is a convex set, it then follows that it is the convex hull of X, $\langle X \rangle$ (why?). In our next theorem we will establish a general criterion stating precisely at what point we will obtain the convex hull when we construct the sets X_1, X_2, etc.; but first we must establish the following.

5.12. Lemma. *Suppose S is a linear subspace of L having $\dim(S) = m$, let n be a positive integer, and suppose that a point $\mathbf{x} \in S$ is a nonnegative linear combination of $\{\mathbf{x}_1, \ldots, \mathbf{x}_n\} \subseteq S$. Then $\mathbf{x}$ can be represented as a nonnegative linear combination of at most m elements of the set $\{\mathbf{x}_1, \ldots, \mathbf{x}_n\}$.*

Proof. Let

$$\mathbf{x} = \sum_{i=1}^{n} a_i \mathbf{x}_i, \tag{5.4}$$

where $\mathbf{x}_i \in S$ and $a_i \in \mathbf{R}_+$, for $i = 1, \ldots, n$. If $n \leq m$, we have nothing to prove, so suppose $n > m$. Then, since $\dim(S) = m$, $\{\mathbf{x}_1, \ldots, \mathbf{x}_n\}$ must be a linearly dependent set; so that there exist $b_1, \ldots, b_n \in \mathbf{R}$, not all of which are zero, satisfying:

$$\sum_{i=1}^{n} b_i \mathbf{x}_i = \mathbf{0}. \tag{5.5}$$

Furthermore, we can assume, without loss of generality, that

$$b_j > 0 \quad \text{for some } j \in \{1, \ldots, n\};$$

and thus that the set J defined by:

$$J = \{j \in \{1, \ldots, n\} \mid b_j > 0\},$$

is non-empty. Defining

$$\theta = \min\{a_j / b_j \mid j \in J\},$$

we see that $\theta \geq 0$; and by (5.4) and (5.5), we have:

$$\mathbf{x} = \sum_{i=1}^{n} a_i \mathbf{x}_i - \theta \sum_{i=1}^{n} b_i \mathbf{x}_i = \sum_{i=1}^{n} (a_i - \theta b_i) \mathbf{x}_i. \tag{5.6}$$

Since $b_i \leq 0$ for each $i \notin J$, we see that

$$(\forall i \in \{1, \ldots, n\} \setminus J)\colon a_i - \theta b_i \geq 0;$$

while, since $\theta \leq a_j/b_j$ for $j \in J$, we have:

$$(\forall j \in J)\colon a_j - \theta b_j \geq 0,$$

as well. Moreover, for some $k \in J$, we have

$$a_k/b_k = \theta;$$

and thus, for this k,

$$a_k - \theta b_k = 0.$$

From (5.6) we see, therefore, that $\mathbf{x}$ can be expressed as a nonnegative linear combination of at most $n-1$ elements of $\{\mathbf{x}_1, \ldots, \mathbf{x}_n\}$. If $n-1 > m$, we can go through the same procedure a second time to obtain $\mathbf{x}$ as a nonnegative linear combination of at most $n-2$ elements of $\{\mathbf{x}_1, \ldots, \mathbf{x}_n\}$, etc. □.

5.13. Theorem. (Carathéodory's Theorem). *If dim(X) = m, then:*

$$\langle X \rangle = X_{m+1},$$

where X_{m+1} is defined in equation (5.2), above.

Proof. As we have already noted [equation (5.3)], X_{m+1} is contained in the convex hull of X; consequently, it suffices to prove that $\langle X \rangle$ is contained in X_{m+1}. Suppose, therefore, that $\mathbf{x} \in \langle X \rangle$. Then it follows from 5.11 that there exists a positive integer, n, along with $\mathbf{x}_1, \ldots, \mathbf{x}_n \in X$, and a vector $\mathbf{b} \in \Delta_n$, such that:

$$\mathbf{x} = \sum_{i=1}^{n} b_i \mathbf{x}_i. \tag{5.7}$$

If $n \leq m+1$, then we obviously have $\mathbf{x} \in X_{m+1}$. Otherwise, defining

$$\mathbf{y} = \begin{pmatrix} \mathbf{x} \\ 1 \end{pmatrix} \text{ and } \mathbf{y}_i = \begin{pmatrix} \mathbf{x}_i \\ 1 \end{pmatrix} \quad \text{for } i = 1, \ldots, n,$$

it follows from (5.7) and the fact that $\mathbf{b} \in \Delta_n$, that

$$\mathbf{y} = \sum_{i=1}^{n} b_i \mathbf{y}_i.$$

However, since $\dim(X) = m$, at most $m+1$ of the $\mathbf{x}_i$'s are affinely independent, and thus by exercise 5 at the end of the last section of Chapter 4, at most $m+1$ of the $\mathbf{y}_i$'s are linearly independent. Consequently, it follows from Lemma 5.12 that there exists some subset, Y, of $\{\mathbf{y}_1, \ldots, \mathbf{y}_n\}$, containing at most $m+1$ elements, such that $\mathbf{y}$ is a nonnegative linear combination of the vectors in Y. In fact, for notational convenience, we may suppose (renumbering the $\mathbf{y}_i$'s if necessary), that

$$Y = \{\mathbf{y}_1, \ldots, \mathbf{y}_{m+1}\},$$

and that there exists $\mathbf{a} \in \mathbf{R}^{m+1}$ such that:

$$\mathbf{y} = \begin{pmatrix} \mathbf{x} \\ 1 \end{pmatrix} = \sum_{j=1}^{m+1} a_j \mathbf{y}_j = \sum_{j=1}^{m+1} a_j \begin{pmatrix} \mathbf{x}_i \\ 1 \end{pmatrix}. \tag{5.8}$$

From (5.8) we then have:

$$\mathbf{x} = \sum_{j=1}^{m+1} a_j \mathbf{x}_j \quad \text{and} \quad \sum_{j=1}^{m+1} a_j = 1,$$

so that it follows that $\mathbf{x} \in X_{m+1}$; and thus we conclude that

$$\langle X \rangle \subseteq X_{m+1}. \quad \square$$

The following is then an immediate consequence of 5.13.

5.14. Corollary. *If X is a non-empty subset of $\mathbf{R}^n$, then:*

$$\langle X \rangle = \left\{\mathbf{x} \in \mathbf{R}^n \mid (\exists \mathbf{x}_1, \ldots, \mathbf{x}_{n+1} \;\&\; \mathbf{a} \in \Delta_{n+1}) \colon \mathbf{x} = \sum_{i=1}^{n+1} a_i \mathbf{x}_i\right\}.$$

Many uses have been found in economic theory for the following result; one of which is in some of the proofs that the 'core of the economy shrinks to the set of competitive equilibria as the number of agents becomes large.' I will leave the proof as an exercise, since it follows fairly readily from Propositions 5.11 and 5.4.

5.15. Proposition. *If C_i is a non-empty convex subset of L, for $i = 1, \ldots, n$, then the convex hull of the union of the C_i's is given by:*

$$\left\langle \bigcup_{i=1}^n C_i \right\rangle = \{\mathbf{x} \in L \mid (\exists \mathbf{a} \in \Delta_n, \mathbf{x}_i \in C_i, \textit{ for } i = 1, \ldots, n) \colon \mathbf{x} = \sum_{i=1}^n a_i \mathbf{x}_i\}. \tag{5.9}$$

5.16. Proposition. *If $X_i \subseteq L$, for $i = 1, \ldots, n$, then:*

$$\sum_{i=1}^n \langle X_i \rangle = \left\langle \sum_{i=1}^n X_i \right\rangle;$$

that is, the sum of convex hulls is equal to the convex hull of the sum.

Proof. We begin by noting that, by Proposition 5.8, $\sum_{i=1}^n \langle X_i \rangle$ is a convex set; and, since it clearly contains $\sum_{i=1}^n X_i$, it follows that

$$\left\langle \sum_{i=1}^n X_i \right\rangle \subseteq \sum_{i=1}^n \langle X_i \rangle; \tag{5.10}$$

We will use an induction argument to prove the reverse inclusion, as follows.

Our result is trivial if $n = 1$. To show that

$$\sum_{i=1}^{n} \langle X_i \rangle \subseteq \left\langle \sum_{i=1}^{n} X_i \right\rangle; \tag{5.11}$$

for $n = 2$, let X and Y be subsets of L. If either X or Y is the empty set, then both of the sums in (5.11) are empty, and the inclusion obviously holds. Suppose, therefore, that both X and Y are non-empty, and suppose

$$\mathbf{z} \in X + \langle Y \rangle.$$

Then it follows from our definitions and from Proposition 5.11 that there exist a positive integer, m, $\mathbf{x} \in X, \mathbf{y}_1, \dots, \mathbf{y}_m \in Y$, and $\mathbf{a} \in \Delta_m$ such that:

$$\mathbf{z} = \mathbf{x} + \sum_{i=1}^{m} a_i \mathbf{y}_i. \tag{5.12}$$

However, we obviously have

$$\mathbf{x} + \mathbf{y}_i \in X + Y \quad \text{for } i = 1, \dots, m;$$

and therefore

$$\sum_{i=1}^{m} a_i(\mathbf{x} + \mathbf{y}_i) = \mathbf{x} + \sum_{i=1}^{m} a_i \mathbf{y}_i \in \langle X + Y \rangle.$$

Thus we see that

$$X + \langle Y \rangle \subseteq \langle X + Y \rangle. \tag{5.13}$$

Now suppose that

$$\mathbf{z} \in \langle X \rangle + \langle Y \rangle.$$

Using Proposition 5.11 once again, it follows that there exist $\mathbf{x}_1, \dots, \mathbf{x}_q \in X, \mathbf{b} \in \Delta_q$, and $\mathbf{y} \in \langle Y \rangle$, such that

$$\mathbf{z} = \sum_{j=1}^{q} b_j \mathbf{x}_j + \mathbf{y}.$$

But then, proceeding as before,

$$\mathbf{x}_j + \mathbf{y} \in X + \langle Y \rangle \subseteq \langle X + Y \rangle \quad \text{for } j = 1, \dots, q;$$

[where the inclusion is by (5.13)] and, consequently,

$$\sum_{j=1}^{q} b_j(\mathbf{x}_j + \mathbf{y}) = \sum_{j=1}^{q} b_j \mathbf{x}_j + \mathbf{y} = \mathbf{z} \in \langle X + Y \rangle.$$

We conclude, therefore, that

$$\langle X \rangle + \langle Y \rangle \subseteq \langle X + Y \rangle. \tag{5.14}$$

Now suppose the inclusion in equation (5.11) holds for $n = k$. Then, making use of our inductive hypothesis and (5.14) in turn, we have:

$$\sum_{i=1}^{k+1}\langle X_i\rangle = \sum_{i=1}^{k}\langle X_i\rangle + \langle X_{k+1}\rangle \subseteq \left\langle \sum_{i=1}^{k} X_i \right\rangle + \langle X_{k+1}\rangle$$
$$\subseteq \left\langle \left(\sum_{i=1}^{k} X_i\right) + X_{k+1}\right\rangle = \left\langle \sum_{i=1}^{k+1} X_i\right\rangle. \quad \square$$

Exercises.

1. Prove Proposition 5.5.
2. Prove Proposition 5.6.
3. Prove Proposition 5.7.
4. Prove Proposition 5.8. It may be useful for you to try proving it by using Propositions 5.6 and 5.7; although admittedly it is nearly as easy to prove it directly.
5. Prove Proposition 5.15.

5.2 Relative Interiors of Convex Sets

In considering the following definitions, you may wish to review the notions of relatively open and closed sets. See Section 6 of Chapter 1, particularly 1.71 - 1.77. *In this section, unless otherwise stated, we shall take L to be an arbitrary normed real linear space, with norm '$\|\cdot\|$.'* As in Section 4 of Chapter 4, I will denote the spherical neighborhood of a point $\mathbf{x} \in L$ with radius $\epsilon > 0$ by '$N(\mathbf{x}, \epsilon)$;' that is,

$$N(\mathbf{x}, \epsilon) = \{\mathbf{y} \in L \mid \|\mathbf{y} - \mathbf{x}\| < \epsilon\}.$$

5.17. Definitions. If X is a convex subset of L, we define riX, the **relative interior of X**, by

$$riX = \{\mathbf{x} \in X \mid (\exists \epsilon \in \mathbf{R}_{++})\colon N(\mathbf{x}, \epsilon) \cap \mathit{aff}\, X \subseteq X\};$$

that is, riX is the interior of X relative to $\mathit{aff}\, X$.[1] We also define the **relative boundary of X**, or **frontier of X**, denoted by '$Fr(X)$,' by

$$Fr(X) = \overline{X} \setminus riX.$$

Finally, we shall say that X is **relatively open** iff $riX = X$.

Before proceeding further, I should warn you that the definition of relative interior just presented is not very satisfactory when applied to infinite-dimensional linear spaces, and we will consider a modified definition in Chapter 10. However, the results involving relative interiors which are presented

[1] Where, remember, '$\mathit{aff}\, X$' denotes the affine hull of X; the smallest affine set containing X. See Definition 4.100.

in the remainder of this chapter are valid, unless specifically stated otherwise, for any normed real linear space.

As one would hope, if a convex subset of $\mathbf{R}^n$ has dimension n, then its relative interior coincides with its interior in the normal sense; as is formally set out in the next result.

5.18. Proposition. *If $X \subseteq \mathbf{R}^n$ is convex and non-empty, then* $\dim(X) = n$ *if, and only if,* $\text{int}(X) = riX$.

Proof. Suppose first that $\dim(X) = n$. Then there exists a maximal affinely independent subset of X containing $n+1$ points, $\{\mathbf{x}_1, \ldots, \mathbf{x}_{n+1}\}$. However, since $\{\mathbf{x}_1, \ldots, \mathbf{x}_{n+1}\}$ is affinely independent, we have

$$aff\, X = aff\{\mathbf{x}_1, \ldots, \mathbf{x}_{n+1}\} = \mathbf{R}^n.$$

It is then obvious from the definitions that

$$riX = int(X).$$

Conversely, suppose that $int(X) = riX$, and let $\mathbf{x}^* \in int(X)$. Then there exists $\epsilon > 0$ such that

$$N(\mathbf{x}^*, \epsilon) \subseteq X. \tag{5.15}$$

But then we see that, if we define $\mathbf{x}_i$ by

$$\mathbf{x}_i = \mathbf{x}^* + (\epsilon/2)\mathbf{e}_i \quad \text{for } i = 1, \ldots, n, \text{and } \mathbf{x}_{n+1} = \mathbf{x}^*,$$

then $X^* \equiv \{\mathbf{x}_1, \ldots, \mathbf{x}_{n+1}\}$ is an affinely independent set;[2] and, by (5.15),

$$X^* \subseteq X.$$

Therefore $\dim(X) = n$. □

Clearly a convex set may have an empty interior. However, one of the many ways in which convex sets are unique and uniquely useful is that a convex set of finite dimension always has a non-empty relative interior; as is established in the following theorem.

5.19. Theorem. *If C is a convex subset of L, and* $\dim(C)$ *is finite, then* $riC \neq \emptyset$.

Proof. Let

$$m = \dim(C) = \dim[aff\, C],$$

let $X = \{\mathbf{x}_1, \ldots, \mathbf{x}_{m+1}\}$ be an affinely independent subset of C, and define:

$$\mathbf{x}^* = \sum_{i=1}^{m+1} [1/(1+m)]\mathbf{x}_i = [1/(1+m)] \sum_{i=1}^{m+1} \mathbf{x}_i. \tag{5.16}$$

[2]See Exercise 3, at the end of this section.

Since C is convex, $\mathbf{x}^* \in C$, and we will show that $\mathbf{x}^*$ is, in fact, an element of riC.

If $m = 0$, $\mathbf{x}^*$ is obviously a relative interior point of C. On the other hand, if $m \geq 1$, then by Proposition 4.108, each element, $\mathbf{x}$, of $\mathit{aff}\, C$ has a unique representation of the form

$$\mathbf{x} = \sum_{i=1}^{m+1} a_i \mathbf{x}_i; \tag{5.17}$$

where $\mathbf{a} \in \mathbf{R}^{m+1}$ is such that

$$\sum_{i=1}^{m+1} a_i = 1.$$

Defining

$$\mathbf{y}_i = \mathbf{x}_i - \mathbf{x}_{m+1} \quad \text{for } i = 1, \ldots, m,$$

we note that if $\mathbf{x}$ can be expressed in the form (5.17), we have:

$$\begin{aligned}\mathbf{x} - \mathbf{x}^* &= \mathbf{x} - \mathbf{x}_{m+1} - (\mathbf{x}^* - \mathbf{x}_{m+1}) \\ &= \sum_{i=1}^{m} a_i \mathbf{x}_i + (a_{m+1} - 1)\mathbf{x}_{m+1} \\ &- \left[\left(\frac{1}{m+1}\right) \sum_{i=1}^{m} \mathbf{x}_i + \left(\frac{1}{m+1} - 1\right)\mathbf{x}_{m+1}\right] \\ &= \sum_{i=1}^{m} a_i \mathbf{y}_i - \left(\frac{1}{m+1}\right) \sum_{i=1}^{m} \mathbf{y}_i \\ &= \sum_{i=1}^{m} [a_i - 1/(m+1)]\mathbf{y}_i.\end{aligned} \tag{5.18}$$

Moreover, by Proposition 4.50, there exists a number $\eta > 0$ such that, for all $\mathbf{b} \in \mathbf{R}^m$,

$$\sum_{i=1}^{m} |b_i| \leq \eta \cdot \left\| \sum_{i=1}^{m} b_i \mathbf{y}_i \right\|. \tag{5.19}$$

Now, define

$$\epsilon^* = \frac{1}{\eta m(m+1)}.$$

Then, if

$$\mathbf{x} \in N(\mathbf{x}^*, \epsilon^*) \cap \mathit{aff}\, C,$$

we have, using (5.18) and (5.19)

$$\sum_{i=1}^{m} \left| a_i - \frac{1}{m+1} \right| \leq \eta \cdot \left\| \sum_{i=1}^{m} \left[a_i - \frac{1}{m+1} \right] \mathbf{y}_i \right\| = \eta \cdot \|\mathbf{x} - \mathbf{x}^*\| < \frac{1}{m(m+1)};$$

where the scalars a_i are from (5.17), above. But then it follows at once from this last inequality that, for each $i \in \{1, \ldots, m\}$,

$$-\frac{1}{m(m+1)} < a_i - \frac{1}{m+1} < \frac{1}{m(m+1)},$$

and thus

$$0 \leq \frac{m-1}{m(m+1)} < a_i < \frac{m+1}{m(m+1)} = 1/m \quad \text{for } i = 1, \ldots, m.$$

It then follows that

$$a_i > 0 \quad \text{for } i = 1, \ldots, m,$$

and

$$\sum_{i=1}^{m} a_i < m \cdot (1/m) = 1;$$

so that

$$a_{m+1} = 1 - \sum_{i=1}^{m} a_i > 0,$$

as well. Consequently, $\mathbf{x} \in C$ (since C is convex); and, since $\mathbf{x}$ was an arbitrary element of $\mathit{aff}\, C \cap N(\mathbf{x}^*, \epsilon^*)$, we conclude that $\mathbf{x}^* \in \mathit{ri} C$. □

An especially useful property of convex sets is that (in finite dimensional space) the closure of a convex set is equal to the closure of its relative interior. This is one of the many consequences of the following theorem, as we will see.

5.20. Theorem. *If X is a convex subset of $L, \mathbf{x} \in \mathit{ri} X$, and $\mathbf{y} \in \overline{X}$, then*

$$(\forall \theta \in [0, 1[)\colon \theta\mathbf{y} + (1 - \theta)\mathbf{x} \in \mathit{ri} X.$$

Proof. Let $\mathbf{x} \in \mathit{ri} X, \mathbf{y} \in \overline{X}$, and $\theta \in [0, 1[$ and define

$$\mathbf{z} = \theta\mathbf{y} + (1 - \theta)\mathbf{x}.$$

Since $\mathbf{x} \in \mathit{ri} X$, there exists $\epsilon > 0$ such that

$$N(\mathbf{x}, \epsilon) \cap \mathit{aff}\, X \subseteq X; \tag{5.20}$$

and notice that, if we can prove that, for $\theta > 0$,

$$N[\mathbf{z}, (1 - \theta)\epsilon] \cap \mathit{aff}\, X \subseteq X,$$

then our result will follow. As you read through the proof, it may be helpful to refer to Figure 5.2, on the next page, which shows the basic schematic relationships among $\mathbf{x}$, $\mathbf{y}$, and $\mathbf{z}$.

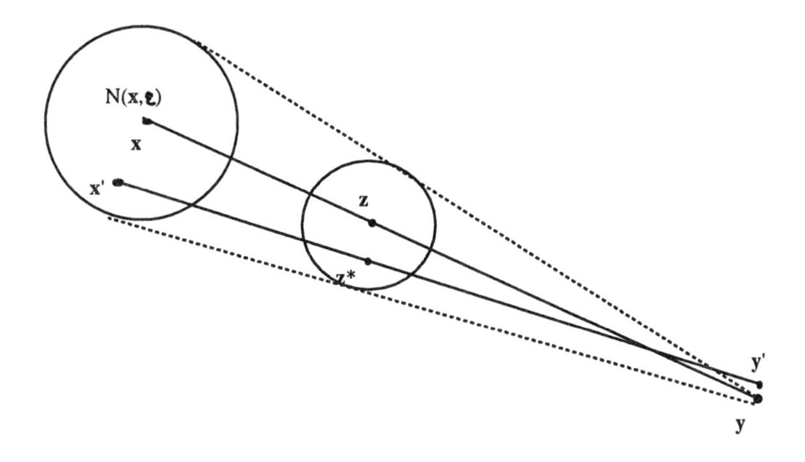

Figure 5.2: Schematic of the Proof of 5.20.

Suppose $\theta > 0$, and let $\mathbf{z}^* \in N[\mathbf{z}, (1-\theta)\epsilon] \cap \mathit{aff}\, X$ be arbitrary. Since $\mathbf{y} \in \overline{X}$, there exists $\mathbf{y}' \in X$ such that:

$$\|\mathbf{y} - \mathbf{y}'\| < (1/\theta)\big[(1-\theta)\epsilon - \|\mathbf{z}^* - \mathbf{z}\|\big].$$

If we now define

$$\mathbf{x}' = [1/(1-\theta)]\mathbf{z}^* - [\theta/(1-\theta)]\mathbf{y}';$$

we see that, since $\mathbf{z}^*$, $\mathbf{y}' \in \mathit{aff}\, X$, we have

$$\mathbf{x}' \in \mathit{aff}\, X.$$

Consequently, since we obviously have

$$\mathbf{z}^* = \theta\mathbf{y}' + (1-\theta)\mathbf{x}';$$

we see that, if we can show that $\mathbf{x}' \in N(\mathbf{x}, \epsilon)$, it will follow from (5.20) and the fact that $\mathbf{y}' \in X$ that $\mathbf{z}^* \in X$. However, we have, making use of our definitions,

$$\begin{aligned}
\|\mathbf{x}' - \mathbf{x}\| &= \big\|[[1/(1-\theta)]\mathbf{z}^* - [\theta/(1-\theta)]\mathbf{y}' - \mathbf{x}\big\| \\
&= \left\|\left(\frac{1}{1-\theta}\right)\mathbf{z}^* - \left(\frac{1}{1-\theta}\right)\mathbf{z} + \left(\frac{1}{1-\theta}\right)\mathbf{z} - \left(\frac{\theta}{1-\theta}\right)\mathbf{y}' - \mathbf{x}\right\| \\
&= \left\|\left(\frac{1}{1-\theta}\right)(\mathbf{z}^* - \mathbf{z}) + \left(\frac{1}{1-\theta}\right)[\theta\mathbf{y} + (1-\theta)\mathbf{x}] - \left(\frac{\theta}{1-\theta}\right)\mathbf{y}' - \mathbf{x}\right\| \\
&\leq [1/(1-\theta)]\|\mathbf{z}^* - \mathbf{z}\| + [\theta/(1-\theta)]\|\mathbf{y} - \mathbf{y}'\| \\
&< [1/(1-\theta)]\|\mathbf{z}^* - \mathbf{z}\| + [\theta/(1-\theta)](1/\theta)\big[(1-\theta)\epsilon - \|\mathbf{z}^* - \mathbf{z}\|\big] = \epsilon.
\end{aligned}$$

Hence, $\mathbf{x}' \in N(\mathbf{x}, \epsilon) \cap \mathit{aff}\, X$, and our result follows. □

The following two results are easy corollaries of Theorem 5.20, and you should have no difficulty in proving them.

5.21. Corollary. *If X is a convex subset of L, then riX is also convex.*

5.22. Corollary. *If X is a convex subset of L, and $riX \neq \emptyset$,[3] then $\overline{riX} = \overline{X}$.*

If $\mathbf{x}, \mathbf{y}$ and $\mathbf{z}$ are points in $\mathbf{R}^2$, and

$$\mathbf{y} = (1-\theta)\mathbf{x} + \theta\mathbf{z} = \mathbf{x} + \theta(\mathbf{z} - \mathbf{x}), \tag{5.21}$$

with $\theta > 1$, then we can think of $\mathbf{y}$ as being on the line, ℓ, determined by $\mathbf{x}$ and $\mathbf{z}$, but as being on the opposite side of $\mathbf{z}$ from $\mathbf{x}$. Clearly, if X is a convex set and $\mathbf{z} \in riX$, while $\mathbf{x} \in X$, then we should find, for values of θ greater than, but close to one, that $\mathbf{y}$, as given by (5.21), remains in X. This intuition holds up nicely, as is shown in the following result.

5.23. Proposition. *Let X be a convex subset of L, and $\mathbf{z} \in riX$. Then, for any $\mathbf{x} \in \text{aff}X$, there exists $\delta > 0$ such that, for all θ satisfying $|\theta - 1| < \delta$,*

$$(1-\theta)\mathbf{x} + \theta\mathbf{z} = \mathbf{x} + \theta(\mathbf{z} - \mathbf{x}) \in X.$$

Proof. If $\mathbf{z} = \mathbf{x}$, our result is trivial. Suppose, therefore, that $\mathbf{z} \neq \mathbf{x}$. Since $\mathbf{z} \in riX$, there exists $\epsilon > 0$ such that

$$N(\mathbf{z}, \epsilon) \cap \mathit{aff}\, X \subseteq X. \tag{5.22}$$

Defining

$$\delta = \epsilon / \|\mathbf{x} - \mathbf{z}\|,$$

we then see that, for θ such that $|\theta - 1| < \delta$, we have

$$(1-\theta)\mathbf{x} + \theta\mathbf{z} \in \mathit{aff}\, X, \tag{5.23}$$

and

$$\|(1-\theta)\mathbf{x} + \theta\mathbf{z} - \mathbf{z}\| = |1-\theta| \cdot \|\mathbf{x} - \mathbf{z}\| < \delta \cdot \|\mathbf{x} - \mathbf{z}\| = \epsilon,$$

so that

$$(1-\theta)\mathbf{x} + \theta\mathbf{z} \in N(\mathbf{z}, \epsilon). \tag{5.24}$$

Our result then follows from equations (5.22) - (5.24). □

The following partial converse of 5.23 is also quite often useful. I will leave the proof as an exercise.

[3]Remember that, if $\dim(X)$ is finite, then the relative interior of X is nonempty.

5.24. Proposition. *If C is a convex subset of L, and $\mathbf{z} \in L$ is such that, for some $\mathbf{x} \in riC$ and $\mu \in \mathbf{R}$, we have*

$$\mu > 1 \text{ and } (1-\mu)\mathbf{x} + \mu\mathbf{z} \in \overline{C},$$

then $\mathbf{z} \in riC$.

Exercises.

1. Prove Corollary 5.21.
2. Prove Corollary 5.22.
3. Show that the vectors, $\mathbf{x}_1, \ldots, \mathbf{x}_{n+1}$, defined in the proof of Proposition 5.18 are affinely independent.
4. Prove Proposition 5.24.

5.3 Extreme Points of a Convex Set

In this section our main concern will be with 'extreme points' of a convex set; however, before introducing this concept, it will be convenient to first study some related definitions, beginning with the following.

5.25. Definitions. Given a point $\mathbf{x}^* \in L$, and $\mathbf{v} \in L \setminus \{\mathbf{0}\}$, we define:

1. the **ray from x* in the direction v, R(x*, v)**, by:

$$R(\mathbf{x}^*, \mathbf{v}) = \{\mathbf{x} \in L \mid (\exists\theta \in \mathbf{R}_+) : \mathbf{x} = \mathbf{x}^* + \theta\mathbf{v}\}$$

and

2. the **line through x* determined by v** by:

$$L(\mathbf{x}^*, \mathbf{v}) = \{\mathbf{x} \in L \mid (\exists\theta \in \mathbf{R}) : \mathbf{x} = \mathbf{x}^* + \theta\mathbf{v}\}.$$

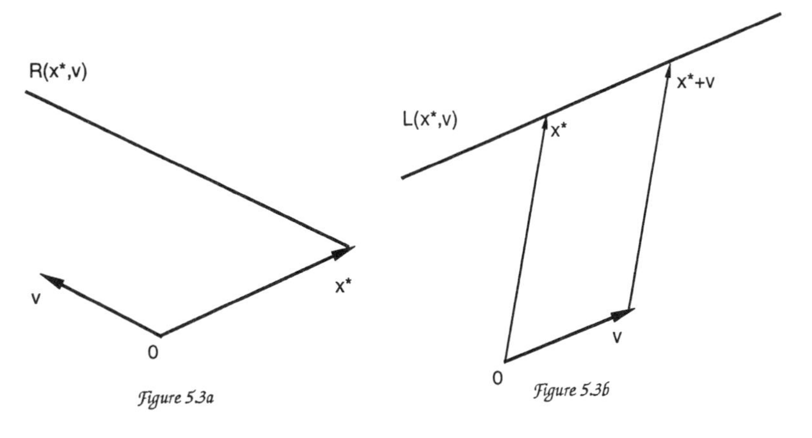

Figure 5.3: Lines and Rays Determined by x* and v.

Notice that the line through $\mathbf{x}^*$ determined by $\mathbf{v}$ will not generally pass through $\mathbf{v}$ (see Figure 5.3, above). In fact, it coincides with the line determined by $\mathbf{x}^*$ and $\mathbf{x}^* + \mathbf{v}$, as we previously defined said line; for notice that if $\mathbf{x}$ is of the form

$$\mathbf{x} = \mathbf{x}^* + \theta\mathbf{v},$$

for some $\theta \in \mathbf{R}$, then we can equally well write

$$\mathbf{x} = \theta(\mathbf{x}^* + \mathbf{v}) + (1 - \theta)\mathbf{x}^*.$$

It will be convenient at several points in the remainder of this text to make use of the following notions: we define the **unit sphere**, S, and the **unit ball**, B, in L, by:

$$S = \{\mathbf{u} \in L \mid \|u\| \leq 1\},$$

and

$$B = \{\mathbf{u} \in L \mid \|u\| = 1\},$$

respectively.

Now suppose X is a convex subset of L, and that $\mathbf{x}^* \in X$. If we write

$$M = \mathit{aff}\, X, \tag{5.25}$$

recall that we can also express M as

$$M = \mathbf{x}^* + K, \tag{5.26}$$

where K is the linear subspace

$$K = M - M. \tag{5.27}$$

Defining the set V as

$$V = B \cap K, \tag{5.28}$$

we then have the following.

5.26. Proposition. *If X is a convex subset of L, $\mathbf{x}^* \in riX$, and we define M, K, and V as in (5.25)-(5.28), above, then:*

1. there is a one-to-one correspondence between the set V and the set of rays in M issuing from $\mathbf{x}^*$, and

2. given any $\mathbf{v}^* \in V$, if we define $A \subseteq \mathbf{R}_+$ by

$$A = \{a \in \mathbf{R}_+ \mid \mathbf{x}^* + a\mathbf{v}^* \in X\}, \tag{5.29}$$

then either

a. $A = \mathbf{R}_+$, *or*

b. there exists $\alpha \in \mathbf{R}_{++}$ *such that*

$$[0, \alpha[\subseteq A \subseteq [0, \alpha], \tag{5.30}$$

$\mathbf{x}^* + \alpha \mathbf{v}^* \in FrX$, *and*

$$(\forall a \in [0, \alpha[) : \mathbf{x}^* + a\mathbf{v}^* \in riX. \tag{5.31}$$

Proof. If $R(\mathbf{x}^*, \mathbf{y}')$ is a ray in M issuing from $\mathbf{x}^*$ in the direction $\mathbf{y}'$, then, since:

$$\mathbf{x}' \stackrel{\text{def}}{=} \mathbf{x}^* + \mathbf{y}',$$

is an element of $R(\mathbf{x}^*, \mathbf{y}')$, we see that $\mathbf{y}' \in K$. Therefore, $\mathbf{v}' \stackrel{\text{def}}{=} (1/\|\mathbf{y}'\|)\mathbf{y}'$ is an element of V, and clearly,

$$R(\mathbf{x}^*, \mathbf{y}') = R(\mathbf{x}^*, \mathbf{v}').$$

Conversely, if $\mathbf{v} \in V$, then, since $V \subseteq K$, we see that

$$R(\mathbf{x}^*, \mathbf{v}) \subseteq \mathbf{x}^* + K = M;$$

and thus $\mathbf{v}$ defines a ray issuing from $\mathbf{x}^*$ in M; which establishes part 1 of our conclusion.

Now let $\mathbf{v}^* \in V$, and define the set A as in (5.29), above. If $A \neq \mathbf{R}_+$, then there exists $b \in \mathbf{R}_{++} \setminus A$. However, it must then be the case that b is an upper bound for A; for suppose, by way of obtaining a contradiction, that there exists $b' > b$ such that $b' \in A$. Then

$$0 < b/b' < 1,$$

and, defining

$$\mathbf{x} = \mathbf{x}^* + b\mathbf{v}^* \text{ and } \mathbf{x}' = \mathbf{x}^* + b'\mathbf{v}^*,$$

it is a matter of simple algebra to show that

$$\mathbf{x} = \left(1 - \frac{b}{b'}\right)\mathbf{x}^* + \left(\frac{b}{b'}\right)\mathbf{x}';$$

implying (by Theorem 5.20) that $\mathbf{x} \in riX$, which contradicts the assumption that $b \notin A$. Thus if $A \neq \mathbf{R}_+$, it follows, using Proposition 5.23, that

$$\alpha \equiv \sup A$$

is positive (and finite). From the definitions of $R(\mathbf{x}^*, \mathbf{v}^*)$ and A, it now follows easily that, defining

$$\mathbf{y}^* = \mathbf{x}^* + \alpha \mathbf{v}^*,$$

we have $\mathbf{y}^* \in Fr(X)$. Equations (5.30) and (5.31) then follow readily from Theorem 5.20. □

We are especially interested in a particular class of lines passing through an element of a convex set, and which are defined as follows.

5.27. Definition. Let $\mathbf{x}^* \in C$, where C is a convex subset of L, and let $\mathbf{v} \in L \setminus \{\mathbf{0}\}$. We shall say that **C is absorbing at x* in the direction v** iff there exists a positive number, κ, such that

$$(\forall \theta \in \mathbf{R})\colon |\theta| < \kappa \Rightarrow \mathbf{x}^* + \theta\mathbf{v} \in C.$$

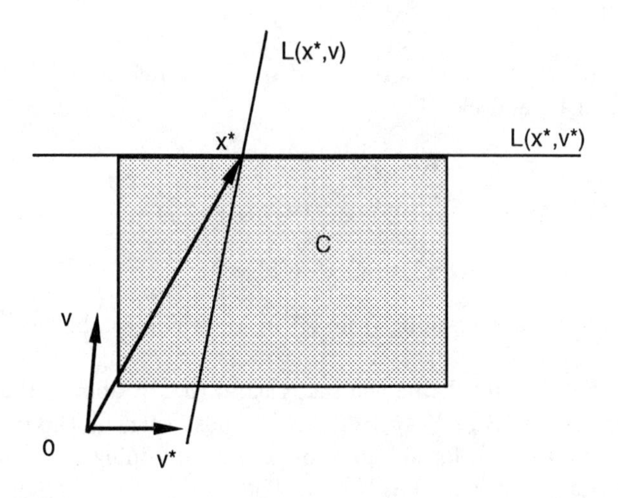

Figure 5.4: C is abosrbing at x* in the direction v*.

Thus, in Figure 5.4, above, the set C is absorbing at $\mathbf{x}^*$ in the direction $\mathbf{v}^*$, whereas C is *not* absorbing at $\mathbf{x}^*$ in the direction $\mathbf{v}$. It is easy to see that, even if $\mathbf{v}$ is a non-zero element of $Sp(C - C)$, the set C may not be absorbing at $\mathbf{x}^*$ in the direction $\mathbf{v}$.[4] We make note of two simple properties involving convex sets and absorbency in the following, the proofs of which will be left as exercises (see Exercise 2 at the end of this section).

5.28. Proposition. *Let C be a convex subset of L, and let $\mathbf{x}^* \in C$. If C is absorbing at $\mathbf{x}^*$ in the direction $\mathbf{v} \in L$, then*

$$\mathbf{v} \in Sp(C - C) = Sp(M - M),$$

where $M = \text{aff}C$.

[4]In fact, in Figure 5.4, $Sp(C - C) = \mathbf{R}^2$. Thus $\mathbf{v}$ in that diagram provides an example of a non-zero element of $Sp(C - C)$ such that C is not absorbing at $\mathbf{x}^*$ in the direction $\mathbf{v}$.

5.29. Proposition. *Let C be a convex subset of L, and let $\mathbf{x}^* \in riC$. Then, for all $\mathbf{v} \in Sp(C - C) \setminus \{\mathbf{0}\}$, C is absorbing at $\mathbf{x}^*$ in the direction $\mathbf{v}$.*

5.30. Proposition. *Let C be a convex subset of L having finite dimension n, and let $\mathbf{x}^* \in L$. Then $\mathbf{x}^* \in riC$ if, and only if, for all $\mathbf{v} \in Sp(C-C)\setminus\{\mathbf{0}\}$, C is absorbing at $\mathbf{x}^*$ in the direction $\mathbf{v}$.*

Proof. The necessity part of this result follows immediately from Proposition 5.29. To prove sufficiency, suppose that for all $\mathbf{v} \in Sp(C - C) \setminus \{\mathbf{0}\}$, C is absorbing at $\mathbf{x}^*$ in the direction $\mathbf{v}$, and let $\mathbf{y}_1, \ldots, \mathbf{y}_n$ be a basis for $T \equiv Sp(C-C)$. Then for each i $(i = 1, \ldots, n)$, there exists $\kappa_i > 0$ such that

$$(\forall \theta \in \mathbf{R}) : |\theta| < \kappa_i \Rightarrow \mathbf{x}^* + \theta \mathbf{y}_i \in C. \tag{5.32}$$

Furthermore, by Proposition 4.50, there exists $\eta > 0$ such that, for all $\mathbf{a} \in \mathbf{R}^n$, we have

$$\sum_{i=1}^{n} |a_i| \leq \eta \cdot \left\| \sum_{i=1}^{n} a_i \mathbf{y}_i \right\|.$$

We then define

$$\epsilon = \min\{\kappa_1/\eta, \kappa_2/\eta, \ldots, \kappa_n/\eta\}.$$

Now suppose

$$\mathbf{z} \in N(\mathbf{x}^*, \epsilon) \cap M,$$

where $M \equiv \mathit{aff}C$. Then, since $\mathbf{z} \in M$, there exists $\mathbf{b} \in \mathbf{R}^n$ such that

$$\mathbf{z} = \mathbf{x}^* + \sum_{i=1}^{n} b_i \mathbf{y}_i;$$

and, since $\mathbf{z} \in N(\mathbf{x}^*, \epsilon)$:

$$\left\| \sum_{i=1}^{n} b_i \mathbf{y}_i \right\| < \epsilon.$$

Thus, for $i \in \{1, \ldots, n\}$ such that $b_i \neq 0$:

$$\left| \left(\frac{b_i}{|b_i|} \right) \sum_{j=1}^{n} |b_j| \right| = \sum_{j=1}^{n} |b_j| \leq \eta \cdot \left\| \sum_{j=1}^{n} b_j \mathbf{y}_j \right\| < \eta \cdot \epsilon \leq \kappa_i;$$

and it follows from (5.32) that:

$$\mathbf{x}^* + \left[\left(\frac{b_i}{|b_i|} \right) \sum_{j=1}^{n} |b_j| \right] \mathbf{y}_i \in C, \text{ for all } i \in I \stackrel{\text{def}}{=} \{k \in \{1, \ldots, n\} \mid b_k \neq 0\}.$$

Consequently, since C is convex:

$$\sum_{i \in I} \left[\left(\frac{|b_i|}{\sum_{j=1}^{n} |b_j|} \right) \left(\mathbf{x}^* + \left[\left(\frac{b_i}{|b_i|} \right) \sum_{j=1}^{n} |b_j| \right] \mathbf{y}_i \right) \right]$$
$$= \mathbf{x}^* + \sum_{i \in I} b_i \mathbf{y}_i = \mathbf{x}^* + \sum_{i=1}^{n} b_i \mathbf{y}_i = \mathbf{z},$$

is an element of C. Therefore, it follows that:

$$N(\mathbf{x}^*, \epsilon) \cap M \subseteq C;$$

and thus $\mathbf{x}^* \in riC$. □

5.31. Definitions. Let C be a convex subset of L, and let $\mathbf{x}^* \in C$. We shall say that $\mathbf{x}^*$ is an **extreme point of C** iff there exists no $\mathbf{v} \in L \setminus \{\mathbf{0}\}$ such that C is absorbing at $\mathbf{x}^*$ in the direction $\mathbf{v}$. The collection of all extreme points of C is called the **profile of C**, and is denoted by '$\ddot{C}$.'

Equivalently, $\mathbf{x}^*$ is an extreme point of C if, and only if, there exist no elements, $\mathbf{x}_1$ and $\mathbf{x}_2$ in C, both of which are distinct from $\mathbf{x}^*$, such that $\mathbf{x}^*$ is a convex combination of $\mathbf{x}_1$ and $\mathbf{x}_2$.[5] In the following discussion, we will develop some material which may be quite helpful to you in terms of developing a feeling for the geometry of relative interior and extreme points.

5.32. Examples/Exercises. Let C be a convex subset of dimension n, and let $M = aff C$. Then if $Y \equiv \{\mathbf{y}_1, \dots, \mathbf{y}_n\}$ is a basis for $M - M$, and $\mathbf{z}$ is any element of M, we know that any element, $\mathbf{x} \in M$, can be uniquely expressed as:

$$\mathbf{x} = \sum_{i=1}^{n} a_i \mathbf{y}_i + \mathbf{z}, \tag{5.33}$$

for some $\mathbf{a} \in \mathbf{R}^n$ (see Exercise 12, at the end of Section 4.6). Because of this uniqueness, we can then define a function, $f\colon M \to \mathbf{R}^n$, by

$$f\left(\sum_{i=1}^{n} a_i \mathbf{y}_i + \mathbf{z}\right) = \mathbf{a} = (a_1, \dots, a_n)'. \tag{5.34}$$

Given any such basis, Y, for $M - M$, and any such $\mathbf{z} \in M$, we shall say that the corresponding function, f, defined as per (5.34), above, is the **pseudo-isomorphism between M and $\mathbf{R}^n$ determined by z and Y.**[6]

Exercise 1. Show that a pseudo isomorphism is an affine function which is one-to-one and onto $\mathbf{R}^n$.

Given any such pseudo-isomorphism, define the set $D \subseteq \mathbf{R}^n$, by

$$D = f(C). \tag{5.35}$$

It then follows at once from Exercise 1, above, and 5.7 that D is a convex set. We can also prove the following.

Exercise 2. Let C be a convex subset of L, let $M \equiv aff C, \mathbf{z} \in M$, and let $Y \equiv \{\mathbf{y}_1, \dots, \mathbf{y}_n\}$ be a basis for $M - M$. We can extend the domain

[5] See Exercise 7, at the end of this section.

[6] Each point in $M - M$ has a unique representation of the form (5.33) with $\mathbf{z} = \mathbf{0}$. If we take $M - M$ as the domain of our function rather than M, then the function f defined by (5.34) [with $\mathbf{z} = \mathbf{0}$] is an isomorphism. See Chapter 4, Section 3, particularly 4.38.

of the pseudo-isomorphism, f, determined by $\mathbf{z}$ and Y in the obvious way; that is, define $f(\mathbf{x})$ for $\mathbf{x} \in M$ as in (5.34), and for $\mathbf{y} \in M - M, \mathbf{y}$ will be of the form

$$\mathbf{y} = \sum_{i=1}^{n} b_i \mathbf{y}_i, \text{ for some } \mathbf{b} \in \mathbf{R}^n;$$

and we define

$$f(\mathbf{y}) = f\left(\sum_{i=1}^{n} b_i \mathbf{y}_i\right) = \mathbf{b}.$$

Defining the set D as in (5.35), above, and given an arbitrary element, $\mathbf{x}^*$, of C, prove the following.

If we define $\mathbf{a}^* = f(\mathbf{x}^*)$*, then* C *is absorbing at* $\mathbf{x}^*$ *in the direction* $\mathbf{v} \in M - M$ *if, and only if,* D *is absorbing at* $\mathbf{a}^*$ *in the direction* $f(\mathbf{v})$.

This exercise is fairly routine to prove if one makes note of the fact that if $\mathbf{v} \in M - M$, then a point on the line through $\mathbf{x}^*$ determined by $\mathbf{v}$ is of the form:

$$\mathbf{x} = \mathbf{x}^* + \theta \mathbf{v};$$

and thus $\mathbf{x}$ has the representation

$$\mathbf{x} = \mathbf{z} + \sum_{i=1}^{n} a_i^* \mathbf{y}_i + \theta\left(\sum_{i=1}^{n} b_i \mathbf{y}_i\right),$$

where

$$\mathbf{v} = \sum_{i=1}^{n} b_i \mathbf{y}_i. \quad \square$$

The following result will often enable us to extend a proof pertaining to a convex subset of $\mathbf{R}^n$ to apply to a convex set of dimension n, but contained in an arbitrary normed real linear space. In Chapter 10 we will further explore the power and applicability of the result.

5.33. Proposition. *Given a convex set,* C*, of dimension* n*, a pseudo-isomorphism,* $f\colon M \to \mathbf{R}^n$*, and* $D = f(C)$*, we have, for* $\mathbf{x}^* \in C$*, and defining* $\mathbf{a}^* = f(\mathbf{x}^*)$*:*

1. *D is a convex set,*
2. *$\mathbf{x}^*$ is an extreme point of C iff $\mathbf{a}^*$ is an extreme point of D.*
3. *$\mathbf{x}^* \in riC$ if, and only if, $\mathbf{a}^* \in int(D)$.*

Proof. Part 1 of our conclusion follows straightforwardly from the definitions, and part 2 is an almost immediate consequence of Exercise 2 of 5.32, above. Details of these proofs will be left as an exercise.

In the proof of part 3, we shall denote spherical neighborhoods in L by '$S(\cdot)$,' while in $\mathbf{R}^n$ such neighborhoods will be denoted by '$N(\cdot)$.' We assume that $Y \equiv \{\mathbf{y}_1, \ldots, \mathbf{y}_n\}$ and $\mathbf{z} \in M$ are the basis for $T \equiv Sp(C - C)$ and the element of $M \equiv \mathit{aff}C$ used to define the pseudo-isomorphism, f.

Let

$$\mathbf{x}^* \in riC \text{ and } \mathbf{a}^* = f(\mathbf{x}^*).$$

Then there exists $\epsilon > 0$ such that

$$S(\mathbf{x}^*, \epsilon) \cap M \subseteq C.$$

Define

$$\alpha = \max\{\|\mathbf{y}_1\|, \|\mathbf{y}_2\|, \ldots, \|\mathbf{y}_n\|, \},$$

and

$$\delta = \epsilon/n\alpha,$$

let

$$\mathbf{b} \in N(\mathbf{a}^*, \delta),$$

and define

$$\mathbf{x} = \sum_{i=1}^n b_i\mathbf{y}_i + \mathbf{z}.$$

Then we note that $\mathbf{x} \in M$, and

$$\|\mathbf{x} - \mathbf{x}^*\| = \left\|\sum_{i=1}^n b_i\mathbf{y}_i + \mathbf{z} - \left(\sum_{i=1}^n a^*{}_i\mathbf{y}_i + \mathbf{z}\right)\right\| = \left\|\sum_{i=1}^n (b_i - a_i^*)\mathbf{y}_i\right\|$$

$$\leq \sum_{i=1}^n |b_i - a_i^*| \cdot \|\mathbf{y}_i\| \leq \alpha \cdot \sum_{i=1}^n |b_i - a_i^*| < \alpha \cdot n\delta = \epsilon;$$

where the last inequality follows from the fact that, for $\mathbf{x} \in \mathbf{R}^n$:[7]

$$\sum_{i=1}^n |x_i| \leq n\|\mathbf{x}\|.$$

Consequently, we see that $\mathbf{x} \in C$, and thus

$$\mathbf{b} = f(\mathbf{x}) \in D.$$

Therefore, $\mathbf{a}^* \in int(D)$.

To prove the converse, suppose $\mathbf{a}^* \in int(D)$. Then there exists $\epsilon > 0$ such that

$$N(\mathbf{a}^*, \epsilon) \subseteq D;$$

[7] See exercise 3, at the end of this section.

and, by Proposition 4.50 there exists a positive number, η, such that, for all $\mathbf{b} \in \mathbf{R}^n$,

$$\sum_{i=1}^{n} |b_i| \leq \eta \cdot \left\| \sum_{i=1}^{n} b_i \mathbf{y}_i \right\|.$$

Define

$$\epsilon^* = \epsilon/\eta,$$

and let

$$\mathbf{x} \in S(\mathbf{x}^*, \epsilon^*) \cap M.$$

Since $\mathbf{x} \in M$, there exists $\mathbf{b} \in \mathbf{R}^n$ such that

$$\mathbf{x} = \sum_{i=1}^{n} b_i \mathbf{y}_i + \mathbf{z};$$

and, since $\mathbf{x} \in S(\mathbf{x}^*, \epsilon^*)$, we have

$$\sum_{i=1}^{n} |b_i - a_i^*| \leq \eta \cdot \left\| \sum_{i=1}^{n} (b_i - a_i^*)\mathbf{y}_i \right\| = \eta \cdot \left\| \sum_{i=1}^{n} b_i \mathbf{y}_i + \mathbf{z} - \left(\sum_{i=1}^{n} a_i^* \mathbf{y}_i + \mathbf{z} \right) \right\|$$

$$= \eta \cdot \|\mathbf{x} - \mathbf{x}^*\| < \eta \cdot \epsilon^* = \epsilon.$$

Thus we see that $\mathbf{b} \in D$, so that $\mathbf{x} = f^{-1}(\mathbf{b}) \in C$. Since $\mathbf{x}$ was an arbitrary element of $S(\mathbf{x}^*, \epsilon^*) \cap M$, it then follows that $\mathbf{x}^* \in riC$. □

Exercises.

1. Show that if X is a non-empty subset of L, and we define $M = \mathit{aff}\, X$, then

$$Sp(X - X) = Sp(M - M).$$

2. Prove (a) Proposition 5.28, and (b) Proposition 5.29.

3. Prove that, for that, for $\mathbf{x} \in \mathbf{R}^n$,

$$\sum_{i=1}^{n} |x_i| \leq n \max_i |x_i| \leq n \|\mathbf{x}\|.$$

4. Prove Parts 1 and 2 of Proposition 5.33.

5. Recall that a set is said to be a **proper** subset of L iff it is neither empty nor equal to L itself. Show that any proper convex subset of L has at least one frontier point. (Hint: try making use of Proposition 5.26.)

6. Prove that a point $\mathbf{x} \in C$ is a non-extreme point of C if, and only if, there exist $\mathbf{y}, \mathbf{z} \in C$ and $\theta \in]0, 1[$ satisfying:

$$\mathbf{y} \neq \mathbf{z} \text{ and } \mathbf{x} = \theta \mathbf{y} + (1 - \theta)\mathbf{z}.$$

7. Let M be an affine subset of an inner product space, L, such that $\dim M = n$. Show that there exists a 'pseudo isomorphism,' $f \colon M \to \mathbf{R}^n$, and a constant $\alpha \geq 0$, such that, for all $\mathbf{x}, \mathbf{y} \in M$:

$$\mathbf{x} \cdot \mathbf{y} = f(\mathbf{x}) \cdot f(\mathbf{y}) + \alpha.$$

(Hint: see exercise 8 at the end of Section 4.5.)

5.4 Minimum Distance and Projection Theorems

In this section our main concern will be with the examination of what is really a minimization problem: given a closed convex subset, C, of a Hilbert space, L, and a point $\mathbf{x} \in L$, we will seek to find a point $\mathbf{y} \in L$ which minimizes the distance between $\mathbf{x}$ and C; that is, we seek $\mathbf{y} \in C$ (if such a point exists) satisfying:

$$(\forall \mathbf{z} \in C): \|\mathbf{y} - \mathbf{x}\| \leq \|\mathbf{z} - \mathbf{x}\|.$$

It turns out that a great deal can be said about the solution to such a problem, and we shall find that the results characterizing the solution are useful in many applications; the first of which will be to the derivation of 'separating hyperplane' results, which will be the topic taken up in the next section of this chapter (although we will there deal only with the case in which $L = \mathbf{R}^n$).

Before beginning our study of these minimum distance problems, however, we need to take a bit closer look at the notion of closed sets; since the closure (or lack thereof) of the set C will play a key role in our minimum distance problem. As was noted in Section 4.4, and will be proved in Chapter 7, a subset, X, of a normed linear space, L, is closed if, and only if, $X = \overline{X}$. In a Banach space, this is in turn equivalent to saying that X is complete; where we define this as follows.

5.34. Definition. Let L be a normed linear space, and let X be a nonempty subset of L. We will say that X is **complete** iff, given any Cauchy sequence, $\langle \mathbf{x}_n \rangle \subseteq X$, there exists $\mathbf{x}^* \in X$ such that $\mathbf{x}_n \to \mathbf{x}^*$.

Recall that a normed linear space, L, is said to be a **Banach space** iff L itself is complete.

5.35. Proposition. *In a Banach space a subset is complete if, and only if, it is closed.*

Proof. Suppose first that X is closed, and let $\langle \mathbf{x}_n \rangle \subseteq X$ be a Cauchy sequence. Then, since L is complete, there exists a vector, $\mathbf{x}^* \in L$ such that $\mathbf{x}_n \to \mathbf{x}^*$. However, it is then clear that $\mathbf{x}^* \in \overline{X}$, and since X is closed (and thus $X = \overline{X}$), it follows that $\mathbf{x}^* \in X$.

Conversely, suppose that X is complete, and let $\mathbf{x}^* \in \overline{X}$. Then, for each positive integer, n, there exists $\mathbf{x}_n \in X$ satisfying:

$$\|\mathbf{x}_n - \mathbf{x}^*\| < 1/n.$$

But it is then apparent that $\langle \mathbf{x}_n \rangle$ is a Cauchy sequence, and, in addition, that $\mathbf{x}_n \to \mathbf{x}^*$. Consequently, since X is complete, it follows that $\mathbf{x}^* \in X$. We conclude, therefore, that:

$$\overline{X} \subseteq X;$$

and, since the opposite inclusion is always true, it follows that $X = \overline{X}$, and thus that X is closed. □

It is an unfortunate fact of life that a linear subspace of a normed linear space is not necessarily closed, as the following example demonstrates.

5.36. Example. Let 'ℓ_1' denote that subset of $\mathbf{R}^\infty$ consisting of all absolutely convergent series; that is, $\mathbf{x} = (x_1, x_2, \dots) \in \ell_1$ iff:

$$\sum_{i=1}^{\infty} |x_i| < \infty,$$

and define the norm '$\|\cdot\|_1$' on ℓ_1 by:

$$\|\mathbf{x}\|_1 = \sum_{i=1}^{\infty} |x_i|.$$

We then define $f\colon \ell_1 \to \ell_1$ by:

$$f(\mathbf{x}) = (x_1, (1/2)x_2, (1/3)x_3, \dots, (1/n)x_n, \dots).$$

Notice that it follows immediately from Example/Exercise 2.44.4 that f is well-defined; that is, if $\mathbf{x} \in \ell_1$, then $f(\mathbf{x}) \in \ell_1$ as well. Moreover, it is easily seen that f is a linear function; so we know from Proposition 4.30 that $S \equiv f(\ell_1)$ is a linear subspace of ℓ_1. However, consider the sequence $\langle \mathbf{x}_n \rangle$ defined by:

$$\begin{aligned}
\mathbf{x}_1 &= (1, 0, 0, \dots),\\
\mathbf{x}_2 &= (1, 1/2, 0, \dots),\\
&\dots\\
\mathbf{x}_n &= (1, 1/2, 1/3, \dots, 1/n, 0, \dots)\\
&\dots;
\end{aligned}$$

that is:

$$x_{in} = \begin{cases} 1/i & \text{for } i = 1, \dots, n,\\ 0 & \text{for } i = n+1, \dots. \end{cases}$$

Clearly $\mathbf{x}_n \in \ell_1$, for $n = 1, 2, \dots$. However, defining:

$$\mathbf{x}^* = (1, 1/4, 1/9, \dots, 1/n^2, \dots),$$

it is easily seen that (See 2.44.2):

$$\lim_{n\to\infty} \|f(\mathbf{x}_n) - \mathbf{x}^*\|_1 = \lim_{n\to\infty} \left[\sum_{i=n+1}^{\infty} (1/i)^2\right] = 0,$$

and thus $f(\mathbf{x}_n) \to \mathbf{x}^*$, so that $\mathbf{x}^* \in \overline{S}$. However, since the harmonic series does not converge (see 2.37.5), and thus is not an element of ℓ_1, we see that $\mathbf{x}^* \notin S$. □

In Chapter 10 we will study the conditions under which a subspace is closed at some length. However, in the next section of this chapter, we will be concerned with the case in which $L = \mathbf{R}^n$. Since $\mathbf{R}^n$ is a Banach space (see Theorem 2.66), one implication of the following result is that any subspace in $\mathbf{R}^n$ is closed.

5.37. Theorem. *In a normed linear space, any finite-dimensional subspace is complete.*

Proof. Denote the subspace by 'S.' Our proof will proceed by induction on the dimension, n, of S.

For $n = 1$, we note that a sequence $\langle \mathbf{x}_m \rangle$ from S can be written in the form:

$$\mathbf{x}_m = \theta_m \mathbf{z},$$

where $\mathbf{z}$ is a non-zero element of S. Since, given any positive integers, m and p, we have:

$$\|\mathbf{x}_m - \mathbf{x}_p\| = |\theta_m - \theta_p| \cdot \|\mathbf{z}\|,$$

it is easily seen that $\langle \mathbf{x}_m \rangle$ is a Cauchy sequence if, and only if, $\langle \theta_m \rangle$ is a Cauchy sequence. As a consequence, it follows from the completeness of $\mathbf{R}$ that S is complete in this case.

Now suppose our result holds for $n = k \geq 1$, and let $n = k + 1$. Letting $Z = \{\mathbf{z}_1, \ldots, \mathbf{z}_{k+1}\}$ be a basis for S, define the subspaces S_j $(j = 1, \ldots, k + 1)$, by:

$$S_j = Sp[Z \setminus \mathbf{z}_j];$$

that is, S_j is the k-dimensional subspace spanned by deleting the vector $\mathbf{z}_j$ from Z. Defining

$$\delta_j = \inf_{\mathbf{x} \in S_j} \|\mathbf{z}_j - \mathbf{x}\|, \text{ for } j = 1, \ldots, k + 1;$$

suppose, by way of obtaining a contradiction, that, for some j we have $\delta_j = 0$. In this case it is clear that there exists a sequence $\langle \mathbf{z}_m \rangle \subseteq S_j$ which converges to $\mathbf{z}_j$. However, from our induction hypothesis, S_j is complete; so that the existence of such a sequence implies $\mathbf{z}_j \in S_j$, contradicting the linear independence of Z. Therefore we must have:

$$\delta_j > 0 \text{ for } j = 1, \ldots, k + 1;$$

and we define the positive number δ by:

$$\delta = \min\{\delta_1, \ldots, \delta_{k+1}\}.$$

Now let $\langle \mathbf{x}_m \rangle$ be a sequence of vectors from S. Then there exists a sequence $\langle \mathbf{a}_m \rangle$ from $\mathbf{R}^{k+1}$ such that:

$$\mathbf{x}_m = \sum_{i=1}^{k+1} a_{im} \mathbf{z}_i \text{ for } m = 1, 2, \ldots.$$

Thus, for any positive integers, m and p, we have, if $a_{jm} \neq a_{jp}$:

$$\begin{aligned} \|\mathbf{x}_m - \mathbf{x}_p\| &= \left\| \sum_{i=1}^{k+1} (a_{im} - a_{ip}) \mathbf{z}_i \right\| \\ &= |a_{jm} - a_{jp}| \cdot \left\| \mathbf{z}_j - \sum_{i \neq j} \left(\frac{a_{im} - a_{ip}}{a_{jp} - a_{jm}} \right) \mathbf{z}_i \right\| \geq |a_{jm} - a_{jp}| \cdot \delta; \end{aligned}$$

and the inequality obviously holds if $a_{jm} = a_{jp}$. Consequently, it follows that if $\langle \mathbf{z}_m \rangle$ is a Cauchy sequence, then $\langle a_{jm} \rangle$ is a Cauchy sequence as well, for each j; in which event it follows from the fact that $\mathbf{R}$ is complete that there exists $\mathbf{a}^* \in \mathbf{R}^{k+1}$ such that $\mathbf{a}_m \to \mathbf{a}^*$. If we now define:

$$\mathbf{x}^* = \sum_{i=1}^{k+1} a_i^* \mathbf{z}_i,$$

we, of course, have $\mathbf{x}^* \in S$. Moreover, from the inequality:

$$\|\mathbf{x}_m - \mathbf{x}^*\| = \left\| \sum_{i=1}^{k+1} (a_{im} - a_i^*) \mathbf{z}_i \right\| \leq \sum_{i=1}^{k+1} |a_{im} - a_i^*| \cdot \|\mathbf{z}_i\|,$$

and the fact that $a_{im} \to a_i^*$, for $i = 1, \ldots, k+1$, it is clear that $\mathbf{x}_m \to \mathbf{x}^*$. □

We can now turn to the derivation of our minimum norm results, the first and most fundamental of which is as follows.

5.38. Theorem. *Let $\mathbf{x}^* \in L$, where L is a Hilbert space, and let C be a nonempty, closed, and convex subset of L. Then there exists a unique point $\mathbf{y} \in C$ satisfying:*

$$(\forall \mathbf{z} \in C) \colon \|\mathbf{y} - \mathbf{x}^*\| \leq \|\mathbf{z} - \mathbf{x}^*\|. \tag{5.36}$$

Moreover, a necessary and sufficient condition that $\mathbf{y}$ be the unique minimizing vector is that:

$$(\forall \mathbf{z} \in C) \colon (\mathbf{y} - \mathbf{x}^*) \cdot \mathbf{z} \geq (\mathbf{y} - \mathbf{x}^*) \cdot \mathbf{y}. \tag{5.37}$$

Proof. If $\mathbf{x}^* \in C$, then the first part of the result is trivial. Suppose, therefore, that $\mathbf{x}^* \notin C$. Then we note that, defining:

$$\delta = \inf_{\mathbf{z} \in C} \|\mathbf{z} - \mathbf{x}^*\|,$$

we must have $\delta > 0$. Moreover, from the definition of an infimum, we see that there exists a sequence $\langle \mathbf{z}_n \rangle \subseteq C$ satisfying:

$$\delta^2 \leq \|\mathbf{z}_n - \mathbf{x}^*\|^2 < \delta^2 + 1/n \quad \text{for } n = 1, 2, \ldots. \tag{5.38}$$

Now suppose $\mathbf{x}_1$ and $\mathbf{x}_2$ are any two elements of C. Then by the Parallelogram Law (Proposition 4.57):

$$\|\mathbf{x}_1 - \mathbf{x}_2\|^2 + \|(\mathbf{x}_1 + \mathbf{x}_2) - 2\mathbf{x}^*\|^2 = 2 \cdot \|\mathbf{x}_1 - \mathbf{x}^*\|^2 + 2 \cdot \|\mathbf{x}_2 - \mathbf{x}^*\|^2,$$

so that:

$$\begin{aligned}\|\mathbf{x}_1 - \mathbf{x}_2\|^2 &= 2\|\mathbf{x}_1 - \mathbf{x}^*\|^2 + 2\|\mathbf{x}_2 - \mathbf{x}^*\|^2 - 4\left\|\mathbf{x}^* - (1/2)(\mathbf{x}_1 + \mathbf{x}_2)\right\|^2 \\ &\leq 2\|\mathbf{x}_1 - \mathbf{x}^*\|^2 + 2\|\mathbf{x}_2 - \mathbf{x}^*\|^2 - 4\delta^2, \quad (5.39)\end{aligned}$$

where the last inequality is by the definition of δ and the fact that C is a convex set.

Now let $\epsilon > 0$ be given. Then by (5.38) there exists an integer, p, such that, for all $n \geq p$:

$$\|\mathbf{z}_n - \mathbf{x}^*\|^2 < \delta^2 + \epsilon/4. \quad (5.40)$$

But then, for $m, n \geq p$, we have, from (5.39) and (5.40):

$$\begin{aligned}\|\mathbf{z}_n - \mathbf{z}_m\|^2 &\leq 2\|\mathbf{z}_n - \mathbf{x}^*\|^2 + 2\|\mathbf{z}_m - \mathbf{x}^*\|^2 - 4\delta^2 \\ &< 2(\delta^2 + \epsilon/4) + 2(\delta^2 + \epsilon/4) - 4\delta^2 = \epsilon.\end{aligned}$$

Thus we see that $\langle \mathbf{z}_n \rangle$ is a Cauchy sequence. Since L is a Hilbert space, and C is a closed set, it then follows that there exists $\mathbf{y} \in C$ such that:

$$\mathbf{z}_n \to \mathbf{y} \text{ and } \|\mathbf{y} - \mathbf{x}^*\| = \delta.$$

To show that $\mathbf{y}$ is unique, suppose $\overline{\mathbf{y}} \in C$ is such that:

$$\|\overline{\mathbf{y}} - \mathbf{x}^*\| = \delta = \|\mathbf{y} - \mathbf{x}^*\|.$$

Then by inequality (5.39):

$$\|\mathbf{y} - \overline{\mathbf{y}}\|^2 \leq 2\|\mathbf{y} - \mathbf{x}^*\|^2 + 2\|\overline{\mathbf{y}} - \mathbf{x}^*\|^2 - 4\delta^2 = 0.$$

Therefore $\|\mathbf{y} - \overline{\mathbf{y}}\|^2 = 0$, and it follows that:

$$\mathbf{y} = \overline{\mathbf{y}}.$$

In order to prove that inequality (5.37) is a necessary and sufficient condition that $\mathbf{y}$ satisfy (5.36), suppose first that $\mathbf{y}$ satisfies (5.36), let $\mathbf{z} \in C$ be arbitrary, and, for $t \in [0, 1]$, define:

$$\mathbf{x}_t = t\mathbf{z} + (1 - t)\mathbf{y}.$$

Since C is convex, it follows from (5.36) that the function $f(t)$ defined by:

$$\begin{aligned}f(t) &= \|\mathbf{x}_t - \mathbf{x}^*\|^2 = \|t(\mathbf{z} - \mathbf{x}^*) + (1 - t)(\mathbf{y} - \mathbf{x}^*)\|^2 \\ &= t^2\|\mathbf{z} - \mathbf{x}^*\|^2 + (1 - t)^2\|\mathbf{y} - \mathbf{x}^*\|^2 + 2t(1 - t)(\mathbf{z} - \mathbf{x}^*) \cdot (\mathbf{y} - \mathbf{x}^*),\end{aligned}$$

is minimized at $t = 0$. Consequently, since f is differentiable, we see that we must have $f'(0) \geq 0$. Differentiating f and evaluating the derivative at zero, we obtain:

$$\begin{aligned} 0 \leq f'(0) &= -2(\mathbf{y} - \mathbf{x}^*) \cdot (\mathbf{y} - \mathbf{x}^*) + 2(\mathbf{y} - \mathbf{x}^*) \cdot (\mathbf{z} - \mathbf{x}^*) \\ &= 2(\mathbf{y} - \mathbf{x}^*) \cdot [(\mathbf{z} - \mathbf{x}^*) - (\mathbf{y} - \mathbf{x}^*)] = 2(\mathbf{y} - \mathbf{x}^*) \cdot (\mathbf{z} - \mathbf{y}); \end{aligned}$$

so that:

$$(\mathbf{y} - \mathbf{x}^*) \cdot \mathbf{z} \geq (\mathbf{y} - \mathbf{x}^*) \cdot \mathbf{y}.$$

Conversely, suppose $\mathbf{y} \in C$ satisfies (5.37), and let $\mathbf{z} \in C \setminus \{\mathbf{y}\}$ be arbitrary. Then:

$$\begin{aligned} \|\mathbf{z} - \mathbf{x}^*\|^2 &= \|\mathbf{z} - \mathbf{y} + \mathbf{y} - \mathbf{x}^*\|^2 \\ &= \|\mathbf{z} - \mathbf{y}\|^2 + \|\mathbf{y} - \mathbf{x}^*\|^2 + 2(\mathbf{y} - \mathbf{x}^*) \cdot (\mathbf{z} - \mathbf{y}) > \|\mathbf{y} - \mathbf{x}^*\|^2, \end{aligned}$$

where the last inequality is by (5.37) and the fact that $\mathbf{y} \neq \mathbf{z}$. □

A careful study of the last portion of the proof of the above theorem also yields the following (details of the proof are left to the reader).

5.39. Corollary. *Let $\mathbf{x}^*$ be an element of the real inner product space, L, and let C be a nonempty subset of L. If a point $\mathbf{y} \in C$ satisfies inequality (5.37), then $\mathbf{y}$ is the unique solution of (5.36).*

It is important to carefully distinguish between what is being stated in the above corollary and the statement of the theorem which it follows. The corollary tells us that if we are working in an inner product space, and can find a point $\mathbf{y} \in C$ satisfying (5.37), then it is the unique solution of (5.36). What it does not tell us is (a) whether such a $\mathbf{y}$ exists, or (b) whether a solution of (5.36) necessarily satisfies (5.37). The theorem answers both (a) and (b) in the affirmative, but with a cost of some additional assumptions; namely that L is complete (and thus is a Hilbert space), and that the set C is both closed and convex. However, a second review of the last two paragraphs of the proof of the theorem will confirm that we need the set C to be closed and L to be complete only to be sure of the *existence* of a solution to (5.36); that is, you should easily be able to verify the following result.

5.40. Corollary. *Let $\mathbf{x}^*$ be an element of the real inner product space, L, and let C be a nonempty convex subset of L. Then a point $\mathbf{y} \in C$ satisfies inequality (5.37) if, and only if, $\mathbf{y}$ is the unique solution of (5.36).*

While the above three results cover the applications in which we will be most interested in this chapter, we will at a later stage be very interested in a solution of (5.36) for the case in which the set C is *affine*, and not just convex. Since it will be quite easy to prove the pertinent result for this case now, let's do so.

5.41. Theorem. *Let $\mathbf{x}^*$ be an element of the real inner product space, L, let M be a nonempty affine subset of L, and let S be the subspace of L parallel to M; that is, let $S = M - M$. Then a point $\mathbf{y} \in M$ is the unique solution of (5.36) if, and only if, the vector $\mathbf{y} - \mathbf{x}^*$ is an element of $S^{\perp}$.*

Proof. Suppose first that $\mathbf{y} \in M$ satisfies (5.36), and let $\mathbf{x}$ be an arbitrary element of S. Then there exists an element, $\mathbf{z} \in M$ such that:

$$\mathbf{x} = \mathbf{z} - \mathbf{y}$$

(see Corollary 4.96); and by Corollary 5.40, we have:

$$(\mathbf{y} - \mathbf{x}^*) \cdot \mathbf{x} = (\mathbf{y} - \mathbf{x}^*) \cdot (\mathbf{z} - \mathbf{y}) \geq 0.$$

However, since M is affine, the point $\mathbf{z}'$ defined by:

$$\mathbf{z}' = 2\mathbf{y} + (1 - 2)\mathbf{z} = 2\mathbf{y} - \mathbf{z},$$

is also an element of M; and, thus, using (5.37) a second time:

$$(\mathbf{y} - \mathbf{x}^*) \cdot (\mathbf{z}' - \mathbf{y}) = (\mathbf{y} - \mathbf{x}^*) \cdot (\mathbf{y} - \mathbf{z}) = -(\mathbf{y} - \mathbf{x}^*) \cdot \mathbf{x} \geq 0.$$

Thus we see that:

$$(\mathbf{y} - \mathbf{x}^*) \in S^{\perp}. \tag{5.41}$$

Conversely, if $\mathbf{y} \in M$ satisfies (5.41), then it follows from Corollary 5.39 and the fact that $S = M - \mathbf{y}$ (see 4.96) that $\mathbf{y}$ is the unique solution of (5.36) □

Notice also that, as an immediate application of Theorem 5.38, we can conclude that if the hypotheses of Theorem 5.41 are strengthened to require that L be a Hilbert space, and that M be closed (as well as being nonempty and affine), then there exists a unique $\mathbf{y} \in M$ satisfying (5.36).

Our final result of this section is concerned with operations involving relative interiors. It is the only such result we will deal with in this book; however, Rockafellar [1976] has a very extensive treatment of this sort of operation for $\mathbf{R}^n$.[8]

5.42. Proposition. *Suppose A and B are convex subsets of a real normed linear space, L; suppose both A and B have nonempty relative interiors, and suppose further that $ri(A - B) \neq \emptyset$. Then:*

$$ri(A - B) = riA - riB.$$

[8]It should perhaps be noted that this chapter once contained a section entitled 'Operations Involving Relative Interiors, Convex Hulls, and Affine Hulls.' However, I deleted the section in order to reduce the excessive length of the book a bit.

Proof. Suppose first that $\widehat{\mathbf{z}} \in ri(A - B)$; and, using the assumption that both A and B have non-empty interiors, let $\mathbf{x}^\dagger \in riA$ and $\mathbf{y}^\dagger \in riB$. If we define:

$$\mathbf{z}^\dagger = \mathbf{x}^\dagger - \mathbf{y}^\dagger,$$

it then follows from Proposition 5.23 that there exists $\mu > 1$ such that:

$$\mathbf{z}^* \stackrel{\text{def}}{=} (1-\mu)\mathbf{z}^\dagger + \mu\widehat{\mathbf{z}} \in A - B. \tag{5.42}$$

Thus there exists $\mathbf{x}^* \in A$ and $\mathbf{y}^* \in B$ such that:

$$\mathbf{z}^* = \mathbf{x}^* - \mathbf{y}^*.$$

But then, if we define:

$$\mathbf{x}' = \left(\frac{1}{\mu}\right)\mathbf{x}^* + \left(1 - \frac{1}{\mu}\right)\mathbf{x}^\dagger$$

and:

$$\mathbf{y}' = \left(\frac{1}{\mu}\right)\mathbf{y}^* + \left(1 - \frac{1}{\mu}\right)\mathbf{y}^\dagger,$$

it follows from Theorem 5.20 that:

$$\mathbf{x}' \in riA \quad \text{and} \quad \mathbf{y}' \in riB,$$

and, using (5.42), we have:

$$\begin{aligned}\mathbf{x}' - \mathbf{y}' &= \left(\frac{1}{\mu}\right)\mathbf{x}^* + \left(1 - \frac{1}{\mu}\right)\mathbf{x}^\dagger - \left(\frac{1}{\mu}\right)\mathbf{y}^* - \left(1 - \frac{1}{\mu}\right)\mathbf{y}^\dagger \\ &= \left(\frac{1}{\mu}\right)\mathbf{z}^* + \left(1 - \frac{1}{\mu}\right)\mathbf{z}^\dagger = \widehat{\mathbf{z}}.\end{aligned}$$

Therefore, $\widehat{\mathbf{z}} \in riA - riB$, and we conclude that:

$$ri(A - B) \subseteq riA - riB.$$

Conversely, suppose:

$$\mathbf{z}^* \stackrel{\text{def}}{=} \mathbf{x}^* - \mathbf{y}^* \in riA - riB;$$

and, using the assumption that $ri(A - B) \neq \emptyset$, let $\overline{\mathbf{z}} \in ri(A - B)$. Since $ri(A - B) \subseteq A - B$, there exist $\overline{\mathbf{x}} \in A$ and $\overline{\mathbf{y}} \in B$ such that:

$$\overline{\mathbf{z}} = \overline{\mathbf{x}} - \overline{\mathbf{y}};$$

and, by Proposition 5.23, there exists $\mu > 1$ such that:

$$(1-\mu)\overline{\mathbf{x}} + \mu\mathbf{x}^* \in A,$$

and

$$(1-\mu)\overline{\mathbf{y}} + \mu\mathbf{y}^* \in B.$$

But then:

$$(1-\mu)\overline{\mathbf{x}} + \mu\mathbf{x}^* - [(1-\mu)\overline{\mathbf{y}} + \mu\mathbf{y}^*]$$
$$= (1-\mu)(\overline{\mathbf{x}} - \overline{\mathbf{y}}) + \mu(\mathbf{x}^* - \mathbf{y}^*) = (1-\mu)\overline{\mathbf{z}} + \mu\mathbf{z}^* \in A - B,$$

and it follows from Proposition 5.24 that $\mathbf{z}^* \in ri(A-B)$. We then conclude that:

$$riA - riB \subseteq ri(A-B). \quad \square$$

Recall that it follows from Theorem 5.19 that if the dimensions of A and B are both finite, then we can dispense with the explicit requirement in Proposition 5.42 that the three sets all have nonempty relative interiors. The modification for $\mathbf{R}^n$ is sufficiently important that we will formally state it, as follows (the proof is immediate).

5.43. Corollary. *If A and B are both nonempty convex subsets of $\mathbf{R}^n$, then:*

$$ri(A-B) = riA - riB.$$

Exercises.

1. Complete the proof of (a) Corollary 5.39 and (b) Corollary 5.40.

2. Suppose that L_1 and L_2 are real normed linear spaces, $f\colon L_1 \to L_2$ is a continuous linear function, and that $\mathbf{z} \in f(L_1)$. Show that, if we define $M \subseteq L_1$ by:

$$M = f^{-1}(\{\mathbf{z}\}) = \{\mathbf{x} \in L_1 \mid f(\mathbf{x}) = \mathbf{z}\},$$

then: (a) M is affine, and (b) M is closed.

3. Prove the **Projection Theorem**: If L is a (real) Hilbert space, and S is a closed linear subspace of L, then, given any $\mathbf{x} \in L$, there exist unique points $\mathbf{y} \in S$ and $\mathbf{z} \in S^{\perp}$ such that $\mathbf{x} = \mathbf{y} + \mathbf{z}$.

5.5 Basic Separation Theorems in $\mathbf{R}^n$

In this section we shall return to the context of n-dimensional Euclidean space. Many of the definitions and propositions to follow can be generalized to arbitrary real linear spaces, but we will postpone the consideration of such generalizations until Chapter 10.

5.44. Definitions. Let $\mathbf{a} \in \mathbf{R}^n \setminus \{\mathbf{0}\}$, and $b \in \mathbf{R}$. We define the:

1. **hyperplane determined by (a, b)**, $H_{(\mathbf{a},b)}$, by:

$$H_{(\mathbf{a},b)} = \{\mathbf{x} \in \mathbf{R}^n \mid \mathbf{a} \cdot \mathbf{x} = b\},$$

2. **(closed) upper** [respectively, **lower**] **half-space determined by (a, b)**, $H^{\geq}_{(\mathbf{a},b)}$, by

$$H^{\geq}_{(\mathbf{a},b)} = \{\mathbf{x} \in \mathbf{R}^n \mid \mathbf{a} \cdot \mathbf{x} \geq b\}$$

[respectively, $H^{\leq}_{(\mathbf{a},b)} = \{\mathbf{x} \in \mathbf{R}^n \mid \mathbf{a} \cdot \mathbf{x} \leq b\}$],

3. **open upper** [respectively, **lower**] **half-space determined by (a, b)**, $H^{>}_{(\mathbf{a},b)}$, by

$$H^{>}_{(\mathbf{a},b)} = \{\mathbf{x} \in \mathbf{R}^n \mid \mathbf{a} \cdot \mathbf{x} > b\}$$

[respectively, $H^{<}_{(\mathbf{a},b)} = \{\mathbf{x} \in \mathbf{R}^n \mid \mathbf{a} \cdot \mathbf{x} < b\}$].

The proof of the following result is quite easy, and will be left as an exercise.

5.45. Proposition. *If* $\mathbf{a} \in \mathbf{R}^n \setminus \{\mathbf{0}\}$, *and* $b \in \mathbf{R}$, *then the sets,* $H_{(\mathbf{a},b)}$, $H^{\geq}_{(\mathbf{a},b)}$, $H^{\leq}_{(\mathbf{a},b)}$, $H^{>}_{(\mathbf{a},b)}$, *and* $H^{<}_{(\mathbf{a},b)}$ *are all convex.*

5.46. Definition. A subset, C, of $\mathbf{R}^n$ is said to be a **polyhedral convex set** iff C can be expressed as the intersection of a finite number of closed half-spaces; that is, iff there exist a positive integer, m, and $\mathbf{a}_i \in \mathbf{R}^n \setminus \{\mathbf{0}\}$ and $b_i \in \mathbf{R}$, for $i = 1, \ldots, m$, such that:[9]

$$C = \{\mathbf{x} \in \mathbf{R}^n \mid \mathbf{a}_i \cdot \mathbf{x} \geq b_i \quad \text{for } i = 1, \ldots, m\}. \tag{5.43}$$

Notice that inequality (5.43) can equivalently be expressed as:

$$C = \{\mathbf{x} \in \mathbf{R}^n \mid \mathbf{A}\mathbf{x} \geq \mathbf{b}\},$$

where the $m \times n$ matrix $\mathbf{A}$ and the m-vector $\mathbf{b}$ are defined in the obvious way.

5.47. Definition. A set which is the convex hull of a finite set of points from $\mathbf{R}^n$ is called a **polytope**. If $X = \langle\{\mathbf{x}_1, \ldots, \mathbf{x}_m\}\rangle$, and $\{\mathbf{x}_1, \ldots, \mathbf{x}_m\}$ is an affinely independent set, then X is called a [$(m-1)$-dimensional] **simplex** (with vertices $\{\mathbf{x}_1, \ldots, \mathbf{x}_m\}$).

Example. $\Delta_n = \{\mathbf{p} \in \mathbf{R}^n_+ \mid \sum_{i=1}^n p_i = 1\}$ is an $(n-1)$-dimensional simplex with vertices $\{\mathbf{e}_1, \ldots, \mathbf{e}_n\}$.

---

[9] Notice that there is no loss in generality in supposing that all of these half-spaces are closed *upper* half-spaces.

5.48. Definitions. Let C_1 and C_2 be non-empty convex subsets of $\mathbf{R}^n$, let $\mathbf{a} \in \mathbf{R}^n \setminus \{\mathbf{0}\}, b \in \mathbf{R}$, and let H be the hyperplane determined by $(\mathbf{a}, b)$. Then H is said to:

1. **separate $\mathbf{C_1}$ and $\mathbf{C_2}$** iff C_1 is contained in one of the closed half-spaces determined by H, and C_2 lies in the opposite closed half-space.
2. **properly separate $\mathbf{C_1}$ and $\mathbf{C_2}$** iff H separates C_1 and C_2, and at least one of the C_i is not contained in H.
3. **strictly separate $\mathbf{C_1}$ and $\mathbf{C_2}$** iff H separates C_1 and C_2, and $C_i \cap H = \emptyset$, for $i = 1, 2$.
4. **strongly separate $\mathbf{C_1}$ and $\mathbf{C_2}$** iff either:

$$\inf_{\mathbf{x} \in C_1} \mathbf{a} \cdot \mathbf{x} > b > \sup_{\mathbf{x} \in C_2} \mathbf{a} \cdot \mathbf{x},$$

or

$$\sup_{\mathbf{x} \in C_1} \mathbf{a} \cdot \mathbf{x} < b < \inf_{\mathbf{x} \in C_2} \mathbf{a} \cdot \mathbf{x}.$$

At first glance, it probably appears that strict separation and strong separation are logically equivalent notions. However, the following example shows that two sets may be strictly separated by a hyperplane, H, in a situation where there exists no hyperplane which strongly separates the two sets.

5.49. Example. Let C_1 and C_2 be the subsets of $\mathbf{R}^2$ defined by

$$C_1 = \{(x, y) \in \mathbf{R}^2 \mid x < 0 \ \& \ y \geq -1/x\},$$

and

$$C_2 = \{(x, y) \in \mathbf{R}^2 \mid x > 0 \ \& \ y \geq 1/x\};$$

and consider the hyperplane, H, defined by

$$H = \{(x, y) \in \mathbf{R}^2 \mid \mathbf{e}_1 \cdot (x, y) = x = 0\}$$

(that is, H is the vertical axis). Then H strictly separates C_1 and C_2; and, in fact, is the *only* hyperplane in $\mathbf{R}^2$ which separates the two sets. However, H does *not* strongly separate C_1 and C_2, since

$$\sup_{(x,y) \in C_1} \mathbf{e}_1 \cdot (x, y) = 0 = \inf_{(x,y) \in C_2} \mathbf{e}_1 \cdot (x, y). \quad \square$$

We will defer a proof of the following result until Chapter 8 (see 8.32).

5.50. Lemma. *If C is a convex and non-empty subset of $\mathbf{R}^n$, if $\mathbf{0} \notin riC$, and if we define $M = \text{aff}C$, then there exists $\mathbf{a} \in M \setminus \{\mathbf{0}\}$, satisfying:*

$$\inf_{\mathbf{x} \in \overline{C}} \mathbf{a} \cdot \mathbf{x} \geq 0 \text{ and } (\forall \mathbf{x} \in riC) \colon \mathbf{a} \cdot \mathbf{x} > 0.$$

5.51. Theorem. *Let C and D be convex and non-empty subsets of $\mathbf{R}^n$. Then there exists a hyperplane, H, properly separating C and D if, and only if, $riC \cap riD = \emptyset$.*

Proof.
1. Suppose $riC \cap riD = \emptyset$, and define the set B by

$$B = C - D.$$

It follows from Proposition 5.8 that B is convex, and, since $riC \cap riD = \emptyset$, it follows from Corollary 5.43 that $\mathbf{0} \notin riB$. Since $riB \subseteq B$, it then follows from Lemma 5.50 that, defining $M = \mathit{aff}\, B$, there exists $\mathbf{a} \in M \setminus \{\mathbf{0}\}$ such that:

$$(\forall \mathbf{z} \in riB)\colon \mathbf{a} \cdot \mathbf{z} > 0, \tag{5.44}$$

and

$$(\forall \mathbf{z} \in B)\colon \mathbf{a} \cdot \mathbf{z} \geq 0. \tag{5.45}$$

Now, let $\mathbf{x} \in C$ and $\mathbf{y} \in D$ be arbitrary. Then it follows at once from inequality (5.45) that:

$$\mathbf{a} \cdot (\mathbf{x} - \mathbf{y}) \geq 0,$$

and thus that

$$\mathbf{a} \cdot \mathbf{x} \geq \mathbf{a} \cdot \mathbf{y}.$$

Consequently, we have from Proposition 1.15 that, defining

$$b = \inf\{\mathbf{a} \cdot \mathbf{x} \mid \mathbf{x} \in C\}$$

(b is finite, and),

$$\inf_{\mathbf{x} \in C} \mathbf{a} \cdot \mathbf{x} \geq b \geq \sup_{\mathbf{y} \in D} \mathbf{a} \cdot \mathbf{y}.$$

Thus we see that the hyperplane, H, determined by $(\mathbf{a}, b)$, separates C and D. Moreover, from (5.44) and Corollary 5.43, we see that if $\mathbf{x}^* \in riC$ and $\mathbf{y}^* \in riD$, then

$$\mathbf{a} \cdot \mathbf{x}^* > \mathbf{a} \cdot \mathbf{y}^*;$$

and thus either $\mathbf{a} \cdot \mathbf{x}^* > b$ or $\mathbf{a} \cdot \mathbf{y}^* < b$; and so we see that H cannot contain both C and D. It follows that H properly separates C and D.

2. Now suppose the hyperplane $H_{(\mathbf{a},b)}$ properly separates C and D; and suppose for the sake of convenience (re-labeling if necessary) that

$$\inf_{\mathbf{x} \in C} \mathbf{a} \cdot \mathbf{x} \geq b \geq \sup_{\mathbf{y} \in D} \mathbf{a} \cdot \mathbf{y}, \tag{5.46}$$

and

$$(\exists \mathbf{x}^* \in C): \mathbf{a} \cdot \mathbf{x}^* > b. \tag{5.47}$$

Let $\mathbf{x} \in riC$. Then, by Proposition 5.23 there exists $\theta > 1$ such that

$$\theta \mathbf{x} + (1 - \theta)\mathbf{x}^* \in C$$

[where $\mathbf{x}^*$ is from (5.47)], and thus by (5.46),

$$\theta \mathbf{a} \cdot \mathbf{x} + (1 - \theta)\mathbf{a} \cdot \mathbf{x}^* \geq b.$$

Therefore, using (5.47) and the fact that $\theta > 1$, we have:

$$\mathbf{a} \cdot \mathbf{x} \geq \left(\frac{\theta - 1}{\theta}\right)\mathbf{a} \cdot \mathbf{x}^* + \left(\frac{1}{\theta}\right)b > \left(\frac{\theta - 1}{\theta}\right)b + \left(\frac{1}{\theta}\right)b = b, \tag{5.48}$$

and we conclude that:

$$(\forall \mathbf{x} \in riC): \mathbf{a} \cdot \mathbf{x} > b. \tag{5.49}$$

On the other hand, since $riD \subseteq D$, it follows from (5.46) that

$$(\forall \mathbf{y} \in riD): \mathbf{a} \cdot \mathbf{y} \leq b;$$

and, combining this with (5.49), we conclude that

$$riC \cap riD = \emptyset. \quad \square$$

Our next result, the proof of which will be left as an exercise, is an easy consequence of Theorem 5.51. In it we will use the following notation.

Notation. In the remainder of this chapter, we shall denote the unit ball in $\mathbf{R}^n$ by 'B_n;' that is,

$$B_n = \{\mathbf{x} \in \mathbf{R}^n \mid \|\mathbf{x}\| = 1\}.$$

5.52. Corollary. *Let C be a convex subset of $\mathbf{R}^n$, other that $\mathbf{R}^n$ itself. Then there exists $\mathbf{a} \in B_n$, and $b \in \mathbf{R}$, such that C is contained in the closed upper half-space determined by $(\mathbf{a}, b)$. Moreover, we can choose b so that $H_{(\mathbf{a},b)} \cap C = \emptyset$.*

The following result is an easy consequence of Theorem 5.38. Its proof will be left as an exercise. See Figure 5.5, on the next page.

5.53. Lemma. *Suppose C is a closed, convex, and non-empty subset of $\mathbf{R}^n$, and that $\mathbf{0} \notin C$. Then there exists $\mathbf{a} \in C$ such that:*

$$\inf_{\mathbf{x} \in C} \mathbf{a} \cdot \mathbf{x} = \mathbf{a} \cdot \mathbf{a} > 0.$$

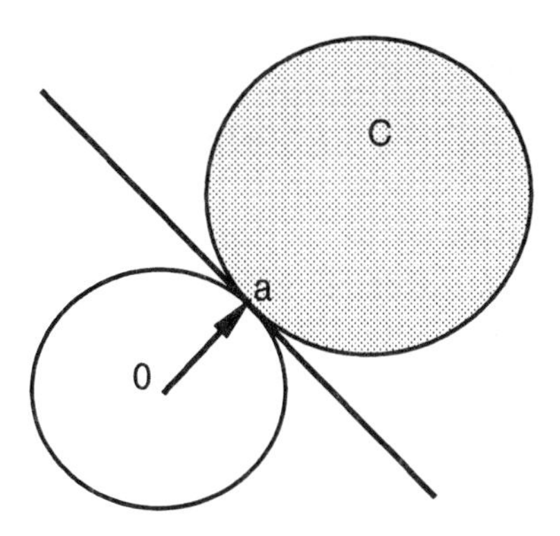

Figure 5.5: Separating a Convex Set from the Origin.

5.54. Theorem. *Let C and D be non-empty convex subsets of $\mathbf{R}^n$, and suppose that*

$$\overline{C} \cap \overline{D} = \emptyset.$$

If either C or D (or both) is bounded, then there exist $\mathbf{a} \in (\overline{C} - \overline{D}) \setminus \{\mathbf{0}\}$ and $b \in \mathbf{R}$ such that

$$\inf_{\mathbf{x} \in C} \mathbf{a} \cdot \mathbf{x} > b > \sup_{\mathbf{y} \in D} \mathbf{a} \cdot \mathbf{y}$$

(that is, $H_{(\mathbf{a},b)}$ strongly separates C and D).

Proof. Defining B by

$$B = \overline{C} - \overline{D},$$

we see that B is closed,[10] convex, and does not contain the origin. Consequently it follows from Lemma 5.53 that there exists $\mathbf{a} \in B$ and $b^* \in \mathbf{R}_{++}$ such that ($\mathbf{a} \neq \mathbf{0}$ and):

$$\inf_{\mathbf{x} \in B} \mathbf{a} \cdot \mathbf{x} = \mathbf{a} \cdot \mathbf{a} \stackrel{\text{def}}{=} b^* > 0.$$

It is then easy to show (exercise) that

$$\inf_{\mathbf{x} \in C} \mathbf{a} \cdot \mathbf{x} \geq b^* + \sup_{\mathbf{y} \in D} \mathbf{a} \cdot \mathbf{y}; \tag{5.50}$$

and consequently, if we define

$$b = \sup_{\mathbf{y} \in D} \mathbf{a} \cdot \mathbf{y} + \frac{b^*}{2},$$

[10]The fact that B is closed will follow as a special case of a result which we will prove in Chapter 10.

we have from (5.50) that

$$\inf_{\mathbf{x}\in C} \mathbf{a}\cdot\mathbf{x} > b > \sup_{\mathbf{y}\in D} \mathbf{a}\cdot\mathbf{y}. \quad \square$$

5.55. Examples/Exercises.

1. The purpose of this example is to show that if C and D satisfy all the hypotheses of 5.54 *except* that neither is bounded, then there may not exist a hyperplane strictly separating C and D, let alone a hyperplane which *strongly* separates the two sets.

Let

$$C = \{(x,y) \in \mathbf{R}^2 \mid x > 0 \,\&\, y \geq 1/x\} = \{(x,y) \in \mathbf{R}^2_+ \mid xy \geq 1\},$$

and

$$D = \{(x,y) \in \mathbf{R}^2 \mid x = 0\}.$$

We will show that D itself is the only hyperplane (line in this case) separating C and D; and, of course, D does not strictly separate C and D.

Accordingly, suppose $\mathbf{a} \in \mathbf{R}^2_+ \setminus \{\mathbf{0}\}$ and $b \in \mathbf{R}$ are such that

$$\inf_{(x,y)\in C} \mathbf{a}\cdot(x,y) \geq b \geq \sup_{(x',y')\in D} \mathbf{a}\cdot(x',y'). \tag{5.51}$$

Since the point $(1,1)$ is an element of C, and we have:

$$(\forall y \in \mathbf{R})\colon (0,y) \in D,$$

it follows from (5.51) that

$$(\forall y \in \mathbf{R})\colon \mathbf{a}\cdot(0,y) = a_2 y \leq \mathbf{a}\cdot(1,1) = a_1 + a_2. \tag{5.52}$$

However, it is obvious that (5.52) can hold only if

$$a_2 = 0; \tag{5.53}$$

and if $\mathbf{a}$ satisfies (5.51) and (5.53), and if $a_1 \neq 0$ (as must be the case if $\mathbf{a} \neq \mathbf{0}$), then

$$D = \{(x,y) \in \mathbf{R}^2 \mid x = 0\} = H_{(\mathbf{a},0)}. \tag{5.54}$$

Now, we have from (5.51) and (5.54) that $b \geq 0$, and thus from (5.51),

$$\inf_{(x,y)\in C} \mathbf{a}\cdot(x,y) = \inf_{(x,y)\in C} a_1 x \geq 0.$$

Since C contains the points $(n, 1/n)$ and $(1/n, n)$ for each positive integer n, it then follows easily that

$$a_1 > 0 \text{ and } b = 0.$$

Thus, $D = H_{(\mathbf{a},0)}$ is the only hyperplane separating C and D.

2. **Theorem of the Alternative.** Let $\mathbf{A}$ be a real $m \times n$ matrix, and let $\mathbf{b} \in \mathbf{R}^m$. We will show that *exactly one* of the following conditions holds. Either (a) there exists $\mathbf{x} \in \mathbf{R}^n_+$ such that:

$$\mathbf{A}\mathbf{x} = \mathbf{b},$$

or (b) there exists $\mathbf{y} \in \mathbf{R}^m$ such that:

$$\mathbf{y}'\mathbf{A} \geq \mathbf{0}' \;\;\&\;\; \mathbf{b} \cdot \mathbf{y} < 0.$$

I will leave as an exercise the proof that not both of conditions (a) and (b) can hold. To prove that at least one of the two conditions must hold, define the set C by:

$$C = \{\mathbf{u} \in \mathbf{R}^m \mid (\exists \mathbf{x} \in \mathbf{R}^n_+) \colon \mathbf{u} = \mathbf{A}\mathbf{x}\}.$$

It is easily seen that C is a convex set, and it can be shown that C is closed as well. Thus it follows from Theorem 5.54 that there exists $\mathbf{y} \in \mathbf{R}^m$ satisfying:

$$\mathbf{y} \cdot \mathbf{b} < \inf_{\mathbf{x} \in \mathbf{R}^n_+} \mathbf{y}'\mathbf{A}\mathbf{x}.$$

It is then easily shown that we must have $\mathbf{b} \cdot \mathbf{y} < 0$ and:

$$(\forall \mathbf{u} \in C) \colon \mathbf{y} \cdot \mathbf{u} \geq 0.$$

Thus, in particular, for the jth unit coordinate vector, $\mathbf{e}_j$, we must have:

$$0 \leq \mathbf{y}'\mathbf{A}\mathbf{e}_j = \mathbf{y} \cdot \mathbf{a}_{\cdot j} \quad \text{for } j = 1, \ldots, n,$$

where '$\mathbf{a}_{\cdot j}$' denotes the jth column of $\mathbf{A}$. □

5.56. Theorem. *A closed, convex, proper subset of* $\mathbf{R}^n$ *is the intersection of the closed half-spaces which contain it.*

Proof. Let C be a closed, convex, proper subset of $\mathbf{R}^n$, and define

$$\mathcal{H} = \{H \subseteq \mathbf{R}^n \mid H \text{ is a closed half-space} \;\&\; C \subseteq H\},$$

and

$$X = \bigcap_{H \in \mathcal{H}} H.$$

It follows from 5.52 that $\mathcal{H} \neq \emptyset$, and clearly X is a closed convex set containing C. To prove that $X \subseteq C$, suppose $\mathbf{x}^* \notin C$. Then, by Theorem 5.54, there exists $\mathbf{a} \in \mathbf{R}^n_+ \setminus \{\mathbf{0}\}$, and $b \in \mathbf{R}$ such that

$$\mathbf{a} \cdot \mathbf{x}^* < b < \inf_{\mathbf{x} \in C} \mathbf{a} \cdot \mathbf{x}.$$

But then we see that, defining:

$$H^* \equiv \{\mathbf{x} \in \mathbf{R}^n \mid \mathbf{a} \cdot \mathbf{x} \geq b\},$$

we have $H^* \in \mathcal{H}$ and $\mathbf{x}^* \notin H^*$. Therefore $\mathbf{x}^* \notin X$, and it follows that

$$C^c \subseteq X^c;$$

or, equivalently,

$$X \subseteq C. \quad \square$$

5.57. Definitions. Let C be a subset of $\mathbf{R}^n$. A **supporting half-space to C** is a closed half-space which contains C and contains an element of $\overline{C}$ in its boundary. A **support plane for C** (or **supporting hyperplane for C**), is a hyperplane which is the boundary of a supporting half-space to C. A support plane for C, H, is said to be a **non-trivial support plane for C** (or a **non-trivial supporting hyperplane for C**) iff $C \nsubseteq H$.

5.58. Theorem. *Let C be a convex subset of $\mathbf{R}^n$, and let $\mathbf{x}^* \in \overline{C}$. Then there exists a non-trivial supporting hyperplane to C which contains $\mathbf{x}^*$ if, and only if, $\mathbf{x}^* \notin riC$ [so that $\mathbf{x}^* \in \mathrm{Fr}(C)$].*

Proof. Suppose that H is a non-trivial supporting hyperplane to C containing $\mathbf{x}^*$, and define $D = \{\mathbf{x}^*\}$. Then H properly separates C and D, and thus it follows from Theorem 5.51 that:

$$riC \cap riD = \emptyset.$$

However, $riD = \{\mathbf{x}^*\}$, and therefore we see that $\mathbf{x}^* \notin riC$.

Conversely, suppose $\mathbf{x}^* \in Fr(C) = \overline{C} \setminus riC$. Defining $D = \{\mathbf{x}^*\}$, we have from 5.52 that there exists $\mathbf{a} \in \mathbf{R}^n \setminus \{\mathbf{0}\}$ and $b \in \mathbf{R}$ such that

$$\inf_{\mathbf{x} \in C} \mathbf{a} \cdot \mathbf{x} \geq b \geq \sup_{\mathbf{y} \in D} \mathbf{a} \cdot \mathbf{y} = \mathbf{a} \cdot \mathbf{x}^*, \tag{5.55}$$

and not both C and D are contained in $H_{(\mathbf{a},b)}$. However,

$$H^{\geq}_{(\mathbf{a},b)} \stackrel{\text{def}}{=} \{\mathbf{x} \in \mathbf{R}^n \mid \mathbf{a} \cdot \mathbf{x} \geq b\}$$

is a closed set, which by (5.55), contains C. Therefore,

$$\overline{C} \subseteq H^{\geq}_{(\mathbf{a},b)};$$

and, since $\mathbf{x}^* \in \overline{C}$, it follows, using (5.55), that

$$\mathbf{x}^* \in H_{(\mathbf{a},b)}$$

[and that $b = \mathbf{a} \cdot \mathbf{x}^*$ and $D \subseteq H_{(\mathbf{a},b)}$]. Since $H_{(\mathbf{a},b)}$ properly separates C and D, it then follows that we must have $C \nsubseteq H_{(\mathbf{a},b)}$. $\square$

It is intuitively apparent, and easy to prove,[11] that any proper convex subset[12] of $\mathbf{R}^n$ has at least one frontier point. Thus any proper convex subset of $\mathbf{R}^n$ has a non-trivial support plane. As a special case of this, we see that any non-empty compact and convex subset of $\mathbf{R}^n$ has a non-trivial support plane. We make use of this fact in proving the following result.

5.59. Theorem. *If C is a non-empty compact, convex subset of $\mathbf{R}^n$, then C has an extreme point. In fact, every supporting hyperplane to C contains an extreme point of C.*

Proof. We prove this by induction on n, the dimension of the space.

1. For $n = 1$, that is, for $\mathbf{R}^n = \mathbf{R}$, a compact convex set is necessarily a closed and bounded interval of the form:

$$C = [a, b],$$

and a supporting plane is necessarily of the form

$$H = \{a\} \text{ or } H' = \{b\}$$

In the first case, H contains the point a, which is an extreme point; and in the second case, H' contains b, which is also an extreme point.

2. Now suppose that the result holds for $1 \leq n \leq k$. Then we note that, using 5.58, it also follows that every compact convex set in $\mathbf{R}^n$, for $1 \leq n \leq k$, has an extreme point.

Let C be a compact convex set in $\mathbf{R}^{k+1}$. By Theorem 5.58, C has a support plane, $H = H_{(\mathbf{a},b)}$, satisfying

$$b = \inf\{\mathbf{a} \cdot \mathbf{x} \mid \mathbf{x} \in C\}. \tag{5.56}$$

Since C is closed, it also follows that:

$$D \stackrel{\text{def}}{=} H \cap C \neq \emptyset.$$

The set D is convex, being the intersection of two convex sets; and, since it is contained in H, which has dimension k, we see that the dimension of D is at most k. Letting $f\colon H \to \mathbf{R}^k$ be a pseudo-isomorphism defined as per Examples 5.32, and

$$D^* = f(D),$$

it follows from our inductive hypothesis that D^* has at least one extreme point; call it $\mathbf{a}^*$. Furthermore, we then have from Theorem 5.33 that D has an extreme point,

$$\mathbf{x}^* \stackrel{\text{def}}{=} f^{-1}(\mathbf{a}^*).$$

[11]See exercise 5 at the end of Section 3.

[12]That is, any convex subset of $\mathbf{R}^n$ which is neither empty nor $\mathbf{R}^n$ itself.

Since $\mathbf{x}^*$ is an element of D, it follows from the definition of D that $\mathbf{x}^*$ is also an element of C. Moreover, and also from the definition of D and the fact that $\mathbf{x}^*$ is an extreme point of that set, it follows that there exists no $\mathbf{v} \in H - H$ such that C is absorbing at $\mathbf{x}^*$ in the direction $\mathbf{v}$. Now suppose, by way of obtaining a contradiction, that there exists a point $\mathbf{v}^*$ satisfying

$$\mathbf{v}^* \in \mathbf{R}^{k+1} \setminus (H - H), \tag{5.57}$$

and such that C is absorbing at $\mathbf{x}^*$ in the direction $\mathbf{v}^*$. Then there exists $\kappa > 0$ such that

$$(\forall \theta \in \mathbf{R}) \colon |\theta| < \kappa \Rightarrow \mathbf{x}^* + \theta \mathbf{v}^* \in C. \tag{5.58}$$

However, from (5.57) and the definition of $H = H_{(\mathbf{a},b)}$, we can suppose, without loss of generality, that

$$\mathbf{a} \cdot \mathbf{v}^* < 0; \tag{5.59}$$

while from (5.58) we have that the vector $\mathbf{y}^*$ defined by

$$\mathbf{y}^* = \mathbf{x}^* + (\kappa/2)\mathbf{v}^*,$$

is an element of C. But this is impossible, for it follows from (5.59) and the fact that $\mathbf{x}^*$ is an element of H, that

$$\begin{aligned} \mathbf{a} \cdot \mathbf{y}^* &= \mathbf{a} \cdot [\mathbf{x}^* + (\kappa/2)\mathbf{v}^*] = \mathbf{a} \cdot \mathbf{x}^* + (\kappa/2)\mathbf{a} \cdot \mathbf{v}^* \\ &= b + (\kappa/2)\mathbf{a} \cdot \mathbf{v}^* < b; \end{aligned}$$

contradicting (5.56). □

Our next definition will, as we shall discover shortly, turn out to be somewhat redundant. It is nonetheless useful in the statement of the Krein-Milman Theorem, which we will be proving shortly.

5.60. Definition. Given a non-empty set, $X \subseteq \mathbf{R}^n$, we define the **closed convex hull of X**, denoted by '$c(X)$,' as the smallest closed convex set containing X.

I will be asking you to prove the legitimacy of the preceding definition at the end of this section (Exercise 4). As to what I mean by calling the definition 'somewhat redundant,' consider the following result.

5.61. Proposition. *If X is a nonempty subset of $\mathbf{R}^n$, then $c(X) = \overline{\langle X \rangle}$; that is, $c(X)$ is the closure of the convex hull of X.*

Proof. It follows from Proposition 5.9 that $\overline{\langle X \rangle}$ is convex; and, since it is closed, it is necessarily a closed convex set containing X. Therefore:

$$c(X) \subseteq \overline{\langle X \rangle}.$$

On the other hand, $c(X)$ is a convex set containing X, and therefore:

$$\langle X \rangle \subseteq c(X).$$

Consequently,

$$\overline{\langle X \rangle} \subseteq \overline{c(X)} = c(X),$$

and it follows that the two sets are equal. □

In our next result, we make reference to the 'profile' of a set. Recall that this is the collection of all extreme points of the set, and that the profile of a set X is denoted by '$\ddot{X}$' (see Definition 5.31).

5.62. Theorem. (Krein-Milman Theorem) *If X is a nonempty compact convex subset of $\mathbf{R}^n$, then $X = c(\ddot{X})$; that is, X is the closed convex hull of its profile.*

Proof. Suppose, by way of obtaining a contradiction, that:

$$X \nsubseteq c(\ddot{X}).$$

Then there exists $\mathbf{x}^* \in X$ such that $\mathbf{x}^* \notin c(\ddot{X})$; and thus, by Theorem 5.54, there exists a non-null vector $\mathbf{p} \in \mathbf{R}^n$ such that:

$$\mathbf{p} \cdot \mathbf{x}^* < \alpha \stackrel{\text{def}}{=} \inf_{\mathbf{x} \in c(\ddot{X})} \mathbf{p} \cdot \mathbf{x}. \tag{5.60}$$

Now, the function f defined by:

$$f(\mathbf{x}) = \mathbf{p} \cdot \mathbf{x} \quad \text{for } \mathbf{x} \in \mathbf{R}^n,$$

is obviously continuous, and thus there exists a point $\mathbf{y} \in X$ satisfying:

$$(\forall \mathbf{x} \in X) \colon \mathbf{p} \cdot \mathbf{x} \geq \mathbf{p} \cdot \mathbf{y}.$$

But then, defining $\beta = \mathbf{p} \cdot \mathbf{y}$, we see that, since $\mathbf{x}^* \in X$, we must have:

$$\beta \leq \mathbf{p} \cdot \mathbf{x}^*. \tag{5.61}$$

On the other hand, consider the hyperplane, H, defined by:

$$H = \{\mathbf{x} \in \mathbf{R}^n \mid \mathbf{p} \cdot \mathbf{x} = \beta\}.$$

Obviously H is a supporting hyperplane to X, and thus by Theorem 5.59 must contain an extreme point, $\mathbf{y}^*$. Since $\mathbf{y}^* \in \ddot{X}$, we then see that we must have:

$$\beta = \mathbf{p} \cdot \mathbf{y}^* \geq \alpha, \tag{5.62}$$

where α is from (5.60). However, combining (5.60)–(5.62), we then have:

$$\alpha \leq \beta \leq \mathbf{p} \cdot \mathbf{x}^* < \alpha;$$

and from this contradiction, we infer that:

$$X \subseteq c(\ddot{X}).$$

To establish the opposite inclusion, note that, since X is a closed convex set containing $\ddot{X}$, we must have:

$$c(\ddot{X}) \subseteq X. \quad \square$$

Interestingly enough the following result, although due to Shapley and Folkman, was first reported in Starr [1969]. The proof presented here is, essentially, an adaptation of that published in Green and Heller [1981, pp. 41–2], who attribute the proof to J. P. Aubin and I. Ekeland.

5.63. Theorem. (Shapley-Folkman) *Let $Y_1, \ldots, Y_m$ be non-empty subsets of $\mathbf{R}^n$, define:*

$$Y = \sum_{i=1}^{m} Y_i,$$

and let $\mathbf{x} \in \langle Y \rangle$. Then there exist $\mathbf{x}_i \in \langle Y_i \rangle$ $(i = 1, \ldots, m)$ such that:

$$\mathbf{x} = \sum_{i=1}^{m} \mathbf{x}_i,$$

and at most *n of the $\mathbf{x}_i$'s satisfy:*

$$\mathbf{x}_i \in \langle Y_i \rangle \setminus Y_i.$$

Proof. Letting $\mathbf{x} \in \langle Y \rangle$, it follows from Carathéodory's Theorem (Theorem 5.13) that there exists $k \leq n + 1$, $\mathbf{a} \in \Delta_k$, and $\mathbf{y}_j \in Y$, for $j = 1, \ldots, k$, such that:

$$\mathbf{x} = \sum_{j=1}^{k} a_j \mathbf{y}_j. \tag{5.63}$$

Moreover, from the definition of Y, we see that for each j, there exists $\{\mathbf{y}_{1j}, \ldots, \mathbf{y}_{mj}\} \subseteq \mathbf{R}^n$ such that:

$$\mathbf{y}_{ij} \in Y_i \quad \text{for } i = 1, \ldots, m \text{ and } \mathbf{y}_j = \sum_{i=1}^{m} \mathbf{y}_{ij}.$$

Now, define the sets $Z_i \subseteq Y_i$ by:

$$Z_i = \{\mathbf{y}_{i1}, \ldots, \mathbf{y}_{ik}\} \quad \text{for } i = 1, \ldots, m,$$

and notice that:

$$\mathbf{x} = \sum_{i=1}^{m} \sum_{j=1}^{k} a_j \mathbf{y}_{ij},$$

and thus it follows that:

$$\mathbf{x} \in \sum_{i=1}^{m} \langle Z_i \rangle.$$

Therefore, defining the set $P \subseteq \mathbf{R}^{mn}$ by:

$$P = \left\{ (\mathbf{x}_1, \ldots, \mathbf{x}_m) \in \mathbf{R}^{mn} \mid \mathbf{x}_i \in \langle Z_i \rangle \text{ for } i = 1, \ldots, m \ \& \ \sum_{i=1}^{m} \mathbf{x}_i = \mathbf{x} \right\},$$

we see that $P \neq \emptyset$.

It will follow from results in Chapter 8 that P is a compact set, and thus (by Theorem 5.59) P has at least one extreme point. We will show that any extreme point of P satisfies the conclusion of our result.

Accordingly, let $\overline{\mathbf{x}}$ be an extreme point of P. Then $\overline{\mathbf{x}}$ is of the form:

$$\overline{\mathbf{x}} = (\overline{\mathbf{x}}_1, \ldots, \overline{\mathbf{x}}_m),$$

where:

$$\overline{\mathbf{x}}_i \in \langle Z_i \rangle \quad \text{for } i = 1, \ldots, m.$$

I claim that, for all but at most n of these vectors, $\overline{\mathbf{x}}_i$ must be an extreme point of $\langle Z_i \rangle$. For suppose not. Then we may suppose, without loss of generality, that $\overline{\mathbf{x}}_1, \ldots, \overline{\mathbf{x}}_{n+1}$ are all non-extreme points; and thus, for each i ($i = 1, \ldots, n+1$), there exists $\mathbf{z}_i \in \mathbf{R}^n \setminus \{\mathbf{0}\}$ such that $\langle Z_i \rangle$ is absorbing at $\overline{\mathbf{x}}_i$ in the direction $\mathbf{z}_i$. However, since $\{\mathbf{z}_1, \ldots, \mathbf{z}_{n+1}\}$ is a subset of $\mathbf{R}^n$, it must be a linearly dependent set; so that there exists $\mathbf{b} \in \mathbf{R}^{n+1} \setminus \{\mathbf{0}\}$ such that:

$$\sum_{i=1}^{n+1} b_i \mathbf{z}_i = \mathbf{0}.$$

Moreover, defining:

$$\overline{\mathbf{z}}_i = b_i \mathbf{z}_i \quad \text{for } i = 1, \ldots, n+1,$$

it is obvious that $\langle Z_i \rangle$ is also absorbing at $\overline{\mathbf{x}}_i$ in the direction $\overline{\mathbf{z}}_i$, for $i = 1, \ldots, n+1$. Consequently, there exists $\delta > 0$ such that:

$$\big(\forall t \in \,]-\delta, \delta\,[\big) : \overline{\mathbf{x}}_i + t\overline{\mathbf{z}}_i \in \langle Z_i \rangle \quad \text{for } i = 1, \ldots, n+1; \tag{5.64}$$

and we also note that:

$$\sum_{i=1}^{n+1} \overline{\mathbf{z}}_i = \mathbf{0}. \tag{5.65}$$

But then we see that, defining $\widehat{\mathbf{z}} \in \mathbf{R}^{kn}$ by:

$$\widehat{\mathbf{z}}_i = \begin{cases} \overline{\mathbf{z}}_i & \text{for } i = 1, \ldots, n+1, \\ \mathbf{0} & \text{for } i = n+2, \ldots, m, \end{cases}$$

$\widehat{\mathbf{z}} \neq \mathbf{0}$ and, for all $t \in]-\delta, \delta[$:

$$\overline{\mathbf{x}} + t\widehat{\mathbf{z}} \in \prod_{i=1}^{m} \langle Z_i \rangle;$$

by (5.64); while by (5.65):

$$\sum_{i=1}^{m} (\overline{\mathbf{x}}_i + t\widehat{\mathbf{z}}_i) = \sum_{i=1}^{m} \overline{\mathbf{x}}_i + t \sum_{i=1}^{n+1} \widehat{\mathbf{z}}_i = \sum_{i=1}^{m} \overline{\mathbf{x}}_i = \mathbf{x}.$$

Consequently, P is absorbing at $\overline{\mathbf{x}}$ in the direction $\widehat{\mathbf{z}}$; contradicting the assumption that $\overline{\mathbf{x}}$ is an extreme point of P.

It follows, therefore, that $\overline{\mathbf{x}}_i$ must be an extreme point of $\langle Z_i \rangle$ for all but *at most n* of these sets. However, it is an easy consequence of Proposition 5.11 (see also Exercise 6, at the end of Section 5.3) that any extreme point of $\langle Z_i \rangle$ must be an element of Z_i, and our result now follows. □

To illustrate the meaning and significance of the Shapley-Folkman Theorem, consider the subsets of $\mathbf{R}^2$ defined by:

$$Y_i = \{(i, 0), (i, 1)\} \quad \text{for } i = 1, \ldots, 4.$$

In this case, the summation set, Y, is given by:

$$Y = \sum_{i=1}^{4} Y_i = \{\mathbf{x} \in \mathbf{R}^2 \mid x_1 = 10 \;\&\; x_2 \in \{0, 1, 2, 3, 4\}\},$$

whose convex hull, $\langle Y \rangle$, is given by:

$$\langle Y \rangle = \{\mathbf{x} \in \mathbf{R}^2 \mid x_1 = 10 \;\&\; x_2 \in [0, 4]\}.$$

If, for example, we consider $\mathbf{y} = (10, 3.5)$, we can write:

$$\mathbf{y} = (1, .5) + (2, 1) + (3, 1) + (4, 1),$$

where the point $(1, .5)$ is in the convex hull of Y_1, while the last three points are in Y_2, Y_3, and Y_4, respectively. In fact, you should have no difficulty in convincing yourself that if $\mathbf{y} \in \langle Y \rangle$, then one can find points $\mathbf{y}_i \in \langle Y_i \rangle$, for $i = 1, \ldots, 4$, such that:

$$\mathbf{y} = \sum_{i=1}^{4} \mathbf{y}_i,$$

and *at most two* of the $\mathbf{y}_i$'s are not elements of the corresponding Y_i sets.

In the following definition we make use of the function $\delta(\cdot, Y)$, defined for a non-empty set $Y \subseteq \mathbf{R}^n$ by:

$$\delta(\mathbf{x}, Y) = \inf_{\mathbf{y} \in Y} \|\mathbf{y} - \mathbf{x}\| \quad \text{for } \mathbf{x} \in \mathbf{R}^n$$

(see Example 4.46.2), which is a measure of the distance from $\mathbf{x}$ to the set Y. We use the function $\delta(\cdot, Y)$ to define a measure of the non-convexity of Y as follows.

5.64. Definition. Let Y be a non-empty subset of $\mathbf{R}^n$. We define the **inner measure of Y, $\rho(Y)$**, by:

$$\rho(Y) = \sup_{\mathbf{x} \in \langle Y \rangle} \delta(\mathbf{x}, Y).$$

It is easy to show that $\rho(Y) = 0$ if, and only if, $\overline{Y}$ is convex; so that, in particular, $\rho(Y) = 0$ if Y is a convex set. In general, it measures the maximum distance from a point in the convex hull to a point in the original set, and thus the larger is $\rho(Y)$, the more non-convex Y is, in some sense.[13] Our final result of this section, which is a refinement by Heller of a corollary of the Shapley-Folkman Theorem developed by Starr [1969], makes use of the inner measure and is central to most of the approximate equilibrium results in the economics literature. We will not include a proof here, since a very accessible and elegant proof is included in Green and Heller [1981, p. 43].

5.65. Corollary. *Let $S_1, \ldots, S_m$ be subsets of $\mathbf{R}^n$ such that there exists a positive number, a, satisfying:*

$$\rho(S_i) \leq a \quad \text{for } i = 1, \ldots, m.$$

Then, given any element, $\mathbf{x} \in \left\langle \sum_{i=1}^m S_i \right\rangle$, there exist elements $\mathbf{y}_i \in S_i$, for $i = 1, \ldots, m$ such that:

$$\left\| \mathbf{x} - \sum_{i=1}^{m} \mathbf{y}_i \right\| \leq na.$$

Exercises.

1. Prove: (a) Corollary 5.52, and (b) Lemma 5.53.

2. Show that if C is a proper convex subset of $\mathbf{R}^n$, then there exists a non-constant linear function, $f \colon \mathbf{R}^n \to \mathbf{R}$, which is bounded above on C.

3. Supposing that C is a convex subset of $\mathbf{R}^n$, prove the following statement. A point $\mathbf{x}^* \in \mathbf{R}^n$ is a relative boundary point of C [that is, $\mathbf{x}^* \in Fr(C)$] if, and only if, there exists a linear function, $f \colon \mathbf{R}^n \to \mathbf{R}$, such that f is not constant on C, and f attains a maximum over C at $\mathbf{x}^*$.

4. Prove the legitimacy of the definition of 'closed convex hull;' that is, prove that, given any non-empty subset, X, of $\mathbf{R}^n$, there exists a smallest (in the sense of set inclusion) closed convex set containing X.

5. *Theorem of the Alternative.* If $\mathbf{A}$ is a real $m \times n$ matrix, and $\mathbf{c} \in \mathbf{R}^m$, show that *exactly one* of the following alternatives holds. Either (a) there exists $\mathbf{y} \in \mathbf{R}^m$ such that:

$$\mathbf{y}'\mathbf{A} \geq \mathbf{c}',$$

[13] For a more detailed discussion, see Heller [1972].

or (b) there exists $\mathbf{x} \in \mathbf{R}^n_+$ such that:

$$\mathbf{A}\mathbf{x} = \mathbf{0} \;\; \& \;\; \mathbf{x} \cdot \mathbf{c} = 1.$$

Try applying 5.55.2 to the matrix $\mathbf{B}$ and the vector $\mathbf{b}$ given by:

$$\mathbf{B} = \begin{bmatrix} \mathbf{A} \\ \mathbf{c}' \end{bmatrix} \;\; \text{and} \;\; \mathbf{b} = \begin{pmatrix} \mathbf{0} \\ 1 \end{pmatrix},$$

respectively.

A number of additional useful results of the type established in this exercise are set out in Gale [1960, pp. 47–9]. See also Nikaido [1968].

5.6 Convex and Concave Functions

The theory of concave and convex functions is, arguably, as important to modern economic theory as is the theory of convex sets. In this section we shall present a brief introduction to this theory. Once again, unless specifically stated otherwise, we will suppose that we are working with an arbitrary real linear space, L.

5.66. Definitions. Let $X \subseteq L$ be a convex set, and suppose $f \colon X \to \mathbf{R}$. If for all $\theta \in [0, 1]$, and all $\mathbf{x}, \mathbf{y} \in X$, we have:

1. $f[\theta\mathbf{x} + (1-\theta)\mathbf{y}] \leq \theta f(\mathbf{x}) + (1-\theta) f(\mathbf{y})$, we say that f is **convex** (or **convex on X**).
2. $f[\theta\mathbf{x} + (1-\theta)\mathbf{y}] \geq \theta f(\mathbf{x}) + (1-\theta) f(\mathbf{y})$, we say that f is **concave** (or **concave on X**).
3. $f[\theta\mathbf{x} + (1-\theta)\mathbf{y}] < \theta f(\mathbf{x}) + (1-\theta) f(\mathbf{y})$ whenever $0 < \theta < 1$ and $\mathbf{x} \neq \mathbf{y}$, we say that f is **strictly convex** (or **strictly convex on X**).
4. $f[\theta\mathbf{x} + (1-\theta)\mathbf{y}] > \theta f(\mathbf{x}) + (1-\theta) f(\mathbf{y})$, whenever $0 < \theta < 1$ and $\mathbf{x} \neq \mathbf{y}$, we say that f is **strictly concave** (or **strictly concave on X**).

Notice that a function, f, is convex if, and only if, its negative, $-f$, is concave (the student should prove this). Consequently, any result we might prove for concave functions will have an immediate analogue for convex functions; and, since in economic theory concave functions are probably encountered more often than are convex functions, we will generally concentrate our attention upon concavity in the remainder of this section.

5.67. Examples/Exercises.

1. Suppose $g \colon A \to L_2$, where A is an affine subset of L_1. Prove that g is both concave and convex if, and only if, it is affine.
2. Let X be a convex subset of a normed real linear space, L, and define the function $f \colon L \to \mathbf{R}_+$ by:

$$f(\mathbf{x}) = \delta(\mathbf{x}, X) \stackrel{\text{def}}{=} \inf\{\rho(\mathbf{x} - \mathbf{y}) \mid \mathbf{y} \in X\},$$

where $\rho(\cdot)$ is the norm for L. Then f is a convex function.

3. In the notation and context of the immediately preceding example, if we define $X = \{\mathbf{0}\}$, then

$$\rho(\mathbf{x}) = \delta(\mathbf{x}, X) \qquad \text{for each } \mathbf{x} \in L.$$

Thus it follows from example 2 that $\rho(\cdot)$ is a convex function.

Note. Until further notice, we will always be dealing with functions of the form $f\colon X \to \mathbf{R}$, where X is a convex subset of a real linear space, L. I will leave the proof of the following result as an exercise.

5.68. Proposition. *If f is concave on X, then, for all positive integers, m, all $\mathbf{a} \in \Delta_m$, and all $\mathbf{x}_1, \ldots, \mathbf{x}_m \in X$, we have*

$$f\left(\sum_{i=1}^{m} a_i \mathbf{x}_i\right) \geq \sum_{i=1}^{m} a_i f(\mathbf{x}_i).$$

5.69. Proposition. *The function f is concave [respectively, convex] on X if, and only if, for all $\mathbf{x}, \mathbf{y} \in X$, the function $g(\cdot)$ defined on $[0,1]$ by:*

$$g(\lambda) = f[\lambda \mathbf{x} + (1-\lambda)\mathbf{y}] = f[\mathbf{y} + \lambda(\mathbf{x} - \mathbf{y})] \quad \text{for } \lambda \in [0,1], \tag{5.66}$$

is concave [respectively, convex] on $[0,1]$.

Proof. We shall prove this result for the concave case only. The respective statements follow from the fact that f is convex if, and only if, $-f$ is concave.

Suppose first that g is concave, for each $\mathbf{x}, \mathbf{y} \in X$; let $\mathbf{x}^*$ and $\mathbf{y}^*$ be elements of X, and let $\theta \in [0,1]$. Then

$$\begin{aligned} f[\theta \mathbf{x}^* + (1-\theta)\mathbf{y}^*] &\equiv g(\theta) = g[\theta 1 + (1-\theta)0] \\ &\geq \theta g(1) + (1-\theta)g(0) = \theta f(\mathbf{x}^*) + (1-\theta) f(\mathbf{y}^*). \end{aligned}$$

Conversely, suppose f is concave on X, let $\mathbf{x}^*$ and $\mathbf{y}^*$ be elements of X, define g on $[0,1]$ as in (5.66) (with $\mathbf{x} = \mathbf{x}^*$ and $\mathbf{y} = \mathbf{y}^*$), and let λ_1, λ_2 and $\theta \in [0,1]$ be given. If we then define:

$$\mathbf{z}_1 = \lambda_1 \mathbf{x}^* + (1-\lambda_1)\mathbf{y}^* \quad \text{and} \quad \mathbf{z}_2 = \lambda_2 \mathbf{x}^* + (1-\lambda_2)\mathbf{y}^*;$$

we have

$$\begin{aligned} &\theta \mathbf{z}_1 + (1-\theta)\mathbf{z}_2 \\ &\quad = \theta\lambda_1 \mathbf{x}^* + \theta \mathbf{y}^* - \theta\lambda_1 \mathbf{y}^* + \lambda_2(1-\theta)\mathbf{x}^* + (1-\theta)\mathbf{y}^* - \lambda_2(1-\theta)\mathbf{y}^* \\ &\quad = [\theta\lambda_1 + (1-\theta)\lambda_2]\mathbf{x}^* + [1 - \theta\lambda_1 - (1-\theta)\lambda_2]\mathbf{y}^*. \end{aligned}$$

Since

$$0 \leq \theta\lambda_1 + (1-\theta)\lambda_2 \leq \theta + (1-\theta) = 1,$$

and f is concave, we then see that:

$$g[\theta\lambda_1 + (1-\theta)\lambda_2] = f\left(\left[\theta\lambda_1 + (1-\theta)\lambda_2\right]\mathbf{x}^* + \left[1 - [\theta\lambda_1 + (1-\theta)\lambda_2]\right]\mathbf{y}^*\right)$$
$$= f[\theta\mathbf{z}_1 + (1-\theta)\mathbf{z}_2] \geq \theta f(\mathbf{z}_1) + (1-\theta)f(\mathbf{z}_2) = \theta g(\lambda_1) + (1-\theta)g(\lambda_2);$$

where the last equality is an immediate consequence of the definitions of g and the $\mathbf{z}_i$'s. □

5.70. Definition. If $f\colon X \to \mathbf{R}$, where X is a convex subset of L, we define the **disposable hull of f**, abbreviated '*hull*(f),' and the **epigraph of f**, abbreviated '*epi*(f),' by:

$$hull(f) = \{(\mathbf{x}, y) \in X \times \mathbf{R} \mid \mathbf{x} \in X \text{ and } y \leq f(\mathbf{x})\},$$

and

$$epi(f) = \{(\mathbf{x}, y) \in X \times \mathbf{R} \mid \mathbf{x} \in X \text{ and } y \geq f(\mathbf{x})\},$$

respectively.

It should be admitted that I have never encountered a definition of *hull*(f) in the mathematical literature. The definition given here arises often and in a very natural way in economic applications, however; for example, in the theory of production. Given the function, f, whose graph is denoted by '$\mathcal{G}_f$' in Figure 5.6, below, the hull of the function, *hull*(f), is as indicated in the figure.

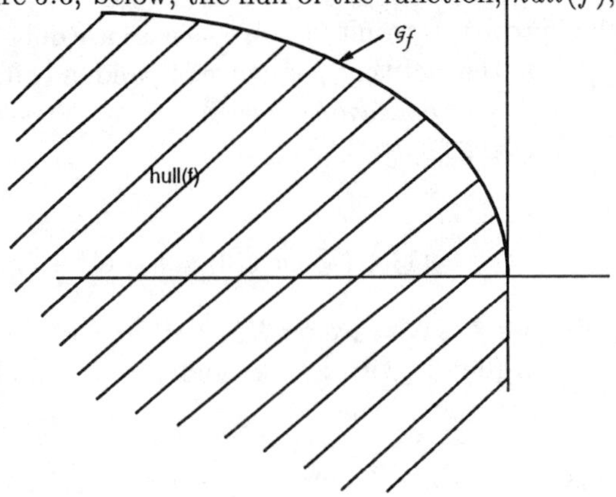

Figure 5.6: The Hull of a Concave Function.

Notice that if we define $G\colon X \times \mathbf{R} \to \mathbf{R}$ by:

$$G(\mathbf{x}, y) = f(\mathbf{x}) - y, \tag{5.67}$$

then

$$hull(f) = \{(\mathbf{x}, y) \in X \times \mathbf{R} \mid G(\mathbf{x}, y) \geq 0\},$$

and

$$epi(f) = \{(\mathbf{x}, y) \in X \times \mathbf{R} \mid G(\mathbf{x}, y) \leq 0\}.$$

We also have the following.

5.71. Theorem. *The function f is concave [respectively, convex] if, and only if,* hull(f) *[respectively,* epi(f)*] is a convex subset of $L \times \mathbf{R}$.*

Proof. We will provide a proof for the concave case only. The convex case can be established by a very similar argument; or, alternatively, by making use of the fact that a function f is convex if, and only if, $-f$ is concave.

1. Suppose $hull(f)$ is a convex subset of $L \times \mathbf{R}$, let $\mathbf{x}, \mathbf{y} \in X$, and $\theta \in [0, 1]$. It follows at once from the definition of $hull(f)$ that:

$$(\mathbf{x}, f(\mathbf{x})) \in hull(f) \quad \text{and} \quad (\mathbf{y}, f(\mathbf{y})) \in hull(f);$$

and thus, since $hull(f)$ is convex,

$$\theta(\mathbf{x}, f(\mathbf{x})) + (1-\theta)(\mathbf{y}, f(\mathbf{y})) = (\theta\mathbf{x} + (1-\theta)\mathbf{y},\ \theta f(\mathbf{x}) + (1-\theta)f(\mathbf{y}))$$

is in $hull(f)$. It then follows from the definition of $hull(f)$ that

$$\theta f(\mathbf{x}) + (1-\theta)f(\mathbf{y}) \leq f[\theta\mathbf{x} + (1-\theta)\mathbf{y}];$$

and we see that f is a concave function.

2. Suppose now that f is concave. If $(\mathbf{x}_i, y_i) \in hull(f)$ for $i = 1, 2$, and $\theta \in [0, 1]$, we have:

$$f(\mathbf{x}_i) \geq y_i \quad \text{for } i = 1, 2;$$

and thus

$$\theta f(\mathbf{x}_1) + (1-\theta)f(\mathbf{x}_2) \geq \theta y_1 + (1-\theta)y_2. \tag{5.68}$$

However, since f is concave, we have:

$$f[\theta\mathbf{x}_1 + (1-\theta)\mathbf{x}_2)] \geq \theta f(\mathbf{x}_1) + (1-\theta)f(\mathbf{x}_2); \tag{5.69}$$

and from (5.68) and (5.69) we see that:

$$\Big(\theta\mathbf{x}_1 + (1-\theta)\mathbf{x}_2, \theta y_1 + (1-\theta)y_2\Big) \in hull(f).$$

Therefore, $hull(f)$ is a convex set. □

5.72. Theorem. *Suppose $\mathfrak{F}$ is a family of concave functions, where*

$$(\forall f \in \mathfrak{F}): \ f : X \to \mathbf{R},$$

and X is a convex subset of L, and suppose there exists $\varphi\colon X \to \mathbf{R}$ which satisfies:[14]

$$(\forall \mathbf{x} \in X)(\forall f \in \mathcal{F})\colon f(\mathbf{x}) \geq \varphi(\mathbf{x}). \tag{5.70}$$

Then the function $g\colon X \to \mathbf{R}$ defined by:

$$g(\mathbf{x}) = \inf_{f \in \mathcal{F}} f(\mathbf{x}) \quad \mathit{for}\, \mathbf{x} \in X,$$

is concave on X.

Proof. Suppose $(\mathbf{x}, y) \in hull(g)$, and let $f \in \mathcal{F}$ be arbitrary. Then

$$f(\mathbf{x}) \geq g(\mathbf{x}) \geq y,$$

and hence

$$(\mathbf{x}, y) \in hull(f).$$

It follows that

$$hull(g) \subseteq \bigcap_{f \in \mathcal{F}} hull(f). \tag{5.71}$$

Conversely, suppose that

$$(\mathbf{x}, y) \in \bigcap_{f \in \mathcal{F}} hull(f).$$

Then we have

$$(\forall f \in \mathcal{F})\colon f(\mathbf{x}) \geq y,$$

and we see that y is a lower bound for the set $F(\mathbf{x})$ defined by:

$$F(\mathbf{x}) = \{f(\mathbf{x}) \mid f \in \mathcal{F}\}.$$

Consequently,

$$y \leq g(\mathbf{x}) \equiv \inf F(\mathbf{x}),$$

and we see that $(\mathbf{x}, y) \in hull(g)$. Therefore,

$$\bigcap_{f \in \mathcal{F}} hull(f) \subseteq hull(g). \tag{5.72}$$

Combining (5.71) and (5.72), we see that

$$\bigcap_{f \in \mathcal{F}} hull(f) = hull(g).$$

---

[14]Condition (5.70) in the hypotheses of Theorem 5.72 is included only to guarantee that the function g is finite-valued (that is, real-valued). If we were willing to include extended real-valued functions (functions which might be equal to minus or plus infinity at some points) in our discussion, then we could delete condition (5.70) from the hypotheses of the Theorem.

However, it follows from Theorem 5.71 that $hull(f)$ is a convex set, for each $f \in \mathcal{F}$, and therefore we see that $hull(g)$ is a convex set. Using Theorem 5.71 once again, it then follows that g is a concave function. □

In our next two results, we state some conditions under which the composition of two concave functions is itself a concave function. We begin with the following general result.

5.73. Proposition. *If*

1. $f \colon X \to L_2$ *is affine, where* X *is a convex subset of* L_1,
2. $g \colon Y \to \mathbf{R}$ *is concave [respectively, convex], where*

$$f(X) \subseteq Y \subseteq L_2,$$

and Y *is a convex set,*

then the composition of f *and* g, $h \equiv g \circ f$, *is concave [respectively, convex] on* X.

Proof. If $\mathbf{x}$ and $\mathbf{y}$ are elements of X, and $\theta \in [0,1]$, we have, since f is affine:

$$f[\theta\mathbf{x} + (1-\theta)\mathbf{y}] = \theta f(\mathbf{x}) + (1-\theta) f(\mathbf{y}) \in f(X).$$

Therefore,

$$\begin{aligned} h[\theta\mathbf{x} + (1-\theta)\mathbf{y}] = g[\theta f(\mathbf{x}) + (1-\theta) f(\mathbf{y})] &\geq \theta g[f(\mathbf{x})] + (1-\theta) g[f(\mathbf{y})] \\ &= \theta h(\mathbf{x}) + (1-\theta) h(\mathbf{y}), \end{aligned}$$

and we see that h is concave.

The respective statements follow by a symmetric argument. □

Our next result concerns the composition of concave functions, the domain of one of which is a subset of a finite-dimensional Euclidean space. In it we shall need some new definitions, as follows.

5.74. Definition. We shall say that a function $\mathbf{f} \colon X \to \mathbf{R}^n$ is **concave** iff

$$(\forall \mathbf{x}, \mathbf{y} \in X)(\forall \theta \in [0,1]) \colon \mathbf{f}[\theta\mathbf{x} + (1-\theta)\mathbf{y}] \geq \theta\mathbf{f}(\mathbf{x}) + (1-\theta)\mathbf{f}(\mathbf{y}).$$

5.75. Definitions. Let $f \colon X \to \mathbf{R}$, where X is a non-empty subset of $\mathbf{R}^n$. We shall say that f is:

1. **non-decreasing** (on X) iff:

$$(\forall \mathbf{x}, \mathbf{y} \in X) \colon \mathbf{x} \geq \mathbf{y} \Rightarrow f(\mathbf{x}) \geq f(\mathbf{y}).$$

2. **increasing** (on X) iff f is non-decreasing and satisfies:

$$(\forall \mathbf{x}, \mathbf{y} \in X) \colon \mathbf{x} \gg \mathbf{y} \Rightarrow f(\mathbf{x}) > f(\mathbf{y}).$$

3. **strictly increasing** (on X) iff:

$$(\forall \mathbf{x}, \mathbf{y} \in X) \colon \mathbf{x} > \mathbf{y} \Rightarrow f(\mathbf{x}) > f(\mathbf{y}).$$

Notice that it is an immediate consequence of Definition 5.74 that $\mathbf{f} = (f_1, \ldots, f_n)$ is concave if, and only if, f_i is concave, for $i = 1, \ldots, n$. Consequently, the basic theory of concave functions mapping into $\mathbf{R}^n$ is readily deduced from the theory of real-valued concave functions; and, of course, similar remarks apply to convex functions mapping into $\mathbf{R}^n$. I will, however, leave the verification of these remarks as exercises for the interested student. The proof of our next result will also be left as an exercise.

5.76. Proposition. *If*

1. $\mathbf{f}: X \to \mathbf{R}^n$ *is concave, where* X *is a convex subset of a real linear space,* L, *and*
2. $g: Y \to \mathbf{R}$ *is a non-decreasing concave function, where*

$$\mathbf{f}(X) \subseteq Y \subseteq \mathbf{R}^n,$$

and Y *is a convex set,*

then the composition of $\mathbf{f}$ *and* g, $h \equiv g \circ \mathbf{f}$, *is concave on* X.

Notice that, as a special case of the above result, we have the following: if $\mathbf{f}: X \to \mathbf{R}^n$ is concave, and $\mathbf{a} \in \mathbf{R}^n_+$, then the function $g: X \to \mathbf{R}$ defined by

$$g(\mathbf{x}) = \sum\nolimits_{i=1}^{n} a_i f_i(\mathbf{x}),$$

is concave. Analogous results follow, of course, for the case of convex functions.

In economic theory, we are often interested in maximizing functions. Consequently, the following result is of particular interest for us.

5.77. Theorem. *Let* C *be a convex set, and let* $f: C \to \mathbf{R}$ *be a convex [respectively, concave] function. If* f *attains a maximum [respectively, a minimum] over* C *at some* $\mathbf{x}^* \in riC$, *then* f *is constant on* C.

Proof. Suppose f is not constant on C. Then, defining

$$\alpha = \sup\{f(\mathbf{x}) \mid \mathbf{x} \in C\},$$

we note that, if $\alpha = +\infty$, then f does not attain a maximum anywhere on C. Suppose, therefore, that $\alpha \in \mathbf{R}$, and let $\mathbf{x}$ be an arbitrary element of riC. Since f is not constant on C, there exists $\mathbf{y} \in C$ such that

$$f(\mathbf{y}) < \alpha;$$

and, it follows easily from Proposition 5.23 and Theorem 5.20 (see Proposition 5.87, below) that there exists $\theta \in]0, 1[$ and $\mathbf{z} \in riC$ such that

$$\mathbf{x} = \theta\mathbf{z} + (1 - \theta)\mathbf{y}.$$

But then we have

$$f(\mathbf{x}) \leq \theta f(\mathbf{z}) + (1 - \theta)f(\mathbf{y}) \leq \theta\alpha + (1 - \theta)f(\mathbf{y}) < \theta\alpha + (1 - \theta)\alpha = \alpha;$$

and we see that f is not maximized over C at $\mathbf{x}$. □

5.78. Definitions. We shall say that a set K is a:

1. **cone** iff, for each $\mathbf{x} \in K$ and each $\lambda \in \mathbf{R}_{++}$, we have

$$\lambda \mathbf{x} \in K.$$

2. **convex cone** iff K is both a cone and a convex set.

Notice that the above definition does not require K to contain the origin in order to be a cone. Occasionally, however, we shall want to require that K both be a cone and contain the origin, in which case we shall say that K is a **nonnegative cone**, and if we want to specifically require that $\mathbf{0} \notin K$ we shall say that K is a **semipositive cone.**[15] If K is both convex and a nonnegative [respectively, semipositive] cone, we shall, of course, say that K is a **nonnegative** [respectively, **semipositive**] **convex cone**. I will leave the proof of the following useful fact as an exercise.

5.79. Proposition. *If K is a convex cone, then K satisfies:*

$$(\forall \mathbf{x}, \mathbf{y} \in K): \mathbf{x} + \mathbf{y} \in K.$$

5.80. Definition. If $f \colon K \to \mathbf{R}$, where K is a cone, we shall say that f is **positively homogeneous of degree r** ≥ 0 iff for all $\mathbf{x} \in K$, and all $\lambda \in \mathbf{R}_{++}$, we have:

$$f(\lambda \mathbf{x}) = \lambda^r f(\mathbf{x}).$$

5.81. Definition. If $f \colon K \to \mathbf{R}$, where K is a convex cone, then we shall say that f is **super-** [respectively, **sub-**] **additive** iff, for all $\mathbf{x}, \mathbf{y} \in K$, we have

$$f(\mathbf{x} + \mathbf{y}) \geq f(\mathbf{x}) + f(\mathbf{y}) \text{ [respectively, } f(\mathbf{x} + \mathbf{y}) \leq f(\mathbf{x}) + f(\mathbf{y})].$$

5.82. Proposition. *Suppose $f \colon K \to \mathbf{R}$ is positively homogeneous of degree one on K, where K is a convex cone. Then f is concave [respectively, convex] on K if, and only if, f is super- [respectively, sub-] additive on K.*

Proof. Suppose first that f is concave on K, and let $\mathbf{x}, \mathbf{y} \in K$. Then, since f is positively homogeneous of degree one,

$$\begin{aligned} f(\mathbf{x} + \mathbf{y}) = f\left[2\left(\frac{\mathbf{x}}{2} + \frac{\mathbf{y}}{2}\right)\right] &= 2f[(1/2)\mathbf{x} + (1/2)\mathbf{y}] \\ &\geq 2[(1/2)f(\mathbf{x}) + (1/2)f(\mathbf{y})] = f(\mathbf{x}) + f(\mathbf{y}). \end{aligned}$$

Therefore, f is super-additive on K.

Now suppose that f is super-additive on K, and let $\mathbf{x}, \mathbf{y} \in K$, and $\lambda \in]0, 1[$. Then

$$f[\lambda \mathbf{x} + (1 - \lambda)\mathbf{y}] \geq f(\lambda \mathbf{x}) + f[(1 - \lambda)\mathbf{y}] = \lambda f(\mathbf{x}) + (1 - \lambda) f(\mathbf{y}).$$

[15] The terminology here is suggested by the nonnegative and semipositive orthants in $\mathbf{R}^n$.

Therefore f is concave on K.

The respective statements follow by similar reasoning. □

We conclude this section by reviewing some facts regarding differentiable convex and concave functions whose domains are subsets of $\mathbf{R}$ or $\mathbf{R}^n$. Since this material is, presumably, largely review, we will not provide proofs here. For more complete discussions of the properties of differentiable concave and convex functions, see Berge [1963, pp. 194-200], Nikaido [1968, pp. 44-51], or Novshek [1993, pp. 95-8].

5.83. Theorem. *Suppose $f\colon I \to \mathbf{R}$, where I is a non-degenerate open interval of real numbers. Then the following statements hold:*

1. If f is differentiable on I, then f is convex on I if, and only if, f' is non-decreasing on I.

2. If f is twice-differentiable on I, then f is convex on I if, and only if, f'' is nonnegative on I.

5.84. Theorem. *Suppose $f\colon C \to \mathbf{R}$, where X is a convex subset of $\mathbf{R}^n$ having a nonempty interior, and suppose f has continuous second partial derivatives on $int(C)$. Then f is convex on $int(C)$ if, and only if, for all $\mathbf{x} \in int(C)$,*

$$H_f(\mathbf{x}) \stackrel{def}{=} [f_{ij}(\mathbf{x})],$$

is nonnegative definite.

5.85. Proposition. *If $f\colon I \to \mathbf{R}$, where I is a non-degenerate open interval of real numbers, and if f'' exists, for all $x \in I$, then f is strictly convex [respectively, concave] if, and only if, for all $x \in I$:*

$$f''(x) \geq 0 \text{ [respectively, } f''(x) \leq 0],$$

and f'' is not identically zero on any non-degenerate sub-interval of I.

5.86. Examples/Exercises.

1. Define $f\colon \mathbf{R}_+ \to \mathbf{R}$ by:

$$f(x) = x^\beta,$$

where β is a fixed real number. Show that f is:

a. convex, if $\beta \leq 0$ or $\beta \geq 1$.

b. concave if $0 \leq \beta \leq 1$.

2. Define $f\colon \mathbf{R}_{++} \to \mathbf{R}$ by

$$f(x) = \beta \log x,$$

where β is a fixed real number. Show that f is:

a. convex, if $\beta \leq 0$, and

b. concave if $\beta \geq 0$.

3. Suppose $\mathbf{p} \in \Delta_n$, and $\mathbf{x} \in \mathbf{R}^n_+$. Then $\mathbf{p} \cdot \mathbf{x} \geq \prod_{i=1}^n (x_i)^{p_i}$.

Proof. From exercise 2, above, it follows that the function f defined by

$$f(x) = \log x,$$

is concave on $\mathbf{R}_{++}$. Consequently, using Proposition 5.68, it follows that

$$\log(\mathbf{p} \cdot \mathbf{x}) \geq \sum_{i=1}^{n} p_i \log x_i = \log \left[\prod_{i=1}^{n} (x_i)^{p_i} \right].$$

Since $\exp(\cdot)$ is an increasing function, it then follows that

$$\mathbf{p} \cdot \mathbf{x} \geq \prod_{i=1}^{n} (x_i)^{p_i}. \quad \square$$

4. Suppose $f\colon X \to \mathbf{R}$, where X is a convex subset of $\mathbf{R}^n$, that f is differentiable on $int(X)$, let $\mathbf{x}^* \in int(X)$, and let $\mathbf{x}$ be an arbitrary element of X. If f is concave on X, then for all $\theta \in]0, 1[$,

$$f[\mathbf{x}^* + \theta(\mathbf{x} - \mathbf{x}^*)] \geq \theta f(\mathbf{x}) + (1 - \theta) f(\mathbf{x}^*) = f(\mathbf{x}^*) + \theta[f(\mathbf{x}) - f(\mathbf{x}^*)];$$

so that

$$\frac{f[\mathbf{x}^* + \theta(\mathbf{x} - \mathbf{x}^*)] - f(\mathbf{x}^*)}{\theta} \geq f(\mathbf{x}) - f(\mathbf{x}^*). \tag{5.73}$$

However, since f is differentiable at $\mathbf{x}^*$, the limit as $\theta \to 0^+$ of the left-hand-side of (5.73) exists, and is equal to the directional derivative of f at $\mathbf{x}^*$ in the direction $\mathbf{x} - \mathbf{x}^*$; in fact:

$$\lim_{\theta \to 0^+} \left(\frac{f[\mathbf{x}^* + \theta(\mathbf{x} - \mathbf{x}^*)] - f(\mathbf{x}^*)}{\theta} \right) = \nabla f(\mathbf{x}^*) \cdot (\mathbf{x} - \mathbf{x}^*).$$

But then it follows from (5.73) that:

$$\nabla f(\mathbf{x}^*) \cdot (\mathbf{x} - \mathbf{x}^*) \geq f(\mathbf{x}) - f(\mathbf{x}^*),$$

and, since $\mathbf{x}$ was an arbitrary element of X, we have:

$$(\forall \mathbf{x} \in X)\colon \nabla f(\mathbf{x}^*) \cdot \mathbf{x} - f(\mathbf{x}) \geq \nabla f(\mathbf{x}^*) \cdot \mathbf{x}^* - f(\mathbf{x}^*). \tag{5.74}$$

From (5.74) and the definition of *hull*(f) [Definition 5.70], we can then see that the vector

$$(\nabla f(\mathbf{x}^*), -1) \in \mathbf{R}^{n+1},$$

defines a supporting hyperplane to *hull*(f) at $(\mathbf{x}^*, f(\mathbf{x}^*))$. $\square$

Exercises.

1. Prove the following Propositions: (a) 5.68, (b) 5.76, and (c) 5.79.
2. Show that if $\mathbf{a} \in \mathbf{R}^n$, then the function $f\colon \mathbf{R}^n \to \mathbf{R}$ defined by

$$f(\mathbf{x}) = (\mathbf{a} \cdot \mathbf{x})^2 \quad \text{for } \mathbf{x} \in \mathbf{R}^n,$$

is convex on $\mathbf{R}^n$.

3. Let $X \subseteq \mathbf{R}^n_+$ be a consumption set, let $\succsim$ be a preference (binary) relation on X, define

$$\Omega = \{(\mathbf{p}, m) \in \mathbf{R}^n_+ \times \mathbf{R} \mid \mathbf{p} \neq \mathbf{0} \ \& \ (\exists \mathbf{x} \in X) \colon \mathbf{p} \cdot \mathbf{x} \leq m\},$$

define h on Ω by:

$$h(\mathbf{p}, m) = \{\mathbf{x} \in X \mid (\forall \mathbf{y} \in X) \colon \mathbf{x} \succsim \mathbf{y}\},$$

and define $\Omega^* \subseteq \Omega$ by:

$$\Omega^* = \{(\mathbf{p}, m) \in \Omega \mid h(\mathbf{p}, m) \neq \emptyset\}.$$

Show that Ω and Ω^* are both cones. Is Ω necessarily a convex cone? If not, can you derive conditions on X and/or $\succsim$ under which it is?

4. Prove directly from the definition of concavity (in particular, without using Theorem 5.72) that if f and g are both concave on X, then the function h defined on X by:

$$h(\mathbf{x}) = \min\{f(\mathbf{x}), g(\mathbf{x})\},$$

is also concave.

5. Suppose $\varphi \colon C \to \mathbf{R}$, where C is a convex set, and that φ is bounded on C. Prove that, if we define the set $\mathcal{F}$ by

$$\mathcal{F} = \{f \colon C \to \mathbf{R} \mid f \text{ is convex and } (\forall \mathbf{x} \in C) \colon f(\mathbf{x}) \leq \varphi(\mathbf{x})\},$$

and define $g \colon C \to \mathbf{R}$ by

$$g(\mathbf{x}) = \sup\{f(\mathbf{x}) \mid f \in \mathcal{F}\},$$

then g is convex on X, and

$$epi(g) = \bigcap_{f \in \mathcal{F}} epi(f) = \langle epi(\varphi) \rangle.$$

6. Show that if K is a convex cone in a normed linear space, then riK is also a convex cone.

5.7 Continuity of Convex Functions

It is almost the case that a convex or concave function is necessarily also continuous. In this section we will investigate the issue of whether and to what extent continuity is a necessary condition for convexity of a function. We will generally state our results only for the case of convex functions; where this is the case, however, the student should keep in mind that an analogous result holds for concave functions, although we will leave the statements and proofs of the analogous results as exercises.

In this section, except where otherwise specifically stated, we shall suppose that the domains of all the functions with which we are dealing are contained in a normed real linear space having norm $\|\cdot\|$.

The following proposition represents a bit of a digression from our present topic, but will be useful in establishing some of the results of this and the next section.

5.87. Proposition. *If C is convex, $\mathbf{x} \in riC$, and $\mathbf{y} \in \mathrm{aff}C$, then there exists $\mathbf{z} \in riC$ and $\theta \in]0,1[$ such that*

$$\mathbf{x} = \theta\mathbf{z} + (1-\theta)\mathbf{y}.$$

Proof. Since $\mathbf{x} \in riC$ and $\mathbf{y} \in \mathit{aff}C$, it follows from Proposition 5.23 that there exists $\mu > 1$ such that

$$\mathbf{z} \stackrel{\text{def}}{=} (1-\mu)\mathbf{y} + \mu\mathbf{x} \in riC.$$

However, if we now define $\theta = 1/\mu$, we see that

$$0 < \theta < 1,$$

and

$$\mathbf{x} = \theta\mathbf{z} + (1-\theta)\mathbf{y}. \quad \square$$

The principal result of this section is the following. After proving it, we will investigate the extent of the need for the boundedness hypothesis in the context of $\mathbf{R}^n$.

5.88. Theorem. *Suppose $f\colon D \to \mathbf{R}$ is a convex function, where D is a convex set. If there exists a point $\mathbf{a} \in riD$, and a positive real number, δ, such that f is bounded above on $N(\mathbf{a},\delta) \cap D$, then f is continuous on riD.*

We will begin our proof of 5.88 by first establishing the following.

5.89. Lemma. *Suppose C is a convex subset of the normed real linear space, L (with norm $\|\cdot\|$), and that $\mathbf{0} \in riC$. If $f\colon C \to \mathbf{R}$ is a convex function satisfying:*

1. *there exists $\eta > 0$ such that f is bounded above on $N(\mathbf{0},\eta) \cap C$, and*
2. *$f(\mathbf{0}) = 0$,*

then f is continuous at $\mathbf{0}$.

Proof. Let $\epsilon > 0$ be given, and let $\alpha > 0$ be such that

$$(\forall \mathbf{x} \in N(\mathbf{0},\eta) \cap C)\colon f(\mathbf{x}) < \alpha.$$

Since $\mathbf{0} \in riC$, we can suppose, without loss of generality, that

$$N(\mathbf{0}, \eta) \cap aff C \subseteq C; \tag{5.75}$$

and obviously we can also suppose, without loss of generality, that $\epsilon < 1$. By way of completing our preliminary observations, we also note that, since $\mathbf{0} \in C$, $aff C$ is necessarily a linear subspace of L.

Now, define

$$\delta = \epsilon\eta/(1+\alpha),$$

and

$$\mu = \epsilon/(1+\alpha) = \delta/\eta;$$

and note that, since $\epsilon < 1$,

$$0 < \mu < 1.$$

Suppose now that:

$$\mathbf{x} \in N(\mathbf{0}, \delta) \cap C.$$

Then we have

$$\|(1/\mu)(-\mathbf{x})\| = \|(1/\mu)\mathbf{x}\| = (1/\mu)\|\mathbf{x}\| < (1/\mu)\delta = \eta; \tag{5.76}$$

and, since $aff C$ is a linear space, and $\mathbf{x} \in C$, we see that both $(1/\mu)\mathbf{x}$ and $(1/\mu)(-\mathbf{x}) = (-1/\mu)\mathbf{x}$ are elements of $aff C$. Thus we have from (5.75) and (5.76) that both $(1/\mu)\mathbf{x}$ and $(1/\mu)(-\mathbf{x})$ are elements of C. Furthermore, we have:

$$\begin{aligned} f(\mathbf{x}) &= f[(1-\mu)\mathbf{0} + \mu(1/\mu)\mathbf{x}] \\ &\leq (1-\mu)f(\mathbf{0}) + \mu f[(1/\mu)\mathbf{x}] = \mu f[(1/\mu)\mathbf{x}] < \mu\alpha < \epsilon; \end{aligned} \tag{5.77}$$

where the next-to-the-last inequality in (5.77) is from (5.76) and (5.75) [as well as the definitions of μ and δ]. Furthermore, since

$$\mathbf{0} = \left[\frac{1}{1+\mu}\right]\mathbf{x} + \left[\frac{\mu}{1+\mu}\right] \cdot \left[\left(\frac{1}{\mu}\right)(-\mathbf{x})\right];$$

we also have

$$\begin{aligned} 0 = f(\mathbf{0}) &\leq \left(\frac{1}{1+\mu}\right) f(\mathbf{x}) + \left(\frac{\mu}{1+\mu}\right) f\left[\left(\frac{1}{\mu}\right)(-\mathbf{x})\right] \\ &< \left(\frac{1}{1+\mu}\right) f(\mathbf{x}) + \left(\frac{\mu}{1+\mu}\right) \alpha; \end{aligned}$$

and thus:

$$f(\mathbf{x}) > -\mu\alpha = -\epsilon\alpha/(1+\alpha) > -\epsilon. \tag{5.78}$$

Combining (5.77) and (5.78), we see that:

$$-\epsilon < f(\mathbf{x}) = f(\mathbf{x}) - f(\mathbf{0}) < \epsilon;$$

and thus:

$$|f(\mathbf{x}) - f(\mathbf{0})| < \epsilon \quad \text{for } \mathbf{x} \in N(\mathbf{0}, \delta) \cap C,$$

and we conclude that f is continuous at $\mathbf{0}$. □

Proof of Theorem 5.88. Let $\mathbf{x}^* \in riD$ be arbitrary. We will first show that there exists a positive real number, η, such that f is bounded above on $N(\mathbf{x}^*, \eta)$, and we then show that this implies that f is continuous at $\mathbf{x}^*$.

We proceed as follows. From Proposition 5.87 we see that there exists $\mathbf{z} \in riD$, and $\theta \in]0, 1[$ such that

$$\mathbf{x}^* = \theta\mathbf{z} + (1 - \theta)\mathbf{a}; \tag{5.79}$$

and we note that it follows from the fact that $\mathbf{a} \in riD$ that we can assume, without loss of generality, that

$$N(\mathbf{a}, \delta) \cap \mathit{aff}\, D \subseteq D. \tag{5.80}$$

Let $\alpha \in \mathbf{R}$ be such that

$$(\forall \mathbf{x} \in N(\mathbf{a}, \delta) \cap D)\colon f(\mathbf{x}) \leq \alpha,$$

and define

$$\eta = (1 - \theta)\delta,$$

and

$$\beta = \theta f(\mathbf{z}) + (1 - \theta)\alpha.$$

We will prove that

$$(\forall \mathbf{y} \in N(\mathbf{x}^*, \eta) \cap D)\colon f(\mathbf{y}) \leq \beta. \tag{5.81}$$

Accordingly, let $\mathbf{y}^* \in N(\mathbf{x}^*, \eta) \cap D$ be arbitrary, and define $\mathbf{y}$ by

$$\mathbf{y} = [1/(1 - \theta)]\mathbf{y}^* - [\theta/(1 - \theta)]\mathbf{z}.$$

Since $\mathbf{y}$ is an affine combination of $\mathbf{y}^*$ and $\mathbf{z}$, both of which are elements of D, we see that

$$\mathbf{y} \in \mathit{aff}\, D; \tag{5.82}$$

and, using the fact [from (5.79)] that

$$\mathbf{a} = [1/(1 - \theta)]\mathbf{x}^* - [\theta/(1 - \theta)]\mathbf{z},$$

we have:

$$\begin{aligned} \|\mathbf{y} - \mathbf{a}\| &= \left\| \left(\frac{1}{1-\theta}\right)\mathbf{y}^* - \left(\frac{\theta}{1-\theta}\right)\mathbf{z} - \left(\frac{1}{1-\theta}\right)\mathbf{x}^* + \left(\frac{\theta}{1-\theta}\right)\mathbf{z} \right\| \\ &= [1/(1-\theta)] \cdot \|\mathbf{y}^* - \mathbf{x}^*\| < [1/(1-\theta)] \cdot \eta = \delta. \end{aligned} \tag{5.83}$$

From (5.80), (5.82), and (5.83), we see that $\mathbf{y} \in D$, and $f(\mathbf{y}) \leq \alpha$. Consequently,

$$f(\mathbf{y}^*) = f[\theta \mathbf{z} + (1-\theta)\mathbf{y}] \leq \theta f(\mathbf{z}) + (1-\theta) f(\mathbf{y}) \leq \beta;$$

and (5.81) follows.

Next, we define the set C as the displacement of D by $\mathbf{x}^*$; that is,

$$C = D - \{\mathbf{x}^*\}.$$

Then C is convex, and by Corollary 5.43, $\mathbf{0} \in riC$. Now define $h\colon C \to D$ by:

$$h(\mathbf{y}) = \mathbf{y} + \mathbf{x}^* \quad \text{for } \mathbf{y} \in C,$$

and note that h is affine. Consequently, it follows from Proposition 5.73 that the function $g\colon C \to \mathbf{R}$ defined by

$$g(\mathbf{y}) = f[h(\mathbf{y})] - f(\mathbf{x}^*)$$

is convex, and $g(\mathbf{0}) = 0$. We then have from Lemma 5.89 that g is continuous at $\mathbf{0}$.

Now define $\varphi\colon D \to C$ by

$$\varphi(\mathbf{x}) = \mathbf{x} - \mathbf{x}^*,$$

and note that φ is continuous on D. Therefore the function

$$g[\varphi(\mathbf{x})] + f(\mathbf{x}^*),$$

is continuous at $\varphi^{-1}(\mathbf{0})$. However, $\varphi^{-1}(\mathbf{0}) = \mathbf{x}^*$, and, for any $\mathbf{x} \in D$, we have:

$$\begin{aligned} g[\varphi(\mathbf{x})] + f(\mathbf{x}^*) &= \left[f\left(h[\varphi(\mathbf{x})]\right) - f(\mathbf{x}^*)\right] + f(\mathbf{x}^*) \\ &= f[\varphi(\mathbf{x}) + \mathbf{x}^*] = f(\mathbf{x} - \mathbf{x}^* + \mathbf{x}^*) = f(\mathbf{x}). \end{aligned}$$

Consequently, we see that f is continuous at $\mathbf{x}^*$. □

The following is an immediate consequence of 5.88 and the definition of upper semi-continuity.

5.90. Corollary. *If $f: D \to \mathbf{R}$ is convex, where D is a convex set, and if f is upper semi-continuous at some $\mathbf{a} \in riD$, then f is continuous on riD.*

If the domain of the function is of finite dimension, then we can dispense with the assumption that the function is bounded above in Theorem 5.88. This is because a convex function on such a domain is necessarily bounded above in some neighborhood of each relative interior point; as we shall demonstrate shortly. We begin with the following lemma.

5.91. Lemma. *If $f : C \to \mathbf{R}$ is a convex funtion, where C is a convex subset of $\mathbf{R}^n$, and if $\mathbf{0} \in int(C)$, then there exists $\epsilon > 0$ such that f is bounded on $N(\mathbf{0}, \epsilon)$.*

Proof. Since $\mathbf{0} \in int(C)$, there exists $\delta > 0$ such that for all $\mathbf{x} \in \mathbf{R}^n$:

$$\sum_{i=1}^n |x_i| < \delta \Rightarrow \mathbf{x} \in C. \tag{5.84}$$

Define $\mathbf{y}_1, \ldots, \mathbf{y}_{2n}$ by:

$$\mathbf{y}_i = \begin{cases} \delta \mathbf{e}_i & \text{for } i = 1, \ldots, n, \\ -\delta \mathbf{e}_{i-n} & \text{for } i = n+1, \ldots, 2n; \end{cases}$$

where '$\mathbf{e}_i$' denotes the ith unit coordinate vector, and note that, from (5.84):

$$\mathbf{y}_i \in C \quad \text{for } i = 1, \ldots, 2n.$$

Now suppose $\mathbf{x}$ is such that:

$$\sum_{i=1}^n |x_i| < \delta.$$

Then by (5.84), $\mathbf{x} \in C$, and obviously:

$$\mathbf{x} = \sum_{i=1}^n x_i \mathbf{e}_i.$$

However, if we define $\mathbf{a} \in \mathbf{R}_+^{2n}$ by:

$$a_i = \begin{cases} x_i/\delta & \text{if } x_i \geq 0 \\ 0 & \text{if } x_i < 0, \quad \text{for } i = 1, \ldots, n, \\ -x_{i-n}/\delta & \text{if } x_{i-n} < 0 \\ 0 & \text{if } x_{i-n} \geq 0 \quad \text{for } i = n+1, \ldots, 2n; \end{cases}$$

we have:

$$\sum_{i=1}^{2n} a_i \mathbf{y}_i = \sum_{i=1}^n \left(\frac{x_i}{\delta}\right) \mathbf{y}_i = \sum_{i=1}^n x_i \mathbf{e}_i = \mathbf{x},$$

and:

$$\sum_{i=1}^{2n} a_i = \left(\frac{1}{\delta}\right) \sum_{i=1}^n |x_i| < 1.$$

Consequently, the point $\mathbf{x}$ is a convex combination:

$$\mathbf{x} = \sum_{i=1}^{2n} a_i \mathbf{y}_i + \left(1 - \sum_{i=1}^{2n} a_i\right) \mathbf{0},$$

and it follows from Proposition 5.68 that:

$$\begin{aligned} f(\mathbf{x}) &= f\left[\sum_{i=1}^{2n} a_i \mathbf{y}_i + \left(1 - \sum_{i=1}^{2n} a_i\right) \mathbf{0}\right] \\ &\leq \sum_{i=1}^{2n} a_i f(\mathbf{y}_i) + \left(1 - \sum_{i=1}^{2n} a_i\right) f(\mathbf{0}) \\ &\leq \max\{f(\mathbf{y}_1), \ldots, f(\mathbf{y}_{2n}), f(\mathbf{0})\}. \end{aligned}$$

We see, therefore, that if we define $\epsilon = \delta/n$ (see Exercise 4, at the end of Section 5.2), then f is bounded above on $N(\mathbf{0}, \epsilon)$. □

5.92. Proposition. *If $f: C \to \mathbf{R}$ is a convex function, where C is a convex subset of L such that* $\dim C = n$, *then f is continuous on riC.*

Proof. Let $\mathbf{x}^* \in riC$, and let $\mathbf{y}_1, \ldots, \mathbf{y}_n$ be a basis for $Sp(C - C)$. Then the function $h: \mathbf{R}^n \to affC$ defined by:

$$h(\boldsymbol{a}) = \sum\nolimits_{i=1}^{n} a_i \mathbf{y}_i + \mathbf{x}^* \quad \text{for } \boldsymbol{a} \in \mathbf{R}^n,$$

is a one-to-one affine function which maps onto $affC$. Moreover, defining the set D by:

$$D = h^{-1}(C),$$

we know that D is convex, and (from Proposition 5.33) that $\mathbf{0} \in int(D)$. Since the composition function,

$$g(\boldsymbol{a}) = f[h(\boldsymbol{a})],$$

is convex (by Proposition 5.73), it now follows from Lemma 5.91 that g is bounded in a neighborhood, $N(\mathbf{0})$ of $\mathbf{0}$; and (since $h[N(\mathbf{0})]$ is a neighborhood of $\mathbf{x}^*$) it then follows from Theorem 5.88 that f is continuous at $\mathbf{x}^*$. □

It is easy to construct examples of convex functions satisfying all of the conditions of Proposition 5.92, but which are *not* lower semi-continuous at boundary points.[16] One might speculate, however, that such functions must necessarily be upper semi-continuous even at boundary points. However, such is not the case, as is shown by the following example; on the other hand, see also Proposition 5.94, which follows this example.

5.93. Example. Define the sets C, D, and B in $\mathbf{R}^2_+$ by:

$$C = \{(x, y) \in \mathbf{R}^2_+ \mid 0 \leq x \leq 1 \ \& \ (x-1)^2 + y^2 \leq 1\},$$

and:

$$D = \{(x, y) \in \mathbf{R}^2_+ \mid 0 < x \leq 1 \ \& \ (x-1)^2 + y^2 = 1\},$$

and:

$$B = C \setminus D.$$

We then define the function $f: C \to \mathbf{R}_+$ by:

$$f(x, y) = \begin{cases} x & \text{for } (x, y) \in B, \\ 1/y^2 & \text{for } (x, y) \in D. \end{cases}$$

[16] See Exercise 2 at the end of this section.

Define the function $g\colon \mathbf{R}^2_+ \to \mathbf{R}_+$ by:

$$g(x,y) = x,$$

and notice that for all $(x,y) \in B$, we have:

$$f(x,y) = g(x,y) = x = g(x,0) = f(x,0). \tag{5.85}$$

Next, we note that if $(x,y) \in D$, then:

$$(x-1)^2 + y^2 = x^2 - 2x + 1 + y^2 = 1,$$

so that:

$$y^2 = 2x - x^2. \tag{5.86}$$

To complete our preliminary observations, consider the function $\sigma\colon \mathbf{R} \to \mathbf{R}$ defined by:

$$\sigma(x) = 2x^2 - x^3. \tag{5.87}$$

It is easy to prove (see Exercise 2, at the end of this section) that σ is strictly increasing on the interval [0, 1], and that:

$$\sigma(0) = 0 \text{ and } \sigma(1) = 1.$$

Thus we have:

$$0 < 2x^2 - x^3 < 1 \quad \text{for } 0 < x < 1,$$

so that:

$$0 < x < \frac{1}{2x - x^2} \quad \text{for } 0 < x < 1. \tag{5.88}$$

Comparing (5.85), (5.86), and (5.88), we see that:

$$f(x,y) = 1/y^2 > x = g(x,0) = f(x,0) \quad \text{for } (x,y) \in D. \tag{5.89}$$

To prove that f is convex, suppose $\mathbf{z} \equiv (x,y), \mathbf{z}' \equiv (x',y') \in C$ are such that $(x,y) \neq (x',y')$, let $\theta \in\,]0,1[$, and define:

$$\mathbf{z}^* = \theta\mathbf{z} + (1-\theta)\mathbf{z}' \stackrel{\text{def}}{=} (x^*, y^*).$$

In order to show that:

$$f(\mathbf{z}^*) \leq \theta f(\mathbf{z}) + (1-\theta) f(\mathbf{z}'), \tag{5.90}$$

we begin by noting that, since each element of D is an extreme point of C, we necessarily have $\mathbf{z}^* \in B$, and thus:

$$f(\mathbf{z}^*) = g(\mathbf{z}^*) = f(x^*, 0) = g(x^*, 0) = x^*. \tag{5.91}$$

We can then distinguish three cases.

a. $\mathbf{z}, \mathbf{z}' \in B$. In this case, the inequality (5.90) follows more or less immediately from the convexity of g. I will leave the details as an exercise.

b. $\mathbf{z} \in B$ and $\mathbf{z}' \in D$. Here we note that, using (5.91), the definition of $\mathbf{z}^*$, (5.85), and (5.89) in turn, we have:

$$\begin{aligned} f(\mathbf{z}^*) = x^* &= \theta x + (1-\theta)x' = \theta g(x,0) + (1-\theta)g(x',0) \\ &= \theta f(\mathbf{z}) + (1-\theta)f(x',0) < \theta f(\mathbf{z}) + (1-\theta)f(\mathbf{z}'). \end{aligned}$$

c. Both $\mathbf{z}$ and $\mathbf{z}'$ are in D. Inequality (5.90) can be established in this case by an argument quite similar to that used in case b. Details will be left as an exercise.

It is easy to establish that our function f defined in this example has a number of most inconvenient properties: (i) it is not upper semicontinuous at the origin, and in fact, it is not bounded above in any neighborhood of the origin; (ii) the point $(1,1)$ is the only point in D at which the function is lower semicontinuous, and (iii) despite the fact that the domain of the function is compact, the function is not bounded above on C. □

We will change format slightly to state the following result for concave functions, since it seems to be more useful in economic applications than is the analogous result for convex functions.

5.94. Proposition. *If $f\colon \mathbf{R}^n_+ \to \mathbf{R}$ is non-decreasing and concave on $\mathbf{R}^n_+$, then f is lower semicontinuous on $\mathbf{R}^n_+$.*[17]

Proof. Notice, first of all, that it suffices to prove this result for the case in which $f(\mathbf{0}) = 0$, which we will assume to be the case in the following argument. Let $\mathbf{x}^* \in \mathbf{R}^n_+$ and let $\epsilon > 0$ be given. We distinguish two cases.

a. $f(\mathbf{x}^*) = 0$. In this case, it is very easy to prove that f is lower semicontinuous at $\mathbf{x}^*$, and I will leave the details as an exercise.

b. $f(\mathbf{x}^*) > 0$. If we define:

$$\theta = \max\{1/2, [f(\mathbf{x}^*) - \epsilon/2]/f(\mathbf{x}^*)\},$$

we note that we have $1/2 \leq \theta < 1$, and, by the concavity of f:

$$\begin{aligned} f(\theta\mathbf{x}^*) = f[\theta\mathbf{x}^* + (1-\theta)\mathbf{0}] &\geq \theta f(\mathbf{x}^*) + (1-\theta)f(\mathbf{0}) \\ &= \theta f(\mathbf{x}^*) \geq f(\mathbf{x}^*) - \epsilon/2. \end{aligned} \tag{5.92}$$

Next, define:

$$I = \{i \in \{1, \ldots, n\} \mid x_i^* > 0\},$$

[17] It is also upper semi-continuous on $\mathbf{R}^n_{++}$, as follows easily from Proposition 5.92. However, see Exercise 2, at the end of this section.

and:

$$\delta = (1-\theta)\min\{x_i^* \mid i \in I\}.$$

If $\mathbf{x} \in N(\mathbf{x}^*, \delta) \cap \mathbf{R}_+^n$, we have:

$$|x_i - x_i^*| \leq \|\mathbf{x} - \mathbf{x}^*\| < \delta;$$

so that, for $i \in I$:

$$-\delta < x_i - x_i^* < \delta,$$

and thus:

$$x_i > x_i^* - \delta = x_i^* - (1-\theta)\min\{x_i \mid i \in I\} \geq x_i^* - (1-\theta)x_i^* = \theta x_i^*.$$

Thus it follows that:

$$(\forall \mathbf{x} \in N(\mathbf{x}^*, \delta) \cap \mathbf{R}_+^n) \colon \mathbf{x} \geq \theta \mathbf{x}^*;$$

and thus from (5.92) and the fact that f is nondecreasing, it follows that:

$$(\forall \mathbf{x} \in N(\mathbf{x}^*, \delta) \cap \mathbf{R}_+^n) \colon f(\mathbf{x}) \geq f(\mathbf{x}^*) - \epsilon/2 > f(\mathbf{x}^*) - \epsilon. \quad \square$$

Our final result of this section may appear to be somewhat out of place. However, it can be proved by an argument very similar to the proof just completed; and consequently is included here. The details of the proof will be left as an exercise.[18]

5.95. Proposition. *If $f \colon \mathbf{R}_+^n \to \mathbf{R}$ is non-decreasing and positively homogeneous of degree one on $\mathbf{R}_+^n$, then f is lower semicontinuous on $\mathbf{R}_+^n$.*

Exercises.

1. Prove Corollary 5.90.

2. Construct an example of a function mapping $\mathbf{R}_+^n$ into $\mathbf{R}_+$ which is non-decreasing and concave, but which is *not* everywhere upper semicontinuous. (Try taking $n = 1$; that is, let the domain of the function be $\mathbf{R}_+$.)

3. Prove the stated properties of the function $\sigma(\cdot)$ defined in equation (5.87) of Example 5.93.

4. Establish inequality (5.90) for cases (a) and (c) of Example 5.93.

5. Prove Proposition 5.95.

6. Define $C \subseteq \mathbf{R}^3$ by:

$$C = \{(x, y, z) \in \mathbf{R}^3 \mid z \geq 0 \;\&\; x^2 + y^2 \leq z^2\},$$

and define f on C by:

$$f(x,y,z) = \begin{cases} z & \text{if } x^2 + y^2 < z^2 \text{ or } y^2 = z^2 \;\&\; x = 0, \\ 2z & \text{if } x^2 + y^2 = z^2 \;\&\; x \neq 0. \end{cases}$$

Prove that f is positively homogeneous of degree one and convex, but is *neither* upper-, nor lower semicontinuous.

---

[18]See also Exercise 6, at the end of this section.

5.8 Quasi-Concave and Quasi-Convex Functions

In intermediate microtheory texts, indifference maps for consumers always have indifference curves which are convex to the origin; that is, such that the preferred-or-indifferent sets are always convex. Mathematically this will be so if the utility function involved is (non-decreasing and) quasi-concave. We will study a few of the properties of such functions in this section, along with the symmetric concept of quasi-convex functions.

5.96. Definitions. Let $f\colon X \to \mathbf{R}$, where X is a convex subset of L. We shall say that f is:

1. **quasi-convex** (on X) iff, for all $\mathbf{x} \in X$,

$$\{\mathbf{y} \in X \mid f(\mathbf{y}) \leq f(\mathbf{x})\}$$

is a convex set.

2. **quasi-concave** (on X) iff, for all $\mathbf{x} \in X$,

$$\{\mathbf{y} \in X \mid f(\mathbf{y}) \geq f(\mathbf{x})\}$$

is a convex set.

Until further notice, we will suppose that the domains of all functions with which we are dealing are convex subsets of some arbitrary real linear space, L.

5.97. Proposition. *The following two conditions are equivalent to quasi-concavity, for* $f\colon X \to \mathbf{R}$:

1. for all $a \in \mathbf{R}$, *the set*

$$X_a \equiv \{\mathbf{x} \in X \mid f(\mathbf{x}) \geq a\},$$

is a convex set.

2. for all $\mathbf{x}, \mathbf{y} \in X$, *and all* $\theta \in [0, 1]$,

$$f[\theta\mathbf{x} + (1-\theta)\mathbf{y}] \geq \min\{f(\mathbf{x}), f(\mathbf{y})\}.$$

Proof. I will leave most of this proof as an exercise. However, let me note that you may find it easiest to prove these equivalences by establishing the following implications as the steps in your proof: (a) quasi-concavity implies condition 2, (b) condition 2 implies condition 1, and (c) condition 1 implies quasi-concavity. □

Notice that if a function is concave, then it is also necessarily quasi-concave; and similarly, a convex function is quasi-convex. The converse statements are not correct, however; in particular, any non-decreasing real-valued function whose domain is an interval of real numbers is quasi-concave. Thus, for example, the function $\exp(x)$ is quasi-concave; even though it is obviously not concave. Other simple properties of quasi-concave and quasi-convex functions are set out in the following.

5.98. Examples/Exercises.

1. If $f\colon X \to \mathbf{R}^n$ is concave, and $g\colon f(X) \to \mathbf{R}$ is nondecreasing, then the composition function, $h \equiv g \circ f$ is quasi-concave.

2. Suppose $f\colon C \to \mathbf{R}$ is quasi-concave on C, where $C \subseteq \mathbf{R}^n$ is convex and has a non-empty interior. Let $\mathbf{x}^* \in int(C)$ be such that f is differentiable at $\mathbf{x}^*$, and define $X_{\mathbf{x}^*}$ by:

$$X_{\mathbf{x}^*} = \{\mathbf{x} \in C \mid f(\mathbf{x}) \geq f(\mathbf{x}^*)\}.$$

Then, if $\mathbf{x}$ is an arbitrary element of $X_{\mathbf{x}^*}$, it follows easily from the definition of quasi-concavity that, for $\theta \in]0,1[$:

$$\frac{f[\mathbf{x}^* + \theta(\mathbf{x} - \mathbf{x}^*)] - f(\mathbf{x}^*)}{\theta} \geq 0;$$

and thus:

$$\lim_{\theta \to 0+} \left[\frac{f[\mathbf{x}^* + \theta(\mathbf{x} - \mathbf{x}^*)] - f(\mathbf{x}^*)}{\theta}\right] = \nabla f(\mathbf{x}^*) \cdot (\mathbf{x} - \mathbf{x}^*) \geq 0.$$

Consequently, $\nabla f(\mathbf{x}^*)$ defines a supporting hyperplane to $X_{\mathbf{x}^*}$ at $\mathbf{x}^*$. □

In economic theory we often assume that a function is both quasi-concave and positively homogeneous of degree one. Roughly speaking, the conjunction of these two conditions implies that the function is actually concave. In most of the remainder of this section, we will concern ourselves with deriving precise conditions under which this statement is correct. Our first result of this kind is the following.

5.99. Theorem. *Suppose $f\colon C \to \mathbf{R}$ is positively homogeneous of degree one, where $C \subseteq L$ is a convex cone, and suppose f satisfies:*

$$(\forall \mathbf{x} \in C \setminus \{\mathbf{0}\})\colon f(\mathbf{x}) > 0. \tag{5.93}$$

Then f is quasi-concave [respectively, quasi-convex] if, and only if, f is concave [respectively, convex] on C.

Proof. The sufficiency portion of the result is, of course, obvious. Suppose now that f is quasi-concave on C, and let $\mathbf{x}$ and $\mathbf{y}$ be arbitrary elements of C. If, say, $\mathbf{x} = \mathbf{0}$, then we have from the definition of positive homogeneity that

$$f(\mathbf{x}) = 0;$$

and thus

$$f(\mathbf{x} + \mathbf{y}) = f(\mathbf{y}) = f(\mathbf{x}) + f(\mathbf{y}).$$

Suppose now that $\mathbf{x}, \mathbf{y} \in C \setminus \{\mathbf{0}\}$, and define:

$$\mathbf{x}^* = [1/f(\mathbf{x})]\mathbf{x} \ \text{ and } \ \mathbf{y}^* = [1/f(\mathbf{y})]\mathbf{y}.$$

Since f is positively homogeneous of degree one on C, it follows that

$$f(\mathbf{x}^*) = f(\mathbf{y}^*) = 1.$$

Therefore, since f is quasi-concave, we have that, for all $\theta \in [0,1]$:

$$f[\theta\mathbf{x}^* + (1-\theta)\mathbf{y}^*] \geq \min\{f(\mathbf{x}^*), f(\mathbf{y}^*)\} = 1. \tag{5.94}$$

In particular, therefore, equation (5.94) must hold if we set

$$\theta = \frac{f(\mathbf{x})}{f(\mathbf{x}) + f(\mathbf{y})}.$$

However, for this value of θ, we have

$$f[\theta\mathbf{x}^* + (1-\theta)\mathbf{y}^*] = f\Big[\Big(1/[f(\mathbf{x}) + f(\mathbf{y})]\Big)(\mathbf{x}+\mathbf{y})\Big];$$

and, combining this with (5.94) and the homogeneity of f, we obtain:

$$f(\mathbf{x}+\mathbf{y}) \geq f(\mathbf{x}) + f(\mathbf{y}).$$

Thus we see that f is super-additive on C, and it then follows from Proposition 5.82 that f is concave.

The respective statements follow either (a) by symmetric arguments, or (b) by the fact that f is quasi-convex iff $-f$ is quasi-concave. □

It should be noted that the above result does not remain correct if we drop the requirement expressed in equation (5.93) from its hypotheses; even if we retain the assumption that $f\colon C \to \mathbf{R}_+$. Thus, for example, let $C = \mathbf{R}^2_+$ and define f on C by:

$$f(\mathbf{x}) = \begin{cases} 0 & \text{if } x_1 > 0, \\ x_2 & \text{if } x_1 = 0. \end{cases} \tag{5.95}$$

Here it is easy to show that f is positively homogeneous of degree one on C and quasi-concave. However, for $\mathbf{x} = (0,1)$, $\mathbf{y} = (1,0)$, and $\theta = 1/2$, we have:

$$f[\theta\mathbf{x} + (1-\theta)\mathbf{y}] = f(1/2, 1/2) = 0 < (1/2)f(\mathbf{x}) + (1/2)f(\mathbf{y}) = 1/2.$$

On the other hand, we can prove the following.

5.100. Theorem. *Suppose $f\colon C \to \mathbf{R}_+$ is positively homogeneous of degree one, where C is a convex cone with $riC \neq \emptyset$, and suppose f satisfies:*

$$(\forall \mathbf{x} \in riC)\colon f(\mathbf{x}) > 0 \quad \textit{and} \quad (\forall \mathbf{x} \in \mathrm{Fr}(C))\colon f(\mathbf{x}) = 0.$$

If, in addition, f is quasi-concave [respectively, quasi-convex] on riC, then f is concave [respectively, convex] on C.

Proof. Since riC is a convex cone (Exercise 4, at the end of Section 6), it follows from 5.99 that f is concave on riC. To prove that f is concave on all of C, let $\mathbf{x}$ and $\mathbf{y}$ be arbitrary elements of C, and let $\theta \in]0,1[$. We distinguish three cases.

Case 1. $\mathbf{x}, \mathbf{y} \in Fr(C)$. Here we have:

$$f(\mathbf{x}) = 0 = f(\mathbf{y}),$$

and thus we obviously have:

$$f[\theta\mathbf{x} + (1-\theta)\mathbf{y}] \geq 0 = \theta f(\mathbf{x}) + (1-\theta)f(\mathbf{y}).$$

Case 2. $\mathbf{x}, \mathbf{y} \in riC$. Here, as we have already noted, it follows at once from Theorem 5.99 that:

$$f[\theta\mathbf{x} + (1-\theta)\mathbf{y}] \geq \theta f(\mathbf{x}) + (1-\theta)f(\mathbf{y}).$$

Case 3. One of the two points is in riC, and the other in $Fr(C)$. Here we can suppose, without loss of generality, that:

$$\mathbf{x} \in riC \text{ and } \mathbf{y} \in Fr(C).$$

Define:

$$\mathbf{z} = \theta\mathbf{x} + (1-\theta)\mathbf{y};$$

and, for each positive integer, m, define:

$$\mu_m = \frac{1}{m(1-\theta)+1},$$

and

$$\mathbf{y}_m = \mu_m\mathbf{x} + (1-\mu_m)\mathbf{y}.$$

For each m, we have:

$$0 < \mu_m < 1,$$

and therefore, by Theorem 5.20, we see that

$$\mathbf{y}_m \in riC \quad \text{for } n = 1, 2, \ldots .$$

Consequently, using Theorem 5.99, we see that, for each m, and all $\lambda \in]0,1[$, we have

$$f[\lambda\mathbf{x} + (1-\lambda)\mathbf{y}_m] \geq \lambda f(\mathbf{x}) + (1-\lambda)f(\mathbf{y}_m). \tag{5.96}$$

Now, defining λ_m by:

$$\lambda_m = \theta - 1/m,$$

we have, for all positive integers, m, satisfying $m > 1/\theta$,

$$0 < \lambda_m < 1;$$

and thus, from (5.96):

$$f[\lambda_m \mathbf{x} + (1 - \lambda_m)\mathbf{y}_m] \geq \lambda_m f(\mathbf{x}) + (1 - \lambda_m) f(\mathbf{y}_m). \tag{5.97}$$

However, for any such m,

$$\begin{aligned} &\lambda_m \mathbf{x} + (1 - \lambda_m)\mathbf{y}_m \\ &= [(\theta m - 1)/m]\mathbf{x} + [(m - \theta m + 1)/m)]\mu_m \mathbf{x} + [(m - \theta m + 1)/m][1 - \mu_m]\mathbf{y} \\ &\qquad = \theta \mathbf{x} + (1 - \theta)\mathbf{y} = \mathbf{z}. \end{aligned}$$

Therefore, using (5.97) and the fact that $f(\mathbf{y}) = 0$, we have

$$\begin{aligned} f[\theta \mathbf{x} + (1 - \theta)\mathbf{y}] &\geq (\theta - 1/m) f(\mathbf{x}) + (1 - \lambda_m) f(\mathbf{y}_m) \\ &\geq (\theta - 1/m) f(\mathbf{x}) + (1 - \theta) f(\mathbf{y}), \end{aligned}$$

for all $m > 1/\theta$. Since $\theta - 1/m \to \theta$ as $m \to +\infty$, it then follows that

$$f[\theta \mathbf{x} + (1 - \theta)\mathbf{y}] \geq \theta f(\mathbf{x}) + (1 - \theta) f(\mathbf{y}).$$

The respective statements follow from similar considerations. □

A similar argument, based upon inequality (5.97) in the above proof, establishes the following result.

5.101. Corollary. *Suppose $f\colon C \to \mathbf{R}_+$ is positively homogeneous of degree one and continuous on C, where C is a convex cone having $riC \neq \emptyset$, and suppose C satisfies*

$$(\forall \mathbf{x} \in riC)\colon f(\mathbf{x}) > 0.$$

If, in addition, f is quasi-concave [respectively, quasi-convex] on riC, then f is concave [respectively, convex] on C.

We can make use of 5.100 to derive a very useful application, as follows.

5.102. Proposition. *If $\alpha_j \geq 0$ for each j, then the function $f\colon \mathbf{R}^n_+ \to \mathbf{R}_+$ defined by:*

$$f(\mathbf{x}) = \prod_{j=1}^{n} x_j^{\alpha_j},$$

is quasi-concave on $\mathbf{R}^n_+$. If, in addition,

$$\sum_{j=1}^{n} \alpha_j \leq 1, \tag{5.98}$$

then f is concave on $\mathbf{R}^n_+$.

Proof. We begin by noting that if

$$\sum_{j=1}^{n} \alpha_j = 0,$$

then f is constant, and is then obviously concave. Consequently, we suppose throughout the following that

$$\sum_{j=1}^{n} \alpha_j > 0.$$

From Exercise 1 at the end of Section 6 of this chapter, and 5.86.2, the function

$$\log[f(\mathbf{x})] = \sum_{j=1}^{n} \alpha_j \log x_j,$$

is concave on $\mathbf{R}^n_{++}$. From 5.98.1 it then follows that

$$\exp\Big(\log[f(\mathbf{x})]\Big) = f(\mathbf{x})$$

is quasi-concave on $\mathbf{R}^n_{++} = ri\mathbf{R}^n_+$. Since f is obviously continuous, it then follows easily that f is quasi-concave on $\mathbf{R}^n_+$.

Now suppose that

$$\sum_{j=1}^{n} \alpha_j = 1. \tag{5.99}$$

Then f is obviously positively homogeneous of degree one on $\mathbf{R}^n_+$, and, using Theorem 5.100, it follows that for values of the α_j's satisfying (5.99), f is concave on $\mathbf{R}^n_+$. However, for the general case satisfying (5.98), if we define:

$$\alpha = \sum_{j=1}^{n} \alpha_j,$$

and

$$\beta_j = \alpha_j/\alpha \quad \text{for } j = 1, \ldots, n,$$

the function g defined by

$$g(\mathbf{x}) = \prod_{j=1}^{n} x_j^{\beta_j},$$

is concave on $\mathbf{R}^n_+$. Thus we have from Proposition 5.76 and 5.86.1 that

$$f(\mathbf{x}) = [g(\mathbf{x})]^{\alpha},$$

is concave on $\mathbf{R}^n_+$. □

In the remainder of this section, we will continue our investigation of functions which are positively homogeneous of degree one. We begin with the following result.

5.103. Proposition. *Suppose $f_i\colon K \to \mathbf{R}_+$, for $i = 1, 2$, and where $K \subseteq L$ is a nonempty cone and suppose that f_1 and f_2 satisfy the following conditions:*

1. *f_i is positively homogeneous of degree one, for $i = 1, 2$, and*

2. for some $a_1, a_2 \in \mathbf{R}$, we have $a_1 > 0$; and, defining:

$$X_{a_i} = \{\mathbf{x} \in K \mid f_i(\mathbf{x}) \geq a_i\} \quad \textit{for } i = 1, 2,$$

we have $X_{a_1} = X_{a_2}$. Then:

$$(\forall \mathbf{x} \in K)\colon f_1(\mathbf{x}) = (a_2/a_1) f_1(\mathbf{x}).$$

Proof. Suppose, by way of obtaining a contradiction, that $a_2 \leq 0$. Then, since $f_2 \colon K \to \mathbf{R}_+$, we see that we must have $X_{a_2} = K$, and hypothesis 2 then implies that $X_{a_1} = K$ as well. But this is impossible; for the fact that $a_1 > 0$ now implies that:

$$(\forall \mathbf{x} \in K)\colon f(\mathbf{x}) > 0.$$

However, if $\mathbf{x}^* \in K$ is such that:

$$f_1(\mathbf{x}^*) \geq a_1,$$

we have, for $\mu = a_1/[2f_1(\mathbf{x}^*)]$:

$$f_1(\mu\mathbf{x}^*) = \frac{a_1}{2f_1(\mathbf{x}^*)} \cdot f_1(\mathbf{x}^*) = \frac{a_1}{2};$$

which implies $\mu\mathbf{x}^* \notin X_{a_1}$; contradicting the conclusion that $X_{a_1} = K$. Thus, we conclude that $a_2 > 0$.

Now let $\mathbf{x}^* \in K$ be arbitrary. We distinguish two cases, as follows.

Case 1. $f_1(\mathbf{x}^*) = 0$. In this case, it follows from the homogeneity of f_1 that:

$$(\forall \mu \in \mathbf{R}_+)\colon f_1(\mu\mathbf{x}^*) = \mu f_1(\mathbf{x}^*) = 0 < a_1;$$

and thus it follows from hypothesis 2 that:

$$(\forall \mu \in \mathbf{R}_+)\colon f_2(\mu\mathbf{x}^*) < a_2.$$

Since f_2 is positively homogeneous of degree one and nonnegative-valued, it then follows that $f_2(\mathbf{x}^*) = 0$; and, therefore, that:

$$f_2(\mathbf{x}^*) = (a_2/a_1) f_1(\mathbf{x}^*).$$

Case 2. $f_1(\mathbf{x}^*) > 0$. Defining $\mu^* = a_1/f_1(\mathbf{x}^*)$, we see that:

$$f_1(\mu^*\mathbf{x}^*) = \Big(\frac{a_1}{f_1(\mathbf{x}^*)}\Big) f_1(\mathbf{x}^*) = a_1;$$

so that, from hypothesis 2, we have:

$$a_2 \leq f_2(\mu^*\mathbf{x}^*) = [a_1/f_1(\mathbf{x}^*)] f_2(\mathbf{x}^*),$$

and therefore:

$$f_2(\mathbf{x}^*) \geq (a_2/a_1)f_1(\mathbf{x}^*). \tag{5.100}$$

Conversely [and using the fact that, by (5.100), $f_2(\mathbf{x}^*) > 0$], let $\bar{\mu} = a_2/f_2(\mathbf{x}^*)$. Then:

$$f_2(\bar{\mu}\mathbf{x}^*) = a_2,$$

so that by hypothesis 2:

$$a_1 \leq f_1(\bar{\mu}\mathbf{x}^*) = [a_2/f_2(\mathbf{x}^*)]f_1(\mathbf{x}^*);$$

and therefore:

$$f_2(\mathbf{x}^*) \leq (a_2/a_1)f_1(\mathbf{x}^*). \tag{5.101}$$

Combining (5.100) and (5.101), we then have:

$$f_2(\mathbf{x}^*) = (a_2/a_1)f_1(\mathbf{x}^*) \quad \square$$

5.104. Definition. If K is a cone, we define the **ordering on L induced by K**, $\geq_K$, by:

$$\mathbf{x} \geq_K \mathbf{y} \iff \mathbf{x} - \mathbf{y} \in K.$$

While the above definition may appear to be a bit strange at first glance, notice that one can define the weak inequality on $\mathbf{R}^n$ in a manner equivalent to the usual definition, by:

$$\mathbf{x} \geq \mathbf{y} \iff \mathbf{x} - \mathbf{y} \in \mathbf{R}^n_+;$$

and, of course, $\mathbf{R}^n_+$ is a convex cone. It is easy to show that if $\mathbf{0} \in K$, then $\geq_K$ is reflexive; and if K is **pointed**; that is, if K satisfies the condition:

$$(\forall \mathbf{x} \in L): [\mathbf{x} \in K \ \& \ -\mathbf{x} \in K] \Rightarrow \mathbf{x} = \mathbf{0},$$

then $\geq_K$ is antisymmetric. It is also easy to show that, if K is convex, then $\geq_K$ is transitive.

5.105. Definition. Suppose K is a cone, and that Y is a nonempty subset of L. We shall say that **Y is unbounded with respect to** $\geq_K$ iff, for all $\mathbf{x}, \mathbf{y} \in L$,

$$[\mathbf{y} \in Y \text{ and } \mathbf{x} \geq_K \mathbf{y}] \Rightarrow \mathbf{x} \in Y.$$

If K is a convex cone, then K itself is unbounded with respect to $\geq_K$. In the elementary theory of consumer preferences, where we take the consumption set to be equal to $\mathbf{R}^n_+$, if the consumer's utility function is nondecreasing, then upper countour sets are always unbounded with respect to the usual weak inequality on $\mathbf{R}^n$.

In our next result, we will be interested in functions which are nondecreasing with respect to the ordering $\geq_K$, where we define this as follows.

5.106. Definition. Suppose K is a cone, that Y is a nonempty subset of L which is unbounded with respect to $\geq_K$, and that $f\colon Y \to \mathbf{R}$. We will say that **f is nondecreasing with respect to $\geq_K$** iff we have:

$$(\forall \mathbf{x}, \mathbf{y} \in Y)\colon \mathbf{x} \geq_K \mathbf{y} \Rightarrow f(\mathbf{x}) \geq f(\mathbf{y}).$$

5.107. Proposition. *Suppose $f\colon K \to \mathbf{R}_+$ is nondecreasing with respect to $\geq_K$ and positively homogeneous of degree one on the convex cone K. Then f is quasi-concave [respectively, quasi-convex] if, and only if f is concave [respectively, convex] on K.*

Proof. The sufficiency portion of this result is obvious. Suppose, therefore, that f is quasi-concave, and let $\mathbf{x}$ and $\mathbf{y}$ be arbitrary elements of K. If $f(\mathbf{x}) \cdot f(\mathbf{y}) = 0$, and, for example, $f(\mathbf{x}) = 0$, then it follows from the fact that f is nondecreasing on K and the fact that $\mathbf{x} + \mathbf{y} \geq_K \mathbf{y}$, that:

$$f(\mathbf{x} + \mathbf{y}) \geq f(\mathbf{y}) = f(\mathbf{x}) + f(\mathbf{y}).$$

On the other hand, if $f(\mathbf{x}) \cdot f(\mathbf{y}) > 0$, then we can proceed by an argument identical to that used in the proof of Theorem 5.99 to show that:

$$f(\mathbf{x} + \mathbf{y}) \geq f(\mathbf{x}) + f(\mathbf{y}).$$

The remaining details of the proof are left as an exercise. □

In the theoretical literature in economics one often encounters strictly quasi-concave functions, which we define as follows.

5.108. Definition. Let $f\colon X \to \mathbf{R}$, where X is a convex subset of L. We shall say that f is:

1. **strictly quasi-concave** (on X) iff, for each $\mathbf{x}, \mathbf{y} \in X$ such that $\mathbf{x} \neq \mathbf{y}$, and each $\theta \in]0,1[$, we have:

$$f[\theta\mathbf{x} + (1-\theta)\mathbf{y}] > \min\{f(\mathbf{x}), f(\mathbf{y})\}.$$

2. **strictly quasi-convex** (on X) iff, for each $\mathbf{x}, \mathbf{y} \in X$ such that $\mathbf{x} \neq \mathbf{y}$, and each $\theta \in]0,1[$, we have:

$$f[\theta\mathbf{x} + (1-\theta)\mathbf{y}] < \max\{f(\mathbf{x}), f(\mathbf{y})\}.$$

If f is strictly concave, then f is strictly quasi-concave; and, of course, stict convexity implies strict quasi-convexity. The property set out in the following result constitutes one of the reasons strictly quasi-concave functions are of particular interest in economic theory. The proof of the proposition is more or less immediate, and will be left as an exercise.

5.109. Proposition. *Suppose $f\colon X \to \mathbf{R}$ is strictly quasi-concave [respectively, stricIty quasi-convex], where X is a convex subset of L, and suppose $\mathbf{x}', \mathbf{x}^* \in X$ satisfy:*

$$(\forall \mathbf{x} \in X)\colon f(\mathbf{x}^*) \geq f(\mathbf{x})\ \&\ f(\mathbf{x}') \geq f(\mathbf{x})$$

[respectively, $f(\mathbf{x}^*) \leq f(\mathbf{x})\ \&\ f(\mathbf{x}') \leq f(\mathbf{x})$]. Then $\mathbf{x}^* = \mathbf{x}'$.

Exercises.

1. Show that the function defined in equation (5.95) of this section is quasi-concave and positively homogeneous of degree one on $\mathbf{R}^2_+$.

2. Suppose $f: C \to \mathbf{R}$, where C is a convex subset of $\mathbf{R}^n$ having a non-empty interior, and that f is quasi-concave and differentiable on C. Show that if $\mathbf{x}^* \in C$ is such that

$$(\exists \mathbf{z} \in C): \nabla f(\mathbf{x}^*) \cdot \mathbf{z} < \nabla f(\mathbf{x}^*) \cdot \mathbf{x}^*,$$

then $\mathbf{x}^*$ maximizes f subject to

$$\nabla f(\mathbf{x}^*) \cdot (\mathbf{x} - \mathbf{x}^*) = 0.$$

[Hint: recall that if f is differentiable, then it is continuous, and make use of 5.98.2.]

3. Prove Corollary 5.101.

4. Prove Proposition 5.97.

5. Prove the statements made in the text regarding the properties of $\geq_K$.

6. Complete the details of the proof of Proposition 5.107.

7. Show that if $f: \mathbf{R}^n_+ \to \mathbf{R}_+$ and is non-decreasing and positively homogeneous of degree one, then f is lower semi-continuous. To prove that f is lower semi-continuous at a point $\mathbf{x}^*$ at which $f(\mathbf{x}^*) > 0$, try first proving that, for any $\mu \in [0, 1[$, there exists $\delta > 0$ such that:

$$(\forall \mathbf{x} \in N(\mathbf{x}^*, \delta) \cap \mathbf{R}^n_+): \mathbf{x} \geq \mu \mathbf{x}^*.$$

Chapter 6

Applications of Convexity

6.1 Introduction

In this chapter we will develop several examples from economic theory which draw heavily upon the material concerning convex sets and functions which we studied in the last chapter. We begin with the theory of cost and production, and the duality between cost and production functions. In Section 2, we present the basics of the theory of production and cost with which we will be dealing. In Section 4, our main topic will be to develop the duality relationship between cost and 'technology functions,' while in Section 5 we will look at two notions of homotheticity for general technology and cost functions.[1] The intervening section 3 will be devoted to the development of a theory of distance and support functions; which, we shall find, are the basic mathematical tools which can be utilized to construct a very general and efficient development of the production and cost theory being presented in the other four of the first five sections.

In Section 6, we develop a version of the 'Two Fundamental Theorems of Welfare Economics;' showing the relationship between Pareto efficiency and competitive equilibrium in the case of finite numbers of consumers and commodities. Finally, in Section 7, we will develop a very brief treatment of a general theory of constrained maximization.

6.2 Basic Production and Cost Theory

The notion of a technology set, T, is probably the simplest general characterization of the production possibilities available to a 'firm' which is used in economic theory. In this section, we will develop the basics of this theory; supposing that T is a subset of $\mathbf{R}_+^{n+m}$, and that T consists of the collection

[1]Sections 1–5 of this chapter can be regarded as companions to and extensions of Sections 5 and 6 of Chapter 3. The student may wish to review the latter material quickly before continuing with this chapter.

of all input-output pairs, $(\mathbf{v}, \mathbf{x})$, where

$\mathbf{v} \in \mathbf{R}^n_+$ is a vector of available input quantities, and

$\mathbf{x} \in \mathbf{R}^m_+$ is a vector of output quantities,

such that $\mathbf{v}$ can produce $\mathbf{x}$. In dealing with such a set, T, of technologically feasible input-output pairs, $(\mathbf{v}, \mathbf{x})$, we define the **extended producible set for T**, X_0, by:

$$X_0 = \{\mathbf{x} \in \mathbf{R}^m_+ \mid (\exists \mathbf{v} \in \mathbf{R}^n_+) \colon (\mathbf{v}, \mathbf{x}) \in T\}, \tag{6.1}$$

the **producible set for T**, X, by:[2]

$$X = X_0 \setminus \{\mathbf{0}\};$$

and, for each $\mathbf{x} \in X_0$, the **input requirement set for x**, $V(\mathbf{x})$, by:

$$V(\mathbf{x}) = \{\mathbf{v} \in \mathbf{R}^n_+ \mid (\mathbf{v}, \mathbf{x}) \in T\}. \tag{6.2}$$

In principle, it would be appropriate to refer to any non-empty subset of $\mathbf{R}^{n+m}_+$ which is to be interpreted as defining a firm's technologicially feasible input-output pairs as a 'technology set.' However, it will be more convenient for us to reserve this terminology for a set which satisfies the following conditions.

6.1. Definition. We shall say that a nonempty subset, T, of $\mathbf{R}^{n+m}_+$ is a **technology set** iff T satisfies the following conditions:

1. for all $(\mathbf{v}, \mathbf{x}) \in T$, we have $\mathbf{v} > \mathbf{0}_n$ and $\mathbf{x} > \mathbf{0}_m$, where '$\mathbf{0}_t$' denotes the origin in $\mathbf{R}^t$, for $t = m, n$.[3]
2. T satisfies **weak disposability**, that is, if $(\mathbf{v}, \mathbf{x}) \in T$, and $(\mathbf{v}', \mathbf{x}')$ is an element of $\mathbf{R}^{n+m}_+$ satisfying:

$$\mathbf{v}' \geq \mathbf{v} \text{ and } \mathbf{0} < \mathbf{x}' \leq \mathbf{x}, \tag{6.3}$$

then $(\mathbf{v}', \mathbf{x}') \in T$;

and, for each $\mathbf{x} \in X$, the input requirement set for $\mathbf{x}$, $V(\mathbf{x})$, is:

3. closed, and
4. convex.

As was the case in Chapter 3, we do not consider the possibility of zero production in our definition of a technology set. The reasons for this are essentially the same as were set out in Section 5 of Chapter 3; the most important of which is that this omission allows us to develop a theory of cost functions which is equally applicable to total and variable cost.

[2]Strictly speaking, we should denote these two sets by something like '$X_0(T)$' and '$X(T)$,' respectively; but in practice our omission of the indication of the dependence of these sets upon the set T should cause no confusion.

[3]Where it is clear that we are dealing with a particular Euclidean space, $\mathbf{R}^m$ or $\mathbf{R}^n$, or whatever, we will simply use the notation '$\mathbf{0}$' to denote the origin in that space.

It is easy to show that if T is a closed set, then it satisfies property 3 of the above definition; however, there exist sets satisfying condition 3 which are not closed. Similarly, if T is a convex set, then it satisfies condition 4 of 6.1; but a set may satisfy said condition 4, yet not be convex.

If T is a technology set, then T generates an 'input requirement correspondence' via equation (6.2). We shall find that the properties of the correspondence so generated are those set out in the following definition.

6.2. Definition. Let X be a nonempty subset of $\mathbf{R}^m_+ \setminus \{\mathbf{0}\}$. We shall say that a nonempty-valued correspondence, $V\colon X \mapsto \mathbf{R}^n_+$ is an **input-requirement correspondence** iff V satisfies:

1. for all $\mathbf{x} \in X, \mathbf{0} \notin V(\mathbf{x})$.
2. a. if $\mathbf{x}$ is an element of X, and $\mathbf{x}'$ is such that $\mathbf{0} < \mathbf{x}' \leq \mathbf{x}$, then $\mathbf{x}' \in X$ as well, and

   b. for each $\mathbf{x}_1,\ \mathbf{x}_2 \in X$,

$$\mathbf{x}_1 \geq \mathbf{x}_2 \Rightarrow V(\mathbf{x}_1) \subseteq V(\mathbf{x}_2).$$

   c. for each $\mathbf{x} \in X$, and each $\mathbf{v}, \mathbf{v}' \in \mathbf{R}^n_+$, we have:
      if $\mathbf{v} \in V(\mathbf{x})$, and $\mathbf{v}' \geq \mathbf{v}$, then $\mathbf{v}' \in V(\mathbf{x})$ as well; and:

for each $\mathbf{x} \in X, V(\mathbf{x})$ is a:

3. closed, and
4. convex set.

We will use condition 2.a of Definition 6.2 to define a **producible set**; that is, hereafter, when we say that X is a **producible set**, we shall mean that X is a non-empty subset of $\mathbf{R}^m_+ \setminus \{\mathbf{0}\}$ satisfying:

$$(\forall \mathbf{x}, \mathbf{x}' \in \mathbf{R}^m_+)\colon [\mathbf{x} \in X \;\&\; \mathbf{0} < \mathbf{x}' \leq \mathbf{x}] \Rightarrow \mathbf{x}' \in X. \tag{6.4}$$

Suppose now that T is a technology set, and let $V = \iota(T)$ be the correspondence $V\colon X \mapsto \mathbf{R}^n_+$ implicitly defined by equations (6.1) and (6.2), above. It is easy to see that V is then an input-requirement correspondence (satisfying Definition 6.2). Moreover, if V is an input-requirement correspondence, and we define $T = \tau(V)$ by

$$\tau(V) \equiv T \equiv \{(\mathbf{v}, \mathbf{x}) \in \mathbf{R}^{n+m}_+ \mid \mathbf{x} \in X \;\&\; \mathbf{v} \in V(\mathbf{x})\},$$

then T is a technology set. This is also 'easy to see,' albeit a trifle tedious to prove formally. It is nonetheless a good exercise for the student to prove this equivalence.

Since the technology set and the input-requirement representations of production possibilities are formally equivalent, we can see that a theory of cost and production which is based upon the concept of a technology set is exactly equivalent to one which is based upon the notion of an input-requirement correspondence. As a matter of fact, instead of developing the idea of an input-requirement correspondence, we could instead have defined the **production correspondence**, $P\colon \mathbf{R}^n_+ \mapsto \mathbf{R}^m_+$, generated by a technology set T, as:

$$P(\mathbf{v}) = \{\mathbf{x} \in \mathbf{R}^m_+ \mid (\mathbf{v}, \mathbf{x}) \in T\}.$$

If we take the production correspondence as the starting point for our analysis, then the corresponding technology set is simply the graph of the correspondence; and as this discussion suggests, it is very natural in some contexts to take a production correspondence as the focal point of our analysis. However, we will not develop this idea further here. Instead, in the remainder of this section we shall develop yet a fourth characterization of available technological possibilities; one which is more convenient in applied work (as well as some theoretical work) than any of these other representations, despite being logically equivalent to the two modes of analysis already presented.

This fourth method of characterizing production possibilities involves the use of an implicit function. The trick, insofar as obtaining a particularly useful and convenient representation is concerned, is to choose the right implicit function; and the representation which we will be using here is developed from ideas introduced by Ronald Shephard [1953], and further developed and generalized by S. E. Jacobsen [1970] and D. McFadden [1970, 1978]. We define the technology function for a given input-requirement correspondence via the two steps set out in the following definition.

6.3. Definition. Given an input-requirement correspondence, $V\colon X \mapsto \mathbf{R}^n_+$, we define:

$$\Lambda(\mathbf{v}, \mathbf{x}) = \{\lambda \in \mathbf{R}_{++} \mid (1/\lambda)\mathbf{v} \in V(\mathbf{x})\} \quad \text{for } (\mathbf{v}, \mathbf{x}) \in \mathbf{R}^n_+ \times X.$$

We then define the **technology function corresponding to V**, $\psi = \delta(V)$, by:

$$\psi(\mathbf{v}, \mathbf{x}) = \left\{ \begin{array}{cl} \max \Lambda(\mathbf{v}, \mathbf{x}) & \text{if } \Lambda(\mathbf{v}, \mathbf{x}) \neq \emptyset \\ 0 & \text{if } \Lambda(\mathbf{v}, \mathbf{x}) = \emptyset \end{array} \right\} \quad \text{for } (\mathbf{v}, \mathbf{x}) \in \mathbf{R}^n_+ \times X. \quad (6.5)$$

Since $V(\mathbf{x})$ is closed, it is easy to see that the technology function, as defined in equation (6.5), above, is well-defined. Notice that in Figure 6.1, on the next page, $\Lambda(\mathbf{v}', \mathbf{x}^*)$ is empty, while $\Lambda(\mathbf{v}^*, \mathbf{x}^*)$ is non-empty. It should be apparent that if $\mathbf{x} \in X$, and $\mathbf{v} \gg \mathbf{0}$, then it is necessarily the case that $\Lambda(\mathbf{v}, \mathbf{x}) \neq \emptyset$; and that, as is indicated in the diagram, if $\Lambda(\mathbf{v}^*, \mathbf{x}^*) \neq \emptyset$, then $[1/\psi(\mathbf{v}^*, \mathbf{x}^*)]\mathbf{v}^* \in V(\mathbf{x}^*)$.

A consideration of the diagram and properties of an input-requirement set should make it clear that we will have $\psi(\mathbf{v}, \mathbf{x}) \geq 1$ if, and only if, $\mathbf{v} \in V(\mathbf{x})$; a fact which you will be asked to prove in the exercises at the end of this section. In the meantime, let's take a look at an analytic example.

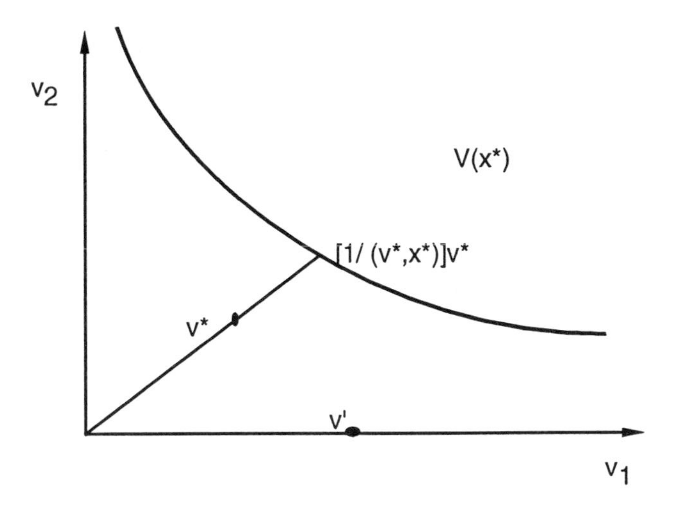

Figure 6.1: Defining the Technology Function.

6.4. Example. Suppose a production unit produces a single output with a technology which can be characterized by the production function $\varphi \colon \mathbf{R}^n_+ \to \mathbf{R}_+$. Suppose further that φ is homothetic (Definition 3.69) so that we can write

$$\varphi(\mathbf{v}) = F[\sigma(\mathbf{v})],$$

where F is a transformation function (Definition 3.50), and σ is a production function which is positively homogeneous of degree one. If, following our usual practice, we denote the producible set by X, then recall that there exists a unique input-requirement function, $f \colon X \to \mathbf{R}_{++}$, dual to F, and we proved in Theorem 3.70 that, for $x \in X$,

$$V(x) = \{\mathbf{v} \in \mathbf{R}^n_+ \mid \sigma(\mathbf{v}) \geq f(x)\}. \tag{6.6}$$

In order to derive the technology function corresponding to V [and, of course, to φ], let $x^* \in X$ and $\mathbf{v}^* \in \mathbf{R}^n_+$. We distinguish two cases, as follows.

a. $\sigma(\mathbf{v}^*) > 0$. In this case, it follows from (6.6) that, for $\lambda \in \mathbf{R}_{++}$:

$$\begin{aligned}(1/\lambda)\mathbf{v}^* \in V(x^*) &\iff f(x^*) \leq \sigma[(1/\lambda)\mathbf{v}^*] = (1/\lambda)\sigma(\mathbf{v}^*) \\ &\iff \lambda \leq \frac{\sigma(\mathbf{v}^*)}{f(\mathbf{x}^*)}.\end{aligned} \tag{6.7}$$

Consequently, it is apparent that:

$$\Lambda(\mathbf{v}^*, x^*) =]0, \sigma(\mathbf{v}^*)/f(x^*)],$$

from which it follows, using Definition 6.3, that

$$\psi(\mathbf{v}^*, \mathbf{x}^*) \stackrel{\text{def}}{=} \max \Lambda(\mathbf{v}^*, \mathbf{x}^*) = \sigma(\mathbf{v}^*)/f(x^*).$$

b. $\sigma(\mathbf{v}^*) = 0$. Here the homogeneity of σ implies that for all $\lambda \in \mathbf{R}_{++}$,

$$\sigma[(1/\lambda)\mathbf{v}^*] = (1/\lambda)\sigma(\mathbf{v}^*) = 0;$$

and thus it follows from (6.7) that:

$$\Lambda(\mathbf{v}^*, x^*) = \emptyset.$$

But we then have, from the definition of ψ that:

$$\psi(\mathbf{v}^*, x^*) = 0 = \sigma(\mathbf{v}^*)/f(x^*)$$

in this case as well. □

In the remainder of this section we will state and discuss some of the most important properties of the technology function and its relationship to the corresponding cost function. *All* proofs, except for some very simple arguments, will, in effect, be deferred to Section 3 of this chapter; where we will present an introduction to distance functions and support functions, of which the theory of technology and cost functions is a special case.

The following sets forth several properties of $\Lambda(\mathbf{v}, \mathbf{x})$, and $\psi(\mathbf{v}, \mathbf{x})$ which we will frequently find to be useful.

6.5. Proposition. *Let $V\colon X \mapsto \mathbf{R}^n_+$ be an input-requirement correspondence, and let $\Lambda(\cdot)$ and $\psi = \delta(V)$ be as defined in 6.3, above. Then we have:*

1. $(\forall \mathbf{v} \in \mathbf{R}^n_{++})(\forall \mathbf{x} \in X)\colon \Lambda(\mathbf{v}, \mathbf{x}) \neq \emptyset$,

2. for each $(\mathbf{v}, \mathbf{x}) \in \mathbf{R}^n_+ \times X$ such that $\Lambda(\mathbf{v}, \mathbf{x}) \neq \emptyset$, we have:

$$\Lambda(\mathbf{v}, \mathbf{x}) =]0, \psi(\mathbf{v}, \mathbf{x})],$$

3. $(\forall \mathbf{x} \in X)\colon V(\mathbf{x}) = \{\mathbf{v} \in \mathbf{R}^n_+ \mid \psi(\mathbf{v}, \mathbf{x}) \geq 1\}$.

The results of the next section will establish that the function $\psi = \delta(V)$ defined in 6.3, above, actually is a technology function, as now defined.

6.6. Definition. Let X be a producible set, as specified in equation (6.4), above. A function $\psi\colon \mathbf{R}^n_+ \times X \to \mathbf{R}_+$ will be said to be a **technology** (or **distance) function** iff ψ satisfies:

1. for each $\mathbf{v} \in \mathbf{R}^n_+$, $\psi(\mathbf{v}, \cdot)$ is non-increasing in $\mathbf{x}$, and

for each $\mathbf{x} \in X$, $\psi(\cdot, \mathbf{x})$ is:

2. increasing,
3. continuous,
4. concave, and
5. positively homogeneous of degree one,

in $\mathbf{v}$.

Before proceeding further, we should take a moment to consider a couple of aspects of the above definition. First of all, the stipulation that X be a producible set, as we have defined same, plays almost no role in the analysis of this and Section 4 of this chapter. On the other hand, considerations similar to those spelled out at the beginning of Section 3.6 [the discussion centered around equations (3.72) and (3.73)] imply that we can incorporate this assumption in our analysis essentially without loss of generality. Secondly, notice that the assumption that $\psi(\cdot, \mathbf{x})$ is increasing in $\mathbf{v}$, together with the fact that it is positively homogeneous of degree one, implies that:

$$(\forall \mathbf{x} \in X)(\forall \mathbf{v} \in \mathbf{R}^n_{++})\colon \psi(\mathbf{v}, \mathbf{x}) > \psi(\mathbf{0}, \mathbf{x}) = 0.$$

Finally, it should also be mentioned that if we were to include the zero vector in X, we would necessarily have to define:

$$\psi(\mathbf{v}, \mathbf{0}) = +\infty, \quad \text{for all } \mathbf{v} \in \mathbf{R}^n_+.$$

This turns out to be rather a pain, since it would require most of the proofs involving technology functtions to be broken down into two parts: one for the case in which $\psi(\mathbf{v}, \mathbf{x})$ is finite, and the second for the case in which it is infinite-valued.

We can use a technology function to characterize feasible input-output pairs in the following way: if ψ is a technology function characterizing a firm's technology, then an input-output pair, $(\mathbf{v}, \mathbf{x})$, with $\mathbf{x} \neq \mathbf{0}$, will be technologically feasible for the firm if, and only if, $\psi(\mathbf{v}, \mathbf{x}) \geq 1$. Thus, given a technology function, ψ, we define the **technology set corresponding to ψ** by:

$$T = \{(\mathbf{v}, \mathbf{x}) \in \mathbf{R}^n_+ \times X \mid \psi(\mathbf{v}, \mathbf{x}) \geq 1\}.$$

We will present the definition of the input-requirement correspondence generated by a technology function a bit more formally, as follows.

6.7. Definition. Given a technology function, $\psi\colon \mathbf{R}^n_+ \times X \to \mathbf{R}_+$, we define the **input-requirement correspondence for** $\psi, V = \nu(\psi)$, by:

$$V(\mathbf{x}) = \{\mathbf{v} \in \mathbf{R}^n_+ \mid \psi(\mathbf{v}, \mathbf{x}) \geq 1\} \quad \text{for } \mathbf{x} \in X.$$

I will leave the proof of the following result for the 'interested reader.' It is more likely to challenge such a reader's perserverance and energy than his or her imagination or mathematical prowess.

6.8. Theorem. *If $\psi\colon \mathbf{R}^n_+ \times X \to \mathbf{R}_+$ is a technology function, then the correspondence $V = \nu(\psi)$ defined in 6.7, above, is an input-requirement correspondence.*

One can also prove that if $V\colon X \mapsto \mathbf{R}^n_+$ is an input-requirement correspondence, then the function defined from V by Definition 6.3 is a technology function.[4] This fact and Theorem 6.8 can be used as the basis of

[4]This will follow from the results in Section 3.

the proof of a one-to-one correspondence between the collection, $\mathcal{V}$, of all input-requirement correspondences and the collection, Ψ, of all technology functions. Thus, in particular, it follows that a theory of production which is based upon the concept of a technology function is equivalent to one which is based upon the notion of an input-requirement correspondence.

Turning our attention now to the subject of cost functions, the following definition will be shown to set forth the properties which will be satisfied by a cost function which is derived from a technology function. The student may find it instructive to compare this definition with the conclusion of Theorem 3.65.

6.9. Definition. A function $c\colon \mathbf{R}^n_+ \times X \to \mathbf{R}_+$ will be said to be a **cost function** iff X is a producible set, and c satisfies:

1. for each $\mathbf{w} \in \mathbf{R}^n_+, c(\mathbf{w}, \cdot)$ is non-decreasing in $\mathbf{x}$, and:

for each $\mathbf{x} \in X, c(\cdot, \mathbf{x})$ is:

2. increasing,
3. continuous,
4. concave, and
5. positively homogeneous of degree one,

in $\mathbf{w}$.

6.10. Definition. If $\psi\colon \mathbf{R}^n_+ \times X \to \mathbf{R}_+$ is a technology function, we define the **cost function dual to** $\psi, c = \kappa(\psi)$, by:

$$c(\mathbf{w}, \mathbf{x}) = \inf\{\mathbf{w} \cdot \mathbf{v} \mid \mathbf{v} \in \mathbf{R}^n_+ \ \&\ \psi(\mathbf{v}, \mathbf{x}) \geq 1\} \qquad \text{for } (\mathbf{w}, \mathbf{x}) \in \mathbf{R}^n_+ \times X.$$

While it is a bit unconventional to define a cost function as an infimum, rather than as a minimum, it will be considerably more convenient to do so. For a more complete explanation of the reasons for this, see Example 3.66, Proposition 3.67, and the discussion which follows the proof of 3.67. The results of the next section will establish the propriety of referring to the function defined in 6.10 as a *cost function*.

As it turns out, one can find the technology function corresponding to a given cost function by exactly the same methods as are used to find the cost function corresponding to a given technology function. Once again this will follow as an easy consequence of the results of the next section.

Exercises.

1. Show that if production possibilities can be characterized by the production function, φ, given by:

$$\varphi(\mathbf{v}) = \prod_{i=1}^{n} v_i^{a_i},$$

where

$$a_i > 0, \quad \text{for } i = 1, \ldots, n;$$

then the corresponding technology function is given by:

$$\psi(\mathbf{v}, x) = \prod_{i=1}^{n} v_i^{b_i}/x^{1/a},$$

where:

$$b_i \stackrel{\text{def}}{=} a_i/a \quad \text{for } i = 1, \ldots, n \text{ and } a \stackrel{\text{def}}{=} \sum_{i=1}^{n} a_i.$$

2. Show that if production possibilities can be characterized by a production function, $\varphi\colon \mathbf{R}_+^n \to \mathbf{R}_+$, which is positively homogenous of degree one, then the corresponding technology function is given by:

$$\psi(\mathbf{v}, x) = \sigma(\mathbf{v})/x^{1/k} \quad \text{for } (\mathbf{v}, x) \in \mathbf{R}_+^n \times \mathbf{R}_{++};$$

where $\sigma\colon \mathbf{R}_+^n \to \mathbf{R}_+$ is defined by:

$$\sigma(\mathbf{v}) = [\varphi(\mathbf{v})]^{1/k} \quad \text{for } \mathbf{v} \in \mathbf{R}_+^n.$$

3. Show that if $V\colon X \mapsto \mathbf{R}_+^n$ is an input-requirement correspondence, then:

$$T = \{(\mathbf{v}, \mathbf{x}) \in \mathbf{R}_+^n \times X \mid \mathbf{v} \in V(\mathbf{x})\},$$

is a technology set.

4. Show that if $\psi\colon \mathbf{R}_+^n \times X \to \mathbf{R}_+$ is a technology function, then the set:

$$T = \{(\mathbf{v}, \mathbf{x}) \in \mathbf{R}_+^n \times X \mid \psi(\mathbf{v}, \mathbf{x}) \geq 1\},$$

is a technology set.

5. Prove Proposition 6.5.
6. Prove Theorem 6.8.

6.3 Distance and Support Functions

In this section we will be developing some of the theory of 'distance functions' and support functions; all of which material will be applicable to the study of technology and cost functions which we began in the previous section. *Until further notice, we will be assuming that all sets under discussion are subsets of a real normed linear space, L.*

The property established in the following proposition will often be useful to us in the remainder of this, and some later chapters. The ordering $\geq_K$, used here, and the definition of 'unbounded with respect to $\geq_K$,' which is used in the next definition, are set out in 5.104 and 5.105, respectively.

6.11. Proposition. *If K is a cone and $\mathbf{x} \in ri(K)$, then, given any $\mathbf{y} \in K$, there exists $\mu > 0$ such that $\mu\mathbf{x} \geq_K \mathbf{y}$.*

Proof. Let $\mathbf{x}^* \in riK$. Then there exists $\delta > 0$ such that

$$N(\mathbf{x}^*, \delta) \cap \mathit{aff}\, K \subseteq K.$$

Now let $\mathbf{y}^* \in K$. If $\mathbf{x}^* = \mathbf{y}^*$, our result is trivial. Otherwise,

$$\|\mathbf{x}^* - \mathbf{y}^*\| > 0;$$

and if we define θ by:

$$\theta = \frac{\delta}{2\|\mathbf{x}^* - \mathbf{y}^*\|},$$

we have $\theta > 0$ and

$$(1+\theta)\mathbf{x}^* - \theta\mathbf{y}^* \in \mathit{aff}\, K.$$

Furthermore,

$$\begin{aligned}\|(1+\theta)\mathbf{x}^* - \theta\mathbf{y}^* - \mathbf{x}^*\| &= |\theta| \cdot \|\mathbf{x}^* - \mathbf{y}^*\| \\ &= \left(\frac{\delta}{2\|\mathbf{x}^* - \mathbf{y}^*\|}\right) \cdot \|\mathbf{x}^* - \mathbf{y}^*\| = \delta/2 < \delta.\end{aligned}$$

Hence,

$$(1+\theta)\mathbf{x}^* - \theta\mathbf{y}^* \in K.$$

But then

$$[1/\theta] \cdot [(1+\theta)\mathbf{x}^* - \theta\mathbf{y}^*] = \left[\frac{1+\theta}{\theta}\right]\mathbf{x}^* - \mathbf{y}^* \in K,$$

and it follows that:

$$[(1+\theta)/\theta]\mathbf{x}^* \geq_K \mathbf{y}^*. \quad \square$$

If the function we are about to define reminds you of a technology function, let me assure you that the resemblance is not coincidental!

6.12. Definition. Let K be a cone, and let Y be a closed subset[5] of K which is unbounded with respect to $\geq_K$ and such that $\mathbf{0} \notin Y$. Then we define the **distance function for Y on K**, $f\colon K \to \mathbf{R}_+$, by:

$$f(\mathbf{z}) = \begin{cases} \sup \Lambda(\mathbf{z}) & \text{if } \Lambda(\mathbf{z}) \neq \emptyset \\ 0 & \text{if } \Lambda(\mathbf{z}) = \emptyset. \end{cases}$$

where:

$$\Lambda(\mathbf{z}) \stackrel{\text{def}}{=} \{\lambda \in \mathbf{R}_{++} \mid (1/\lambda)\mathbf{z} \in Y\}.$$

[5] So that $Y = \overline{Y}$.

6.13. Theorem. (Properties of the Distance Function) *Under the conditions of the preceding definition, the distance function for Y on K satisfies the following properties:*

1. *for all $\mathbf{z} \in riK$, $\Lambda(\mathbf{z}) \neq \emptyset$, and for all $\mathbf{z}$ such that $\Lambda(\mathbf{z}) \neq \emptyset$, $f(\mathbf{z}) > 0$.*
2. *f is non-decreasing with respect to $\geq_K$.*
3. *f is positively homogeneous of degree one on K.*
4. *f is upper semi-continuous on K.*
5. *for all $\mathbf{z}$ such that $\Lambda(\mathbf{z}) \neq \emptyset$,*

$$\Lambda(\mathbf{z}) =]0, f(\mathbf{z})];$$

and thus we also have:

$$f(\mathbf{z}) = \max \Lambda(\mathbf{z}),$$

6. *and, finally:*

$$Y = \{\mathbf{z} \in K \mid f(\mathbf{z}) \geq 1\}.$$

Proof. In order to prove the first of the six properties listed in the conclusion, we note first that it follows easily from Proposition 6.11 and the fact that Y is unbounded with respect to $\geq_K$, that, for any $\mathbf{z} \in riK, \Lambda(\mathbf{z}) \neq \emptyset$. On the other hand, since $\mathbf{0} \notin Y$, and Y is a closed set, there exists $\delta > 0$ such that

$$N(\mathbf{0}, \delta) \cap Y = \emptyset.$$

Consequently, as you can easily show, if $\Lambda(\mathbf{z}) \neq \emptyset$ and $\lambda \in \Lambda(\mathbf{z})$, we must have:

$$\lambda \leq \|\mathbf{z}\|/\delta.$$

Thus, if $\Lambda(\mathbf{z}) \neq \emptyset$, $\sup \Lambda(\mathbf{z})$ exists (is a finite real number), and it follows at once that if $\Lambda(\mathbf{z}) \neq \emptyset$, then:

$$f(\mathbf{z}) \equiv \sup \Lambda(\mathbf{z}) > 0.$$

2. Suppose $\mathbf{z}, \mathbf{z}' \in K$ are such that $\mathbf{z} \geq_K \mathbf{z}'$. If $\Lambda(\mathbf{z}') = \emptyset$, then it is immediate that $f(\mathbf{z}) \geq f(\mathbf{z}')$. Suppose, therefore, that $\Lambda(\mathbf{z}') \neq \emptyset$, and let $\lambda \in \mathbf{R}_{++}$ be such that:

$$(1/\lambda)\mathbf{z}' \in Y.$$

Then, since K is a cone and $\mathbf{z} \geq_K \mathbf{z}'$, it follows that:

$$(1/\lambda)\mathbf{z} \geq_K (1/\lambda)\mathbf{z}',$$

and thus, since Y is unbounded with respect to $\geq_K$:

$$(1/\lambda)\mathbf{z} \in Y.$$

Consequently we see that $\lambda \in \Lambda(\mathbf{z})$, and it follows that:

$$\Lambda(\mathbf{z}') \subseteq \Lambda(\mathbf{z});$$

and thus that $f(\mathbf{z}) = \sup \Lambda(\mathbf{z})$ is an upper bound for $\Lambda(\mathbf{z}')$. Since $f(\mathbf{z}')$ is the *least* upper bound for $\Lambda(\mathbf{z}')$, we then conclude that:

$$f(\mathbf{z}') \leq f(\mathbf{z}).$$

3. I will leave the proof that f is positively homogeneous of degree one on K as an exercise.

4. To prove that f is upper semi-continuous on K, let $\mathbf{z}^*$ be arbitrary, and let $\epsilon > 0$ be given. If we define $\theta^* = f(\mathbf{z}^*) + \epsilon/2$, we see that $\theta^* > f(\mathbf{z}^*)$, and thus:

$$(1/\theta^*)\mathbf{z}^* \in K \setminus Y.$$

Since Y is closed, it then follows that there exists $\delta > 0$ such that:

$$N[(1/\theta^*)\mathbf{z}^*, \delta] \cap K \subseteq K \setminus Y.$$

Now, let $\mathbf{z} \in N(\mathbf{z}^*, \theta^*\delta) \cap K$. Then

$$(1/\theta^*)\mathbf{z} \in N[(1/\theta^*)\mathbf{z}^*, \delta] \cap K,$$

and thus:

$$(1/\theta^*)\mathbf{z} \in K \setminus Y.$$

Consequently,

$$f(\mathbf{z}) \leq \theta^* = f(\mathbf{z}^*) + \epsilon/2 < f(\mathbf{z}^*) + \epsilon.$$

5. Let $\mathbf{z}^* \in K$ be such that $\Lambda(\mathbf{z}^*) \neq \emptyset$, and let $\lambda \in \mathbf{R}_{++}$ be such that $0 < \lambda < f(\mathbf{z}^*)$. Then there exists $\lambda' \in \Lambda(\mathbf{z}^*)$ such that:

$$\lambda < \lambda' \leq f(\mathbf{z}^*)$$

[since otherwise λ is an upper bound for $\Lambda(\mathbf{z}^*)$]. But then we see that $(1/\lambda')\mathbf{z}^* \in Y$; and, since

$$(1/\lambda)\mathbf{z}^* \geq_K (1/\lambda')\mathbf{z}^*,$$

it then follows from the fact that Y is unbounded with respect to $\geq_K$ that $(1/\lambda)\mathbf{z}^* \in Y$. Moreover, since Y is closed, it is easy to see that $[1/f(\mathbf{z}^*)]\mathbf{z}^* \in Y$. This argument establishes that

$$]0, f(\mathbf{z}^*)] \subseteq \Lambda(\mathbf{z}^*);$$

and, since the opposite inclusion is obvious, it follows that

$$\Lambda(\mathbf{z}^*) =]0, f(\mathbf{z}^*)].$$

6. I will let the proof that $Y = \{\mathbf{z} \in K \mid f(\mathbf{z}) \geq 1\}$ be an exercise. □

The following result completes our catalog of the properties of a distance function.

6.14. Proposition. *If, in addition to the assumptions of Theorem 6.13, Y and K are both convex sets, then the distance function for Y on K is also super-additive and concave, in addition to the other properties listed in 6.13.*

Proof. Using Theorem 6.13, Proposition 5.82, and Proposition 5.107, we see that it suffices to prove that f is quasi-concave.

Accordingly, let $\mathbf{z}_1, \mathbf{z}_2 \in K$ and $\theta \in [0,1]$ be given, define $\mathbf{z}^*$ by:

$$\mathbf{z}^* = \theta\mathbf{z}_1 + (1-\theta)\mathbf{z}_2,$$

and suppose, without loss of generality, that:

$$f(\mathbf{z}_1) \geq f(\mathbf{z}_2). \tag{6.8}$$

If $f(\mathbf{z}_2) = 0$, then we obviously have:

$$f(\mathbf{z}^*) \equiv f[\theta\mathbf{z}_1 + (1-\theta)\mathbf{z}_2] \geq 0 = \min\{f(\mathbf{z}_1), f(\mathbf{z}_2)\}.$$

Suppose, therefore, that $f(\mathbf{z}_2) > 0$. Then we note that it follows from equation (6.8) and Proposition 6.13.5 that:

$$\mathbf{y}_i \stackrel{\text{def}}{=} \left[\frac{1}{f(\mathbf{z}_2)}\right] \mathbf{z}_i \in V \quad \text{for } i = 1, 2.$$

Therefore, since Y is a convex set,

$$\theta\mathbf{y}_1 + (1-\theta)\mathbf{y}_2 \in Y.$$

However, since

$$\theta\mathbf{y}_1 + (1-\theta)\mathbf{y}_2 = \left(\frac{1}{f(\mathbf{z}_2)}\right)\mathbf{z}^*,$$

it then follows that $f(\mathbf{z}_2) \in \Lambda(\mathbf{z}^*)$, and thus that:

$$f(\mathbf{z}^*) \equiv \sup\Lambda(\mathbf{z}^*) \geq f(\mathbf{z}_2) = \min\{f(\mathbf{z}_1), f(\mathbf{z}_2)\}. \quad \square$$

The other class of functions which we will be studying in this section is introduced in the following.

6.15. Definition. Let $K \subseteq L$ be a convex cone. We shall say that a function $f : K \to \mathbf{R}_+$ is a **support function on K** iff f is:

1. not identically zero,[6]
2. non-decreasing with respect to $\geq_K$,
3. positively homogeneous of degree one,
4. super-additive, and
5. upper semi-continuous,

on K.

Notice that, under the conditions of Proposition 6.14, a distance function is also a support function. While we have only required that a support function not be identically zero, such a function must satisfy a stronger property; as set out in the following.

6.16. Proposition. *If $K \subseteq L$ is a convex cone, and $f : K \to \mathbf{R}_+$ is a support function, then f satisfies:*

$$(\forall \mathbf{z} \in riK) \colon f(\mathbf{z}) > 0.$$

Proof. If f is a support function, then there must exist a point $\mathbf{z}^* \in K$ such that $f(\mathbf{z}^*) > 0$. However, if $\mathbf{z} \in riK$, then by Proposition 6.11, there exists $\mu > 0$ such that:

$$\mu\mathbf{z} \geq_K \mathbf{z}^*;$$

and, thus, since f is non-decreasing with respect to $\geq_K$, it follows that:

$$f(\mu\mathbf{z}) \geq f(\mathbf{z}^*) > 0.$$

But then the homogeneity of f implies that we also have:

$$f(\mathbf{z}) = (1/\mu) f(\mu\mathbf{z}) > 0. \quad \square$$

In the remainder of this section, we will suppose that L is a real inner product space (see Section 4.5). In the context of such a space, we will make use of the following definition.

6.17. Definition. We shall say that a cone, K, is **acute** iff, for all $\mathbf{y}, \mathbf{z} \in K$, we have $\mathbf{y} \cdot \mathbf{z} \geq \mathbf{0}$.

6.18. Proposition. *If K is an acute convex cone, and Z is a closed, convex, and nonempty subset of K which does not contain the origin ($\mathbf{0} \notin Z$), then the function $f : K \to \mathbf{R}_+$ defined by:*

$$f(\mathbf{y}) = \inf\{\mathbf{y} \cdot \mathbf{z} \mid \mathbf{z} \in Z\}$$

is a support function.

---

[6] The function $f : K \to \mathbf{R}_+$ given by $f(\mathbf{z}) = 0$ for all $\mathbf{z} \in K$ satisfies properties 2–5 of Definition 6.15; and might, therefore, also be called a 'support function.' In practice, however, all of the applications with which we will be concerned either imply, or require the condition that f not be identically zero.

Proof. By Theorem 5.38, there exists $\mathbf{z}^* \in Z$ satisfying:

$$(\forall \mathbf{z} \in Z)\colon \|\mathbf{z}\| \geq \|\mathbf{z}^*\| > 0 \text{ and } \mathbf{z} \cdot \mathbf{z}^* \geq \mathbf{z}^* \cdot \mathbf{z}^* > 0.$$

It follows immediately that $f(\mathbf{z}^*) > 0$.

To prove that f is non-decreasing, let $\mathbf{y}', \mathbf{y} \in K$ be such that $\mathbf{y}' \geq_K \mathbf{y}$. Then $\mathbf{y}' - \mathbf{y} \in K$, and if $\mathbf{z} \in Z$, it follows from the fact that K is acute that:

$$(\mathbf{y}' - \mathbf{y}) \cdot \mathbf{z} \geq 0.$$

Consequently,

$$\mathbf{y}' \cdot \mathbf{z} \geq \mathbf{y} \cdot \mathbf{z} \geq f(\mathbf{y}).$$

Since $\mathbf{z} \in Z$ was arbitrary, it then follows that:

$$f(\mathbf{y}') \stackrel{\text{def}}{=} \inf\{\mathbf{y}' \cdot \mathbf{z} \mid \mathbf{z} \in Z\} \geq f(\mathbf{y}).$$

It is an easy exercise to establish the fact that f is positively homogeneous of degree one on K. Moreover, since K is a convex cone, an argument quite similar to that of the above paragraph can be used to show that f is super-additive. I will leave the details as an exercise.

To prove that f is upper semi-continuous, let $\mathbf{y}^* \in K$, and let $\epsilon > 0$ be given. From the definition of f, there exists $\mathbf{z}^* \in Z$ satisfying:

$$\mathbf{y}^* \cdot \mathbf{z}^* < f(\mathbf{y}^*) + \epsilon/2; \tag{6.9}$$

and by the continuity of the inner product function, there exists $\delta > 0$ such that:

$$(\forall \mathbf{y} \in N(\mathbf{y}^*, \delta) \cap K)\colon \mathbf{y} \cdot \mathbf{z}^* < \mathbf{y}^* \cdot \mathbf{z}^* + \epsilon/2. \tag{6.10}$$

Combining (6.9) and (6.10), we then see that:

$$(\forall \mathbf{y} \in N(\mathbf{y}^*, \delta) \cap K)\colon f(\mathbf{y}) \leq \mathbf{y} \cdot \mathbf{z}^* < f(\mathbf{y}^*) + \epsilon;$$

and it follows that f is upper semi-continuous at $\mathbf{y}^*$. □

6.19. Definition. If K and Z satisfy the assumptions of Proposition 6.18, and if the function $f\colon K \to \mathbf{R}_+$ satisfies:

$$f(\mathbf{y}) = \inf\{\mathbf{y} \cdot \mathbf{z} \mid \mathbf{z} \in Z\} \quad \text{for } \mathbf{y} \in K,$$

then we shall say that f is the **support function for Z on K**.

The rationale for the term 'support function' stems from the fact that if Z, K, and f are related as in Definition 6.19, then for any $\mathbf{y} \in K \setminus \{\mathbf{0}\}$, the hyperplane, H defined by:

$$H = \{\mathbf{z} \in \mathbf{R}^n \mid \mathbf{y} \cdot \mathbf{z} = f(\mathbf{y})\}$$

is a support plane for Z.

We will be particularly interested in the subset of K defined in the following proposition.

6.20. Proposition. *If $K \subseteq L$ is an acute convex cone, if $f: K \to \mathbf{R}_+$ is a support function, and if we define the set Y by:*

$$Y = \{\mathbf{y} \in K \mid f(\mathbf{y}) \geq 1\},$$

then Y is nonempty, closed, convex, unbounded with respect to $\geq_K$, and does not contain the origin.

Proof. The fact that Y satisfies the first four of the properties listed in the conclusion follows more or less immediately from the definition of a support function. The fact that $\mathbf{0} \notin Z$ follows immediately from the homogeneity of degree one of the function f. □

6.21. Definition. If $K \subseteq L$ is an acute convex cone and $f: K \to \mathbf{R}_+$ is a support function, then the support function for the set Y, defined by:

$$Y = \{\mathbf{y} \in K \mid f(\mathbf{y}) \geq 1\},$$

is called the **dual support function for** (or is **the support function dual to**) f; that is, the function $g: K \to \mathbf{R}_+$ defined by:

$$g(\mathbf{z}) = \inf\{\mathbf{z} \cdot \mathbf{y} \mid \mathbf{y} \in Y\},$$

is the support function dual to f.

It of course follows at once from 6.18 and 6.20 that if f is a support function on K, then its dual, g, is also a support function on K.

6.22. Definition. We shall say that a nonempty cone, K, is **self-dual** iff $K = K^*$; that is, iff K and satisfies:

1. for all $\mathbf{x}, \mathbf{y} \in K, \mathbf{x} \cdot \mathbf{y} \geq \mathbf{0}$ (so that K is acute), and:
2. if $\mathbf{x} \in L$ is such that for all $\mathbf{y} \in K$, we have $\mathbf{x} \cdot \mathbf{y} \geq 0$, then $\mathbf{x} \in K$.

The particularly interesting property of dual support functions, from our point of view, is set out in the following.

6.23. Theorem. *If $K \subseteq L$ is a self-dual convex cone, $f: K \to \mathbf{R}_+$ is a support function, and $g: K \to \mathbf{R}_+$ is its dual, and if we define:*

$$Z = \{\mathbf{z} \in K \mid g(\mathbf{z}) \geq 1\},$$

then f is the support function for Z on K (and thus f is the support fuction dual to g as well).

Proof. Defining the set Z as in the above statement:

$$Z = \{\mathbf{z} \in K \mid g(\mathbf{z}) \geq 1\},$$

let f^* be the support function for Z on K, and define:

$$Y^* = \{\mathbf{y} \in K \mid f^*(\mathbf{y}) \geq 1\}.$$

Since both f and f^* are positively homogeneous of degree one, it follows from Proposition 5.103 that, in order to establish the fact that $f \equiv f^*$, it suffices to prove that $Y = Y^*$; where the set Y is defined by:

$$Y = \{\mathbf{y} \in K \mid f(\mathbf{y}) \geq 1\}.$$

Accordingly, let $\mathbf{y}^* \in Y$. Then, for an arbitrary $\mathbf{z} \in Z$, we have:

$$\mathbf{y}^* \cdot \mathbf{z} \geq \inf\{\mathbf{y} \cdot \mathbf{z} \mid \mathbf{y} \in Y\} \equiv g(\mathbf{z}) \geq 1;$$

and thus, since $\mathbf{z} \in Z$ was arbitrary:

$$\inf\{\mathbf{y}^* \cdot \mathbf{z} \mid \mathbf{z} \in Z\} \equiv f^*(\mathbf{y}^*) \geq 1.$$

It follows that $\mathbf{y}^* \in Y^*$, and, consequently, that $Y \subseteq Y^*$.

To prove the converse, suppose $\widehat{\mathbf{y}} \in K \setminus Y$. Then it follows from Lemma 5.53 and Theorem 5.54 that there exists $\mathbf{z}^* \in Y - \widehat{\mathbf{y}}$ satisfying:

$$\widehat{\mathbf{y}} \cdot \mathbf{z}^* < a \overset{\text{def}}{=} \inf\{\mathbf{z}^* \cdot \mathbf{y} \mid \mathbf{y} \in Y\}. \tag{6.11}$$

Since Y is unbounded with respect to $\geq_K$, it follows easily that $\mathbf{z}^* \in K^*$; and, since $\widehat{\mathbf{y}} \in K$, that:

$$a > 0.$$

Moreover, since K is self-dual (and thus $K^* \subseteq K$), we see that the vector $\overline{\mathbf{z}}$ defined by:

$$\overline{\mathbf{z}} = (1/a)\mathbf{z}^*,$$

is an element of K. From (6.11), it then follows that:

$$g(\overline{\mathbf{z}}) = 1,$$

so that $\overline{\mathbf{z}} \in Z$. However, it then follows that:

$$f^*(\widehat{\mathbf{y}}) \leq \widehat{\mathbf{y}} \cdot \overline{\mathbf{z}} = (1/a)\widehat{\mathbf{y}} \cdot \mathbf{z}^* < 1.$$

and thus $\widehat{\mathbf{y}} \notin Y^*$. Therefore:

$$K \setminus Y \subseteq K \setminus Y^*,$$

and thus $Y^* \subseteq Y$; so that we can now conclude that $Y^* = Y$. □

In all probability, the above result remains correct even if the assumption that K is self-dual is dropped from its hypotheses. In all the applications which we will make of Theorem 6.23, however, it will turn out that the cone involved is always self-dual in any event.

Appendix/Exercises.

1. Prove parts 3 and 6 of Theorem 6.13.

2. Show that, under the conditions of Definition 6.16, the support function f is actually **increasing**, in the sense that if we define $\gg_k$ on L by:

$$\mathbf{x} \gg_K \mathbf{y} \iff \mathbf{x} - \mathbf{y} \in riK,$$

then, for all $\mathbf{x}, \mathbf{y} \in K$,

$$\mathbf{x} \gg_K \mathbf{y} \Rightarrow f(\mathbf{x}) > f(\mathbf{y}).$$

3. A given support function, $f \colon K \to \mathbf{R}_+$, may be the support function for more than one subset of K. There is, however, a unique largest set for which f is the support function, defined as follows.

6.24. Definition. If $K \subseteq L$ is a convex cone, and the function $f \colon K \to \mathbf{R}_+$ is a support function, we define U_f, the **implicit support set determined by f** as:

$$U_f = \{\mathbf{z} \in K \mid (\forall \mathbf{y} \in K) \colon \mathbf{y} \cdot \mathbf{z} \geq f(\mathbf{y})\}.$$

Show that if f is a support function, then U_f satisfies the following properites:

1. U_f is non-empty, closed, convex, and unbounded with respect to $\geq_K$,
2. $\mathbf{0} \notin U_f$, and
3. if $D \subseteq K$ is such that f is the support function for D, then $D \subseteq U_f$.

6.4 Duality of Cost and Production

In this section, our primary concern will be with the applications of the results of Section 2 to derive what seem to me to be some of the more interesting and useful properties of technology and cost functions, and their relationships to one another. The results to be developed here will assume that the technological possibilities under study can be represented by a subset of finite-dimensional euclidean space, as was assumed to be the case in Section 2 of this chapter. Those who have read Section 3 carefully, however, will probably already have realized that this theory can easily be extended to some infinite-dimensional spaces; however, this extension will be left as a project for the interested reader.

All of the results regarding the properties of technology and cost functions, and the relationships among them to which we alluded in Section 2 can now be established as very simple consequences of the results of Section 3; the proofs being almost immediate consequences of the following considerations. First, if we compare Definitions 6.3 and 6.12, it is easy to see that, for a fixed $\mathbf{x} \in X$, the technology function $\psi(\cdot, \mathbf{x})$ is the distance function for $V(\mathbf{x})$ on $\mathbf{R}^n_+$. This yields a very easy proof that the function defined in 6.3 is, in fact, a technology function, as defined in 6.6. Secondly, it is equally easy to see that the cost function derived from a given technology function is, for each $\mathbf{x} \in X$, the support function dual to $\psi(\cdot, \mathbf{x})$. This allows us to deduce the properties of the cost function as a nearly immediate consequence of 6.18. Moreover, we then obtain the very important duality theorem as a more or less immediate consequence of Theorem 6.23 (notice that $\mathbf{R}^n_+$ is a self-dual, closed, convex cone). The details follow.

While it is perhaps somewhat surprising that the representation of production possibilities via a technology function is as general as the representation of technology by an input-requirement correspondence, this is so; as is established in the following result.

6.25. Theorem. *If $V\colon X \mapsto \mathbf{R}^n_+$ is an input-requirement correspondence, then $\psi = \delta(V)$, as defined in 6.3, is a technology function.*

Proof. Condition 2.a of Definition 6.2 obviously implies that X is a producible set [as specified by equation (6.4) of Section 1]. In order to prove that ψ is non-increasing in $\mathbf{x}$, let $\mathbf{v}^* \in \mathbf{R}^n_+$ be arbitrary, and let $\mathbf{x}_1, \mathbf{x}_2 \in X$ be such that $\mathbf{x}_1 \geq \mathbf{x}_2$. Then by 6.2.2b, we have:

$$V(\mathbf{x}_1) \subseteq V(\mathbf{x}_2),$$

and thus we must have:

$$\Lambda(\mathbf{v}^*, \mathbf{x}_1) \subseteq \Lambda(\mathbf{v}^*, \mathbf{x}_2). \tag{6.12}$$

If $\Lambda(\mathbf{v}^*, \mathbf{x}_1) = \emptyset$, then we obviously have:

$$0 = \psi(\mathbf{v}^*, \mathbf{x}_1) \leq \psi(\mathbf{v}^*, \mathbf{x}_2).$$

On the other hand, if $\Lambda(\mathbf{v}^*, \mathbf{x}_1) \neq \emptyset$, then from (6.12) it follows that $\Lambda(\mathbf{v}^*, \mathbf{x}_2) \neq \emptyset$, and that:

$$\psi(\mathbf{v}^*, \mathbf{x}_1) \equiv \max \Lambda(\mathbf{v}^*, \mathbf{x}_1) \leq \max \Lambda(\mathbf{v}^*, \mathbf{x}_2) \equiv \psi(\mathbf{v}^*, \mathbf{x}_2).$$

To prove that $\psi(\cdot, \mathbf{x})$ satisfies the properties specified in Definition 6.6 as a function of $\mathbf{v}$, let $\mathbf{x}^* \in X$ be given. Then we note that $\psi(\cdot, \mathbf{x}^*)$ is simply the distance function for $V(\mathbf{x}^*)$ on $\mathbf{R}^n_+$. The fact that properties 2–5 of Definition 6.6 are satisfied, apart from the lower semi-continuity of ψ, is then an immediate consequence of Theorem 6.13 and Proposition 6.14. The fact that ψ is also lower semi-continuous in $\mathbf{v}$ follows from the fact that it is non-decreasing and positively homogeneous of degree one (see Exercise 7, at the end of Section 5.8.) □

The primary business with which we will be concerned in the remainder of this section is the establishment of the duality between technology and cost functions; which I believe to be one of the more important and useful relationships developed in modern theoretical economics. The original version of this duality relationship was first stated by Shephard [1953] for single-output production functions. Generalized versions were later developed by Uzawa [1964], Jacobsen [1972], and McFadden [1970, 1978]. We begin with the following definition.

6.26. Definition. If $c\colon \mathbf{R}^n_+ \times X \to \mathbf{R}_+$ is a cost function, we define the **technology function dual to c**, $\psi = d(c)$, by:

$$\psi(\mathbf{v}, \mathbf{x}) = \inf\{\mathbf{w} \cdot \mathbf{v} \mid \mathbf{w} \in \mathbf{R}^n_+ \ \&\ c(\mathbf{w}, \mathbf{x}) \geq 1\} \quad \text{for } (\mathbf{v}, \mathbf{x}) \in \mathbf{R}^n_+ \times X.$$

6.27. Theorem. *If ψ is a technology function, then the function $c = \kappa(\psi)$, defined in 6.10, above, is a cost function, and ψ is the technology function dual to c [that is, in terms of Definition 6.26, $\psi = d(c)$].*

Proof. Let $\mathbf{w}^* \in \mathbf{R}^n_+$ be arbitrary, and $\mathbf{x}_1, \mathbf{x}_2 \in X$ be such that $\mathbf{x}_1 \geq \mathbf{x}_2$. Then, since $\psi(\mathbf{v}, \cdot)$ is non-increasing in $\mathbf{x}$:

$$\{\mathbf{v} \in \mathbf{R}^n_+ \mid \psi(\mathbf{v}, \mathbf{x}_1) \geq 1\} \equiv V(\mathbf{x}_1) \subseteq V(\mathbf{x}_2) \equiv \{\mathbf{v} \in \mathbf{R}^n_+ \mid \psi(\mathbf{v}, \mathbf{x}_2) \geq 1\}.$$

Thus it is apparent that:

$$c(\mathbf{w}^*, \mathbf{x}_1) \equiv \inf\{\mathbf{w}^* \cdot \mathbf{v} \mid \mathbf{v} \in V(\mathbf{x}_1)\} \geq c(\mathbf{w}^*, \mathbf{x}_2) \equiv \inf\{\mathbf{w}^* \cdot \mathbf{v} \mid \mathbf{v} \in V(\mathbf{x}_2)\}.$$

To prove the remaining properties of Definition 6.9, we note that, for a fixed $\mathbf{x}^* \in X$, $c(\cdot, \mathbf{x}^*)$ is the support function for:

$$V(\mathbf{x}^*) \stackrel{\text{def}}{=} \{\mathbf{v} \in \mathbf{R}^n_+ \mid \psi(\mathbf{v}, \mathbf{x}^*) \geq 1\}.$$

Consequently, the required properties are a more or less immediate consequence of Proposition 6.18.[7] I will leave the verification of the details of this step as an exercise.

In the above two paragraphs, we have established that c is a cost function; and we have done so by noting that, for each $\mathbf{x} \in X$, $c(\cdot, \mathbf{x})$ is the support function on $\mathbf{R}^n_+$ for the set $V(\mathbf{x})$ defined by:

$$V(\mathbf{x}) = \{\mathbf{v} \in \mathbf{R}^n_+ \mid \psi(\mathbf{v}, \mathbf{x}) \geq 1\}.$$

It follows that, for each $\mathbf{x} \in X$, $c(\cdot, \mathbf{x})$ is the support function dual to $\psi(\cdot, \mathbf{x})$. Since $\mathbf{R}^n_+$ is a convex cone which is self-dual, it then follows from Theorem 6.23 that $\psi(\cdot, \mathbf{x})$ is dual to $c(\cdot, \mathbf{x})$ as well. Upon comparing Definition 6.26, it then follows at once that $\psi = d(c)$. □

As set out in Definition 6.10 (and as now partially justified by Theorem 6.27), we will refer to the cost function $c = \kappa(\psi)$, as the **cost function dual to ψ**. The justification for this terminology is then completed by the following result. The proof is quite symmetric to the proof of Theorem 6.27, and will be left as an exercise.

[7] See also Exercise 7, at the end of Section 5.8.

6.28. Theorem. *If* $c\colon \mathbf{R}^n_+ \times X \to \mathbf{R}_+$ *is a cost function, then the function* $\psi = d(c)$ *defined in 6.26, above, is a technology function; and* $c(\cdot)$ *is the cost function dual to* ψ *[that is,* $c = \kappa(\psi)$.

The relationships and properties established by the last two results are worth some additional discussion. Let 'Ψ' denote the collection of all technology functions defined on $\mathbf{R}^n_+ \times X$, for a given producible set $X \subseteq \mathbf{R}^n_+ \backslash \{\mathbf{0}\}$; and, similarly, let '$\mathcal{C}$' denote the collection of all cost functions defined on $\mathbf{R}^n_+ \times X$. If we use the notation of the two results just established, we have shown the following. First, let $\psi \in \Psi$. Then, letting $c = \kappa(\psi)$ be the cost function dual to ψ, it follows from Theorem 6.27 that $c \in \mathcal{C}$ and that $\psi = d(c)$. Thus we see that:

$$(\forall \psi \in \Psi)\colon d[\kappa(\psi)] = \psi; \tag{6.13}$$

in words, if we take the cost function dual to ψ, and use Definition 6.26 to find the technology function dual to c, we will recover ψ. Now suppose we begin with an arbitrary cost function, $c \in \mathcal{C}$, and use Definition 6.27 to find the technology function, $\psi = d(c)$, dual to c. Then it follows from Theorem 6.28 that c is the cost function dual to ψ; that is, $c = \kappa(\psi)$. Consequently, we see that:

$$(\forall c \in \mathcal{C})\colon \kappa[d(c)] = c. \tag{6.14}$$

Using equations (6.13) and (6.14), it now follows from Proposition 3.54 that both functions $\kappa(\cdot)$ and $d(\cdot)$ are one-to-one and onto, and that one is the inverse of the other. In particular, it follows that we have established a *one-to-one correspondence between* Ψ *and* $\mathcal{C}$. Among the many useful implications of this fact is the following. Suppose we have a function, $c\colon \mathbf{R}^n_+ \times X \to \mathbf{R}_+$, which satisfies all of the properties specified in 6.9. Definition 6.26 tells us how to find the technology function dual to c (and from which c could be derived); the derivation of which involves solving a constrained minimization problem. However, it may be that the constrained minimization problem involved is impossible, or at least infeasible, to solve. This need not worry us particularly, however; that is, it should never prevent our making use of such cost functions in estimation work, because Theorem 6.28 and the considerations just discussed tell us that *there exists a unique technology function dual to c whether or not we can actually find it!*

The following result, the proof of which will be left as an exercise, is both very simple and often quite useful. The set $W(\mathbf{x})$ to which reference is made in the result (and which is used in Theorem 6.30, which follows it) is defined, for a given cost function, c, by:

$$W(\mathbf{x}) = \{\mathbf{w} \in \mathbf{R}^n_+ \mid c(\mathbf{w}, \mathbf{x}) \geq 1\} \quad \text{for } \mathbf{x} \in X.$$

6.29. Proposition. *Let* $\psi\colon \mathbf{R}^n_+ \times X \to \mathbf{R}_+$ *be a technology function, and* $c\colon \mathbf{R}^n_+ \times X \to \mathbf{R}_+$ *be its dual cost function. Then for all* $\mathbf{x} \in X$, *and all*

$\mathbf{w}, \mathbf{v} \in \mathbf{R}^n_+$, *we have:*

$$\mathbf{w} \cdot \mathbf{v} \geq c(\mathbf{w}, \mathbf{x})\psi(\mathbf{v}, \mathbf{x}).$$

Furthermore, if $\mathbf{x}^* \in X$ *and* $\mathbf{v}^*, \mathbf{w}^* \in \mathbf{R}^n_+$ *are such that* $c(\mathbf{w}^*, \mathbf{x}^*)\psi(\mathbf{v}^*, \mathbf{x}^*) > 0$, *then*

$$\mathbf{w}^* \cdot \mathbf{v}^* = c(\mathbf{w}^*, \mathbf{x}^*)\psi(\mathbf{v}^*, \mathbf{x}^*),$$

if, and only if, we have both:

1. the vector $[1/\psi(\mathbf{v}^*, \mathbf{x}^*)]\mathbf{v}^*$ *minimizes the cost of producing* $\mathbf{x}^*$, *given* $\mathbf{w}^*$, *and*

2. $[1/c(\mathbf{w}^*, \mathbf{x}^*)]\mathbf{w}^*$ *minimizes* $\mathbf{w} \cdot \mathbf{v}^*$, *subject to* $\mathbf{w} \in W(\mathbf{x}^*)$.

The following result is somewhat more fundamental than it may appear at first. We will discuss this idea further following its proof.

6.30. Theorem. *The following conditions are all equivalent:*

1. T *is a convex set.*

2. X *is convex, and* $c(\mathbf{w}, \cdot)$ *is convex in* $\mathbf{x}$, *for each* $\mathbf{w} \in \mathbf{R}^n_+$.

3. X *is convex and* ψ *is quasi-concave on* $\mathbf{R}^n_+ \times X$.

Proof.

1. If T is convex, then it is obvious that X, the producible set, is convex. In order to prove that $c(\mathbf{w}, \cdot)$ is convex in $\mathbf{x}$, for each $\mathbf{w}$, let $\mathbf{w}^* \in \mathbf{R}^n_+, \mathbf{x}_1, \mathbf{x}_2 \in X, \theta \in [0, 1]$, and $\epsilon > 0$ be given. Then, for each i, there exists $\mathbf{v}_i \in \mathbf{R}^n_+$ such that:

$$(\mathbf{v}_i, \mathbf{x}_i) \in T \text{ and } \mathbf{w}^* \cdot \mathbf{v}_i < c(\mathbf{w}^*, \mathbf{x}_i) + \epsilon \quad \text{for } i = 1, 2. \tag{6.15}$$

However, since T is convex, we have:

$$\theta(\mathbf{v}_1, \mathbf{x}_1) + (1 - \theta)(\mathbf{v}_2, \mathbf{x}_2) = (\theta\mathbf{v}_1 + (1 - \theta)\mathbf{v}_2, \theta\mathbf{x}_1 + (1 - \theta)\mathbf{x}_2) \in T;$$

and therefore, using (6.15):

$$\begin{aligned} &c[\mathbf{w}^*, \theta\mathbf{x}_1 + (1 - \theta)\mathbf{x}_2] \leq \mathbf{w}^* \cdot [\theta\mathbf{v}_1 + (1 - \theta)\mathbf{v}_2] \\ &= \theta\mathbf{w}^* \cdot \mathbf{v}_1 + (1 - \theta)\mathbf{w}^* \cdot \mathbf{v}_2 < \theta[c(\mathbf{w}^*, \mathbf{x}_1) + \epsilon] + (1 - \theta)[c(\mathbf{w}^*, \mathbf{x}_2) + \epsilon] \\ &= \theta c(\mathbf{w}^*, \mathbf{x}_1) + (1 - \theta)c(\mathbf{w}^*, \mathbf{x}_2) + \epsilon. \end{aligned}$$

Since $\epsilon > 0$ was arbitrary, we then conclude that:

$$c[\mathbf{w}^*, \theta\mathbf{x}_1 + (1 - \theta)\mathbf{x}_2] \leq \theta c(\mathbf{w}^*, \mathbf{x}_1) + (1 - \theta)c(\mathbf{w}^*, \mathbf{x}_2).$$

2. Now suppose that X is convex, and that for each $\mathbf{w}$, $c(\mathbf{w}, \cdot)$ is convex in $\mathbf{x}$; let $\theta \in [0, 1]$ and $(\mathbf{v}_i, \mathbf{x}_i) \in \mathbf{R}^n_+ \times X$ be given, and define:

$$\mathbf{v}_3 = \theta\mathbf{v}_1 + (1 - \theta)\mathbf{v}_2 \text{ and } \mathbf{x}_3 = \theta\mathbf{x}_1 + (1 - \theta)\mathbf{x}_2.$$

If either

$$\psi(\mathbf{v}_1, \mathbf{x}_1) = 0 \text{ or } \psi(\mathbf{v}_2, \mathbf{x}_2) = 0,$$

then obviously:

$$\psi(\mathbf{v}_3, \mathbf{x}_3) \geq \beta \stackrel{\text{def}}{=} \min\{\psi(\mathbf{v}_1, \mathbf{x}_1), \psi(\mathbf{v}_2, \mathbf{x}_2)\};$$

so suppose that $\beta > 0$, and define:

$$\mathbf{v}_i^* = (1/\beta)\mathbf{v}_i \quad \text{for } i = 1, 2, 3.$$

Since $\psi(\cdot, \mathbf{x})$ is positively homogeneous of degree one in $\mathbf{v}$, we have:

$$\psi(\mathbf{v}_i^*, \mathbf{x}_i) \geq 1 \quad \text{for } i = 1, 2. \tag{6.16}$$

Now let $\mathbf{w}^* \in W(\mathbf{x}_3)$. Then, from (6.16) and the convexity of $c(\mathbf{w}^*, \cdot)$ in $\mathbf{x}$, we see that:

$$\begin{aligned}\mathbf{w}^* \cdot \mathbf{v}_3^* = \theta \mathbf{w}^* \cdot \mathbf{v}_1^* + (1-\theta)\mathbf{w}^* \cdot \mathbf{v}_2^* \\ \geq \theta c(\mathbf{w}^*, \mathbf{x}_1) + (1-\theta)c(\mathbf{w}^*, \mathbf{x}_2) \geq c(\mathbf{w}^*, \mathbf{x}_3) \geq 1.\end{aligned}$$

Since ψ is the technology function dual to c, and $\mathbf{w}^* \in W(\mathbf{x}_3)$ was arbitrary, it now follows that:

$$\psi(\mathbf{v}_3^*, \mathbf{x}_3) = \inf\{\mathbf{w} \cdot \mathbf{v}_3^* \mid \mathbf{w} \in W(\mathbf{x}_3)\} \geq 1;$$

and, using the homogeneity of $\psi(\cdot, \mathbf{x}_3)$ in $\mathbf{v}$, and the definitions of β, $\mathbf{v}_3^*$ and $\mathbf{v}_3$, we then have:

$$\psi(\mathbf{v}_3, \mathbf{x}_3) \geq \beta = \min\{\psi(\mathbf{v}_1, \mathbf{x}_1), \psi(\mathbf{v}_2, \mathbf{x}_2)\}.$$

3. The proof that if ψ is quasi-concave on $\mathbf{R}_+^n \times X$, then T is convex, is very straightforward, and will be left as exercise. □

I have not stated the above result as exactly as it perhaps should be stated. More completely set out, condition 1 should read: 'if T is convex, then the cost function generated by T is convex in $\mathbf{x}$, and the technology function generated by T is quasi-concave,' while condition 2 should read 'if X is convex, and $c(\mathbf{w}, \cdot)$ is convex in $\mathbf{x}$, then the technology function dual to c is quasi-concave, and the technology set corresponding to c is convex,[8]' and so on. Presumably, you get the idea.

A number of further results regarding the relationship between quasi-concavity of ψ in output and the convexity of the production possibility correspondence and/or the quasi-convexity in $\mathbf{x}$ of the cost function, etc., can be found in Jacobsen [1970]. However, we will turn now to a discussion of homotheticity.

[8]Formally, the technology set corresponding to the technology function dual to c.

6.5 Homotheticity

In this section, we will consider some extensions of the notion of homotheticity, which was developed for single-output production functions in Section 3.6, to the case in which any finite number of goods are being produced. In Example 6.4 of Section 1 of this chapter, we showed that the technology function corresponding to a homothetic (single-output) production function of the form $\varphi(\mathbf{v}) = F[\sigma(\mathbf{v})]$ took the form:

$$\psi(\mathbf{v}, x) = \sigma(\mathbf{v})/f(x),$$

where f is the input-requirement function dual to F. This functional form easily generalizes to the case of m outputs, as follows.

6.31. Definition. Let $X \subseteq \mathbf{R}^m_+ \setminus \{\mathbf{0}\}$ be a producible set. We shall say that a technology function $\psi\colon \mathbf{R}^n_+ \times X \to \mathbf{R}_+$ is **homothetic** iff there exist a support function $\sigma\colon \mathbf{R}^n_+ \to \mathbf{R}_+$ and a non-decreasing function, $f\colon X \to \mathbf{R}_{++}$, such that for every $(\mathbf{v}, \mathbf{x}) \in \mathbf{R}^n_+ \times X$:

$$\psi(\mathbf{v}, \mathbf{x}) = \sigma(\mathbf{v})/f(\mathbf{x}).$$

McFadden [1978] uses the term 'separable' to refer to the above property. Giora Hanoch has introduced a weaker notion of homotheticity, which, given the terminology of Definition 6.31, might be called 'weak homotheticity.' We will not consider Hanoch's definition here, but we will consider a property which we will call 'strong homotheticity,' later on in this section. Before proceeding further, let's consider some examples, all of which are borrowed, virtually intact, from McFadden [1978, pp. 38–9].

6.32. Examples.

1. **The Cobb-Douglas Family of Technology Functions.** We shall say that a technology function is of the Cobb-Douglas form iff $\psi\colon \mathbf{R}^n_+ \times X \to \mathbf{R}_+$ takes the form:

$$\psi(\mathbf{v}, \mathbf{x}) = A \cdot \left[\prod_{i=1}^n v_i^{a_i}\right]/f(\mathbf{x}), \tag{6.17}$$

where X is a producible set, $f\colon X \to \mathbf{R}_{++}$ is non-decreasing, and:

$$a_i \geq 0, \text{ for } i = 1, \ldots, n; \sum_{i=1}^n a_i = 1, \text{ and } A > 0.$$

One usually assumes that X is a cone, and that:

$$\mathbf{R}^m_{++} \subseteq X,$$

when working with the Cobb-Douglas technology function.

It can be shown that the cost function corresponding to the Cobb-Douglas technology function (the **Cobb-Douglas cost function**) is given by:

$$c(\mathbf{w}, \mathbf{x}) = B \cdot \left[\prod_{i=1}^{n} w_i^{a_i}\right] \cdot f(\mathbf{x}), \tag{6.18}$$

where:

$$B \stackrel{\text{def}}{=} \left[A \cdot \prod_{i=1}^{n} (a_i)^{a_i}\right]^{-1}.$$

2. **The Leontief Technology Function.** We shall say that a technology function is of the **Leontief form** iff $\psi \colon \mathbf{R}_+^n \times X \to \mathbf{R}_+$ takes the form:

$$\psi(\mathbf{v}, \mathbf{x}) = \min\{v_1/D_1(\mathbf{x}), v_2/D_2(\mathbf{x}), \ldots, v_n/D_n(\mathbf{x})\}, \tag{6.19}$$

where X is a producible set and $D_i \colon X \to \mathbf{R}_{++}$ is non-decreasing, for $i = 1, \ldots, n$. It is easy to show that, if there exist positive constants, β_i ($i = 1, \ldots, n$) and a non-decreasing function $f \colon X \to \mathbf{R}_{++}$ such that $D_i(\cdot)$ is given by:

$$D_i(\mathbf{x}) = \beta_i f(\mathbf{x}) \quad \text{for } i = 1, \ldots, n;$$

then ψ is homothetic. It is also easy to show that the cost function corresponding to this technology function (the **Leontief Cost Function**) is given by:

$$c(\mathbf{w}, \mathbf{x}) = \sum_{i=1}^{n} w_i D_i(\mathbf{x}). \tag{6.20}$$

3. The **CES (Constant Elasticity of Substitution) Technology Function.** We shall say that a technology function is of the **CES form** iff $\psi \colon \mathbf{R}_+^n \times X \to \mathbf{R}_+$ takes the form:

$$\psi(\mathbf{v}, \mathbf{x}) = \left(\sum_{i=1}^{n} [v_i/D_i(\mathbf{x})]^{-\rho}\right)^{-1/\rho}, \tag{6.21}$$

where X is a producible set, $D_i \colon X \to \mathbf{R}_{++}$ is non-decreasing, for $i = 1, \ldots, n$; and:[9]

$$-1 \le \rho < 0 \ \text{ or } \ 0 < \rho.$$

This particular form, as opposed to the form:

$$\psi(\mathbf{v}, x) = \left(\sum_{i=1}^{n} a_i v_i^{-\rho}\right)^{-1/\rho} / x,$$

[9]It can be shown that the Cobb-Douglas form can be regarded as the limiting form of the CES as $\rho \to 0$, and the Leontief form can be regarded as the limiting form as $\rho \to +\infty$.

is apparently due to McFadden [1978]. The one-output, two-input version of this function was introduced into the economics literature by Solow [1956]. Its properties, for this case, were first investigated by Arrow, Chenery, Minhas, and Solow [1961]; who also did some extensive estimation work with the function. The one-output, n-input version of the function was first presented and analyzed by Uzawa [1962].

It can be shown that the cost function corresponding to the CES technology function (the **CES cost function**) is given by:

$$c(\mathbf{w}, \mathbf{x}) = \left(\sum_{i=1}^{n} [w_i D_i(\mathbf{x})]^{\rho/(1+\rho)}\right)^{(1+\rho)/\rho} \quad \text{for } -1 < \rho < 0 \text{ or } 0 < \rho, \tag{6.22}$$

and:

$$c(\mathbf{w}, \mathbf{x}) = \min\{w_1 D_1(\mathbf{x}), w_2 D_2(\mathbf{x}), \ldots, w_n D_n(\mathbf{x})\} \quad \text{for } \rho = -1. \tag{6.23}$$

It can also be shown that the (constant) elasticity of (factor) substitution is in this case given by:

$$s = 1/(1+\rho) \quad \text{for } -1 < \rho < 0 \text{ or } 0 < \rho,$$

and is defined to be equal to $+\infty$ for $\rho = -1$.

As mentioned earlier, the Cobb-Douglas technology function can be considered to be the special case of the CES technology function in which $\rho = 0$ (and we have $s = 1$ in the Cobb-Douglas case); and the Leontief technology function can be considered to be the special case in which $\rho = +\infty$ (and we have $s = 0$ for the Leontief case). □

As presented here, the CES technology function is not necessarily homothetic; however, we have the following. I will leave the proof as an exercise.

6.33. Proposition. *If $\psi\colon \mathbf{R}^n_+ \times X \to \mathbf{R}_+$ is of the CES form:*

$$\psi(\mathbf{v}, \mathbf{x}) = \left(\sum_{i=1}^{n} [v_i / D_i(\mathbf{x})]^{-\rho}\right)^{-1/\rho},$$

and if there exist a non-decreasing function $f\colon X \to \mathbf{R}_{++}$ and constants $d_i > 0$, for $i = 1, \ldots, n$, such that the functions $D_i(\cdot)$ take the form:

$$D_i(\mathbf{x}) = d_i f(\mathbf{x}) \quad \textit{for } i = 1, \ldots, n; \tag{6.24}$$

then ψ is homothetic.

If you work out the details of the proof of Proposition 6.33, you can show that if ψ is of the CES form and satisfies equation (6.24), then, defining:

$$a_i = d_i^{\rho} \quad \text{for } i = 1, \ldots, n,$$

we have:

$$\psi(\mathbf{v}, \mathbf{x}) = \left(\sum_{i=1}^{n} a_i v_i^{-\rho}\right)^{-1/\rho} / f(\mathbf{x}). \tag{6.25}$$

Conversely, we shall refer to a function of the form (6.25) [and for which $a_i > 0$ for $i = 1, \ldots, n$, and $-1 \leq \rho < 0$ or $0 < \rho$] as being of the **homothetic CES form**.

6.34. Definition. We shall say that a cost function $c\colon \mathbf{R}^n_+ \times X \to \mathbf{R}_+$, is **homothetic** iff there exist a non-decreasing function, $f\colon X \to \mathbf{R}_{++}$ and a support function $\gamma\colon \mathbf{R}^n_+ \to \mathbf{R}_+$ such that, for each $(\mathbf{w}, \mathbf{x}) \in \mathbf{R}^n_+ \times X$:

$$c(\mathbf{w}, \mathbf{x}) = \gamma(\mathbf{w}) f(\mathbf{x}).$$

We can now make use of Theorems 6.27 and 6.28 to prove the following.

6.35. Theorem. *A technology function, ψ, is of the homothetic form:*

$$\psi(\mathbf{v}, \mathbf{x}) = \sigma(\mathbf{v}) / f(\mathbf{x}),$$

if, and only if the cost function, c, dual to ψ has the homothetic form:

$$c(\mathbf{w}, \mathbf{x}) = \gamma(\mathbf{w}) f(\mathbf{x});$$

where the support functions, σ and γ are dual to one another.

Proof. Suppose ψ is homothetic, so that we can write:

$$\psi(\mathbf{v}, \mathbf{x}) = \sigma(\mathbf{v}) / f(\mathbf{x}).$$

Define:

$$U = \{\mathbf{v} \in \mathbf{R}^n_+ \mid \sigma(\mathbf{v}) \geq 1\},$$

and let $\gamma(\cdot)$ be the support function for U; so that γ is dual to σ.

Now, let $(\mathbf{w}^*, \mathbf{x}^*) \in \mathbf{R}^n_+ \times X$, and suppose first that:

$$\mathbf{v} \in V(\mathbf{x}^*) \equiv \{\mathbf{v} \in \mathbf{R}^n_+ \mid \psi(\mathbf{v}, \mathbf{x}^*) \geq 1\}.$$

Then:

$$1 \leq \sigma(\mathbf{v}) / f(\mathbf{x}^*) = \sigma\big[\big(1/f(\mathbf{x}^*)\big)\mathbf{v}\big],$$

and thus $\big(1/f(\mathbf{x}^*)\big)\mathbf{v} \in U$. Since γ is the support function for U, we then have:

$$\mathbf{w}^* \cdot \big(1/f(\mathbf{x}^*)\big)\mathbf{v} = \big(1/f(\mathbf{x}^*)\big)\mathbf{w}^* \cdot \mathbf{v} \geq \gamma(\mathbf{w}^*),$$

or:

$$\mathbf{w}^* \cdot \mathbf{v} \geq \gamma(\mathbf{w}^*) f(\mathbf{x}^*).$$

It now follows that $\gamma(\mathbf{w}^*)f(\mathbf{x}^*)$ is a lower bound for:

$$\{\mathbf{w}^* \cdot \mathbf{v} \mid \mathbf{v} \in V(\mathbf{x}^*)\};$$

and, since $c(\mathbf{w}^*, \mathbf{x}^*)$ is the infimum for this set, we conclude that:

$$c(\mathbf{w}^*, \mathbf{x}^*) \geq \gamma(\mathbf{w}^*)f(\mathbf{x}^*). \tag{6.26}$$

Conversely, if $\mathbf{v} \in U$, then:

$$f(\mathbf{x}^*)\mathbf{v} \in V(\mathbf{x}^*).$$

so that:

$$\mathbf{w}^* \cdot [f(\mathbf{x}^*)\mathbf{v}] = f(\mathbf{x}^*)\mathbf{w}^* \cdot \mathbf{v} \geq c(\mathbf{w}^*, \mathbf{x}^*),$$

and thus:

$$\mathbf{w}^* \cdot \mathbf{v} \geq c(\mathbf{w}^*, \mathbf{x}^*)/f(\mathbf{x}^*).$$

Since $\gamma(\mathbf{w}^*)$ is the infimum of the set of such inner products, it then follows that:

$$\gamma(\mathbf{w}^*) \geq c(\mathbf{w}^*, \mathbf{x}^*)/f(\mathbf{x}^*),$$

or:

$$\gamma(\mathbf{w}^*)f(\mathbf{x}^*) \geq c(\mathbf{w}^*, \mathbf{x}^*) \tag{6.27}$$

Combining equations (6.26) and (6.27), we then have:

$$\gamma(\mathbf{w}^*)f(\mathbf{x}^*) = c(\mathbf{w}^*, \mathbf{x}^*).$$

A similar argument establishes the fact that if c is homothetic, then the technology function dual to c is also homothetic. □

6.36. Definition. We shall say that an input-requirement correspondence, $V \colon X \mapsto \mathbf{R}^n_+$ is **homothetic** iff there exist a nondecreasing function, $f \colon X \to \mathbf{R}_{++}$, and a non-empty, closed, and convex set, $V \subseteq \mathbf{R}^n_+ \setminus \{\mathbf{0}\}$ such that:

$$(\forall \mathbf{x} \in X) \colon V(\mathbf{x}) = f(\mathbf{x})V.$$

It can be shown that an input-requirement correspondence is homothetic if, and only if, the corresponding technology function is homothetic; and, in turn this will be true if, and only if, the dual cost function is homothetic. You are asked to prove these statements in Exercise 4, at the end of this section.

Using Theorems 6.35 and 6.30, you should also find the proof of the following result to be an easy exercise.

6.37. Proposition. (Jacobsen [1970]) *Suppose X is convex, and that ψ is homothetic, so that we can write:*

$$\psi(\mathbf{v}, \mathbf{x}) = \sigma(\mathbf{v})/f(\mathbf{x}).$$

Then the corresponding technology set, T, is convex if, and only if f is a convex function.

A somewhat more stringent homotheticity condition than that which we have been studying will make use of the following.

6.38. Definition. Let $X \subseteq \mathbf{R}_+^m$ be a producible set. We shall say that a function $g\colon X \to \mathbf{R}_{++}$ is a **homothetic production index** iff there exist a producible set, $Z \subseteq \mathbf{R}_{++}$ and functions $f\colon Z \to \mathbf{R}_{++}$ and $\zeta\colon X \to Z$, satisfying:

1. f is an input-requirement function,[10]
2. ζ is positively homogeneous of degree one and non-decreasing on X, and
3. for all $\mathbf{x} \in X, g(\mathbf{x}) = f[\zeta(\mathbf{x})]$.

Notice the similarity between the definitions of a homothetic production index and a homothetic production function, as defined in 3.69. This similarity carries over to the family of homothetic production indexes presented in the following example.

6.39. Example. Suppose the function $g\colon X \to \mathbf{R}_{++}$ is positively homogeneous of degree $k > 0$ on X, where $X \subseteq \mathbf{R}_+^m \setminus \{\mathbf{0}\}$ is a non-empty cone. If we define f and ζ by:

$$f(z) = z^k \quad \text{for } z \in \mathbf{R}_{++} \ \text{ and } \ \zeta(\mathbf{x}) = [g(\mathbf{x})]^{1/k} \quad \text{for } \mathbf{x} \in X,$$

respectively, then we see that we can write:

$$g(\mathbf{x}) = f[\zeta(\mathbf{x})] \quad \text{for } \mathbf{x} \in X,$$

and that ζ is positively homogeneous of degree one, while f is an input-requirement function. Therefore, g is a homothetic production index. □

6.40. Definition. Let $X \subseteq \mathbf{R}_+^m$ be a producible set. We shall say that a technology function $\psi\colon \mathbf{R}_+^n \times X \to \mathbf{R}_+$ is **strongly homothetic** iff there exist a support function $\sigma\colon \mathbf{R}_+^n \to \mathbf{R}_+$ and a homothetic production index, $g\colon X \to \mathbf{R}_{++}$, satisfying:

$$(\forall(\mathbf{v}, \mathbf{x}) \in \mathbf{R}_+^n \times X)\colon \psi(\mathbf{v}, \mathbf{x}) = \sigma(\mathbf{v})/g(\mathbf{x}).$$

---

[10] See Definition 3.56.

If ψ is strongly homothetic, then it is, of course, homothetic; but the converse is not generally true. We will consider the question of why this special case may be of some particular interest shortly; but before we do, let's take a look at another example.

6.41. Example. Consider the function g defined by:

$$g(\mathbf{x}) = \sum_{j=1}^{m} \beta_j x_j^{\delta_j} \quad \text{for } \mathbf{x} \in X \stackrel{\text{def}}{=} \mathbf{R}_+^m \setminus \{\mathbf{0}\}, \tag{6.28}$$

where:

$$\beta_j > 0 \ \& \ \delta_j > 0 \quad \text{for } j = 1, \ldots, m.$$

If $m \geq 2$, and, for some j and k, $\delta_j \neq \delta_k$, then it can be shown that g is *not* a homothetic production index; in other words, the function g is a homothetic production index only if:

$$\delta_1 = \delta_2 = \cdots = \delta_m. \tag{6.29}$$

Conversely, if (6.29) holds, let 'δ' denote the common value of the δ_j, and define the function ζ by:

$$\zeta(\mathbf{x}) = \left(\sum_{j=1}^{m} \beta_j x_j^{\delta}\right)^{1/\delta} \quad \text{for } \mathbf{x} \in X \stackrel{\text{def}}{=} \mathbf{R}_+^m \setminus \{\mathbf{0}\}. \tag{6.30}$$

It is then easily seen that ζ is positively homogeneous of degree one, and that, defining f on $\mathbf{R}_{++}$ by:

$$f(z) = z^{\delta},$$

then, for all $\mathbf{x} \in X$, we have $g(\mathbf{x}) = f[\zeta(\mathbf{x})]$. Therefore, the function g *is* homothetic if (6.29) holds; a conclusion which also follows immediately as a special case of Example 6.39. On the other hand, it is probably of considerably greater interest to note that if f is *any* input requirement function (Definition 3.56) having as its domain $Z = \mathbf{R}_{++}$, then the function g defined by:

$$g(\mathbf{x}) = f[\zeta(\mathbf{x})], \tag{6.31}$$

with ζ given by (6.30), is a homothetic production index. Thus, for example, one could take the function $f\colon Z \to \mathbf{R}_{++}$ to be given by:

$$f(z) = exp(z) \quad \text{with } Z = \mathbf{R}_{++},$$

or:

$$f(z) = z^a \quad \text{for } a \in \mathbf{R}_{++} \text{ and } Z = \mathbf{R}_{++}$$

Using either of these input requirement functions, we can then obtain a strongly homothetic technology function by letting,

$$\psi(\mathbf{v}, \mathbf{x}) = \sigma(\mathbf{v})/g(\mathbf{x}) \quad \text{for } (\mathbf{v}, \mathbf{x}) \in \mathbf{R}^n_+ \times X,$$

where $\sigma\colon \mathbf{R}^n_+ \to \mathbf{R}_+$ is any support function; for example:

$$\sigma(\mathbf{v}) = \left(\sum_{i=1}^n a_i v_i^{-\rho}\right)^{-1/\rho} \quad \text{for } \mathbf{v} \in \mathbf{R}^n_+.$$

The function ζ defined in (6.30) is particularly useful in the case in which $\delta \geq 1$, for then ζ is a convex function. It is then easily shown that if ζ is paired with any input requirement function, f, as in (6.31), then the resultant production possibility correspondence will be convex-valued. (However, see Exercise 3 at the end of this section.) □

Suppose we now set out what we will mean by a strongly homothetic cost function.

6.42. Definition. We shall say that a cost function $c\colon \mathbf{R}^n_+ \times X \to \mathbf{R}_+$ is **strongly homothetic** iff (a) c is homothetic, so that c has the form:

$$c(\mathbf{w}, \mathbf{x}) = \gamma(\mathbf{w})g(\mathbf{x}),$$

and (b) g is a homothetic production index.

The following is then a more or less immediate implication of Theorem 6.35.

6.43. Theorem. *A technology function, ψ, is of the strongly homothetic form:*

$$\psi(\mathbf{v}, \mathbf{x}) = \sigma(\mathbf{v})/g(\mathbf{x}),$$

if, and only if the cost function, c, dual to ψ has the strongly homothetic form:

$$c(\mathbf{w}, \mathbf{x}) = \gamma(\mathbf{w})g(\mathbf{x});$$

where the support functions, σ and γ are dual to one another.

We can obtain a somewhat different perspective on the implications of strong homotheticity from the following result.

6.44. Proposition. *Suppose $\psi\colon \mathbf{R}^n_+ \times X \to \mathbf{R}_+$ is strongly homothetic, so that there exist a set $Z \subseteq \mathbf{R}_{++}$, an input requirement function $f\colon Z \to \mathbf{R}_{++}$, and functions $\zeta\colon X \to Z$ and $\sigma\colon \mathbf{R}^n_+ \to \mathbf{R}_+$ satisfying Definitions 6.38 and 6.40, and such that:*

$$(\forall \mathbf{x} \in X)\colon \psi(\mathbf{v}, \mathbf{x}) = \sigma(\mathbf{v})/f[\zeta(\mathbf{x})]$$

Then the input-requirement correspondence for ψ can be defined by:

$$V(\mathbf{x}) = \{\mathbf{v} \in \mathbf{R}^n_+ \mid F[\sigma(\mathbf{v})] \geq \zeta(\mathbf{x})\} \quad \textit{for}\, \mathbf{x} \in X,$$

where $F\colon \mathbf{R}_+ \to \mathbf{R}_+$ is the transformation function dual to f.

Proof. We know that V is given by:

$$V(\mathbf{x}) = \{\mathbf{v} \in \mathbf{R}^n_+ \mid \psi(\mathbf{v}, \mathbf{x}) = \sigma(\mathbf{v})/g(\mathbf{x}) \geq 1\} \quad \text{for}\, \mathbf{x} \in X;$$

or, equivalently:

$$V(\mathbf{x}) = \{\mathbf{v} \in \mathbf{R}^n_+ \mid \sigma(\mathbf{v}) \geq f[\zeta(\mathbf{x})]\} \quad \text{for}\, \mathbf{x} \in X.$$

Thus, in order to establish our result, it suffices to prove that:

$$(\forall (\mathbf{v}, \mathbf{x}) \in \mathbf{R}^n_+ \times X)\colon \sigma(\mathbf{v}) \geq f[\zeta(\mathbf{x})] \Leftrightarrow F[\sigma(\mathbf{v})] \geq \zeta(\mathbf{x}). \tag{6.32}$$

The equivalence in (6.32) actually follows more or less immediately from our results on transformation functions and the input requirement functions to which they are dual; but for the sake of convenient reference, let's prove the result directly.

Suppose first that $(\mathbf{v}, \mathbf{x})$ satisfies:

$$\sigma(\mathbf{v}) \geq f[\zeta(\mathbf{x})].$$

Since F is non-decreasing, we then have:

$$F[\sigma(\mathbf{v})] \geq F\big(f[\zeta(\mathbf{x})]\big); \tag{6.33}$$

while it follows immediately from the fact that F is the pseudo-inverse of f that [see equation (3.50) of Definition 3.56]:

$$F\big(f[\zeta(\mathbf{x})]\big) \geq \zeta(\mathbf{x}). \tag{6.34}$$

Combining (6.33) and (6.34), we then have:

$$F[\sigma(\mathbf{v})] \geq \zeta(\mathbf{x}). \tag{6.35}$$

Conversely, suppose that $(\mathbf{v}, \mathbf{x})$ is such that (6.35) holds. Then, since f is non-decreasing,

$$f\big(F[\sigma(\mathbf{v})]\big) \geq f[\zeta(\mathbf{x})]; \tag{6.36}$$

and we then have by the fact that F is the pseudo-inverse of f that:

$$\sigma(\mathbf{v}) \geq f\big(F[\sigma(\mathbf{v})]\big). \tag{6.37}$$

Combining (6.36) and (6.37), we then have:

$$\sigma(\mathbf{v}) \geq f[\zeta(\mathbf{x})] \equiv g(\mathbf{x}). \quad \square$$

As it turns out, one of the advantages of being able to characterize the technology available to a firm as being strongly homothetic is that in this case, the function ζ can be interpreted as an index of output. This is easly seen by the consideration of two facts. First, the function ζ is positively homogeneous of degree one; a property we would normally require of an index function. Secondly, notice that the function $\varphi\colon \mathbf{R}^n_+ \to \mathbf{R}_+$ given by:

$$\varphi(\mathbf{v}) = F[\sigma(\mathbf{v})],$$

where F is the transformation function dual to f, is a homothetic production function, as defined in 3.69. If we think of ζ as an index of output, $z = \zeta(\mathbf{x})$, then it follows from (6.37) of the proof of Proposition 6.44 that the input-requirement correspondence for φ is exactly that defined by ψ [taking $z = \zeta(\mathbf{x})$]]. Moreover, it is easily seen that the corresponding production possibility correspondence is defined on $\mathbf{R}^n_+$ by:

$$P(\mathbf{v}) = \{\mathbf{x} \in \mathbf{R}^m_+ \mid \sigma(\mathbf{v}) \geq f[\zeta(\mathbf{x})]\} = \{\mathbf{x} \in \mathbf{R}^m_+ \mid \varphi(\mathbf{v}) \geq \zeta(\mathbf{x})\}.$$

Before closing this section, there is one further point about strongly homothetic technology functions which is worth noting. In Example 6.41, we remarked that we could pair the function ζ defined in equation (6.30), above, with any input requirement function whose domain was $\mathbf{R}_{++}$ to obtain a homothetic production index. We could, however, use an input requirement function whose domain, Z is a proper sub-interval of $\mathbf{R}_{++}$, as long as we modified our specification of the producible set for the technology accordingly. Thus, consider the function:

$$f(z) = \frac{z}{\alpha - z},$$

where α is any positive real number, whose domain is $Z =]0, \alpha[$ (see Example 3.63.2). If we restrict the domain of ζ to the set X defined by:

$$X = \{\mathbf{x} \in \mathbf{R}^m_+ \mid 0 < \sum_{j=1}^m \beta_j x_j^\delta < (\alpha)^\delta\},$$

then $g \equiv f \circ \zeta$ is a homothetic production index.

Exercises.

1. Prove the assertions in:

a. 6.32.1. That is, show that the cost function corresponding to the Cobb-Douglas technology function is the function given by (6.18).

b. 6.32.2. That is, show that the condition indicated there implies homotheticity, and that the cost function given is the correct cost function.

c. 6.32.3. That is, show that the condition indicated there implies homotheticity, and that the cost function given is the correct cost function.

2. Prove Proposition 6.37.

3. Show that if $\psi(\mathbf{v},\mathbf{x}) = \sigma(\mathbf{v})/f(\mathbf{x})$ is homothetic, and f is quasi-convex, then the corresponding production possibility correspondence is convex-valued.

4. Prove that an input-requirement correspondence is homothetic if, and only if, the corresponding technology function is homothetic. If the input-requirement function is of the form $V(\mathbf{x}) = f(\mathbf{x})V$, and the technology function takes the form $\psi(\mathbf{v},\mathbf{x}) = \sigma(\mathbf{v})/f(\mathbf{x})$, what is the relationship between V and σ?

6.6 The 'Fundamentals of Welfare Economics '

We will suppose in our discussion here that there are given finite (integer) numbers of commodities, consumers, and firms; and we will denote these quantities by 'n, m,' and 'ℓ,' respectively. The **commodity space** then becomes $\mathbf{R}^n$, and we will employ the following system of notation:

$X_i \subseteq \mathbf{R}^n$ denotes the ith consumer's consumption set, and we will assume throughout that $X_i \neq \emptyset$;

'P_i' denotes the ith consumer's (strict) preference relation on X_i, and *we will assume throughout that P_i is irreflexive, for* $i = 1, \ldots, m$,

$Y_k \subseteq \mathbf{R}^n$ denotes the kth firm's production set, and we will asssume that $Y_k \neq \emptyset$, for $k = 1, \ldots, \ell$,

We will denote the coordinates of bundles $\mathbf{x}_i$ by 'x_{ij}', $(j = 1, \ldots, n)$; that is, we write:

$$\mathbf{x}_i = (x_{i1}, \ldots, x_{in}),$$

where 'x_{ij}' denotes the quantity of the jth commodity available to (or being made available by) the ith consumer, and where we follow the convention that if:

a. $x_{ij} \geq 0$, then the jth commodity is available for i's consumption in the amount x_{ij}, while if:

b. $x_{ij} < 0$, then the consumer is offering to supply the jth commodity (or service), in the amount $|x_{ij}| = -x_{ij}$.

With this convention, notice that if commodity prices are given by the vector $\mathbf{p} \in \mathbf{R}^n_+$, then the net expenditure necessary for the consumer to obtain the bundle $\mathbf{x}_i \in X_i$ is given by the inner product of $\mathbf{p}$ and $\mathbf{x}_i$, that is, $\mathbf{p} \cdot \mathbf{x}_i$.

A similar convention will be followed with respect to production vectors $\mathbf{y}_k \in Y_k$; which, where necessary for clarity, we will write out as:

$$\mathbf{y}_k = (y_{k1}, \ldots, y_{kn}).$$

In this case, if:

a. $y_{kj} \geq 0$, then the kth producer is producing (or planning to produce) the jth commodity in the net amount y_{kj}; while if

b. $y_{kj} < 0$, then the producer is using the jth commodity as an input[11] in the amount $|y_{kj}| = -y_{kj}$.

It is then the case that, given prices $\mathbf{p} \in \mathbf{R}^n_+$, the profit to the kth producer yielded by the choice of $\mathbf{y}_k \in Y_k$ is given by $\mathbf{p} \cdot \mathbf{y}_k$. We will, in fact, define the function π_k on $\mathbf{R}^n_+$ by:

$$\pi_k(\mathbf{p}) = \sup\{\mathbf{p} \cdot \mathbf{y}_k \mid \mathbf{y}_k \in Y_k\},$$

and assume that the producer attempts to maximize profits, taking the price vector as given.[12] In principle, determining conditions under which $\pi_k(\mathbf{p})$ is finite, and where and when the supremum will actually be achieved is a fairly delicate problem; but in our treatment here, we will not have to worry particularly about either issue, as you will see.

We will be dealing with a closed economy; so that all resources used in production must be supplied by consumers, and all profits from production will be distributed to the consumers. We will let $\mathbf{r}_i \in \mathbf{R}^n$ denote the resource vector held by the ith consumer, for $i = 1, \ldots, m$; and let $s_{ik} \geq 0$, be the share of ownership of the ith consumer in the kth firm, for $i = 1, \ldots, m, k = 1, \ldots, \ell$. Since we are supposing that this is a closed economy, we then assume that:

$$\sum_{i=1}^{m} s_{ik} = 1, \text{ for } k = 1, \ldots, \ell. \tag{6.38}$$

Thus, given a price vector, $\mathbf{p} \in \mathbf{R}^n$, the ith **consumer's income** (or **wealth**) is given by:

$$w_i(\mathbf{p}) = \mathbf{p} \cdot \mathbf{r}_i + \sum_{k=1}^{\ell} s_{ik}\pi_k(\mathbf{p}). \tag{6.39}$$

Notice that if $\mathbf{0} \in Y_k$ (although we will not generally be assuming this in our discussion), then for any price vector, $\mathbf{p}$, we will have $w_i(\mathbf{p}) \geq \mathbf{p} \cdot \mathbf{r}_i$.

To sum up, we will say that $\mathcal{E} = ((X_i, P_i), (Y_k), (\mathbf{r}_i), (s_{ik}))$ is a **(private ownership) economy** iff (X_i, P_i) and $\mathbf{r}_i$, for $i = 1, \ldots, m$; Y_k for $k = 1, \ldots, \ell$; and (s_{ik}) for $i = 1, \ldots, m, k = 1, \ldots, \ell$, satisfy the conditions of the above paragraphs.

In dealing with an economy, $\mathcal{E} = ((X_i, P_i), (Y_k), (\mathbf{r}_i), (s_{ik}))$, we denote the cartesian product of the X_i by '$\mathcal{X}$', or by '$\mathcal{X}(\mathcal{E})$,' if it appears that a reminder might be needed as to the association of the set with the economy; that is,

$$\mathcal{X} \equiv \mathcal{X}(\mathcal{E}) = \prod_{i=1}^{m} X_i.$$

Similarly, we denote the product of the Y_k's by $\mathcal{Y}$, or $\mathcal{Y}(\mathcal{E})$:

$$\mathcal{Y} \equiv \mathcal{Y}(\mathcal{E}) = \prod_{k=1}^{\ell} Y_k.$$

---

[11]For this particular production plan; other technologically feasible production plans for the producer may have $y_{kj} = 0$, or, indeed, have $y_{kj} > 0$.

[12]So that the producers are taken to be pure competitors, that is 'price-takers.'

We use the generic notation, '$(\mathbf{x}_i), (\mathbf{x}'_i), (\mathbf{x}^*_i)$,' and so on to denote elements of $\mathcal{X}$; and, similarly, '$(\mathbf{y}_k), (\mathbf{y}'_k), (\mathbf{y}^*_k)$,' will be used to denote elements of $\mathcal{Y}$. Combining these, we will use '$((\mathbf{x}_i), (\mathbf{y}_k)), ((\mathbf{x}'_i), (\mathbf{y}'_k)), ((\mathbf{x}^*_i), (\mathbf{y}^*_k))$,' and so on, to denote elements of:

$$\mathcal{X} \times \mathcal{Y} \stackrel{\text{def}}{=} \left(\prod_{i=1}^{m} X_i\right) \times \left(\prod_{k=1}^{\ell} Y_k\right).$$

6.45. Definitions. Let $\mathcal{E} = ((X_i, P_i), (Y_k), (\mathbf{r}_i), (s_{ik}))$ be an economy. An $(m+\ell)n$-tuple, $((\mathbf{x}_i), (\mathbf{y}_k)) \in \mathbf{R}^{(m+\ell)n}$ will be said to be an **attainable** (or **feasible**) **allocation for $\mathcal{E}$** iff:

1. $\mathbf{x}_i \in X_i$ for $i = 1, \ldots, m$,
2. $\mathbf{y}_k \in Y_k$ for $k = 1, \ldots, \ell$, and:
3. $\sum_{i=1}^{m} \mathbf{x}_i = \sum_{i=1}^{m} \mathbf{r}_i + \sum_{k=1}^{\ell} \mathbf{y}_k$. We will denote the set of all feasible or attainable allocations for $\mathcal{E}$ by '$A(\mathcal{E})$'.

If $\mathcal{E} = ((X_i, P_i), (Y_k), (\mathbf{r}_i), (s_{ik}))$ is an economy, we will denote the projections of $A(\mathcal{E})$ on $\mathcal{X}(\mathcal{E})$ and $\mathcal{Y}(\mathcal{E})$ by '$\mathcal{X}^*(\mathcal{E})$' and '$\mathcal{Y}^*(\mathcal{E})$,' respectively; that is:

$$\mathcal{X}^*(\mathcal{E}) = \{(\mathbf{x}_i) \in \mathcal{X} \mid (\exists (\mathbf{y}_k) \in \mathcal{Y})\colon ((\mathbf{x}_i), (\mathbf{y}_k)) \in A(\mathcal{E})\}$$

and

$$\mathcal{Y}^*(\mathcal{E}) = \{(\mathbf{y}_k) \in \mathcal{Y} \mid (\exists (\mathbf{x}_i) \in \mathcal{X})\colon ((\mathbf{x}_i), (\mathbf{y}_k)) \in A(\mathcal{E})\}.$$

6.46. Definition. An $(m+\ell+1)n$-tuple, $((\mathbf{x}^*_i), (\mathbf{y}^*_k), \mathbf{p}^*)$ is a **competitive equilibrium for the economy $\mathcal{E}$** iff:

1. $\mathbf{p}^* \neq \mathbf{0}$,
2. $((\mathbf{x}^*_i), (\mathbf{y}^*_k)) \in A(\mathcal{E})$,
3. for each k ($k = 1, \ldots, \ell$), we have: $\mathbf{p}^* \cdot \mathbf{y}^*_k = \pi_k(\mathbf{p}^*)$, and
4. for each i ($i = 1, \ldots, m$), we have:
   a. $\mathbf{p}^* \cdot \mathbf{x}^*_i \leq w_i(\mathbf{p}^*)$, and:
   b. $(\forall \mathbf{x}_i \in X_i)\colon \mathbf{x}_i P_i \mathbf{x}^*_i \Rightarrow \mathbf{p}^* \cdot \mathbf{x}_i > w_i(\mathbf{p}^*)$.

Our goal in this section is to explore the relationship between competitive equilibrium and Pareto efficiency. In order to define precisely what we will mean by 'Pareto efficiency,' we begin with the following definitions.

6.47. Definitions. Let $\mathcal{E} = ((X_i, P_i), (Y_k), (\mathbf{r}_i), (s_{ik}))$ be an economy. We then define:

1. the **unanimity ordering** (the **strong Pareto ordering**), $\boldsymbol{Q}$, on $\mathcal{X}$ by:

$$(\mathbf{x}_i) Q (\mathbf{x}'_i) \iff [\mathbf{x}_i P_i \mathbf{x}'_i \quad \text{for } i = 1, \ldots, m.] \tag{6.40}$$

2. the **Pareto (at-least-as-good-as) ordering, $\boldsymbol{R}$,** on $\mathcal{X}$, by:

$$(\mathbf{x}_i) R (\mathbf{x}'_i) \iff [\neg \mathbf{x}'_i P_i \mathbf{x}_i, \text{ for } i = 1, \ldots, m]. \tag{6.41}$$

3. the **strict Pareto ordering**, P, on $\mathcal{X}$, by:

$$(\mathbf{x}_i)P(\mathbf{x}'_i) \iff [(\mathbf{x}_i)R(\mathbf{x}'_i) \ \& \ \neg(\mathbf{x}'_i)R(\mathbf{x}_i)]. \tag{6.42}$$

We will use the following terminology in dealing with these three Pareto orderings. If:
$(\mathbf{x}^*_i)Q(\mathbf{x}_i)$, we shall say that $(\mathbf{x}^*_i)$ is **unanimously preferred to** $(\mathbf{x}_i)$,
$(\mathbf{x}^*_i)R(\mathbf{x}_i)$, we shall say that $(\mathbf{x}^*_i)$ **(weakly) Pareto dominates** $(\mathbf{x}_i)$,
$(\mathbf{x}^*_i)P(\mathbf{x}_i)$, we shall say that $(\mathbf{x}^*_i)$ **strictly Pareto dominates** $(\mathbf{x}_i)$.

6.48. Definitions. Let $\mathcal{E} = ((X_i, P_i), (Y_k), (\mathbf{r}_i), (s_{ik}))$ be an economy. We shall say that a feasible allocation for $\mathcal{E}$, $((\mathbf{x}^*_i), (\mathbf{y}^*_k))$ is **Pareto efficient for** $\mathcal{E}$ [respectively, **strongly Pareto efficient for** $\mathcal{E}$] iff there exists no alternative feasible allocation for $\mathcal{E}$, $((\mathbf{x}_i), (\mathbf{y}_k))$, satisfying:

$$(\mathbf{x}_i)Q(\mathbf{x}^*_i)[\text{respectively, } (\mathbf{x}_i)P(\mathbf{x}^*_i)];$$

where the orderings Q and P are defined in equations (6.40) and (6.42), above.

In the terminology introduced just preceding the above definition, a feasible allocation, $((\mathbf{x}^*_i), (\mathbf{y}^*_k))$, will be said to be Pareto efficient for $\mathcal{E}$ iff there exists no alternative *feasible* allocation which all consumers prefer to $((\mathbf{x}^*_i), (\mathbf{y}^*_k))$. The feasible allocation $((\mathbf{x}^*_i), (\mathbf{y}^*_k))$ is strongly Pareto efficient for $\mathcal{E}$ iff there exists no alternative *feasible* allocation which strictly Pareto dominates $((\mathbf{x}^*_i), (\mathbf{y}^*_k))$. Since $((\mathbf{x}^*_i), (\mathbf{y}^*_k))$ may be such that, while no feasible alternative allocation is unanimously preferred, there nonetheless is another feasible allocation, $((\mathbf{x}'_i), (\mathbf{y}'_k))$, where no consumer is worse off, and at least one consumer is better off than at $((\mathbf{x}^*_i), (\mathbf{y}^*_k))$, there are in principle more Pareto efficient allocations than there are strongly Pareto efficient allocations for a given economy, $\mathcal{E}$. This is the reason for the terminology used here.

Making use of the terminology introduced by Hurwicz [1960], we will demonstrate that, loosely speaking:

1. the competitive mechanism is **non-wasteful**, in the sense that any competitive equilibrium is Pareto efficient, and
2. the competitive mechanism is **unbiased**, in the sense that (given some additional assumptions) any Pareto efficient allocation can be made a competitive equilibrium.

Roughly speaking, these two results respectively constitute what are known as the 'First' and 'Second Fundamental Theorems of Welfare Economics.'

In the material to be presented here, we will concentrate on Pareto efficient allocations, as opposed to strongly Pareto efficient allocations. There are two principal reasons for this choice. First of all, most published versions of the 'Two Fundamental Theorems' of which I am aware deal with strongly Pareto efficient allocations. Consequently, on just this basis alone, it would appear to be better to deal with the weaker notion of efficiency here than to simply present a repetition of the more usual treatments. As a matter of fact, however, I consider Pareto efficiency to be the more important of the two concepts in any case, for the following reason.

In Section 1.3, we presented several reasons for believing it desirable, wherever possible, to take consumers' strict preference relations, (P_i), as opposed to the (reflexive) weak preferences, (G_i), as the basic primitive of our thoery of consumer behavior. In the present context, it is worthwhile to mention one additional reason. If one imagines polling consumers as to the relative desirability of various policy changes, it is likely to be necessary to ask a consumer which of two bundles is preferred when either or both represent quite different consumption patterns than anything the consumer has ever experienced. Where this is the case, a response that neither is preferred to the other is much more likely to mean that the consumer simply cannot make the comparison *in advance of experiencing one or both alternatives* than that he or she is truly indifferent between the two bundles. Consequently, while a response that one of the two bundles is preferred to the other may be meaningful, a response that neither is preferred to the other may be less so. In any event, if we are to agree that strict preference should be the basic primitive of the theory, and that we should try whenever possible to avoid making use of the assumption that P_i is negatively transitive, then Pareto dominance (as opposed to unanimous preference) no longer seems a good guide to policy choices. Correspondingly, Pareto efficiency, as opposed to strong Pareto efficiency, becomes the relevant efficiency concept.

We begin with a preliminary proposition, and then proceed to state a version of the 'First Fundamental Theorem;' that is, the 'non-wastefulness' property of the competitive mechanism.

6.49. Proposition. *Suppose* $((\mathbf{x}_i^*), (\mathbf{y}_k^*), \mathbf{p}^*)$ *is a competitive equilibrium for a private ownership economy,* $\mathcal{E}$. *Then:*

$$\mathbf{p}^* \cdot \mathbf{x}_i^* = w_i(\mathbf{p}^*) = \mathbf{p}^* \cdot \mathbf{r}_i + \sum\nolimits_{k=1}^{\ell} s_{ik}\pi_k(\mathbf{p}^*) \quad \text{for } i = 1, \ldots, m; \tag{6.43}$$

and, for any $(\mathbf{x}_i) \in \mathcal{X}^*(\mathcal{E})$, *we have:*

$$\begin{aligned}\mathbf{p}^* \cdot \left(\sum\nolimits_{i=1}^{m} \mathbf{x}_i^*\right) &= \sum\nolimits_{i=1}^{m} \mathbf{p}^* \cdot \mathbf{x}_i^* \\ &\geq \mathbf{p}^* \cdot \left(\sum\nolimits_{i=1}^{m} \mathbf{x}_i\right) = \sum\nolimits_{i=1}^{m} \mathbf{p}^* \cdot \mathbf{x}_i.\end{aligned} \tag{6.44}$$

Proof. Since $((\mathbf{x}_i^*), (\mathbf{y}_k^*), \mathbf{p}^*)$ is a competitive equilibrium, $((\mathbf{x}_i^*), (\mathbf{y}_k^*)) \in A(\mathcal{E})$, and thus:

$$\begin{aligned}0 &= \mathbf{p}^* \cdot \left(\sum\nolimits_{i=1}^{m} \mathbf{r}_i + \sum\nolimits_{k=1}^{\ell} \mathbf{y}_k^* - \sum\nolimits_{i=1}^{m} \mathbf{x}_i^*\right) \\ &= \sum\nolimits_{i=1}^{m} \mathbf{p}^* \cdot \mathbf{r}_i + \sum\nolimits_{k=1}^{\ell} \mathbf{p}^* \cdot \mathbf{y}_k^* - \sum\nolimits_{i=1}^{m} \mathbf{p}^* \cdot \mathbf{x}_i^*.\end{aligned} \tag{6.45}$$

However, by the definition of a competitive equilibrium, we have that:

$$\mathbf{p}^* \cdot \mathbf{r}_i + \sum_{k=1}^{\ell} s_{ik}(\mathbf{p}^* \cdot \mathbf{y}_k^*) - \mathbf{p}^* \cdot \mathbf{x}_i^* \\ = w_i(\mathbf{p}^*) - \mathbf{p}^* \cdot \mathbf{x}_i^* \geq 0 \quad \text{for } i = 1, \ldots, m. \quad (6.46)$$

Adding the terms in (6.46) yields:

$$\sum_{i=1}^{m} \left[w_i(\mathbf{p}^*) - \mathbf{p}^* \cdot \mathbf{x}_i^*\right] = \sum_{i=1}^{m} \left[\mathbf{p}^* \cdot \mathbf{r}_i + \sum_{k=1}^{\ell} s_{ik}(\mathbf{p}^* \cdot \mathbf{y}_k^*) - \mathbf{p}^* \cdot \mathbf{x}_i^*\right] \\ = \sum_{i=1}^{m} \mathbf{p}^* \cdot \mathbf{r}_i + \sum_{i=1}^{m} \sum_{k=1}^{\ell} s_{ik}(\mathbf{p}^* \cdot \mathbf{y}_k^*) - \sum_{i=1}^{m} \mathbf{p}^* \cdot \mathbf{x}_i^* \\ = \sum_{i=1}^{m} \mathbf{p}^* \cdot \mathbf{r}_i + \sum_{k=1}^{\ell} \mathbf{p}^* \cdot \mathbf{y}_k^* - \sum_{i=1}^{m} \mathbf{p}^* \cdot \mathbf{x}_i^* = 0;$$

where we have used the fact that:

$$\sum_{i=1}^{m} s_{ik} = 1 \quad \text{for } k = 1, \ldots, \ell,$$

in obtaining the next-to-the-last equality, and equation (6.45) to obtain the last equality. However, the sum of nonnegative terms can only be zero if all of these terms are zero; and thus, recalling (6.46), we see that:

$$\mathbf{p}^* \cdot \mathbf{x}_i^* = w_i(\mathbf{p}^*) = \mathbf{p}^* \cdot \mathbf{r}_i + \sum_{k=1}^{\ell} s_{ik}\pi_k(\mathbf{p}^*) \quad \text{for } i = 1, \ldots, m.$$

Now suppose that $(\mathbf{x}_i) \in \mathcal{X}^*(\mathcal{E})$. Then there exist $(\mathbf{y}_k) \in \mathcal{Y}$ such that:

$$\sum_{i=1}^{m} \mathbf{x}_i = \sum_{i=1}^{m} \mathbf{r}_i + \sum_{k=1}^{\ell} \mathbf{y}_k;$$

and therefore we obviously have:

$$\mathbf{p}^* \cdot \sum_{i=1}^{m} \mathbf{x}_i = \mathbf{p}^* \cdot \sum_{i=1}^{m} \mathbf{r}_i + \mathbf{p}^* \cdot \sum_{k=1}^{\ell} \mathbf{y}_k \\ = \mathbf{p}^* \cdot \sum_{i=1}^{m} \mathbf{r}_i + \sum_{k=1}^{\ell} \mathbf{p}^* \cdot \mathbf{y}_k. \quad (6.47)$$

However, since $((\mathbf{x}_i^*), (\mathbf{y}_k^*), \mathbf{p}^*)$ is a competitive equilibrium for $\mathcal{E}$, we have:

$$\sum_{k=1}^{\ell} \mathbf{p}^* \cdot \mathbf{y}_k \leq \sum_{k=1}^{\ell} \mathbf{p}^* \cdot \mathbf{y}_k^*; \quad (6.48)$$

and combining (6.45), (6.47), and (6.48), we then obtain:

$$\mathbf{p}^* \cdot \sum_{i=1}^{m} \mathbf{x}_i \leq \mathbf{p}^* \cdot \sum_{i=1}^{m} \mathbf{x}_i^*. \quad \square$$

The following is a version of the 'First Fundamental Theorem.' Since it follows almost immediately from Propostion 6.49 and the definition of a competitive equilibrium, its proof will be left as an exercise.

6.50. Theorem. *If* $((\mathbf{x}_i^*), (\mathbf{y}_k^*), \mathbf{p}^*)$ *is a competitive equilibrium for an economy,* $\mathcal{E}$*, then* $((\mathbf{x}_i^*), (\mathbf{y}_k^*))$ *is Pareto efficient for* $\mathcal{E}$.

While our proof of the 'First Fundamental Theorem' was straightforward indeed, establishing the 'Second Fundamental Theorem' is a bit more complicated. A first difficulty is that a reallocation of initial endowments and shares of ownership in firms may be necessary in order to make a competitive equilibrium of a Pareto efficient allocation. One way of dealing with this difficulty, which is the approach we will follow here, is to to seek a system of lump-sum transfers, t_i ($i = 1, \ldots, m$), which enable equilibrium to be achieved.[13] A second difficulty arises in that, unless one uses very stringent (or quite non-standard) assumptions, it is difficult, if not impossible, to prove that an arbitrary Pareto efficient allocation can actually be made a competitive equilibrium. The results that people loosely interpret as establishing this relationship usually actually establish the existence of a weakened form of competitive equilibrium. We will follow this pattern initially; the particular weakened form of competitive equilibrium which we will be using being the following.

6.51. Definition. Let $\mathcal{E} = ((X_i, P_i), (Y_k), (\mathbf{r}_i), (s_{ik}))$ be an economy. We shall say that $((\mathbf{x}_i^*), (\mathbf{y}_k^*), \mathbf{p}^*)$ is a **quasi competitive equilibrium for** $\mathcal{E}$ iff:

1. $\mathbf{p}^* \neq \mathbf{0}$,
2. $((\mathbf{x}_i^*), (\mathbf{y}_k^*)) \in A(\mathcal{E})$,
3. $\mathbf{p}^* \cdot \mathbf{y}_k^* = \pi_k(\mathbf{p}^*)$, for $k = 1, \ldots, \ell$,
4. for each i ($i = 1, \ldots, m$), we have $\mathbf{p}^* \cdot \mathbf{x}_i^* \leq w_i(\mathbf{p}^*)$, and either:

$$w_i(\mathbf{p}^*) = \min\{\mathbf{p}^* \cdot \mathbf{x}_i \mid \mathbf{x}_i \in X_i\} \stackrel{\text{def}}{=} \min \mathbf{p}^* \cdot X_i, \tag{6.49}$$

or:

$$(\forall \mathbf{x}_i \in X_i)\colon \mathbf{x}_i P_i \mathbf{x}_i^* \Rightarrow \mathbf{p}^* \cdot \mathbf{x}_i > w_i(\mathbf{p}^*) \tag{6.50}$$

(or both).

Turning now to what we mean by an equilibrium with a system of lump-sum transfers, we begin by defining $\mathbf{t} \in \mathbf{R}^m$ as a **system of lump-sum transfers** for $\mathcal{E}$ iff:

$$\sum\nolimits_{i=1}^{m} t_i = 0. \tag{6.51}$$

We then modify our previous definition as follows.

6.52. Definition. Let $\mathcal{E} = ((X_i, P_i), (Y_k), (\mathbf{r}_i), (s_{ik}))$ be a private ownership economy. We shall say that $((\mathbf{x}_i^*), (\mathbf{y}_k^*), \mathbf{p}^*)$ is a **quasi competitive**

[13] I will explain what I mean by an equilibrium with lump-sum transfers momentarily.

equilibrium (respectively, **competitive equilibrium**)f **for $\mathcal{E}$ with lump-sum transfers $\mathbf{t} \in \mathbf{R}^m$** iff (a) $\mathbf{t}$ satisfies (6.51), and (b) $((\mathbf{x}_i^*), (\mathbf{y}_k^*), \mathbf{p}^*)$ satisfies the conditions of Definition 6.51 (respectively, 6.46) *except* that $w_i^*(\mathbf{p}^*)$, where:

$$w_i^*(\mathbf{p}^*) \stackrel{\text{def}}{=} w_i(\mathbf{p}^*) + t_i \quad \text{for } i = 1, \ldots, m, \tag{6.52}$$

replaces $w_i(\mathbf{p}^*)$ in Condition 4 of 6.51 (respectively, 6.46).

We need one further definition, as follows.

6.53. Definition. Let X_i be a convex subset of $\mathbf{R}^n$, and let P_i be an irreflexive binary relation on X_i. We shall say that ***P_i* is weakly convex** iff, for each $\mathbf{x}_i^* \in X_i$, the set $P_i\mathbf{x}_i^*$ defined by:

$$P_i\mathbf{x}_i^* = \{\mathbf{x}_i \in X_i \mid \mathbf{x}_i P_i \mathbf{x}_i^*\},$$

is convex.

The initial version of the 'Second Fundamental Theorem' which we will consider is as follows.

6.54. Theorem. *Let $\mathcal{E} = ((X_i, P_i), (Y_k), (\mathbf{r}_i), (s_{ik}))$ be an economy such that:*

a. X_i is convex,

b. P_i is (irreflexive and) weakly convex, locally non-saturating, and lower semi-continuous,[14] *for each $i = 1, \ldots, m$; and suppose that:*

c. $Y \stackrel{def}{=} \sum_{k=1}^{\ell} Y_k$ is a convex set.

Then if $((\mathbf{x}_i^*), (\mathbf{y}_k^*))$ is Pareto efficient for $\mathcal{E}$, there exists a price vector, $\mathbf{p}^* \in \mathbf{R}^n$ and a vector $\mathbf{t}^* \in \mathbf{R}^m$ such that $((\mathbf{x}_i^*), (\mathbf{y}_k^*), \mathbf{p}^*)$ is a quasi competitive equilibrium for $\mathcal{E}$ with the lump-sum transfers $\mathbf{t}^*$.

Proof. We note first that, since $((\mathbf{x}_i^*), (\mathbf{y}_k^*))$ is Pareto efficient for $\mathcal{E}$, we must have:

$$\sum_{i=1}^m \mathbf{x}_i^* = \sum_{i=1}^m \mathbf{r}_i + \sum_{k=1}^{\ell} \mathbf{y}_k^*. \tag{6.53}$$

Define:

$$\mathbf{x}^* = \sum_{i=1}^m \mathbf{x}_i^*, \quad \mathbf{r} = \sum_{i=1}^m \mathbf{r}_i;$$

and the subset, $\mathbf{P}$, of $\mathbf{R}^n$, by:

$$\mathbf{P} = \sum_{i=1}^m P_i\mathbf{x}_i^*;$$

[14] Local non-saturation is defined in Example 1.70; lower semi-continuity in Definition 1.78.2.

where we recall the notation:

$$P_i\mathbf{x}_i^* = \{\mathbf{x}_i \in X_i \mid \mathbf{x}_i P_i \mathbf{x}_i^*\} \quad \text{for } i = 1, \ldots, m.$$

By way of completing our preliminaries, we note also that it follows immediately from Proposition 5.8 and the assumption that each P_i is weakly convex, that $\mathbf{P}$ is a convex set.

Next, we note that, by assumption (c) and Proposition 5.8, the set $\mathbf{r}+Y$ is convex; and it is easy to see that, since $((\mathbf{x}_i^*), (\mathbf{y}_k^*))$ is Pareto efficient for $\mathcal{E}$, we must have:

$$\mathbf{P} \cap [\mathbf{r} + Y] = \emptyset.$$

Thus, by Theorem 5.51, there exists a non-zero $\mathbf{p}^* \in \mathbf{R}^n$ such that:

$$\alpha \overset{\text{def}}{=} \sup\{\mathbf{p}^* \cdot \mathbf{z} \mid \mathbf{z} \in \mathbf{r} + Y\} \le \beta \overset{\text{def}}{=} \inf\{\mathbf{p}^* \cdot \mathbf{x} \mid \mathbf{x} \in \mathbf{P}\}. \tag{6.54}$$

Now, it follows at once from (6.53) and our definition of $\mathbf{x}^*$, that:

$$\mathbf{p}^* \cdot \mathbf{x}^* \le \alpha. \tag{6.55}$$

We are going to prove that we also must have $\mathbf{p}^* \cdot \mathbf{x}^* \ge \beta$. To do this, let $\epsilon > 0$ be given. Then, using the continuity of the inner product function and the fact that each P_i is locally non-saturating, we see that, for each i, there exists $\mathbf{x}_i^\dagger$ satisfying:

$$\mathbf{x}_i^\dagger P_i \mathbf{x}_i^* \;\&\; \mathbf{p}^* \cdot \mathbf{x}_i^\dagger < \mathbf{p}^* \cdot \mathbf{x}_i^* + \epsilon/m \quad \text{for } i = 1, \ldots, m. \tag{6.56}$$

Adding the inequalities on the right in (6.56), we then obtain:

$$\begin{aligned} \beta \le \mathbf{p}^* \cdot \left(\sum\nolimits_{i=1}^m \mathbf{x}_i^\dagger\right) &= \sum\nolimits_{i=1}^m \mathbf{p}^* \cdot \mathbf{x}_i^\dagger \\ &< \sum\nolimits_{i=1}^m (\mathbf{p}^* \cdot \mathbf{x}_i^* + \epsilon/m) = \mathbf{p}^* \cdot \mathbf{x}^* + \epsilon, \end{aligned} \tag{6.57}$$

where the first inequality in (6.57) is by the definitions of β and $\mathbf{P}$ and the left-hand part of (6.56). However, since (6.57) has been shown to hold for any positive real number, ϵ, it follows that $\beta \le \mathbf{p}^* \cdot \mathbf{x}^*$; and, combining this with (6.54) and (6.55), we see that:

$$\alpha = \beta = \mathbf{p}^* \cdot \mathbf{x}^*. \tag{6.58}$$

Now let $k \in \{1, \ldots, \ell\}$, let $\mathbf{y}_k$ be an arbitrary element of Y_k, and consider the production allocation $(\mathbf{y}_k^\dagger) \in \mathcal{Y}$ defined by:

$$(\mathbf{y}_k^\dagger) = (\mathbf{y}_1^*, \ldots, \mathbf{y}_{k-1}^*, \mathbf{y}_k, \mathbf{y}_{k+1}^*, \ldots, \mathbf{y}_\ell^*).$$

From (6.53), (6.54), and (6.58), we have:

$$\mathbf{p}^* \cdot \mathbf{r} + \sum_{j \ne k} \mathbf{p}^* \cdot \mathbf{y}_j^* + \mathbf{p}^* \cdot \mathbf{y}_k \le \mathbf{p}^* \cdot \mathbf{r} + \sum\nolimits_{j=1}^\ell \mathbf{p}^* \cdot \mathbf{y}_j^*;$$

from which we obtain:

$$\mathbf{p}^* \cdot \mathbf{y}_k \leq \mathbf{p}^* \cdot \mathbf{y}_k^*;$$

and we conclude that:

$$\mathbf{p}^* \cdot \mathbf{y}_k^* = \pi_k(\mathbf{p}^*) \quad \text{for } k = 1, \ldots, \ell. \tag{6.59}$$

Now define the vector $\mathbf{t}^* \in \mathbf{R}^m$ by:

$$t_i^* = \mathbf{p}^* \cdot \mathbf{x}_i^* - \mathbf{p}^* \cdot \mathbf{r}_i - \sum_{k=1}^{\ell} s_{ik}\pi_k(\mathbf{p}^*) \quad \text{for } i = 1, \ldots, m. \tag{6.60}$$

It follows easily from (6.53), (6.59), and the fact that the s_{ik} terms sum to one, for each k, that:

$$\sum_{i=1}^{m} t_i^* = 0;$$

and thus we see that $\mathbf{t}^*$ is a system of lump-sum transfers for $\mathcal{E}$. Since we obviously have:

$$\mathbf{p}^* \cdot \mathbf{x}_i^* = w_i(\mathbf{p}^*) + t_i^* = \mathbf{p}^* \cdot \mathbf{r}_i + \sum_{k=1}^{\ell} s_{ik}\pi_k(\mathbf{p}^*) + t_i^* \quad \text{for } i = 1, \ldots, m, \tag{6.61}$$

it can easily be seen that we have shown that $((\mathbf{x}_i^*), (\mathbf{y}_k^*), \mathbf{p}^*)$ satisfies the first three of the conditions defining a quasi competitive equilibrium for $\mathcal{E}$ with lump-sum transfers $\mathbf{t}^*$; and the first part of Condition 4 as well. Thus, to complete our proof, we need only establish that, for each i, either (6.49) or (6.50) must hold.

Accordingly, let $i \in \{1, \ldots, m\}$ be arbitrary, and suppose that:

$$w_i^*(\mathbf{p}^*) \stackrel{\text{def}}{=} w_i(\mathbf{p}^*) + t_i^* \neq \min \mathbf{p}^* \cdot X_i; \tag{6.62}$$

but that there exists a commodity bundle $\mathbf{x}_i' \in X_i$ satisfying:

$$\mathbf{x}_i' P_i \mathbf{x}_i^* \;\;\&\;\; \mathbf{p}^* \cdot \mathbf{x}_i' \leq w_i^*(\mathbf{p}^*) = w_i(\mathbf{p}^*) + t_i^*. \tag{6.63}$$

From (6.62) there exists an element, $\mathbf{x}_i^\dagger \in X_i$, such that:

$$\mathbf{p}^* \cdot \mathbf{x}_i^\dagger < w_i^*(\mathbf{p}^*); \tag{6.64}$$

and, since P_i is lower semi-continuous and $\mathbf{x}_i' P_i \mathbf{x}_i^*$, there exists $\delta > 0$ such that:

$$(\forall \mathbf{x}_i \in N(\mathbf{x}_i', \delta) \cap X_i) : \mathbf{x}_i P_i \mathbf{x}_i^*. \tag{6.65}$$

However, since X_i is convex, there exists $\theta \in]0, 1[$ such that:

$$\overline{\mathbf{x}}_i \stackrel{\text{def}}{=} \theta \mathbf{x}_i^\dagger + (1 - \theta)\mathbf{x}_i' \in N(\mathbf{x}_i', \delta) \cap X_i;$$

and thus, from (6.65), [the second part of] (6.63), and (6.64), we then have:

$$\overline{\mathbf{x}}_i P_i \mathbf{x}_i^* \text{ and } \mathbf{p}^* \cdot \overline{\mathbf{x}}_i < w_i^*(\mathbf{p}^*). \tag{6.66}$$

Next, defining:

$$\epsilon = w_i^*(\mathbf{p}^*) - \mathbf{p}^* \cdot \overline{\mathbf{x}}_i > 0; \tag{6.67}$$

we note that it follows from (6.61), the continuity of the inner product function, and the fact that each P_h is locally non-saturating, that, for each $h \neq i$, there exists $\overline{\mathbf{x}}_h \in X_h$ satisfying:

$$\overline{\mathbf{x}}_h P_h \mathbf{x}_h^* \ \& \ \mathbf{p}^* \cdot \overline{\mathbf{x}}_h < w_h^*(\mathbf{p}^*) + \epsilon/[2(m-1)] \quad \text{for all } h \neq i. \tag{6.68}$$

Notice, however, that:

$$\overline{\mathbf{x}} \stackrel{\text{def}}{=} \sum_{h=1}^m \overline{\mathbf{x}}_h \in \mathbf{P};$$

while, from (6.67), (6.68), and (6.61):

$$\mathbf{p}^* \cdot \overline{\mathbf{x}} = \sum_{h=1}^m \mathbf{p}^* \cdot \overline{\mathbf{x}}_h < w_i^*(\mathbf{p}^*) - \epsilon + \sum_{h \neq i} [w^*(\mathbf{p}^*) + \epsilon/[2(m-1)]]$$
$$= \mathbf{p}^* \cdot \mathbf{x}^* - \epsilon + [(m-1)\epsilon]/[2(m-1)] = \mathbf{p}^* \cdot \mathbf{x}^* - \epsilon/2 < \beta$$

[where the last inequality is from (6.58)]; which contradicts (6.54). Therefore we see that we cannot have both (6.62) and (6.63) holding for any one consumer, i, and it follows that $((\mathbf{x}_i^*), (\mathbf{y}_k^*), \mathbf{p}^*)$ satisfies condition 4 of Definition 6.52; and thus is a quasi competitive equilibrium with lump-sum transfers, $\mathbf{t}^*$. □

This last result can be generalized to the extent of allowing for the commodity space to be infinite-dimensional, as we will demonstrate in Chapter 11. It is also possible that it could be generalized within the context of $\mathbf{R}^n$. However, the conclusion of the result does not hold if any one of the conditions of Theorem 6.54 is simply dropped (see the exercises at the end of this section).

Now let's turn our attention to the issue of finding conditions ensuring that $((\mathbf{x}_i^*), (\mathbf{y}_k^*), \mathbf{p}^*)$, as obtained in the conclusion of Theorem 6.54, is a competitive equilibrium with lump-sum transfers; as opposed to the weaker conclusion established in the theorem. As mentioned earlier, one can easily obtain the stronger conclusion by making some rather stringent additional assumptions; the easiest, though in many ways the least attractive of which, being the additional assumption introduced in the following.

6.55. Corollary. *Suppose, in addition to the assumptions and conditions of Theorem 6.54, that the allocation $((\mathbf{x}_i^*), (\mathbf{y}_k^*))$ satisfies:*

$$\mathbf{x}_i^* \in int X_i \quad \textit{for } i = 1, \ldots, m. \tag{6.69}$$

Then there exists $\mathbf{p}^* \in \mathbf{R}^n$ and $\mathbf{t}^* \in \mathbf{R}^m$ such that $((\mathbf{x}_i^*), (\mathbf{y}_k^*), \mathbf{p}^*)$ is a competitive equilibrium for $\mathcal{E}$ with the lump-sum transfers $\mathbf{t}^*$.

In order to prove the corollary, we need only note that a linear function, in this case $f_i(\mathbf{x}_i) \equiv \mathbf{p}^* \cdot \mathbf{x}_i$, can only be minimized at an interior point of a space if it is identically zero on the space. However, since we know that $\mathbf{p}^* \neq \mathbf{0}$, we see that this cannot be the case; that is, we must have:

$$\mathbf{p}^* \cdot \mathbf{x}_i^* > \min \mathbf{p}^* \cdot X_i \quad \text{for } i = 1, \ldots, m.$$

It then follows from the definitions that the quasi competitive equilibrium obtained in the theorem must, with this additional assumption, actually be a competitive equilibrium (with the lump-sum transfers, $\mathbf{t}^*$).

The problem with this result is, of course, that the sort of Pareto efficient allocation which would satisfy this assumption is strange and rare indeed! Notice, in particular, equation (6.69) implies that, for each consumer, i, there exists a point, $\mathbf{x}_i' \in X_i$ satisfying:

$$\mathbf{x}_i' \ll \mathbf{x}_i^*. \tag{6.70}$$

Consequently, it follows that at the allocation $(\mathbf{x}_i^*)$, and given any commodity, j, each consumer must either possess a strictly positive quantity of the jth commodity, or be supplying less of it than he or she is capable of supplying. It is not at all clear that any such point could ever be a Pareto efficient allocation;[15] and in any case, it is clear that most of the efficient allocations of interest will not satisfy this property.

In the remainder of this section, we will prove a bit better version of the modified 'Second Fundamental Theorem.' In order to do so, however, we will first need to establish the following result.

6.56. Proposition. *If P_i is a lower semi-continuous binary relation on a convex set, X_i, and $\mathbf{x}_i^* \in X_i$ and $\mathbf{p}^* \in \mathbf{R}^n$ satisfy:*

$$(\forall \mathbf{x}_i \in X_i) \colon \mathbf{x}_i P_i \mathbf{x}_i^* \Rightarrow \mathbf{p}^* \cdot \mathbf{x}_i \geq \mathbf{p}^* \cdot \mathbf{x}_i^*, \tag{6.71}$$

and

$$\mathbf{p}^* \cdot \mathbf{x}_i^* > \min \mathbf{p}^* \cdot X_i, \tag{6.72}$$

then:

$$(\forall \mathbf{x}_i \in X_i) \colon \mathbf{x}_i P_i \mathbf{x}_i^* \Rightarrow \mathbf{p}^* \cdot \mathbf{x}_i > \mathbf{p}^* \cdot \mathbf{x}_i^*.$$

Proof. Suppose, by way of obtaining a contradiction, that there exists $\mathbf{x}_i' \in X_i$ such that $\mathbf{x}_i' P_i \mathbf{x}_i^*$, but:

$$\mathbf{p}^* \cdot \mathbf{x}_i' \leq \mathbf{p}^* \cdot \mathbf{x}_i^*.$$

[15]Unless consumer preferences are identical, the satisfaction of equation (6.69) will almost certainly indicate that Pareto-improving trades among consumers are possible; and thus that $((\mathbf{x}_i^*), (\mathbf{y}_k^*))$ cannot be Pareto efficient.

Since P_i is lower semi-continuous, there exists a neighborhood, $N(\mathbf{x}_i')$ such that:

$$(\forall \mathbf{x}_i \in N(\mathbf{x}_i') \cap X_i) \colon \mathbf{x}_i P_i \mathbf{x}_i^*. \tag{6.73}$$

Now, by (6.72) there exists $\overline{\mathbf{x}}_i \in X_i$ such that:

$$\mathbf{p}^* \cdot \overline{\mathbf{x}}_i < \mathbf{p}^* \cdot \mathbf{x}_i^*,$$

and we then have:

$$(\forall \theta \in]0,1]) \colon \mathbf{p}^* \cdot [\theta \overline{\mathbf{x}}_i + (1-\theta)\mathbf{x}_i'] < \mathbf{p}^* \cdot \mathbf{x}_i^*. \tag{6.74}$$

However, it is clear that, since X_i is convex, there exists a value of $\theta > 0$ and small enough so that:

$$\theta \overline{\mathbf{x}}_i + (1-\theta)\mathbf{x}_i' \in N(\mathbf{x}_i') \cap X_i;$$

which, given (6.73) and (6.74), contradicts (6.71). □

We will also need the following condition, in the statement of which we define the set M by:

$$M = \{1, \ldots, m\}.$$

6.57. Definition. If $\mathcal{E} = ((X_i, P_i), (Y_k), (\mathbf{r}_i), (s_{ik}))$ is an economy, we shall say that an allocation $((\mathbf{x}_i^*), (\mathbf{y}_k^*)) \in A(\mathcal{E})$ **satisfies the irreducibility condition** iff, for any partition, $\{I_1, I_2\}$ of M, there exists an allocation $((\overline{\mathbf{x}}_i), (\overline{\mathbf{y}}_k)) \in A(\mathcal{E})$ satisfying:

$$(\forall i \in I_1) \colon \overline{\mathbf{x}}_i P_i \mathbf{x}_i^*.$$

The condition just defined is developed from the 'irreducibility condition' introduced by L. McKenzie [1959, 1961], and was introduced in its present form in Moore [1970a]. Effectively, it implies that at the allocation $((\mathbf{x}_i^*), (\mathbf{y}_k^*))$, there is no definable subgroup of consumers, I_1, who could not make themselves better off, collectively, if they were simply allowed to exploit the remaining consumers. It is, consequently, related to notions of the 'core of a competitive economy.' More pertinent to the present discussion, however, is the fact that one could make the claim that this could be used as a necessary condition that $((\mathbf{x}_i^*), (\mathbf{y}_k^*))$ be deemed a 'good' Pareto efficient allocation. That this is so follows from the fact that the satisfaction of the irreducibility condition guarantees that there is no subgroup of consumers (I_2) so poor that the remaining consumers could not (if allowed to do so) exploit them to make themselves better off. In other words, and somewhat loosely interpreting, the condition guarantees that no subgroup of consumers is so poor as to have nothing which is valued by the remaining consumers.

In any case, our extended Second Fundamental Theorem is as follows.

6.58. Theorem. *Let $\mathcal{E} = ((X_i, P_i), (Y_k), (\mathbf{r}_i), (s_{ik}))$ be an economy such that:*

a. X_i is convex,

b. P_i is (irreflexive and) weakly convex, locally non-saturating, and lower semi-continuous, for each $i = 1, \ldots, m$;

c. $Y \stackrel{def}{=} \sum_{k=1}^{\ell} Y_k$ is a convex set, and suppose that:

d. $riX \cap ri(\mathbf{r} + Y) \neq \emptyset$.

Then we have the following.

If $((\mathbf{x}_i^*), (\mathbf{y}_k^*))$ is Pareto efficient for $\mathcal{E}$ and satisfies the irreducibility condition, then there exists a price vector, $\mathbf{p}^* \in \mathbf{R}^n$ and a vector $\mathbf{t}^* \in \mathbf{R}^m$ such that $((\mathbf{x}_i^*), (\mathbf{y}_k^*), \mathbf{p}^*)$ is a competitive equilibrium for $\mathcal{E}$ with the lump-sum transfers $\mathbf{t}^*$.

Proof. We define the set $\mathbf{P}$, and the vectors $\mathbf{x}^*$, $\mathbf{y}^*$, and $\mathbf{r}$ as in the proof of Theorem 6.54; noting this time that, utilizing the same reasoning as we employed in the earlier result, we can obtain a price vector, $\mathbf{p}^*$, which *properly* separates $(\mathbf{r} + Y)$ and $\mathbf{P}$. As before, we can then show that:

$$\sup_{\mathbf{z} \in \mathbf{r}+Y} \mathbf{p}^* \cdot \mathbf{z} = \mathbf{p}^* \cdot (\mathbf{r} + \mathbf{y}^*) = \mathbf{p}^* \cdot \mathbf{x}^* = \inf_{\mathbf{x} \in \mathbf{R}} \mathbf{p}^* \cdot \mathbf{x}, \tag{6.75}$$

and:

$$\mathbf{p}^* \cdot \mathbf{y}_k^* = \pi_k(\mathbf{p}^*) \quad \text{for } k = 1, \ldots, \ell. \tag{6.76}$$

Moreover, using (6.75) and reasoning similar to that employed in obtaining (6.76) [equation (6.59) in the proof of Theorem 6.54], we obtain:

$$(\forall \mathbf{x}_i \in X_i) \colon \mathbf{x}_i P_i \mathbf{x}_i^* \Rightarrow \mathbf{p}^* \cdot \mathbf{x}_i \geq \mathbf{p}^* \cdot \mathbf{x}_i^* \quad \text{for } i = 1, \ldots, m. \tag{6.77}$$

Now, suppose, by way of obtaining a contradiction, that we have both:

$$\mathbf{p}^* \cdot \mathbf{x}^* = \min \mathbf{p}^* \cdot X \tag{6.78}$$

(where we define $X = \sum_{i=1}^m X_i$), and:

$$ri(\mathbf{r} + Y) \cap H^< = \emptyset,$$

where we define the set $H^<$ by:

$$H^< = \{\mathbf{z} \in \mathbf{R}^n \mid \mathbf{p}^* \cdot \mathbf{z} < \mathbf{p}^* \cdot \mathbf{x}^*\}.$$

Since the separation of the sets $(\mathbf{r} + Y)$ and $\mathbf{P}$ is proper, it is then obvious that there must exist $\widehat{\mathbf{x}} \in X$ such that:

$$\mathbf{p}^* \cdot \widehat{\mathbf{x}} > \mathbf{p}^* \cdot \mathbf{x}^*. \tag{6.79}$$

On the other hand, if $\mathbf{x}^\dagger \in riX$ is such that:

$$\mathbf{x}^\dagger = \mathbf{r} + \mathbf{y}^\dagger,$$

where $\mathbf{r} + \mathbf{y}^\dagger \in ri(\mathbf{r} + Y)$ [and the existence of such a point is assured by hypothesis (d)], then we must obviously have, from (6.75):

$$\mathbf{p}^* \cdot \mathbf{x}^\dagger = \mathbf{p}^* \cdot (\mathbf{r} + \mathbf{y}^\dagger) = \mathbf{p}^* \cdot \mathbf{x}^*. \tag{6.80}$$

However, by Proposition 5.23, there exists $\theta > 1$ such that:

$$(1 - \theta)\widehat{\mathbf{x}} + \theta\mathbf{x}^\dagger \stackrel{\text{def}}{=} \mathbf{x}' \in X, \tag{6.81}$$

where $\widehat{\mathbf{x}}$ is from (6.79). But then by (6.79) and (6.80):

$$\begin{aligned} \mathbf{p}^* \cdot \mathbf{x}' - \mathbf{p}^* \cdot \mathbf{x}^* &= \mathbf{p}^* \cdot [(1 - \theta)\widehat{\mathbf{x}} + \theta\mathbf{x}^\dagger] - \mathbf{p}^* \cdot \mathbf{x}^* \\ &= (1 - \theta)\mathbf{p}^* \cdot \widehat{\mathbf{x}} + \theta\mathbf{p}^* \cdot \mathbf{x}^\dagger - \mathbf{p}^* \cdot \mathbf{x}^* \\ &= (1 - \theta)[\mathbf{p}^* \cdot \widehat{\mathbf{x}} - \mathbf{p}^* \cdot \mathbf{x}^*] < 0; \end{aligned}$$

that is:

$$\mathbf{p}^* \cdot \mathbf{x}' < \mathbf{p}^* \cdot \mathbf{x}^*,$$

which contradicts (6.78).

From the contradiction just obtained, we conclude that we must have either:

$$\mathbf{p}^* \cdot \mathbf{x}^* > \min \mathbf{p}^* \cdot X, \tag{6.82}$$

or:

$$ri(\mathbf{r} + Y) \cap H^< \neq \emptyset.$$

However, if the latter holds, it is easy to see that we must then have:

$$ri(\mathbf{r} + Y) \subseteq H^<. \tag{6.83}$$

But then, since by hypothesis (d), there exists a point $\overline{\mathbf{x}}$ satisfying:

$$\overline{\mathbf{x}} \in riX \cap ri(\mathbf{r} + Y),$$

it follows that in this case (6.82) holds as well. We conclude, therefore, that there exists $j \in M$ such that:

$$\mathbf{p}^* \cdot \mathbf{x}_j^* > \min \mathbf{p}^* \cdot X_j. \tag{6.84}$$

Now, define $\mathbf{t}^* \in \mathbf{R}^m$ as in the proof of Theorem 6.54:

$$t_i^* = \mathbf{p}^* \cdot \mathbf{x}_i^* - \mathbf{p}^* \cdot \mathbf{r}_i - \sum\nolimits_{k=1}^{\ell} s_{ik}\pi_k(\mathbf{p}^*) \quad \text{for } i = 1, \ldots, m,$$

and note that, as before, $\mathbf{t}^*$ is a system of lump-sum transfers for $\mathcal{E}$, and:

$$\mathbf{p}^* \cdot \mathbf{x}_i^* = w_i^*(\mathbf{p}^*) \stackrel{\text{def}}{=} w_i(\mathbf{p}^*) + t_i^* \quad \text{for } i = 1, \ldots, m. \tag{6.85}$$

Next, define:

$$I_1 = \{i \in M \mid \mathbf{p}^* \cdot \mathbf{x}_i^* > \min \mathbf{p}^* \cdot X_i\}.$$

Then, since each P_i is lower semi-continuous, it follows immediately from Proposition 6.58 that, for each $i \in I_1$, we have:

$$(\forall \mathbf{x}_i \in X_i) \colon \mathbf{x}_i P_i \mathbf{x}_i^* \Rightarrow \mathbf{p}^* \cdot \mathbf{x}_i > \mathbf{p}^* \cdot \mathbf{x}_i^* = w_i^*(\mathbf{p}^*). \tag{6.86}$$

Next, define:

$$I_2 = \{i \in M \mid \mathbf{p}^* \cdot \mathbf{x}_i^* = \min \mathbf{p}^* \cdot X_i\},$$

and suppose, by way of obtaining a contradiction, that $I_2 \neq \emptyset$. Since $((\mathbf{x}_i^*), (\mathbf{y}_k^*))$ satisfies the irreducibility condition, there exists an allocation $((\overline{\mathbf{x}}_i), (\overline{\mathbf{y}}_k)) \in A(\mathcal{E})$ satisfying:

$$(\forall i \in I_1) \colon \overline{\mathbf{x}}_i P_i \mathbf{x}_i^*. \tag{6.87}$$

But then we have, using in turn the fact that $((\overline{\mathbf{x}}_i), (\overline{\mathbf{y}}_k)) \in A(\mathcal{E})$, (6.85), (6.86), (6.87), and the fact that $((\mathbf{x}_i^*), (\mathbf{y}_k^*)) \in A(\mathcal{E})$:

$$\begin{aligned} \mathbf{p}^* \cdot \sum_{i \in I_2} \overline{\mathbf{x}}_i &= \mathbf{p}^* \cdot \mathbf{r} + \mathbf{p}^* \cdot \overline{\mathbf{y}} - \mathbf{p}^* \cdot \sum_{i \in I_1} \overline{\mathbf{x}}_i < \mathbf{p}^* \cdot \mathbf{r} + \mathbf{p}^* \cdot \overline{\mathbf{y}} - \mathbf{p}^* \cdot \sum_{i \in I_1} \mathbf{x}_i^* \\ &\leq \mathbf{p}^* \cdot \mathbf{r} + \mathbf{p}^* \cdot \mathbf{y}^* - \mathbf{p}^* \cdot \sum_{i \in I_1} \mathbf{x}_i^* = \mathbf{p}^* \cdot \sum_{i \in I_2} \mathbf{x}_i^*. \end{aligned}$$

But this is impossible; for by definition of I_2, we must have:

$$\mathbf{p}^* \cdot \sum_{i \in I_2} \overline{\mathbf{x}}_i = \sum_{i \in I_2} \mathbf{p}^* \cdot \overline{\mathbf{x}}_i \geq \sum_{i \in I_2} \mathbf{p}^* \cdot \mathbf{x}_i^* = \mathbf{p}^* \cdot \sum_{i \in I_2} \mathbf{x}_i^*.$$

We conclude, therefore, that $I_2 = \emptyset$; and thus $I_1 = M$, and it follows from (6.86) that $((\mathbf{x}_i^*), (\mathbf{y}_k^*), \mathbf{p}^*)$ is a competitive equilibrium for $\mathcal{E}$ with the lump-sum transfers $\mathbf{t}^*$. □

At this point, a student who has been reading this material fairly carefully may be shouting, "Foul!" After all, I belittled the result stated as Corollary 6.55 becuase of the key assumption:

$$\mathbf{x}_i^* \in intX_i \quad \text{for } i = 1, \ldots, m; \tag{6.88}$$

yet in Theorem 6.58 I have included hypothesis (d), which may look very similar. However, while hypothesis (d) does play a role in the proof of Theorem 6.58 which is somewhat similar to that played by (6.69) in the proof of 6.55, it is nonetheless a much less stringent condition than is (6.69). To see this, notice first of all that we are implicitly assuming that:

$$X \cap (\mathbf{r} + Y) \neq \emptyset, \tag{6.89}$$

when we assume that there is a Pareto efficient allocation in $\mathcal{E}$ in the first place! In strengthening the condition in equation (6.89) to hypothesis (d) we are, essentially, assuming that the economy is sufficiently wealthy to provide some choice of feasible allocations.[16]

Appendix. In this appendix, we will take a look at an alternative version of the 'First Fundamental Theorem,' and which is non-comparable to 6.50; establishing that the competitive allocation is strongly Pareto efficient. In this appendix, we will employ the following notation: given a consumer's (strict) preference relation, P_i, we define the consumer's **weak preference relation,** $\boldsymbol{R_i}$, on X_i by:

$$\mathbf{x}_i R_i \mathbf{x}_i^* \iff \neg \mathbf{x}_i^* P_i \mathbf{x}_i.$$

Recall from Section 1.3 that the relation R_i is transitive if, and only if, the relation P_i is negatively transitive.

6.59. Proposition. *Letting $\mathcal{E} = ((X_i, P_i), (Y_k), (\mathbf{r}_i), (s_{ik}))$ be a private ownership economy, suppose that for some $h \in \{1, \ldots, m\}$, P_h is asymmetric, negatively transitive, and locally non-saturating, and that $\mathbf{x}_h^* \in X_h$, $\mathbf{p}^* \in \mathbf{R}^n$, and $w^* \in \mathbf{R}$ are such that:*

$$(\forall \mathbf{x}_h \in X_h) \colon \mathbf{x}_h P_h \mathbf{x}_h^* \Rightarrow \mathbf{p}^* \cdot \mathbf{x}_h > w^*. \tag{6.90}$$

Then it must be the case that, for all $\mathbf{x}_h \in X_h$:

$$\mathbf{x}_h R_h \mathbf{x}_h^* \Rightarrow \mathbf{p}^* \cdot \mathbf{x}_h \geq w^*.$$

Proof. Suppose, by way of contradiction, that, while $\mathbf{x}_h^*$, $\mathbf{p}^*$ and w^* satisfy (6.90), that there exists $\mathbf{x}_h' \in X_h$ satisfying:

$$\mathbf{x}_h' R_h \mathbf{x}_h^* \text{ and } \mathbf{p}^* \cdot \mathbf{x}_h' < w^*. \tag{6.91}$$

Then, by the continuity of the inner product, there exists $\epsilon > 0$ such that:

$$(\forall \mathbf{x}_h \in N(\mathbf{x}_h', \epsilon) \cap X_h) \colon \mathbf{p}^* \cdot \mathbf{x}_h < w^*. \tag{6.92}$$

It also follows from the local non-saturation of P_h that:

$$(\exists \mathbf{x}_h^\dagger \in N(\mathbf{x}_h', \epsilon) \cap X_h) \colon \mathbf{x}_h^\dagger P_h \mathbf{x}_h'. \tag{6.93}$$

Moreover, since P_h is asymmetric and negatively transitive, we also have from (6.91) and (6.93) that $\mathbf{x}_h^\dagger P_h \mathbf{x}_h^*$: which, given (6.92), contradicts (6.90). □

We can then establish the following alternative version of the 'First Fundamental Theorem.'

6.60. Theorem. *If $((\mathbf{x}_i^*), (\mathbf{y}_k^*), \mathbf{p}^*)$ is a competitive equilibrium for an economy, $\mathcal{E}$, and each P_i is asymmetric, negatively transitive, and locally non-saturating, then $((\mathbf{x}_i^*), (\mathbf{y}_k^*))$ is strongly Pareto efficient for $\mathcal{E}$.*

[16] And thus that there is some chance for economists and managers to find gainful employment in the economy!

Proof. Suppose $(\mathbf{x}_i^\dagger) \in \mathcal{X}$ satisfies:

$$(\mathbf{x}_i^\dagger)P(\mathbf{x}_i^*).$$

Then, by definition of the strict Pareto ordering, P, we have:

$$\mathbf{x}_i^\dagger R_i \mathbf{x}_i^* \quad \text{for } i = 1, \ldots, m,$$

and, for some $h \in \{1, \ldots, m\}$:

$$\mathbf{x}_h^\dagger P_h \mathbf{x}_h^*. \tag{6.94}$$

It then follows at once from Proposition 6.59 that:

$$\mathbf{p}^* \cdot \mathbf{x}_i^\dagger \geq w_i(\mathbf{p}^*) \quad \text{for } i = 1, \ldots, m; \tag{6.95}$$

while, from (6.94) and the definition of a competitive equilibrium:

$$\mathbf{p}^* \cdot \mathbf{x}_h^\dagger > w_h(\mathbf{p}^*). \tag{6.96}$$

Combining (6.95) and (6.96) with the fact that $((\mathbf{x}_i^*), (\mathbf{y}_k^*), \mathbf{p}^*)$ is a competitive equilibrium, we have:

$$\sum_{i=1}^{m} \mathbf{p}^* \cdot \mathbf{x}_i^\dagger > \sum_{i=1}^{m} w_i(\mathbf{p}^*) = \sum_{i=1}^{m} \mathbf{p}^* \cdot \mathbf{x}_i^*, \tag{6.97}$$

where we have used Equation (6.43) of Proposition 6.49 in obtaining the last equality in (6.97). It follows from Equation (6.44) of Proposition 6.49 that $(\mathbf{x}_i^\dagger) \notin \mathcal{X}^*(\mathcal{E})$; from which it then follows that $((\mathbf{x}_i^*), (\mathbf{y}_i^*))$ is strongly Pareto efficient for $\mathcal{E}$. □

Exercises/Notes on the Literature.

The notion of Pareto dominance was apparently formally introduced into economics in Pareto [1904]; although it seems to have been Pareto's friend and colleague, Enrico Barone who first stated and proved a version of the 'First Fundamental Theorem' (Barone [1908]). Kenneth Arrow first stated and proved a version of the 'Second Fundamental Theorem' (Arrow [1951]): although Gerard Debreu independently published a closely-related result (Debreu [1951]). Early refinements/generalizations of the 'Second Fundamental Theorem' were done by Debreu [1954], Koopmans [1957], and Koopmans and Bausch [1959]. Hurwicz [1960] was the first to examine the issue of non-wastefulness and unbiasedness of abstract resource allocation mechanisms generally.

As mentioned in the text, the irreducibility condition used in Theorem 6.65, while adapted from an earlier condition introduced by Lionel McKenzie, was first introduced in the present form in Moore [1970a]. It was generalized in various ways in that report, and further generalized and refined in Moore [1973, 1975].

Show that none of the hypotheses of Theorem 6.54 can simply be dropped from the result. That is, do the following.

1. Find an example of an economy, $\mathcal{E}$, in which each P_i is irreflexive, weakly convex, locally non-saturating, and lower semi-continuous, and where Y is convex, but such that there exists a Pareto efficient allocation for $\mathcal{E}$, $((\mathbf{x}_i^*), (\mathbf{y}_k^*))$ for which no $\mathbf{p}^*$ can be found to make $((\mathbf{x}_i^*), (\mathbf{y}_k^*), \mathbf{p}^*)$ a quasi competitive equilibrium. (**Hint:** For this, and the other counterexamples you are being asked to construct here, you can probably find what you want in a situation in which there are two commodities,

one consumer, and one producer. There are three things which make such an economy convenient to work with: First, a quasi competitive equilibrium with lump-sum transfers is simply a quasi competitive equilibrium [since there is only one consumer]. Secondly, a Pareto efficient allocation is simply a point which maximizes the single consumer's preferences over the attainable set. Thirdly, you can draw diagrams of the attainable set and the consumer's preference map.)

2. Find an example of an economy, $\mathcal{E}$, in which each X_i is convex, each P_i is irreflexive, locally non-saturating, and lower semi-continuous, and where Y is convex, but such that there exists a Pareto efficient allocation for $\mathcal{E}$, $((\mathbf{x}_i^*), (\mathbf{y}_k^*))$ for which no $\mathbf{p}^*$ can be found to make $((\mathbf{x}_i^*), (\mathbf{y}_k^*), \mathbf{p}^*)$ a quasi competitive equilibrium.

3. As in the previous examples, except this time we want each X_i to be convex, each P_i to be irreflexive, weakly convex, and lower semi-continuous, and Y to be convex.

4. As in the previous examples, except this time we want each X_i to be convex, each P_i to be irreflexive, weakly convex, and locally non-saturating, and Y to be convex.

5. As in the previous examples, except this time we want each X_i to be convex, each P_i to be irreflexive, weakly convex, locally non-saturating, and lower semi-continuous [and Y to be non-convex].[17]

6. Establish equation (6.83) in the proof of Theorem 6.58. (In fact, there were a number of assertions made in the proof of Theorem 6.58 where I was a bit skimpy on proof. A student who is seriously interested in theory should find it quite useful to go back over said proof, filling in all of the details. It would also be useful to try proving it without making use of Theorem 6.54.)

6.7 Constrained Extrema and Saddle Points

In this section we will be considering the following general problem.

Maximization Problem 1 (MP1). Maximize f, subject to the conditions:

$$\mathbf{x} \in X \text{ and } g(\mathbf{x}) \geq \mathbf{0},$$

where:

$$f\colon D \to \mathbf{R}^p, g\colon D \to \mathbf{R}^m, \text{ and } X \subseteq D \subseteq \mathbf{R}^n.$$

We shall say that $\mathbf{x}^* \in \mathbf{R}^n$ is **feasible** for this problem iff:

$$\mathbf{x}^* \in X \;\&\; g(\mathbf{x}^*) \geq \mathbf{0}; \tag{6.98}$$

and, defining the **feasible set** for the problem, X^*, by:

$$X^* = \{\mathbf{x} \in X \mid g(\mathbf{x}) \geq \mathbf{0}\},$$

[17]The condition that each P_i is irreflexive is really a definitional restriction, and there is no particular point in investigating the effect of dropping this condition.

we shall say that $\mathbf{x}^* \in \mathbf{R}^n$ **solves the problem** iff $\mathbf{x}^* \in X^*$ and:

$$(\forall \mathbf{x} \in X^*)\colon f(\mathbf{x}) \geq f(\mathbf{x}^*) \Rightarrow f(\mathbf{x}) = f(\mathbf{x}^*). \tag{6.99}$$

Notice that if $p = 1$ in MP1, then (6.99) reduces to:

$$(\forall \mathbf{x} \in X^*)\colon f(\mathbf{x}^*) \geq f(\mathbf{x}).$$

Allowing for the case in which $p > 1$ in our statement of the problem and in our development of the results will complicate our statements and proof only very slightly; this is particularly true with respect to the proofs, where the argument for the case in which $p \geq 1$ is likely to be virtually identical to that for the case in which $p = 1$.

In this section we will be considering only necessary and sufficient conditions for constrained extrema which do *not* involve differentials; that is, we will be concentrating on the global, rather than the local, theory of constrained extrema.

6.61. Examples.

1. Suppose $X \subseteq D \subseteq \mathbf{R}^n$, and that $f\colon D \to \mathbf{R}$, $g\colon D \to \mathbf{R}^m$, $\mathbf{b} \in \mathbf{R}^m$, and that X is an open set. The problem:

maximize f subject to $\mathbf{x} \in X$ and $g(\mathbf{x}) = \mathbf{b}$,

is the **classical Lagrangian problem**. If we define the function $h\colon D \to \mathbf{R}^{2m}$ by:

$$h(\mathbf{x}) = (g(\mathbf{x}) - \mathbf{b}, \mathbf{b} - g(\mathbf{x})) \quad \text{for } \mathbf{x} \in D,$$

then we see that we can state this problem in the form of MP1:

maximize f subject to $\mathbf{x} \in X$ and $h(\mathbf{x}) \geq \mathbf{0}$.

The sufficiency results which we will be studying are useful for this case; however, it must be admitted that the necessity results which we will be developing are really only useful for this case if the constraint fuction, g, is an affine function.

2. Consider the problem:

$$\text{maximize } \mathbf{c} \cdot \mathbf{x} \text{ subject to: } \mathbf{A}\mathbf{x} \leq \mathbf{b} \text{ and } \mathbf{x} \geq \mathbf{0},$$

where $\mathbf{c} \in \mathbf{R}^n$, $\mathbf{A}$ is an $m \times n$ matrix, and $\mathbf{b} \in \mathbf{R}^m$. This is the **linear programming problem (in standard form)**. If we define:

$$g(\mathbf{x}) = \mathbf{b} - \mathbf{A}\mathbf{x}, \quad \text{and} \quad X = \mathbf{R}^n_+,$$

then this problem is obviously of the form of MP1. Moreover, this problem has the **dual**:

$$\text{minimize } \mathbf{b} \cdot \mathbf{y} \text{ subject to } \mathbf{y}'\mathbf{A} \geq \mathbf{c}' \text{ and } \mathbf{y} \geq \mathbf{0}.$$

Defining:

$$f^*(\mathbf{y}) = -\mathbf{b} \cdot \mathbf{y}, g^*(\mathbf{y}) = \mathbf{A}'\mathbf{y} - \mathbf{c}, Y = \mathbf{R}^m_+,$$

we see that the dual can also be stated in the form of MP1.

3. Consider an economy with m consumers, in which the ith consumer has a utility function $u_i \colon X_i \to \mathbf{R}$, where $X_i \subseteq \mathbf{R}^n$ denotes the ith consumer's consumption set, for $i = 1, \ldots, m$. Let $Y \subseteq \mathbf{R}^n$ be the aggregate production set for the economy, define:

$$\mathbf{X} = \prod_{i=1}^m X_i,$$

$f \colon \mathbf{X} \times \mathbf{R}^n \to \mathbf{R}^m$ by:

$$f(\mathbf{x}_1, \ldots, \mathbf{x}_m, \mathbf{y}) = (u_1(\mathbf{x}_1), \ldots, u_m(\mathbf{x}_m)),$$

and $g \colon \mathbf{X} \times \mathbf{R}^n \to \mathbf{R}^m$ by:

$$g(\mathbf{x}_1, \ldots, \mathbf{x}_m, \mathbf{y}) = \mathbf{y} + \mathbf{z} - \sum_{i=1}^m \mathbf{x}_i,$$

where $\mathbf{z} \in \mathbf{R}^n$ represents the aggregate resource vector for the economy. If we then consider the problem:

maximize $f(\mathbf{x}_1, \ldots, \mathbf{x}_m, \mathbf{y})$ subject to:

$$(\mathbf{x}_1, \ldots, \mathbf{x}_m, \mathbf{y}) \in \mathbf{X} \times Y \;\&\; g(\mathbf{x}_1, \ldots, \mathbf{x}_m, \mathbf{y}) \geq \mathbf{0};$$

then we see the problem is of the form of MP1, and that $((\mathbf{x}_i^*), \mathbf{y}^*)$ is a solution of the problem if, and only if $((\mathbf{x}_i^*), \mathbf{y}^*)$ is strongly Pareto efficient for the economy. □

Given a maximization problem of the form of MP1, we define the associated **Lagrangian function,** $\varphi \colon D \times \mathbf{R}^m \times \mathbf{R}^p \to \mathbf{R}$ by:

$$\varphi(\mathbf{x}, \boldsymbol{\lambda}, \boldsymbol{\mu}) = \boldsymbol{\mu} \cdot f(\mathbf{x}) + \boldsymbol{\lambda} \cdot g(\mathbf{x}).$$

6.62. Definitions. Suppose $(\mathbf{x}^*, \boldsymbol{\lambda}^*, \boldsymbol{\mu}^*) \in X \times \mathbf{R}^m_+ \times \mathbf{R}^p_+$ satisfies:

$$(\boldsymbol{\lambda}^*, \boldsymbol{\mu}^*) \neq \mathbf{0},$$

and, for all $\mathbf{x} \in X$ and $\boldsymbol{\lambda} \in \mathbf{R}^m_+$:

$$\varphi(\mathbf{x}, \boldsymbol{\lambda}^*, \boldsymbol{\mu}^*) \leq \varphi(\mathbf{x}^*, \boldsymbol{\lambda}^*, \boldsymbol{\mu}^*) \leq \varphi(\mathbf{x}^*, \boldsymbol{\lambda}, \boldsymbol{\mu}^*). \tag{6.100}$$

Then we shall say that the Lagrangian function, φ, has a **(nonnegative) saddle point at** $(\mathbf{x}^*, \boldsymbol{\lambda}^*, \boldsymbol{\mu}^*)$. If, in addition,

1. $\boldsymbol{\mu}^* > \mathbf{0}$, we will say that φ has a **proper saddle point at** $(\mathbf{x}^*, \boldsymbol{\lambda}^*, \boldsymbol{\mu}^*)$.
2. $\boldsymbol{\mu}^* \gg \mathbf{0}$, we will say φ has a **positive saddle point at** $(\mathbf{x}^*, \boldsymbol{\lambda}^*, \boldsymbol{\mu}^*)$.

Notice that the distinction between a proper and a positive saddle point disappears in the case in which $p = 1$ (that is, where f is a real-valued function).

6.63. Proposition. *If φ has a saddle point at $(\mathbf{x}^*, \boldsymbol{\lambda}^*, \boldsymbol{\mu}^*)$, then $g(\mathbf{x}^*) \geq \mathbf{0}$, and $\boldsymbol{\lambda}^* \cdot g(\mathbf{x}^*) = 0$.*

Proof. Since φ has a saddle point at $(\mathbf{x}^*, \boldsymbol{\lambda}^*, \boldsymbol{\mu}^*)$, we have from the right-hand inequality in (6.100) and the definition of the Lagrangian function, that:

$$(\forall \boldsymbol{\lambda} \in \mathbf{R}^m_+)\colon \boldsymbol{\mu}^* \cdot f(\mathbf{x}^*) + \boldsymbol{\lambda}^* \cdot g(\mathbf{x}^*) \leq \boldsymbol{\mu}^* \cdot f(\mathbf{x}^*) + \boldsymbol{\lambda} \cdot g(\mathbf{x}^*);$$

from which it follows at once that:

$$(\forall \boldsymbol{\lambda} \in \mathbf{R}^m_+)\colon \boldsymbol{\lambda}^* \cdot g(\mathbf{x}^*) \leq \boldsymbol{\lambda} \cdot g(\mathbf{x}^*). \tag{6.101}$$

Now suppose, by way of obtaining a contradiction, that, for some $i \in \{1, \ldots, m\}$:

$$g_i(\mathbf{x}^*) < 0;$$

where '$g_i(\mathbf{x}^*)$' denotes the ith coordinate of $g(\mathbf{x}^*)$. Then for $\boldsymbol{\lambda}$ of the form:

$$\boldsymbol{\lambda} = (0, \ldots, 0, \lambda_i, 0 \ldots, 0),$$

we see that:

$$\boldsymbol{\lambda} \cdot g(\mathbf{x}^*) = \lambda_i g_i(\mathbf{x}^*) \to -\infty \text{ as } \lambda_i \to +\infty;$$

which contradicts (6.101). Therefore we see that we must have:

$$g(\mathbf{x}^*) \geq \mathbf{0};$$

and thus we also have, using the fact that $\boldsymbol{\lambda}^* \geq \mathbf{0}$, that:

$$\boldsymbol{\lambda}^* \cdot g(\mathbf{x}^*) \geq 0. \tag{6.102}$$

Letting $\boldsymbol{\lambda} = \mathbf{0}$ in the right-hand side of (6.101), however, we also see that we must have:

$$\boldsymbol{\lambda}^* \cdot g(\mathbf{x}^*) \leq 0;$$

and, using (6.102), it then follows that:

$$\boldsymbol{\lambda}^* \cdot g(\mathbf{x}^*) = 0. \quad \square$$

Under very general conditions, the existence of a positive saddle point for φ at $(\mathbf{x}^*, \boldsymbol{\lambda}^*, \boldsymbol{\mu}^*)$ ensures that $\mathbf{x}^*$ solves the maximization problem, as is shown in the following.

6.64. Theorem. *If the Lagrangian function* φ *has a positive saddle point at* $(\mathbf{x}^*, \boldsymbol{\lambda}^*, \boldsymbol{\mu}^*) \in X \times \mathbf{R}^m_+ \times \mathbf{R}^p_+$, *then* $\mathbf{x}^*$ *is a solution of the problem.*

Proof. Suppose φ has a positive saddle point at $(\mathbf{x}^*, \boldsymbol{\lambda}^*, \boldsymbol{\mu}^*)$. We begin by noting that it follows from Proposition 6.63 that $g(\mathbf{x}^*) \geq \mathbf{0}$. Next, let $\mathbf{x} \in X$ be such that $g(\mathbf{x}) \geq \mathbf{0}$. Then, using the fact that $\boldsymbol{\lambda}^* \geq \mathbf{0}$, the left-hand inequality in (6.100), and Proposition 6.63 in turn, we have:

$$\boldsymbol{\mu}^* \cdot f(\mathbf{x}) \leq \boldsymbol{\mu}^* \cdot f(\mathbf{x}) + \boldsymbol{\lambda}^* \cdot g(\mathbf{x}) \leq \boldsymbol{\mu}^* \cdot f(\mathbf{x}^*) + \boldsymbol{\lambda}^* \cdot g(\mathbf{x}^*) = \boldsymbol{\mu}^* \cdot f(\mathbf{x}^*),$$

so that:

$$\boldsymbol{\mu}^* \cdot \left[f(\mathbf{x}^*) - f(\mathbf{x})\right] \geq 0. \tag{6.103}$$

Since $\boldsymbol{\mu}^* \gg \mathbf{0}$, it then follows at once from (6.103) that we cannot have $f(\mathbf{x}) > f(\mathbf{x}^*)$; that is, if $f(\mathbf{x}) \geq f(\mathbf{x}^*)$, then $f(\mathbf{x}) = f(\mathbf{x}^*)$. □

Presently we will prove a partial converse of the above result; however, before we do, let's consider another auxiliary function defined from MP1. Define $h\colon D \to \mathbf{R}^p_+ \times \mathbf{R}^m_+$ by:

$$h(\mathbf{x}) = (f(\mathbf{x}), g(\mathbf{x})) \quad \text{for } \mathbf{x} \in D, \tag{6.104}$$

and the set $A \subseteq \mathbf{R}^p_+ \times \mathbf{R}^m_+$ by:

$$A = \{(\mathbf{y}, \mathbf{z}) \in \mathbf{R}^p_+ \times \mathbf{R}^m_+ \mid (\exists \mathbf{x} \in X)\colon (\mathbf{y}, \mathbf{z}) \leq h(\mathbf{x})\}. \tag{6.105}$$

Suppose now that φ has a saddle point at $(\mathbf{x}^*, \boldsymbol{\lambda}^*, \boldsymbol{\mu}^*)$. Then, for all $\mathbf{x} \in X$:

$$\boldsymbol{\mu}^* \cdot f(\mathbf{x}) + \boldsymbol{\lambda}^* \cdot g(\mathbf{x}) \leq \boldsymbol{\mu}^* \cdot f(\mathbf{x}^*) + \boldsymbol{\lambda}^* \cdot g(\mathbf{x}^*),$$

or:

$$(\boldsymbol{\mu}^*, \boldsymbol{\lambda}^*) \cdot h(\mathbf{x}) \leq (\boldsymbol{\mu}^*, \boldsymbol{\lambda}^*) \cdot h(\mathbf{x}^*).$$

In fact, since $\boldsymbol{\mu}^* \geq \mathbf{0}$ and $\boldsymbol{\lambda}^* \geq \mathbf{0}$, it is clear that:

$$(\forall (\mathbf{y}, \mathbf{z}) \in A)\colon (\boldsymbol{\mu}^*, \boldsymbol{\lambda}^*) \cdot (\mathbf{y}, \mathbf{z}) \leq (\boldsymbol{\mu}^*, \boldsymbol{\lambda}^*) \cdot h(\mathbf{x}^*);$$

and we see that $(\boldsymbol{\mu}^*, \boldsymbol{\lambda}^*)$ defines a supporting hyperplane for A at $h(\mathbf{x}^*)$. Moreover, we have from Proposition 6.63 that:

$$g(\mathbf{x}^*) \geq \mathbf{0} \;\&\; \boldsymbol{\lambda}^* \cdot g(\mathbf{x}^*) = 0.$$

Consequently, it follows easily that if we define:

$$A^* = A - \{(f(\mathbf{x}^*), \mathbf{0})\}, \tag{6.106}$$

then:

$$(\forall (\mathbf{y}, \mathbf{z}) \in A^*)\colon (\boldsymbol{\mu}^*, \boldsymbol{\lambda}^*) \cdot (\mathbf{y}, \mathbf{z}) \leq 0.$$

It is these considerations which form the basis of the proof of the following result.

6.65. Theorem. *If $\mathbf{x}^* \in X$ solves MP1, if X is a convex set, and if f and g are both concave functions, then there exists $(\boldsymbol{\mu}^*, \boldsymbol{\lambda}^*) \in \mathbf{R}_+^{p+m} \setminus \{\mathbf{0}\}$ such that:*

$$(\forall \mathbf{x} \in X)\colon \boldsymbol{\mu}^* \cdot f(\mathbf{x}) + \boldsymbol{\lambda}^* \cdot g(\mathbf{x}) \leq \boldsymbol{\mu}^* \cdot f(\mathbf{x}^*) + \boldsymbol{\lambda}^* \cdot g(\mathbf{x}^*), \tag{6.107}$$

and:

$$\boldsymbol{\lambda}^* \cdot g(\mathbf{x}^*) = 0; \tag{6.108}$$

and consequently the Lagrangian function has a nonnegative saddle point at $(\mathbf{x}^*, \boldsymbol{\lambda}^*, \boldsymbol{\mu}^*)$.

Proof. If f and g are both concave, then the function h defined in (6.104) is concave, and it then follows easily that the set A defined in (6.105) is convex, and thus that the set A^* defined in (6.106), above, is also convex. Moreover, it is also clear that:

$$\dim A = \dim A^* = m + p,$$

and thus that:

$$riA^* = int(A^*).$$

Now suppose, by way of obtaining a contradiction, that:

$$\mathbf{0} \stackrel{\text{def}}{=} (\mathbf{0}_p, \mathbf{0}_m) \in int(A^*),$$

where '$\mathbf{0}_p$' and '$\mathbf{0}_m$' denote the origins in $\mathbf{R}^p$ and $\mathbf{R}^m$, respectively. Then there exist *strictly positive* vectors, $\mathbf{a} \in \mathbf{R}_+^p$ and $\mathbf{b} \in \mathbf{R}_+^m$ such that $(\mathbf{a}, \mathbf{b}) \in A^*$, and it then follows readily from the definition of A^* that there exists $\mathbf{x}^\dagger \in X$ such that:

$$(f(\mathbf{x}^\dagger) - f(\mathbf{x}^*), g(\mathbf{x}^\dagger)) \geq (\mathbf{a}, \mathbf{b}). \tag{6.109}$$

But it then follows from (6.109) that:

$$f(\mathbf{x}^\dagger) \gg f(\mathbf{x}^*) \;\;\&\;\; g(\mathbf{x}^\dagger) \gg \mathbf{0},$$

which contradicts the assumption that $\mathbf{x}^*$ solves the maximization problem.

From the argument of the above paragraph, we conclude that $\mathbf{0} \notin riA^*$, and it then follows from Theorem 5.51 that there exists a point $(\boldsymbol{\mu}^*, \boldsymbol{\lambda}^*) \in \mathbf{R}^{p+m}$ such that:

$$(\boldsymbol{\mu}^*, \boldsymbol{\lambda}^*) \neq \mathbf{0}, \tag{6.110}$$

and:

$$(\forall (\mathbf{y}, \mathbf{z}) \in A^*)\colon (\boldsymbol{\mu}^*, \boldsymbol{\lambda}^*) \cdot (\mathbf{y}, \mathbf{z}) \leq 0. \tag{6.111}$$

Since $-\mathbf{R}_+^{p+m} \subseteq A^*$, it is easy to see that (6.111) implies that $(\boldsymbol{\mu}^*, \boldsymbol{\lambda}^*) \geq \mathbf{0}$; and thus it follows from (6.110) that:

$$(\boldsymbol{\mu}^*, \boldsymbol{\lambda}^*) > \mathbf{0}.$$

Now, from (6.111) and the definition of A^*, we have:

$$(\forall \mathbf{x} \in X)\colon \boldsymbol{\mu}^* \cdot f(\mathbf{x}) + \boldsymbol{\lambda}^* \cdot g(\mathbf{x}) \leq \boldsymbol{\mu}^* \cdot f(\mathbf{x}^*). \tag{6.112}$$

Setting $\mathbf{x} = \mathbf{x}^*$ in (6.112), we see that:

$$\boldsymbol{\lambda}^* \cdot g(\mathbf{x}^*) \leq 0.$$

However, since $\boldsymbol{\lambda}^*$ and $g(\mathbf{x}^*)$ are both nonnegative,

$$\boldsymbol{\lambda}^* \cdot g(\mathbf{x}^*) \geq 0,$$

and thus we conclude that:

$$\boldsymbol{\lambda}^* \cdot g(\mathbf{x}^*) = 0. \tag{6.113}$$

Combining (6.112) and (6.113) easily yields (6.107) and (6.108). □

With the aid of Theorem 6.65, we can establish two useful corollaries; the proof of the first of which I will leave as an exercise.

6.66. Corollary. *If, in addition to the hypotheses of 6.65, the function g satisfies the* **Slater Constraint Qualification***:*

$$(\exists \mathbf{x}^\dagger \in X)\colon g(\mathbf{x}^\dagger) \gg \mathbf{0}, \tag{6.114}$$

then the vector $\boldsymbol{\mu}^* \in \mathbf{R}^p$ obtained in Theorem 6.65 is semi-positive; and thus the Lagrangian function, φ, has a proper saddle point at $(\mathbf{x}^*, \boldsymbol{\lambda}^*, \boldsymbol{\mu}^*)$.

In our next result, we will suppose that the constraint function, g is affine; so that g can be written in the form:

$$g(\mathbf{x}) = \mathbf{A}\mathbf{x} - \mathbf{b},$$

where $\mathbf{A}$ is an $m \times n$ matrix, and $\mathbf{b} \in \mathbf{R}^m$. We will also suppose that the constraint function, g, is **proper for the maximization problem**, by which we will mean that the coordinate functions g_i form a linearly independent set on X; that is, we will suppose that there exits no $\boldsymbol{a} \in \mathbf{R}^n \setminus \{\mathbf{0}\}$ such that:

$$(\forall \mathbf{x} \in X)\colon \boldsymbol{a} \cdot g(\mathbf{x}) = 0. \tag{6.115}$$

Notice that if there exists some $\boldsymbol{a} \in \mathbf{R}^n \setminus \{\mathbf{0}\}$ satisfying (6.115), then one of the constraint functions would be a linear combination of the other $m - 1$ constraint functions, and thus would either be redundant, or, possibly, would guarantee that the feasible set for the problem is empty.

6.67. Corollary. *If, in addition to the other hyupotheses of 6.65, we assume that g is both affine and proper for the maximization problem, and satisfies:*

$$(\exists \mathbf{x}^\dagger \in riX) \colon g(\mathbf{x}^\dagger) \geq \mathbf{0}; \tag{6.116}$$

then the Lagrangian function for the problem, φ, has a proper saddle point at $(\mathbf{x}^*, \boldsymbol{\lambda}^*, \boldsymbol{\mu}^*)$.

Proof. By Theorem 6.65, φ has a saddle point at $(\mathbf{x}^*, \boldsymbol{\lambda}^*, \boldsymbol{\mu}^*)$. If we suppose, by way of obtaining a contradiction, that:

$$\boldsymbol{\mu}^* = \mathbf{0},$$

then it follows from (6.107) and (6.108) that:

$$(\forall \mathbf{x} \in X) \colon \boldsymbol{\lambda}^* \cdot g(\mathbf{x}) \leq \boldsymbol{\lambda}^* \cdot g(\mathbf{x}^*) = 0. \tag{6.117}$$

However, consider the function $G \colon X \to \mathbf{R}$ defined by:

$$\boldsymbol{\lambda}^* \cdot g(\mathbf{x}) \quad \text{for } \mathbf{x} \in X.$$

Since $\boldsymbol{\lambda}^* \geq \mathbf{0}$ and g is affine, it is easy to see that G is a convex function. Moreover, from (6.116) and (6.117), we see that G attains a maximum at $\mathbf{x}^\dagger$, and that maximum value attained is zero. But then it follows from Theorem 5.77 that G is constant on X; and, from (6.117) that we must have $G(\mathbf{x}) = \mathbf{0}$, for all $\mathbf{x} \in X$. However, since we must have $\boldsymbol{\lambda} \neq \mathbf{0}$, this contradicts the assumption that g is proper for the maximization problem.

From the contradiction it follows that we cannot have $\boldsymbol{\mu}^* = \mathbf{0}$, and thus φ has a proper saddle point at $(\mathbf{x}^*, \boldsymbol{\lambda}^*, \boldsymbol{\mu}^*)$. □

Notice that in both 6.66 and 6.67, we have shown that a proper saddle point exists by showing that all of the vectors defining a supporting hyperplane to the set A^* must have certain properties. In principle, however, our interest is in ensuring that there exists at least one vector having the appropriate properties. This suggests that 6.66 and 6.67 could be usefully generalized, but I will leave this as a 'project for the interested reader.'

Exercises.

1. Prove Corollary 6.66.

2. Consider the special case of MP1 in which $p = 1$; that is, where f is real-valued. Supposing that f and g are both concave and that g satisfies the Slater constraint qualification, show that if $\mathbf{x}^*$ solves the problem, then there exists $\mathbf{y}^* \in \mathbf{R}^m_+$ such that for all $\mathbf{x} \in X$, and all $\mathbf{y} \in \mathbf{R}^m_+$:

$$f(\mathbf{x}) + \mathbf{y}^* \cdot g(\mathbf{x}) \leq f(\mathbf{x}^*) + \mathbf{y}^* \cdot g(\mathbf{x}^*) \leq f(\mathbf{x}^*) + \mathbf{y} \cdot g(\mathbf{x}^*).$$

3. Use the results of this section to prove the following result for the linear programming problem presented in Example 6.61.2. You may suppose

that the constraint function $g(\mathbf{x}) = \mathbf{b} - \mathbf{Ax}$ is proper for the maximization problem, and that there exists $\mathbf{x}^\dagger \in \mathbf{R}^n_{++}$ such that $\mathbf{Ax}^\dagger \leq \mathbf{b}$. Show that if $\mathbf{x}^*$ solves the primal problem, then there exists $\mathbf{y}^* \in \mathbf{R}^m_+$ such that $\mathbf{y}^*$ solves the dual problem, and $\mathbf{b} \cdot \mathbf{y}^* = \mathbf{c} \cdot \mathbf{x}^*$.

Hints: (a) Use Theorem 6.65 to show that there exists $\boldsymbol{\lambda}^* \in \mathbf{R}^m_+$ and $\mu^* \in \mathbf{R}_{++}$ such that the Lagrangian function, φ, has a saddle point at $(\mathbf{x}^*, \boldsymbol{\lambda}^*, \mu^*)$. Defining $\mathbf{y}^* = (1/\mu^*)\boldsymbol{\lambda}^*$, show that $\mathbf{y}^*$ solves the dual.

(b) Show that $\mathbf{x}^*$ and $\mathbf{y}^*$ must satisfy:

$$\mathbf{Ax}^* = \mathbf{b} \ \ \& \ \ \mathbf{A}'\mathbf{y}^* = \mathbf{c}.$$

4. Use the results of this section to prove the following result, for the general equilibrium model presented in Example 6.61.3. Suppose $X_i = \mathbf{R}^n_+$, for $i = 1, \ldots, m$, that each u_i is strictly increasing and concave, and that $((\mathbf{x}_i^*), \mathbf{y}^*)$ is a strongly Pareto efficient allocation satisfying:

a. $\sum_{i=1}^m \mathbf{x}_i^* \gg \mathbf{0}$, and:

b. for each i, there exists $j \in \{1, \ldots, n\}$ such that $x_{ij}^* > 0$.

Then there exists a price vector, $\mathbf{p}^*$, such that $((\mathbf{x}_i^*), \mathbf{y}^*, \mathbf{p}^*)$ is a competitive equilibrium for $\mathcal{E}$.

Bibliography

Aleksandrov, P. S. [1956]: *Combinatorial Topology, Vol. I.* Graylock Press, Rochester.

Aliprantis, Charalambos D., and K. C. Border [1994]: *Infinite Dimensional Analysis: A Hitchhiker's Guide.* Springer-Verlag.

Aliprantis, Charalambos D., D. J. Brown, and O. Burkinshaw [1990]: *Existence and Optimality of Competitive Equilibria.* Springer-Verlag.

Aliprantis, Charalambos D., and O. Burkinshaw [1998]: *Principles of Real Analysis,* 3rd edition. Academic Press.

Aliprantis, Charalambos D., R. Tourky, and N. C. Yannelis [1998]: 'Core Conditions in Economics.' Manuscript.

Apostol, Tom M. [1974]: *Mathematical Analysis, 2nd ed.* Reading, Mass; Addison-Wesley.

Armstrong, W. E. [1939]: 'The Determinateness of the Utility Function,' *Economic Journal;* 49, 453-67.

Arrow, Kenneth J. [1951]: 'An Extension of the Basic Theorems of Classical Welfare Economics,' *in:* Neyman, Jerzy, ed.: *Proceedings of the Second Berkeley Symposium on Mathematical Statistics and Probability* University of California Press, pp. 507–32. Reprinted in *Readings in Mathematical Economics,* Vol. I (edited by Peter Newman). Baltimore: The Johns Hopkins Press, 1968, 365–90.

Arrow, Kenneth J. *et. al.* [1961]: 'Capital-Labor Substitution and Economic Efficiency,' *Review of Economic Statistics;* 43; 225–50.

Arrow, Kenneth J., and G. Debreu [1954]: 'Existence of an Equilibrium for a Competitive Economy,' *Econometrica;* 22; 265–90.

Arrow, Kenneth J., and F. H. Hahn [1971]: *General Competitive Analysis.* Holden-Day (republished by North-Holland).

Arrow, Kenneth J., and M. Intriligator, eds. [1981]: *Handbook of Mathematical Economics, Volume I.* North-Holland.

Arrow, Kenneth J., and M. Intriligator, eds. [1982]: *Handbook of Mathematical Economics, Volume II.* North-Holland.

Arrow, Kenneth J., and M. Intriligator, eds. [1986]: *Handbook of Mathematical Economics, Volume III.* North-Holland.

Aumann, Robert J. [1966] 'Existence of Competitive Equilibria in Markets with a Continuum of Traders,' *Econometrica;* 34; 1–17.

Barone, Enrico [1908]: 'Il ministerio dell produzione nello stato colletivista,' *Giornale degli Economisti* [2]; 327; 267–293, 391–414. English translation: 'The Ministry of Production in the Collectivist State,' in *Collectivist Economic Planning* (edited by F. A. Hayek). London: Routledge & Kegan Paul Ltd., 1935, 245–290. Reprinted in *Readings in Mathematical Economics*, Vol. I (edited by Peter Newman). Baltimore: The Johns Hopkins Press, 1968, 319–364.

Berge, Claude [1957]: *Théorie génerale des jeux à n personnes*, Gauthier-Villars, Paris.

Berge, Claude [1963]: *Topological Spaces.* New York: the Macmillan Company. (English Translation by E. M. Patterson of: *Espace Topologiques, fonctions multivoques,* published by Dunod, Paris, 1959.)

Bergstrom, Theodore C. [1976]: 'How to Discard 'Free Disposability'—At No Cost,' *Journal of Mathematical Economics;* 3, 131–4.

Bergstrom, Theodore C., R. P. Parks, and T. Rader [1976]: 'Preferences which have Open Graphs,' *Journal of Mathematical Economics;* 3, 265–68.

Bewley, Truman F. [1972]: 'Existence of Equilibria in Economies with Infinitely Many Commodities,' *Journal of Economic Theory;* 4; 514–40.

Binmore, Kenneth G. [1982]: *Mathematical Analysis: A Straightforward Approach.* Cambridge University Press.

Blackorby, Charles, D. Primont, and R. Russell [1978]: *Duality, Separability, and Functional Structure: Theory and Economic Applications.* North Holland.

Border, Kim C. [1985]: *Fixed Point Theorems with Applications to Economics and Game Theory.* New York: Cambridge University Press.

Castaing, C., and M. Valadier [1977]: *Convex Analysis and Measurable Multifunctions.* Berlin: Springer-Verlag.

Chipman, John S., L. Hurwicz, M. K. Richter, and H. F. Sonnenschein, eds [1971]: *Preferences, Utility, and Demand.* New York: Harcourt Brace Jovanovich, Inc.

Chipman, John S., and J. C. Moore [1971]: 'The Compensation Principle in Welfare Economics,' *in:* Zarley, Arvid V., ed: *Papers in Quantitative Economics, 2.* The University Press of Kansas, pp. 1–77.

Chipman, John S., and J. C. Moore [1979]: 'On Social Welfare Functions and the Aggregation of Preferences,' *Journal of Economic Theory;* 21: 111–39.

Conway, John B. [1985]: *A Course in Functional Analysis.* Springer-Verlag.

Debreu, Gerard [1951]: 'The Coefficient of Resource Utilization,' *Econometrica;* 19; 273–92.

Debreu, Gerard [1954a]: 'Valuation Equilibrium and Pareto Optimum,' *Proceedings of the National Academy of Sciences;* 40; 588–92.

Debreu, Gerard [1954b]: 'Representation of a Preference Ordering by a Numberical Function,' *in:* Thrall, Robert M., C. H. Coombs, and R. L.

Davis, eds.: *Decision Processes.* (New York: John Wiley & Sons).

Debreu, Gerard [1959]: *Theory of Value.* New York: John Wiley & Sons (re-published by the Yale Univesity Press).

Debreu, Gerard [1962]: 'New Concepts and Techniques for Equilibrium Analysis,' *International Economic Review;* 3, 257–73.

Debreu, Gerard [1967]: 'Integration of Correspondences,' *in:* LeCam, L., J. Neyman, and E. L. Scott, eds.: *Proceedings of the Fifth Berkeley Symposium in Math. Stat. and Probability, II,* Part I, Univ of California Press, 351–72.

Debreu, Gerard, and D. Schmeidler [1972]: The Radon-Nikodym Derivative of a Correspondence,' *in:* LeCam, L., J. Neyman, and E. L. Scott, eds.: *Proceedings of the Sixth Berkeley Symposium in Math. Stat. and Probability,* Univ of California Press, 41–56.

Dixmier, Jacques [1984]: *General Topology.* Springer-Verlag.

Dow, James, and H. Sonnenschein [1986]: 'Samuelson and Chipman-Moore on Utility-Generated Community Demand,' Chapter 1, pp. 1–6, *in:* Peston, M. H., and R. E. Quandt, eds.: *Prices, Competition, and Equilibrium.* Barnes & Noble.

Dugundji, James [1966]: *Topology.* Allyn & Bacon, Inc.

Eggleston, Harold G [1962]: *Elementary Real Analysis.* Cambridge University Press.

Ellickson, Bryan [1993]: *Competitive Equilibrium: Theory and Applications.* Cambridge University Press.

Fishburn, Peter C. [1970]: *Utility Theory for Decision Making.* New York: John Wiley & Sons.

Gale, David [1960]: *The Theory of Linear Economic Models.* New York: McGraw-Hill.

Gale, David, and A. Mas-Colell [1975]: 'An Equilibrium Existence Theorem for a General Model Without Ordered Preferences,' *Journal of Mathematical Economics;* 2, 9–15.

Green, Jerry, and W. P. Heller [1981]: 'Mathematical Analysis and Convexity with Applications to Economics,' *in:* Arrow, Kenneth J., and M. D. Intriligator, eds: *Handbook of Mathematical Economics,* Vol. 1 (North-Holland), Chapter 1, pp. 15–52.

Halmos, Paul R. [1982]: *A Hilbert Space Problem Book,* 2nd edition. Springer-Verlag.

Heller, Walter P. [1972]: 'Transactions with Set-Up Costs, *Journal of Economic Theory;* 4, 465–78.

Hildenbrand, Werner [1974]: *Core and Equilibria of a Large Economy.* Princeton University Press.

Hildenbrand, Werner, and A. P. Kirman [1976]: *Introduction to Equilibrium Analysis.* North-Holland.

Hildenbrand, Werner, and A. P. Kirman [1988]: *Equilibrium Analysis.* North-Holland.

Hildenbrand, Werner, and H. Sonnenschein, eds. [1991]: *Handbook of Mathematical Economics, Volume IV.* North-Holland.

Hocking, John G., and G. S. Young [1961]: *Topology.* Addison-Wesley.

Hu, Shouchuan, and N. S. Papageorgiou [1997]: *Handbook of Multivalued Analysis. Volume I: Theory.* Kluwer Academic Publishers.

Hurwicz, Leonid [1960]: 'Optimality and Informational Efficiency in Resource Allocation Processes,' *in:* Arrow, Kenneth J., S. Karlin, and P. Suppes, eds: *Mathematical Methods in the Social Sciences, 1959.* Stanford University Press.

Hurwicz, Leonid [1972]: 'On Informationally Decentralized Systems,' *in:* McGuire, C. B., and R. Radner, eds: *Decision and Organization: A volume in Honor of Jacob Marschak.* North-Holland.

Hurwicz, Leonid [1986]: 'Incentive Aspects of Decentralization,' pp. 1441–1482, *in:* Arrow and Intriligator [1986].

Hurwicz, Leonid, and M. Majumdar [1988]: 'Optimal Intertemporal Allocation Mechanisms and Decentralization of Decisions,' *Journal of Economic Theory;* 45; 228–61.

Jacobsen, Stephen E. [1968]: *Production Correspondences,* Ph.D. Thesis, University of California, Berkeley.

Jacobsen, Stephen E. [1970]: 'Production Correspondences,' *Econometrica;* 38; 754–71.

Jacobsen, Stephen E. [1972]: 'On Shephard's Duality Theorem,' *Journal of Economic Theory;* 4, 458–64.

Jones, Larry E. [1992]: 'Equilibrium in Competitive, Infinite Dimensional, Settings;' Chapter 7, pp. 413–43, *in* Laffont [1992], Vol. II.

Kelley, John L. [1955]: *General Topology.* van Nostrand.

Klein, Erwin, and A. C. Thompson [1984]: *Theory of Correspondences: Including Applications to Mathematical Economics.* Wiley.

Komiya, Hidetoshi [1994]: 'A Simple Proof of the K-K-M-S Theorem,' *Economic Theory;* 4, 463–66.

Koopmans, Tjalling C. [1957]: *Three Essays on the State of Economic Science* (I. 'Allocation of Resources and the Price System'). McGraw-Hill.

Koopmans, Tjalling C., and A. F. Bausch [1959]: 'Selected Topics in Economics Involving Mathematical Rasoning,' *SIAM Review;* 1; 79–128.

Krasa, Stefan, and N. C. Yannelis [1994]: 'An Elementary Proof of the Knaster-Kuratowski-Mazurkiewicz-Shapley Theorem,' *Economic Theory;* 4, 467–71.

Kuratowski, C. [1932]: *Fund. Math.;*

Laffont, Jean-Jacques, ed.[1992]: *Advances in Economic Theory: Sixth World Congress* (2 volumes). Cambridge University Press.

Lang, Karl R., J. C. Moore, and A. B. Whinston [1995]: 'Computational Systems for Qualitative Economics,' *Computational Economics;* 8, 1–26.

Luce, R. Duncan [1956]: 'Semiorders and a Theory of Utility Discrimination,' *Econometrica;* 24, 178–91.

Majumdar, Mukul [1974]: 'Efficient Programs in Infinite Dimensional Spaces: A Complete Characterization,' *Journal of Economic Theory;* 7; 355–69.

Majumdar, Mukul [1988]: 'Decentralization in Infinite Horizon Economies: An Introduction,' *Journal of Economic Theory;* 45; 217–27.

Mas-Colell, Andreu [1974]: 'An Equilibrium Existence Theorem Without Complete or Transitive Preferences,' *Journal of Mathematical Economics;* 1, 237–46.

Mas-Colell, Andreu [1985]: *The Theory of General Economic Equilibrium: A Differentiable Approach.* Cambridge University Press.

Mas-Colell, Andreu, M. D. Whinston, and J. R. Green [1995]: *Microeconomic Theory.* Oxford University Press.

Mas-Colell, Andreu, and W. R. Zame [1991]: 'Equilibrium Theory in Infinite Dimensional Spaces,' *in:* Hildenbrand and Sonnenschein [1991], Chapter 34, pp. 1835–98.

McFadden, Daniel [1966]: 'Cost, Revenue, and Profit Functions: A Cursory Review,' Working Paper 86, Institute of Business and Economic Research, University of California, Berkeley.

McFadden, Daniel [1978]: 'Cost, Revenue, and Profit Functions,' *in:* Fuss, Melvyn, and D. McFadden, eds.: *Production Economics: A Dual Approach to Theory and Applications. vol 1: The Theory of Production.* North-Holland.

McKenzie, Lionel W. [1959]: 'On the Existence of General Equilibrium for a Competitive Market,' *Econometrica;* 27; 54–71.

McKenzie, Lionel W. [1961]: 'On the Existence of General Equilibrium: Some Corrections,' *Econometrica;* 29; 247–8.

McKenzie, Lionel W. [1981]: 'The Classical Theorem on Existence of Competitive Equilibrium,' *Econometrica;* 49, 819–42.

Michael, Ernest [1956a]: 'Continuous Selections, I,' *Annals of Mathematics;* 63, 361–82.

Michael, Ernest [1956b]: 'Continuous Selections, II,' *Annals of Mathematics;* 64, 562–80.

Moore, James C. [1968]: 'A Note on Point-Set Mappings,' *in:* Quirk, James P., and A. M. Zarley, eds.: *Papers in Quantitative Economics.* University Press of Kansas.

Moore, James C. [1970a]: 'On Pareto Optima and Competitive Equilibria: Part 1, Relationships Among Equilibria and Optima,' *Krannert Graduate School of Industrial Administration, Purdue University: Institute Paper No. 268;* April, 1970.

Moore, James C. [1970b]: 'On Pareto Optima and Competitive Equilibria: Part 2, The Existence of Equilibria and Optima,' *Krannert Graduate School of Industrial Administration, Purdue University: Institute Paper No. 269;* April, 1970.

Moore, James C. [1972a]: 'Axiomatic Characterizations of Consumer

Preferences and the Structure of the Consumption Set,' *Krannert Graduate School of Industrial Administration, Purdue University: Institute Paper No. 370;* September, 1972.

Moore, James C. [1972b]: 'A Note on Preference Orderings Which are Convex to the Origin,' *Krannert Graduate School of Industrial Administration, Purdue University: Institute Paper No. 369;* November, 1972.

Moore, James C. [1973]: 'Pareto Optimal Allocations as Competitive Equilibria,' *Krannert Graduate School of Industrial Administration, Purdue University: Institute Paper No. 386;* January, 1973.

Moore, James C. [1975]: 'The Existence of "Compensated Equilibrium" and the Structure of the Pareto Efficiency Frontier,' *International Economic Review;* 16; 267–300.

Munkres, James R. [1975]: *Topology: A First Course.* Prentice-Hall.

Nikaido, Hukukane [1968]: *Convex Structures and Economic Theory.* New York and London: Academic Press.

Novshek, William [1993]: *Mathematics for Economists.* Academic Press.

Pervin, William J. [1964]: *Foundations of General Topology.* Academic Press.

Pareto, Vilfredo [1894]: 'Il massimo di utilità dato dalla libera concorrenza,' *Giornale degli Economisti;* 9; 48–66.

Radner, Roy [1982]: 'Equilibrium under Uncertainty,' pp. 923–1006 in Arrow and Intriligator [1982].

Reed, Michael, and B. Simon [1980]: *Methods of Mathematical Physics: I: Functional Analysis,* Revised ed. Academic Press.

Reny, Philip J., and Myrna H Wooders [1998]: 'An Extension of the KKMS Theorem,' *Journal of Mathematical Economics;* 29, 125–34.

Rockafellar, R. Tyrrell [1970] *Convex Analysis.* Princeton University Press.

Romer, David [1996]: *Advanced Macroeconomics.* McGraw-Hill.

Royden, R. L. [1988]: *Real Analysis,* 3rd edition. Macmillan.

Samuelson, Paul A. [1947]: *Foundations of Economic Analysis.* Cambridge, Mass.: Harvard University press.

Samuelson, Paul A. [1956]: 'Social Indifference Curves,' *Quarterly Journal of Economics;* 70; 1–22.

Sargent, Thomas J. [1987]: *Dynamic Macroeconomic Theory.* Cambridge, Mass.: Harvard University Press.

Shafer, Wayne J. [1974]: 'The Nontransitive Consumer,' *Econometrica;* 42, 913–19.

Shafer, Wayne J. [1976]: 'Equilibrium in Economies Without Ordered Preferences or Free Disposal,' *Journal of Mathematical Economics;* 135–7.

Shafer, Wayne, and H. Sonnenschein [1975]: 'Some Theorems on the Existence of Competitive Equilibrium,' *Journal of Economic Theory;* 11, 83–93.

Shafer, Wayne, and H. Sonnenschein [1976]: 'Equilibrium with Externalities, Commodity Taxation, and Lump Sum Transfers,' *International Economic Review;* 17, 601–11.

Shephard, Ronald W. [1953]: *Cost and production Functions.* Princeton University Press.

Shephard, Ronald W. [1970]: *Theory of Cost and Production Functions.* Princeton University Press.

Shilov, Georgi E. [1973]: *Elementary Real and Complex Analysis.* M. I. T. Press.

Simon, Carl P., and L. Blume [1994]: *Mathematics for Economists.* W. W. Norton & Company.

Solow, Robert M. [1956]: 'A Contribution to the Theory of Economic Growth,' *Quarterly Journal of Economics;* 70; 65–94.

Starr, Ross M. [1969]: 'Quasi-Equilibria in Markets with Non-Convex Preferences,' *Econometrica;* 37; 25–38.

Torchinsky, A. [1988]: *Real Variables.* Addison-Wesley.

Stokey, Nancy L., and R. E. Lucas, with E. C. Prescott [1989]: *Recursive Methods in Economic Dynamics.* Harvard University Press.

Uzawa, Hirofumi: 'Preference and Rational Choice in the Theory of Consumption,' *in:* Arrow, Kenneth J., S. Karlin, and P. Suppes, eds.: *Mathematical Methods in the Social Sciences, 1959,* pp. 129–48. Stanford University Press, 1960. (This article was re-printed, with some corrections, in Chipman, Hurwicz, Richter, and Sonnenschein [1971, 7–28.])

Uzawa, Hirofumi [1962]: 'Production Functions with Constant Elasticities of Substitution,' *Review of Economic Studies;* 29, 291–99.

Walker, Mark [1979]: 'A Generalization of the Maximum Theorem,' *International Economic Review;* 20, 267–72.

Wheeden, R. L., and A Zygmund [1977]: *Measure and Integral.* Marcel Dekker.

Yannelis, Nicholas C. [1987]: 'Equilibria in Noncooperative Models of Competition,' *Journal of Economic Theory;* 41, 96–111.

Yannelis, Nicholas C., and N. D. Prabhakar [1983]: 'Existence of Maximal Elements and Equilibria in Linear Topological Spaces,' *Journal of Mathematical Economics;* 12, 233–45.

Zhou, Lin [1994]: 'A Theorem on Open Coverings of a Simplex and Scarf's Core Existence Theorem Through Brouwer's Fixed Point Theorem,' *Economic Theory;* 4, 473–77.

Index

Printing: Weihert-Druck GmbH, Darmstadt
Binding: Buchbinderei Schäffer, Grünstadt